BIRDS
OF THE
HAROLD HALL AUSTRALIAN EXPEDITIONS

1962-70

Hall's Babbler, *Pomatostomus halli* (painting by P. R. Colston)

BIRDS
OF THE
HAROLD HALL AUSTRALIAN EXPEDITIONS

1962-70

A report on the collections
made for the British Museum (Natural History)

Edited by
B. P. HALL

Results of the Harold Hall
Australian Expeditions, No. 33

LONDON:
TRUSTEES OF THE BRITISH MUSEUM
(NATURAL HISTORY)

1974

Publication number 745
ISBN 0 565 00745 9

Printed and bound in Great Britain by
Staples Printers Limited at The George Press, Kettering Northamptonshire

FOREWORD

Of the avifaunas of the world, the Australian has until recently been the least well represented in the British Museum (Natural History). Such specimens as there were in the collections were old and had little or no information on the labels. Indeed nowhere in the world was there a comprehensive collection of Australian birds with good biological data. When, in 1961, this fact was brought to the attention of Major Harold Hall, an Australian philanthropist interested in natural history, he very generously offered to finance a sufficient number of expeditions from the Museum to make such a collection. This volume reports on the expeditions and on the collections made and gives me the opportunity of acknowledging the great debt owed to Major Hall by the Museum and indeed by all zoologists interested in evolution, for it is on such collections that theories of speciation are based and can be tested and can thus apply far beyond the bounds of ornithology.

Major Hall's gift was administered by the Committee of the Bird Exploration Fund and was supplemented by donations from two members of that Committee, Mr C. W. Mackworth-Praed and the Rt Rev Monsignor F. O. Cave, to cover the cost of the three Land Rovers required.

With funds assured it remained only to seek the co-operation of the Australian authorities before plans could go ahead. Here the enthusiastic support of Dr D. L. (Dom) Serventy helped to overcome the initial misgivings of many Australians at the prospect of extensive collecting at a time when they were trying to encourage conservation. He was foremost in pointing out that good conservation was wholly dependent on accurate knowledge of the feeding habits and ecology of the species concerned, and one of the main objectives was to obtain this type of information. Throughout the next six years his continuing interest, help and hospitality did much to smooth the path of all the expeditions, but especially those which worked in Western Australia.

Once launched the expeditions owed much to the Directors and staff of all the State Museums, to the officers concerned with fauna protection and to others in government departments, both in London and Australia, who helped in sundry ways; also to the owners and managers of properties on which camps were made and other private individuals who gave help and advice. It is impossible to thank all these individually here, but as far as possible they have been named and tribute paid to them in the narratives of the expeditions.

The expeditions enabled the Bird Room staff to gain most valuable experience in the field but it was never possible to spare more than two or three members for any one period of six months or more and the personnel of the expeditions included members from outside the Museum, both from England and Australia. Three of the expeditions were led by Museum staff but the third expedition was led by the indefatigable Mrs Hall (no relation of Major Hall) who has long been unofficially but most actively associated with the Bird Room of the Museum. Thanks are due to Mrs Hall not only for successfully leading this expedition, but also for playing a very active part, as Secretary and Treasurer of the Bird Exploration Fund, in the organisation of all five expeditions. She has also collated and edited the contributions which make up this report. Special tribute is also due to Major B. D. McDonald Booth who led the fifth and last of the major expeditions. Without the help of many who joined the expeditions for little or no more than their expenses, the project could hardly have been mounted; all played a significant part in its success. Three of the expeditions included the leader's wife as cook.

The move of the ornithological collections from the Bird Room to the new four-storey building at Tring has restricted the compilation of this Report and it aims at no more than the provision of as much factual information as possible from the specimens and collectors' notes, including such data as food, breeding seasons and behaviour. The conclusions to be drawn on evolution and speciation are for the future. The specimens are already at Tring and should be available for study about the time this record is in print.

J. P. HARDING
Keeper of Zoology

October 1971

CONTENTS

LIST OF CONTRIBUTORS

Major B. D. McD. Booth *Fresh Fields, Elsenham, Bishops Stortford, Hertfordshire.*
F. Carline *17 Pond Street, London* N.W.3.
P. R. Colston, *Sub-department of Ornithology, British Museum (Natural History).*
G. S. Cowles *Sub-department of Ornithology, British Museum (Natural History).*
D. J. Freeman *Green Farm, Bentley, Hampshire.*
C. B. Frith *46 The Lindens, Fieldway, New Addington, Croydon, Surrey.*
I. C. J. Galbraith *Sub-department of Ornithology, British Museum (Natural History).*
D. Goodwin *Sub-department of Ornithology, British Museum (Natural History).*
Mrs B. P. Hall *Woodside Cottage, Woodgreen, near Fordingbridge, Hampshire.*
Dr C. J. O. Harrison *Sub-department of Ornithology, British Museum (Natural History).*

LIST OF PLATES

Sericornis keri was collected in nearby undergrowth, where *Orthonyx spaldingi* and *Psophodes olivaceus* were common (I. C. J. Galbraith).

(b) Mount Dryander (57), Q'd. Dry tropical rainforest (softwood forest?) on bouldery lower slopes, above Deep Creek (I. C. J. Galbraith).

Plate 10

(a) Mount Bell (32), W.A. General view. Habitat of *Petrophassa albipennis* on rocks behind, and of *Psitteuteles versicolor* and *Trichoglossus haematodus* in foreground (C. B. Frith).

(b) Mount Elizabeth (35), W.A. The Hann River in savanna woodland (D. J. Freeman).

INTRODUCTION

B. P. Hall

Programme

The origin of Major Hall's most generous gift was explained in the Foreword. It was estimated that five expeditions could be equipped and mounted from London, each costing £5000. Each would consist of about six people travelling in three Land Rovers, some of the members being from the British Museum (Natural History) and some from Australian museums, assisted by amateurs from both countries, and each would last about six months. Routes were planned with three objectives: first to provide the British Museum (Natural History) with a representative collection of Australian birds; second, to explore some areas which were poorly known ornithologically; third, to allow the Head of the Bird Room, Mr J. D. Macdonald to visit on the 1st Expedition as many State Museums as possible to prepare the way for subsequent expeditions. In addition each expedition had to end up and store vehicles and equipment at some place from which it was practicable to start the next expedition six months later.

The basic plan was as follows:

1st Expedition Land Perth; cross quickly to Melbourne; collect in Victoria, New South Wales and southern Queensland; finish Brisbane.

2nd Expedition Start Brisbane; move quickly north to collect in northern Queensland; finish Melbourne.

3rd Expedition Start Melbourne; move quickly into South Australia; collect in the desert areas of South Australia and Western Australia; finish Perth.

4th Expedition Start and finish Perth; collect in extreme southwest and in western districts of Western Australia north to the Hamersley Range.

5th Expedition Start and finish Perth; collect in Kimberleys and Arnhem Land.

This plan was carried out in the six years between November 1962 and 1968, most collecting being done in the Australian winters.

In addition to the five main expeditions, grants were made from the Fund to the South Australian Museum, Adelaide, towards two local expeditions in 1963 and 1964, and to C. B. Frith who accompanied C. A. Walker on a primarily palaeontological expedition to northern Queensland in 1970. The dates and localities of the South Australian Museum expeditions have been included after the narratives of the main expeditions, and those birds collected which came to the British Museum have been listed with those of the main expeditions in the systematic section. Birds collected by Frith are listed in Appendix 3.

Equipment & Collecting Techniques

Equipment

The expeditions were fitted out in London with three long wheel-based Land Rovers, tents and camping equipment, nearly all of which lasted for the whole six years. The work of selecting and assembling the equipment was done by Mr and Mrs Macdonald,

G. S. Cowles and P. R. Colston. Help was received from many sources including the War Office, The Rover Company, Martin Walter Ltd (who fitted two of the Land Rovers with Dormobile roofs), Black & Edgington Ltd, P & B Cow (Industrials) Ltd, F.P.T. Industries Ltd, Melaware Ltd, and Safari Beds. In Australia a Sivia refrigerator was purchased which ran on bottled gas and proved useful (though sometimes temperamental) in preserving birds until time could be found for skinning. The 3rd Expedition, operating in the interior, was fitted with a Traegar transceiver and owed much to Mr Graham Pitts of the Royal Flying Doctor Service, Port Augusta, for instruction in its use.

Data Sheets

In order to preserve some uniformity and to ensure that there was a permanent record for each specimen collected, data sheets were designed on which all observations made in the field, at the skinning table, or in the musuem could be noted. These were used by all expeditions except those from South Australian Museum, and with the specimens provide the basis for the systematic report.

Collecting

The majority of specimens were shot, mist nets only being used to a limited extent where the vegetation was suitable.

Preservation

4709 specimens were made into study skins; 786 were made into rough skeletons, and are in the course of being processed; 910 were preserved as fluid specimens.

Weights

An Ohaus beam balance was carried, which records to the nearest 0·1 g, but it was difficult to use in the field unless conditions were still; good series of weights were only obtained by the 2nd and 5th Expeditions. These are listed in the systematic report.

Food

The stomachs (or crops) of most specimens were examined and their contents identified to the limited degree possible in the field. In a few instances seeds and plants were brought back and sent to Dr R. Melville of the Royal Botanic Gardens, Kew, for identification. Dr Melville's identifications are included in the systematic section. Some other stomachs have been retained but their contents await examination.

Examination of skulls

Skulls were examined for pneumatisation. On the 2nd Expedition in northern Queensland (where a considerable number of young birds were collected) this was assessed on the presence or absence of an inner layer of bone, though solidity and light transmission (used by other expeditions) were sometimes noted in addition.

Tongues

Tongues were preserved from the majority of species collected.

Designation of collecting localities

Much of the collecting was done from camps some distance from well-known towns, often on cattle or sheep stations. For each main camp an appropriate name was selected which might be that of the station, the nearest town or a geographical feature. These names have been given a number in rough geographical sequence which is plotted on the map (inside back cover) (white numerals) and which is quoted invariably in conjunction with the name. Numbers have also been given to some subsidiary collecting localities (black numerals) visited for shorter periods, sometimes for only part of a day, which because of their distinctiveness in habitat or remoteness from a main camp seemed to require a separate identity. The named and numbered collecting localities are entered in bold type in the narratives of the expeditions. In the systematic section these names alone are given as the locality for all specimens collected from each base, although extensive use of the Land Rovers meant that the collecting area might cover over 100 square miles. Precise data on direction and distance from base are on the label and data sheet for each bird.

Results

The main objective of the expeditions was to provide a comprehensive and well-documented series of skins and anatomical specimens. This has largely been achieved (Table 1), though gaps in knowledge inevitably remain. In particular, a better understanding of annual cycles and seasonal movements will require comparable series collected in the Australian summer.

Some results have already been published in a series of papers, entitled 'Results of the Harold Hall Australian Expedition(s)', appearing in various journals and listed in Appendix 2. One new species was discovered, *Pomatostomus halli* (Cowles, 1964), one overlooked species was re-instated, *Sericornis keri* (Galbraith & Parker, 1969), and data were obtained on a number of rare species including the Black Grass-wren *Amytornis housei* (Freeman, 1970) and the White-lined Honeyeater *Meliphaga albilineata* (Freeman, 1971).

Two new subspecies have been based on birds collected by the expeditions, *Petrophassa albipennis boothi* Goodwin 1969 and *Colluricincla rufiventris serventyi* Macdonald 1968a, while the status of a distinctive specimen of *Gerygone chloronota* outside its usual range and habitat requires further study. Greater knowledge of the extent of individual, seasonal and geographical variation has also allowed the status of many Mathewsian subspecies to be elucidated.

New evidence has suggested changes in the status of various forms. Thus *Pachycephala melanura* is separated from *P. pectoralis* (Galbraith, 1967), *Philemon gordoni* from *P. buceroides* (Parker, 1971), and *Gerygone tenebrosa* from *G. magnirostris,* while *Coracina papuensis* and *C. robusta* are combined (Galbraith, 1969). The range of several species has been extended through the collections, for instance *Conopophila whitei* (Cowles, 1967), *Conopophila picta* (Colston, 1968), *Neophema chrysostoma* and several of the pardalotes.

New information has been gained on the plumage sequences of several species, studied particularly by Galbraith in the groups on which he worked. Goodwin especially (1967c, and in the systematic notes) and other members of the expeditions have contributed observations on the behaviour of many species.

Types of new forms described on expedition specimens have been deposited in the appropriate Australian State Museum. The totals deposited in the British Museum (Natural History) are listed in Table 1.

Table 1
Specimens Collected for British Museum (Nat. Hist.)

		Skins	Skeletons	Fluid	Tongues
1st Exp.	WA	8	5	—	279 (WA, SA, Vic, NSW, Qd)
	SA	25	14	9	
	Vic	46	6	27	
	NSW	230	78	63	
	Qd	199	44	53	
2nd Exp.	Qd	1481	23	30	c.1500
3rd Exp.	SA	659	47	35	
	WA	573	56	29	c.350
4th Exp.	WA	377	263	213	c.250
5th Exp.	WA	661	195	281	
	NT	286	55	170	c.300
SA Mus '63	SA	123	—	—	—
'64	SA	41	—	—	—
Totals		4709	786	910	c.2700

Authors

Like the expeditions themselves, this report is a joint project. The systematic work was done in the Bird Room by myself and past and present members of the staff who took part in the expeditions, with some assistance during 1970/71 by an amateur zoologist Mr Francis Carline. P. R. Colston was responsible for much of the preliminary summarising of data from the data sheets. The authors responsible for the various sections are named at the start of each section and in the index of contents, followed, where appropriate, by the name of an assistant (in brackets), though it has not been possible to acknowledge all help received from colleagues in various aspects of the compilation. Dr D. W. Snow and D. J. Freeman have helped me in editing.

The manuscripts were completed during 1971 and have not since been extensively revised.

NARRATIVES OF EXPEDITIONS

1st Expedition. November 1962 – June 1963. Perth to Melbourne, Victoria, New South Wales, and southern Queensland

P. R. Colston & G. S. Cowles

Personnel

From British Museum (Natural History): Mr J. D. Macdonald, leader, Mr G. S. Cowles and Mr P. R. Colston. Doctor and caterer: Dr E. C. Macdonald.

From Australia, for Perth to Adelaide section: Mr A. R. McEvey, Curator of Birds, National Museum of Victoria; Mr W. G. Middleton, District Forester, Forests Commission of Victoria. From 5 March to 16 June: Mr R. Boswell, Chief Preparator, National Museum of Victoria. From 5 March to 23 April: Mr R. Lossin, Preparator, Australian Museum, Sydney.

Diary of Expedition

12 – 30 *November*. After the three specially adapted Land Rovers and the camping and collecting equipment were unloaded at **Perth (7),** camp was made on Dr D. L. Serventy's property at Gooseberry Hill, Kalamunda, 12 miles east of Perth. Here the U.K. party was joined by McEvey and Middleton.

30 *November* – 15 *December*. Perth to Adelaide by Eyre Highway. A little collecting was done at **Eyre (16)** on the coast south of Cocklebiddy (15), at **Eucla (17), Nullarbor Homestead (69)** and **Streaky Bay (74).** For an account of this journey and his previous east to west transect see McEvey & Middleton, 1968.

18 – 19 *December*. **The Coorong (85)** S.A. No specimens were collected as this coastal strip with its salt-water lagoons is a protected area, but a short film sequence was made showing feeding habits of Whiskered and Fairy Terns. Some 10,000 birds of 26 species were observed in two hours observation along the shoreline.

19 *December* – 5 *January*. Property of Mr F. and the late Mr J. Livingstone at **Nelson (88),** Glenelg Estuary, western Victoria. The area was predominantly coastal sand-dunes, with low scrub and marshy grassland. The exposed sand and mud flats of the estuary at low tide attracted many duck, gulls, terns and waders, including many Northern Hemisphere migrants. On the river the unusual escape behaviour of the Musk Duck was observed and filmed in the resident flock of about 50 birds. The Blue-winged Parrot was abundant, and on 26 December the first migrant Spine-tailed Swifts appeared. A Red-capped Dotterel was incubating two eggs. Many species were encountered for the first time, including some flocks of Yellow-tailed Black Cockatoo in pine plantations to the east, Red-browed Finches in the coastal scrub and Southern Emu-wrens in swamp heathland.

5 *January* – 2 *February*. Two members only (GSC, PRC) in the field.

5 – 17 *January*. Deltana Homestead (Mr R. Macdonald), 7 miles north-east of **Furner (86).** The property was grazed by sheep and consisted of open savanna and mixed woodland with a small area of virgin scrub woodland. Species typical of the area and

collected for the first time were Red-rumped Parrot, Striated Pardalote, Buff-rumped and Yellow-rumped Thornbills, Yellow Robin, White-throated Tree-creeper and White-winged Chough.

A day was spent making observations at **Bool Lagoon (87)** with the help and permission of Mr J. Hood. It is a sanctuary for three species of ibis and many water-birds. The main purpose of visiting the area, however, was to collect a small topotypical series of Singing Bushlarks *Mirafra javanica keasti* from the surrounding paddocks.

18 *January* – 2 *February*. Willaroo Homestead (Mr C. N. Austin), **Coleraine (89),** western Victoria. No collecting was done near the homestead but under Mr Austin's guidance trips were made into the surrounding country where different habitats could be found, namely (a) woodland with Red Gums close to the river near Harrow, 29 miles to the north; (b) open country with grassy paddocks for sheep near Nareen, 12 miles to the northwest; (c) mixed woodland with stringy-bark and *Banksia* near Vasey, 24 miles to the northeast; (d) open forest at Rocklands Dam, 30 miles to the northeast; (e) **Victoria Range (90),** dense woodland and open forest on the slopes of the range and also in some open heathland at the foot on the west side; (f) **Little Desert (91),** 10 miles north of Nurcoung, an arid sandy area covered with low scrub with isolated patches of woodland and mallee.

2 *February* – 8 *March*. Melbourne. No collecting. Assembling the party and re-equipping for the next stages of the expedition. Work was carried out in the National Museum of Victoria on the specimens already collected (GSC, PRC).

9 – 16 *March*. Kywanna Homestead (Mr F. Hood), 8 miles southwest of **Albury (92),** N.S.W. The area was pastureland with some mixed woodland (Plate 5(a)). The most interesting birds collected were hybrids between the White-backed and Black-backed Magpies.

17 – 27 *March*. Wheogo Homestead (Mr W. Gaden), 10 miles northwest of **Dunedoo (93),** N.S.W. Except where reclaimed for sheep this was well-wooded country (Yellow Box, ironbark, Apple Gum, cypress pine) with low outcrops of sandstone and basalt. In the thicker woodland on laterite ridges (Plate 5(b)) the Speckled Warbler was fairly common and was observed feeding nestlings. New to us were Red-winged Parrots, Banded Finch at the southern limit of their range, Olive-backed Oriole, White-throated Warbler, Little Thornbill and Peaceful Dove. In all 74 species were observed (PRC).

27 *March* – 9 *April*. Southgate, lower Clarence River, 8 miles northeast of **Grafton (94),** N.S.W. Recent heavy rains had made the ground unpleasantly boggy and with the threat of the Clarence flooding Mr C. S. Harrison kindly arranged for us to camp in the village hall. The moist and thickly vegetated country was a new habitat to us; woodlands were of ironbark and Spotted Gums; slow-running streams and pools choked with the introduced Water Hyacinth *Eichhornia crassipes* were ideal for the Lotus Bird and other water-birds. Despite continuing rain and 69% humidity, 90 different species were observed, 47 were collected, 24 being new to us. A film sequence was made of the Southern Stone Curlew. Through the kindness of Mr P. Dwyer we were taken by motor launch to the mouth of river near Yamba to see the Ground Parrot in one of its few remaining haunts. On the estuarine mud flats Northern Hemisphere migrants were beginning to assume summer plumage.

Other species on this area of swampy heathland included Tawny-crowned and White-cheeked Honeyeaters, the latter feeding chicks, Southern Emu-wren, Tawny Grassbird and the delightful Golden-headed Fantail-warbler.

9 – 18 *April*. Timbarra Homestead (Mr C. Rose), 9 miles east of **Tenterfield (95)**, N.S.W., at 2,800 ft on the eastern side of the Great Dividing Range. Much of the area had been cleared for sheep but some collecting was done in both wet and dry sclerophyll forest on either side of the Range, and Mr M. T. Goddard guided us also to a patch of rainforest where we found such typical species as the Red-browed Tree-creeper, the Large-billed Scrub-wren and the Lewin Honeyeater. Southern Emu-wrens were fairly common in a forest clearing of Button Grass and near the summit of the Range. On the west side of the Range some Singing Bush Larks were found and a single Black-throated Finch at the southern limit of its range. On 16 and 17 April heavy southerly movements of *Melithreptus* were observed (PRC) over the forest, probably *M. brevirostris* or *lunatus*, but whether these were local or migratory movements is not known. In all 77 species were observed.

18 – 25 *April*. Brisbane. No collecting.

26 *April* – 8 *May*. **Banoona (58)** Homestead (Mr D. C. Nason), 26 miles southeast of Roma, S. Queensland, on the Condamine River system. The country was mostly acacia scrub or mixed woodland with low scrub and pastureland. Much of the area was being cleared rapidly to extend the pastureland for sheep. The river area was studied first from a light aircraft and later a collecting trip was made to Horse Shoe Lagoon where we found small parties of various ducks and other water-birds. Near the homestead Apostle Birds, bower birds, the Pale-headed Rosella, Pied Currawong and Bar-shouldered Dove were species new to us. In all 81 different species were observed.

8 – 19 *May*. **Tyrone** Homestead **(59)** (Mr G. R. Akers), 3 miles east of Langlo Crossing S. Queensland. The property consisted of mixed woodland with Mulga, sandalwood, Box, Gidgee and low scrub on red sandy soil. Tall eucalypts lined the edges of dry creeks. This area is classified as "Arid Scrub" (Williams, 1955). It was notable for the discovery of a new species, Hall's Babbler *Pomatostomus halli* (Cowles 1964). The area was in some respects a transition zone, for the Noisy Miner was replaced by the Yellow-throated Miner; and several more species were encountered for the first time. Heavy rain fell just as we were leaving so that we were advised to use the southern route via Quilpie to Thylungra (60).

20 – 31 *May*. **Thylungra** Station **(60)** (Mr E. W. Mildren, manager for Australian Estates Ltd), S. Queensland. A large sheep station sited on Kyabra Creek to the west of the Grey Ranges in the Eyre Basin, in an area often affected by floods. The vegetation was mainly savanna woodland with grassy plains (Plate 6(b)), sandhills, and with tall trees along Kyabra Creek. The rare Painted Honeyeater was in low patches of Mulga where parasitised by mistletoe; young birds were in evidence. Singing Bush Larks were seen again after being absent since Tenterfield (95). Five species were seen for the first time, emphasising the continuing transition from the eastern avifauna to that of the centre. In all 100 species were observed during the eleven days, more than at any other camp; they included ten species of birds of prey.

1 – 11 *June*. **Monkira** Station **(61)** (Mr G. MacQueen, manager for Mr D. Fraser),

Diamantina River, S.W. Queensland. The station is deep in the heart of the Channel Country, an area scarred by dry creeks and channels running mostly northeast to southwest, draining rain water from the higher rainfall areas in the north to Lake Eyre. At Monkira the Diamantina can flood the surrounding country for a width of 14 miles in a matter of hours, though locally no rain may fall. On the flood plain (Plate 6(a)) the growth of a most lush, succulent, low vegetation has a great influence on the breeding seasons of birds in the area. The Australian Pratincole, the Australian Dotterel, and the Banded Plover were seen on the flood plains, the latter with young. It was interesting to see the bright yellow Gibber-bird, which is so conspicuous on the red sandy soil, become almost invisible amongst the yellow blossoms of the flowering herbs. Away from the flood basin the higher plains are arid gibber desert with high red sandhills. Cinnamon Quail-thrush and Plumed Pigeon were common. In all 55 species were observed at this our last camp.
11 – 15 *June*. Monkira (61) to Brisbane. Heavy and late rains prevented us following our original plan to return to Melbourne via the Birdsville Track. Instead we returned to Brisbane where the two Land Rovers and most of the equipment were left in store. The third vehicle was driven to Melbourne and stored there.

2nd Expedition, February – August 1964. Northern Queensland, mainly coastal areas

I. C. J. Galbraith

Personnel

From the British Museum (Natural History): Mr I. C. J. Galbraith, leader, Mr S. A. Parker and Mr T. Andersen. Caterer and cook: Mrs E. H. Galbraith. Mr I. McCallum and Mr R. L. Pink from Casterton, Victoria, keen birdwatchers, banders and hunters.

Diary of Expedition

21 *February* – 2 *March*. Preparations in Brisbane, based on the Queensland Museum by permission of the Director, Mr J. T. Woods. Valuable help was given by staff of the Museum, especially Mr D. P. Vernon, and those of the Australian Estates office, Bureau of Meteorology, Department of Primary Industries, Department of Supply, Royal Flying Doctor Service and U.K. Information Service.
2 – 10 *March*. Brisbane to Glendower (55) via Mackay, Mount Dryander (57), Ayr, Townsville and Hughenden.
10 *March* – 4 *April*. **Glendower** Station **(55)** (Mr Dick Wakeford for Winter, Irving & Robertson), northeast of Hughenden, Queensland. This formerly huge station, from which neighbouring properties have been successively separated, is well known as a fossil locality (see Wilkins, 1928 who also collected birds there). Camp was made on the Flinders River two miles above the present homestead. Glendower lies on the western edge of the Einasleigh uplands (CSIRO, 1964) where their low scarp slope subsides into the wide Carpentaria plains. Having emerged from its gorge, the river at this point winds through a widening valley between basalt plateaux before running out onto the

plain through a fringe of buttes. Apart from a few small pools the wide river bed was dry near the camp, and many birds came to drink at a trough fed from beneath the sands by a windmill, though water-birds were not uncommon at pools further up. The flats and concave slopes of the valley were clothed in low woodland with scrubby bushes in places. Grass cover was scanty in the valley, perhaps partly because of a drought which had already lasted several years, but there was still a moderately rich avifauna in which Peaceful Doves, Crested Pigeons, Red-winged Parrots, Weebills, White-plumed Honeyeaters, Little Friar-birds, Yellow-throated Miners, Magpie Larks, Apostle Birds, Spotted Bower-birds and Australian Ravens were especially common. The grass was denser on the plateau above the camp, especially in treeless areas where bushlarks, pipits and fantail warblers were found. The terraced river bank carried a narrow strip of taller eucalypt woodland with casuarinas and a well-grassed floor, to which such birds as the cuckoo-shrikes, butcher-birds and several honeyeaters were confined.

At the end of this camp we combined to compile a list of species seen, but lists for subsequent camps were compiled by Pink and it has not been possible to obtain these for use in writing the species accounts.*

4 – 6 *April.* Glendower (55) to Glenore (49) via Julia Creek and Taldora Station. The highway westwards from Hughenden runs across the vast level Carpentaria plains, broken only by a few trees along the creek beds. A little north of Julia Creek, in the drainage of the western Gulf rivers, the country was much more fertile and varied, with many lagoons and patches of woodland and scrub. We were helped by Mr Vince Kelly of Australian Estates at Julia Creek, and entertained at Taldora by Mr and Mrs Robert Curr, who gave us permission to collect later at Inkerman (50) on the Gulf of Carpentaria.

6 – 27 *April.* **Glenore** Station **(49)** (Mr Ellis Quiltie), south of Normanton, Queensland. Camp was made 15 miles south of the homestead, at a mustering site beside The Lakes, swampy lagoons in a flood-channel of the Norman River. These were frequented by many water-birds including Darters, cormorants, grebes, egrets, ducks and pigmy geese, while Brolgas frequently displayed beside them. We collected along the river channels, from six miles north of the homestead to 28 miles south of it. The plains (CSIRO, 1964) are very level, and had been deeply flooded a few weeks previously. They were covered in featureless low open woodland with moderate grass cover and some taller trees (Plate 7(b)). Here species new to us included the Black-eared Cuckoo, Yellow and Yellow-tinted Honeyeaters, Black-throated Finch and Pictorella. Vegetation was denser and more varied along the watercourses, including scattered river gums on the lagoons and closed woodland on the main channel. Such birds as Rainbow Lorikeets and Blue-faced Honeyeaters were encountered only in the tall trees of a marshy area around the homestead, and Singing Bushlarks and Ground Cuckoo-shrikes in a small area of grassland beside the Normanton road.

27 – 29 *April.* Glenore (49) to Moonlight Creek (47) via Burketown. As suggested by the vegetation map (Williams 1955), woodland gave way to grassland between the Flinders River and Inverleigh homestead, and again between the Leichardt River and

* They appear to have been used by Dr G. M. Storr for his Queensland list (1973).

Burketown. We were guided along the invisible track beyond Escott homestead by Mr Terry Arnold, crossing many creeks before running through savanna woodland dominated by termite mounds, alongside the salt-flats which fringe the Gulf coast.

29 *April* – 19 *May*. **Moonlight Creek (47)** on Brokera Station (Mr Jack Aires and Mr Terry Arnold), northwest of Burketown, Queensland. Camp was made on this uninhabited station at a mustering camp (astronomical control point C22) on Rocky Creek two miles above its confluence with Moonlight Creek. We collected within eight miles of the mouth of the latter, quoting rather inaccurate bearings from Escott Homestead 30 miles away. The creeks cut picturesque billabongs with rocky banks, surrounded by river gums, through the level stony country, with shallow swampy stretches through alluvial deposits. The woodland a few miles from the coast was uniform, open and low with sparse ground cover (Plate 7(a)), and broken only by creek-side vegetation. There was much more diversity nearer the coast, with areas of denser scrub and of grassland and bare soil. As on the Gulf coast of the Cape York Peninsula (Thomson, 1935: 15), sandy ridges separated from the mainland by salt flats supported surprisingly varied woodland and forest (Plate 8(b)). One of these 'islands', shown on the 1:2500,000 map as running north from Moonlight Creek, carried tall woodland with such birds as Rainbow Lorikeets, cuckoo-shrikes, Blue-faced Honeyeaters and butcher-birds. A complex of ridges south of the main mouth carried more varied and locally denser woodland, with tall closed tea-tree forest in a depression, but no species was encountered here which was not present on the mainland. The salt-water channel nearby ('Moonlight Creek mouth') was lined with dense mangrove forest of no great height, consisting mainly of *Rhizophora* and *Ceriops,* with bushes of *Avicennia* on the fringing mudflats. At a minor mouth of the creek ('coast SE of Moonlight Creek mouth') the mangroves consisted almost entirely of *Avicennia,* ranging from trees alongside a channel to salt-stunted bushes beyond the mudflats. Beyond these again the mangroves formed a closed canopy with open understory, on a bank of shell gravel submerged at high tide. Both these mangrove localities were very rich in mangrove birds: Buff-breasted Warblers and Yellow Silvereyes were abundant, and Red-headed Honeyeaters common, in the canopy, while White-breasted and Mangrove Golden Whistlers, Rufous Fantails and Broad-billed Flycatchers skulked in the substage.

At this camp we were concerned largely with the 'Northern Territory' forms which represent typical Queensland ones west of the Carpentaria Gap formed by the grassland barrier. Although Macdonald (1969b) regards the barrier to gene-flow as more important than present vegetational features would warrant, it was the vegetational map (Williams, 1955) which had led us to expect these western forms in the Burketown area, whereas Keast (1961: 413 & fig. 4(a)) would place the barrier further west. The major western forms encountered at Moonlight Creek were the Black-tailed Tree-creeper, Long-tailed Finch and the distinctive races of the Rainbow Lorikeet, Little Cuckoo-shrike, Sittella and Blue-faced Honeyeater.

19 – 22 *May*. **Wernadinga** Station **(48)** (Mr Allan Richards), east of the Leichardt River, Queensland. Camp was made at the site of the former Punchbowl homestead, still marked on some maps. The country around was lightly wooded, corresponding to the more or less wooded 'island' within the grassland barrier (Williams, 1955), and

we concentrated on the species and superspecies which vary across the Carpentaria gap. Lack of tall trees made the country unsuitable for Rainbow Lorikeets, Little Cuckoo-shrikes and Blue-faced Honeyeaters, but of typically western forms we encountered the white-winged Sittella and Black-tailed Treecreeper, while the Black-throated Finch represented eastern forms.

22 – 24 *May*. Wernadinga (48) to Inkerman (50) via Normanton.

24 *May* – 13 *June*. **Inkerman** Station **(50)** (Mr Con Cahalane for Mr Robert Curr), Gulf of Carpentaria coast north of Normanton, Queensland. This locality must not be confused with the Inkerman Station near Home Hill on the east coast of Queensland, where W. Stalker collected in 1907 (Ingram, 1908b). Camp was made at the homestead where we were the guests of Mr and Mrs Cahalane. We collected within 12 miles of the coast, between the Staaten River and Rocky Creek 30 miles to the north. This coastal plain (CSIRO 1970) is occupied by extensive grasslands, interspersed with large salt-flats towards the coast. The 'islands" characteristic of the coast at Moonlight Creek (47) are here represented by sandy ridges bearing varied and sometimes dense woodland, which extend from the coast (though seldom beyond the salt-flats) to at least five miles inland. One of these ridges ('near mouth of Salt Arm Creek') held a swampy depression overhung by a dense closed canopy, and here and in the mangroves were several species characteristic of the east coast: Fairy Warbler, Lemon-breasted Flycatcher, Rufous Shrike-thrush, Yellow-breasted Sunbird, Yellow Oriole, Yellow Figbird and Spangled Drongo. Mangroves at the mouths of the creeks were like those at Moonlight Creek, while there were tracts of tall *Avicennia* forest along the lower reaches of the Staaten River (forming almost pure stands at 'Staaten/North Branch confluence', but considerably mixed with *Rhizophora, Ceriops, Bruguiera* and *Xylocarpus* at 'mouth of Staaten River' where the soil was above the range of normal tides). These mangroves were poorer in bird species and much poorer in individuals than those at Moonlight Creek: neither Rufous Fantails nor White-breasted Whistlers were encountered, Mangrove Robins were the only common substage birds (at the confluence), while there were no large flocks of either Buff-breasted Warblers nor Yellow Silvereyes. Rufous-banded and Rufous-throated Honeyeaters were common in a tea-tree swamp near the aerodrome.

13 – 18 *June*. Inkerman (50) to Ayton (52) via Cairns and Helenvale. From the coastal plain the road to the east coast climbs steadily through woodland to the fertile Atherton Tableland (now almost entirely cleared of rainforest for tobacco growing, dairy farming and other agriculture), and precipitously descends the steep scarp slope through rainforest reserves, to the narrow coastal plain behind Cairns. Leaving the coast road south of Mossman, the road to Cooktown passes through low open woodland in the rainshadow of the Great Dividing Range; but on doubling back to Helenvale the Ayton road passes into rainforest and tall grassy forest to the east of the range. We were helped by the Department of Forestry at Cairns and by the Cummings family at Helenvale.

18 *June* – 8 *July*. **Ayton (52)** (Mr Herbert Olufson for Paterson Plywoods Pty), south of Cooktown, Queensland. Camp was made on the China Camp road at the Granite Creek crossing, just above its confluence with the Bloomfield River two miles above

the township. We collected near the river (Plate 8(a)), from Weary Bay to five miles inland, with a few specimens from Gap Creek five miles to the north. Though along the road between Helenvale and Gap Creek the 'Bloomfield Scrubs' rainforest was virtually unbroken, at Ayton the ridges were mainly occupied by grassy forest. Artificial clearings were overgrown with rank grass and edged with thickets of introduced *Lantana*. We concentrated on the rainforest birds, of which the commonest were Large-billed Scrub-wren, Pale Yellow Robin, Rufous Fantail, Boat-billed, Black-faced and Spectacled Flycatchers, Brown Whistler, Rufous Shrike-thrush, Dusky, Graceful and Lesser Lewin Honeyeaters, Catbird and Victoria Rifle-bird. The mangroves were disappointing, old stands of *Bruguiera* and *Rhizophora* on the coast being almost devoid of birds, while the Mangrove Robin was the only true mangrove species encountered in *Ceriops* thickets on islets in the mouth of the river.

25 *June*, 1 – 2 *July*. **Big Tableland (51)** (Mr Jack Cartaar and Mr Vick Cummings), near Helenvale, south of Cooktown, Queensland (see Storr, 1953). A shelf on the south-western slope of Mount Amos. The very steep sides were clothed in grassy forest which also occupied the river flats at the base. The shelf at about 2,200 ft was densely forested, apart from artificial grassy clearings and lagoons resulting from the working of alluvial tin, with tree-fern thickets in the gulleys. The forest was partly rainforest like that near sea-level, and partly an almost pure stand of Johnstone Hardwood. Rainforest species which were much commoner on the shelf than in the lowlands near Ayton (52) included the Wompoo Pigeon, Northern Chowchilla, Brown Warbler, Yellow-throated Scrub-wren, Golden Whistler, Eastern Whipbird and Lewin Honey-eater.

At Helenvale we received an unfavourable report on the present state of the roads north of Coen. It proved impossible to obtain confirmation or denial of this from Coen, or to arrange for transport by sea. Rather than risk wasting much time which could certainly be profitably spent further south, we reluctantly abandoned our intention of collecting in the rainforests of the Cape York Peninsula.

8 – 10 *July*. Ayton (52) to Tully (54) via Atherton. We were helped by the Department of Forestry at Atherton.

10 – 29 *July*. **Tully (54)** (Mr Harry Sankey for the Department of Forestry, and Mr Neill Alderman for King Ranch), Queensland. Camp was made beside the Cardstone Road to the Kareeya power station, about 10 miles above the township on the Tully River. The plain of the river and of Jarra Creek was occupied by a patchwork of rain-forest, grassy forest of considerable variety, and grasslands (whose outlines were recognisable from twenty-year-old maps). Much forest was being cleared.

We collected on the flats within six miles of the camp. The avifauna was generally similar to that at Ayton (52).

21 & 23 *July*. **Walter Hill Range (53)** (Mr Allan Seawright for Cairns Regional Electricity Board), between Ravenshoe and Tully, Queensland. These mountains, falling precipitously to the valley of the Tully River and Cochable Creek, form the southwestern face of the Atherton Tableland, which throws the long spur of the Cardwell Range down the far side of the valley. Their flanks were covered in almost unbroken forest. We gained access by the private 'H Road' of the CREB, which runs

obliquely up the range from below the Kareeya power station to Ravenshoe, and collected close to the road at about 2,200 ft. On 21 July we were on a flat-topped ridge where the forest undergrowth was rather open, near a wide swath cut beneath the electricity power line and overgrown with low bushes, and on 23 July in steep forested gulleys with a dense undergrowth of lawyer vine (Plate 9(a)). Though the avifauna was similar to that on Big Tableland (51), we rediscovered the scrub-wren *Sericornis keri* (Galbraith & Parker, 1969).

29 – 31 *July*. Tully (54) to Mount Dryander (57).

31 *July* – 21 *August*. **Mount Dryander (57)** (Mr John Ashton), north of Proserpine, Queensland. Camp was made on Mr Ashton's smallholding beside Deep Creek (a tributary of the Gregory River) at the foot of Mount Dryander. We collected on the lower slopes to about 200 ft near camp (Plate 9(b)), and on the flats and lower slopes near Vine Creek on the western side of the spur. Near camp the flats were mainly cleared for crops or overgrown with rank grass and the forest was edged with dense second-growth. The area near Vine Creek, on the other hand, was grazed by cattle and still covered with woodland and grassy forest. At both points the lower slopes of the mountain, which were bouldery with little exposed soil, supported what we regarded as impoverished rainforest (though it may correspond to the 'softwood forest' of CSIRO, 1953). Of rainforest species we encountered Brush Turkey, White-throated and Fairy Warblers, White-browed and Large-billed Scrub-wrens, Spectacled and White-eared Flycatchers, Rufous Fantail, Golden Whistler, Rufous Shrike-thrush and Lewin Honeyeater. Of these the Large-billed Scrub-wren and Golden Whistler were subspecifically distinct from those at Tully (54).

5 – 8 *August*. **Cape Gloucester (56)** (Mr Billy Dieke of Collinvale Station), north of Proserpine, Queensland. Camp was made at Dingo Beach on the north coast, in huts belonging to Mrs Bobby Blair and Mr John Macpherson. We collected near camp and on an inlet of Sinclair Bay on the west coast. At Dingo Beach the fringing coral reef was backed by sand beach and a narrow strip of dense scrub, with woodland like that at Vine Creek (57) behind. Mangrove Honeyeaters, not seen elsewhere, were abundant in the bushes of a rocky island on the reef, where there were small patches of mangroves. The wide inlet at Sinclair Bay was filled with mangrove thickets of *Ceriops* with *Rhizophora* along the muddy drainage channels, and woodland began directly behind the mangroves. Though a Mangrove Golden Whistler was discovered 700 miles south of its known range (Galbraith, 1967), this was the only true mangrove species encountered.

21 – 31 *August*. Mount Dryander (57) to Melbourne via Brisbane and Sydney, and winding-up. Other commitments compelled McCallum to leave the expedition at Mount Dryander on 11 August, and Pink directly on arrival in Brisbane. We were again given every facility at the Queensland Museum, and equipment was stored in Melbourne at the National Museum of Victoria by courtesy of Mr J. McNally and at the Department of Supply by courtesy of Mr E. A. Wallish. Miss J. E. Thomas as representative of the 3rd Expedition was invaluable in enabling us to complete the final arrangements and to leave Australia on schedule.

3rd Expedition. March – September 1965. South and Western Australia, mainly central areas

B. P. Hall

Personnel

From the British Museum (Natural History): Mrs B. P. Hall, leader, Mr Derek Goodwin and Mr G. S. Cowles. Caterer and cook: Miss J. E. Thomas from Guernsey, who had been working temporarily for the Antarctic Division, Department of External Affairs, Melbourne. For the first four months only: Mr D. G. Nicholls, a zoology graduate of the University of Sydney, and Mr C. D. B. Thomas, both keen naturalists with field experience on Macquarie Island when working for the Antarctic Division. For the last two months: Dr D. L. Serventy of CSIRO Perth, and his former assistant Mr Arthur Mathews. For the first month only: General Sir Gerald Lathbury, a member of the Bird Exploration Fund Committee.

Diary of Expedition

10 – 28 *March.* Preparations first in Melbourne, where Miss Andree Griffin generously housed expedition members, then in Adelaide, assisted especially by Dr W. P. Crowcroft, then Director of the South Australian Museum, who arranged accommodation for us in the Belair National Park, and by the chief preparator for the Museum, Mr Paul Lawson. A visit was paid to the Weapons Research Establishment at Salisbury to obtain permission for the expedition to spend some time during the latter stages at the weather station of Giles (25) in Western Australia.

29 *March* – 5 *April.* **Nonning** Station **(79)** (Mr M. I. McTaggart), Gawler Ranges, S.A. Camp was made about 10 miles south of the homestead. The country was light woodland and scrub, broken by rocky outcrops of the Gawler Ranges in the north and by areas of salt bush and Bluebush (Plate 4(a)). South of the camp was some richer woodland with more eucalypts in which we saw for the only time on the expedition some White-winged Choughs and Dusky Wood-swallows. Several other species which were common we did not see again for several months, including the Jacky Winter and the Hooded Robin.

7 – 21 *April.* **Aroona Valley (82),** north Flinders Range, altitude 1,600 ft near the foot of Mount Hayward (*c.*3,000 ft), S.A. Camp was made near a creek which held some shallow pools of water even in this exceptionally dry year and which was lined with tall eucalypts. There were fairly extensive patches of native pine on the sides of the hills (Plate 4(b)) in which thornbills and Weebills were the commonest birds.

Interesting birds found included some hybrids between the Ringneck and Port Lincoln Parrots, while the Peaceful Dove and the Diamond Dove were found within 400 yards of each other. In retrospect the most interesting thing was that, with the sole exception of the Elegant Parrot, all birds collected were found subsequently in very different types of country, some, like the Redthroat collected near the summit of Mount Hayward, being typical of the mulga of the interior.

Day trips were made to **Parachilna (81)** and **Wirrealpa (83)** on the arid plains to the west and east of the Flinders Range. The only Wedgebills collected on the expedition were obtained on these days, though others were seen later.

21 – 26 *April.* Aroona Valley (82) to Innamincka (64) via Leigh Creek, Blanchewater and the Strzelecki Track. The country after several dry years, was mainly sand or gibber desert with no grass and only some dry scrub; a few trees and bushes along dry creeks.

26 *April* – 5 *May.* **Innamincka** Station **(64)** (Manager, Mr K. Kemp), Cooper's Creek, S.A. Camp was made three miles southeast of the homestead on the bank of the Cooper, which held very little water when we arrived but filled partly before we left, following rains in Queensland. At this point the river runs between high sandy banks lined with thick bush and tall gums and there are numerous floods channels with similar vegetation covering about a quarter of a mile on the south side of the river. In this area a number of birds such as Crested Pigeons, Ringnecked Parrots, Tree-Martins and White-plumed Honeyeaters were very common but there was not the variety of species that we expected in such comparatively rich vegetation. The Chestnut-crowned Babbler had replaced the White-browed Babbler and there was a striking absence of thornbills or whitefaces, even in the more open country away from the river. Pallid Cuckoos were collected for the first time.

Two overnight trips were made back down the Strzelecki Track. Crimson and Orange Chats were fairly common on the sand-dunes and along the dry creeks. A Blue-winged Parrot was found outside its known range.

5 – 9 *May.* Innamincka (64) to Clifton Hills (65), via **Cordillo Downs (63), Cadelga (62),** Birdsville and the Birdsville Track. Some collecting was done before crossing the state boundary into Queensland. The country passed through was still arid but with more bushes and stunted trees than along the Strzelecki. Our only Red-winged Parrot was obtained at Cordillo Downs (63), though others had been seen fleetingly along the Cooper. Our first Brolgas were seen at Cadelga (62).

9 – 13 *May.* **Clifton Hills** Station **(65)** (Manager, Mr F. Wilson), Birdsville Track, S.A. Camp was made in a clump of trees in gibber desert halfway between the homestead and the Diamantina River. Most of the country was gibber but with some trees and vegetation round the artesian bore near the homestead, along the Diamantina itself, and along a few dry creeks. Near the river there were some open sandy areas with low scrub. The effect of the drought was even more marked here than at Innamincka (64) but there was a little water in the river which attracted numbers of waders and ducks. A single Flock Pigeon was seen and the rare Gibber-bird was collected, as well as two species of crow.

15 – 23 *May.* **Etadunna** Station **(66)** (Mrs Oldfield), Birdsville Track, S.A. The property includes Lakes Killalpaninna and Kopperamanna, fresh-water lakes fed seasonally by the Cooper. Camp was made in woodland in the flood plain of the Cooper between the Kopperamanna Bore and the lake. The lake itself was disappointing because the retreating shore-line had left a wide, bare, muddy margin which prevented approach to the pelicans and other water-birds. These proved easier to study at Lake Killalpaninna, which has steeper banks with vegetation closer to the water. Here the ruins of the once-flourishing mission station and farm bear witness to the increasing

desiccation of the country in recent years. Yellow-throated Miners and White-plumed Honeyeaters were abundant round the camp but we found also our first Owlet-nightjar and Blue Bonnets, the latter fairly common. In gums on the edge of a small area of salt bush two species of pardalote were collected close together and we found a small party of Chestnut-rumped Thornbills, the first thornbills seen since leaving the Flinders Range.

23 – 29 *May*. Etadunna (66) to Oodnadatta (67), via Marree. No collecting. When in Adelaide we had been expressly asked not to concern ourselves with the rare Eyrean Grass-wren as some Australian ornithologists were planning a field study of it.

29 *May* – 1 *June*. Cramps Camp, The Neales Creek, 12 miles southwest of **Oodnadatta (67).** Most collecting was done in the gums and bush along the creek and at the water-holes, where there were a number of water-fowl including Pink-eared Duck and Black-tailed Native Hens. The first Horsfield Bronze Cuckoo was encountered, and the first Mulga Parrots since Wirrealpa (83). Sergeant Bruce Evans of the police was particularly helpful in advising us on the best collecting areas here and at our next camp. The drought was broken during our stay by some light rain.

3 – 14 *June*. **Granite Downs** Station **(68),** Alice Springs to Port Augusta main road, S.A. Camp was made 30 miles south of the homestead, at an outstation west of the main road. We were in quite different country of red sand dunes covered in mulga and scrub with some bushes in flower, this being the eastern edge of "The Red Centre". Many birds were different; parties of Turquoise Wrens constantly hopped through the camp; flocks of Bourke Parrot and Common Bronzewings came down to the well after dusk and before dawn; Sittellas, last seen at Nonning (79), were common, as were Robust Thornbills, often found in association with other *Acanthiza* species; a Major Mitchell was seen and a Grey Honeyeater was collected, the first known in South Australia.

16 – 28 *June*. North Well Station (Manager, Mr John Taylor), **Kingoonya (70),** S.A. Mr Taylor made us most comfortable in the shearers' quarters and took a great deal of trouble over showing us the different types of vegetation on the property. These included some small patches of mallee 30 miles southwest of the homestead, large pastures of Bluebush and salt bush, some woodlands and scrub, and a strip of native pine on the shores of Lake Labyrinth. The pine was notable only for its complete lack of bird-life but in the pastures we found several birds new to us, the Samphire Thorn-bill, the Australian Dotterel, the Brown Songlark, and the White-winged Triller. In the mallee we found honeyeaters *Meliphaga plumula* and *M. leucotis* seen only in the early part of the expedition.

29 – 30 *June*. Kingoonya (70) to Eucla (17). No collecting since the South Australian Museum had made collections for us from the Nullarbor Homestead (69).

30 *June* – 6 *July*. Two miles north of **Eucla** Pass **(17),** Eyre Highway, W.A. In the belt of thick mallee which stretches about 15 miles inland (Plate 3(a)). Red Wattle-birds were common, and several Currawongs were seen. For the most part the smaller birds moved in mixed parties. The western form of the Silvereye was found, extending its range eastwards: the Yellow-rumped Pardalote replaced the Striated Pardalote, and we collected our only Golden Whistlers. In the coastal scrub Spotted Scrub-wrens were

fairly common along with Redthroats. Remarkably few non-passerines were seen (no pigeons, few parrots), the Boobook Owl being the only nonpassere noted as "fairly common".

2 & 9 *July*. Nullarbor Plain, north of Eucla (17) and Cocklebiddy (15). On each day a trip was made about 30 miles into the Nullarbor Plain, an almost treeless area of Bluebush and salt bush on limestone. Although on each occasion more than ten square miles were covered on foot as well as the distance travelled in the Land Rovers, only two Nullarbor Quail-thrushes were seen, and they eluded us. Birds were scarce but we collected two White-fronted Chats, some more Samphire Thornbills and a Field-wren. Welcome Swallows were found breeding in the limestone pot-holes.

7 – 10 *July*. Three miles south of **Cocklebiddy (15),** Eyre Highway, W.A. An area of more mixed vegetation than Eucla (17) with some mallee, some mixed woodland and open areas between. Again no pigeons were seen but other non-passerines were more common, in particular a flock of over 300 Major Mitchells were seen on the coastal dunes between the cliff and the sea, Purple-crowned Lorikeets were quite common in the mallee and Brown Hawks nearer the road. A visit was made to the Murraelellevan Cave, four and a half miles west of Cocklebiddy Tank, and skeletons were collected from owl pellets. Most of the bones were of small birds with many honeyeaters, but there were also some of Budgerygahs and quails (probably the Stubble Quail).

10 – 30 *July*. Cocklebiddy (15) to Perth (7), to Mount Shenton (18). No collecting, the time being spent re-fitting, changing the teams and organising petrol and supplies for the last two months. In this Dr W. O. Bazhaw of the Hunt Oil Company was very helpful, arranging for petrol dumps and giving us an introduction to Mr Evan Jenkins, the manager of their desert camp.

30 *July* – 5 *August*. **Mount Shenton** Well **(18),** 75 miles northeast of Laverton, W.A., on land belonging to the United Aborigines Mission, Cosmo Newbery. We were back in red sand and mulga country, not unlike Granite Downs (68) but with more woodland and fewer low bushes and shrubs, which may have accounted for our failure to find any blue wrens in what we had hoped was a transition area between *Malurus splendens* and *M. callainus*. There had been some rain so many flowers were out, some birds were in breeding or post-breeding condition and the Pied Butcher-bird was very vocal in the mornings. The only bird new to us was the Pied Honeyeater.

6 – 8 *August*. Mount Shenton (18) to Warburton (23) on the Warburton Track. A little collecting was done at intervals when looking, unsuccessfully, for transitional blue wrens.

8 – 17 *August*. 18 miles east of **Warburton** Mission **(23),** W.A. The area round the camp was characterised by fairly dense mulga woods (Plate 2(b)) with thorny creepers on the ground, interspersed with spinifex flats, but trips were made out to the more open country of the Gibson Desert (more accurately a semi-desert) west of the Mission, and into the Warburton and Mount Eveline ranges. We were thus collecting in a variety of habitats and for the first time found three species of blue wren present and two whitefaces, the Banded Whiteface being new to the collection. Other birds new to us were Cockatiels and a Grey Falcon which was given to us by children at the Mission.

The ranges were disappointing in producing no species that was not found elsewhere but birds may easily have been overlooked, for throughout our stay the weather was overcast with some rain and a lot of wind, so birds were not active.

20 *August* – 5 *September*. Three miles north of **Giles** Weather Station **(25)** (Mr A. Crowe) Rawlinson Range, W.A. Camp was made in the Pass of the Abencerrages near Sladen Water, which in 1873 had served as a watering hole for Giles and his horses for many weeks. But it was now dry and the spinifex on the ranges and plains was three-quarters dead and held no birds except the occasional pipit. The Rawlinson Range rises above 2,500 feet at Mount Russell and on it we found the Plumed Pigeon *Lophophaps plumifera* quite common but perfectly camouflaged among the red rocks. Also common among the rocks in some parts of the range was the miscalled "grass"-wren *Amytornis textilis,* and one bower-bird was found in a wild fig tree. The Grey-headed Honeyeater was common in the scrub and bushes at the foot of the hills, especially to the north of the range where the vegetation was more varied than to the south (Plate 3 (b)). In the same area we collected our first Western Warbler and two different species of quail-thrush were found in close proximity. Ghost Gums along the creeks provided habitats for some honeyeaters and other birds which depend on eucalypts.

Overnight trips were made out to the **Walter James Range (26)** 41 miles to the northeast, and to the western end of the Rawlinson Range. On the first a mist-net was set overnight at a water-hole, but with disappointing results (perhaps because it was visited too early in the morning in a sudden panic least it should have trapped the last extant Night Parrot! In fact in all our travels through the interior we heard no claims of Night Parrots being found). On the second trip we passed through a big stand of Desert Oaks in which Port Lincoln Parrots were common, but in which the bird-life was not otherwise remarkable.

5 – 7 *September*. **Van der Linden Lakes (24),** Gunbarrel Highway 90 miles west of Giles (25). Although the lakes were dry there was richer vegetation here than at Giles (25) with desert mallee and Desert Oaks on large red sand-dunes. This was first of four two-night camps between Giles (25) and Carnegie (19).

8 – 10 *September*. **Hunt Oil Camp (22)** (Mr Evan Jenkins), Gibson Desert, 45 miles west of Warburton (23) on link road between Warburton Track and Gunbarrel Highway. Most of the collecting was done in an area of half-dead Mulga and scrub on gravelly sand, in which thornbills predominated.

10 – 12 *September*. **Lake Gruszka (21),** Gibson Desert 90 miles northwest of Warburton (23). This fresh-water lake had only been discovered four years before our visit. In the area which we reached we found no standing water but many square miles from which water had only recently receded which were covered with gums, but with no ground vegetation. Birds were scarce among these gums but plentiful along the dry creeks leading to them (Plate 2(a)). We had not time to explore the whole area of the lake and there may well have been water elsewhere.

12 – 14 *September*. **Mungilli Claypan (20),** Gunbarrel Highway 90 miles eastnorth-east of Carnegie (19), W.A. Typical mulga country round the claypan, with the usual birds.

14 – 18 *September*. **Carnegie** Station **(19)** (Manager, Mr Roy Lenke), western end of Gunbarrel Highway, W.A. The presence of stock made the vegetation appear less rich here than in the Gibson Desert; but actually as we had been travelling westwards we had noticed that there were fewer signs of drought, with more flowers about and the spinifex less dead. This amelioration resulted in the discovery of our first genuine "grass"-wren, *Amytornis striatus,* in spinifex 26 miles to the east of the homestead. On the whole, however, there were fewer birds about than at any of the stops in the un-settled parts, perhaps partly due to the attraction of the country to the west which had had good rain, carpeting the roadside with flowers along our route from Carnegie (19) to Perth (7).

At Perth Mr H. J. Frith allowed us to store equipment in a CSIRO store and Faull's Motors undertook storage of the Land Rovers.

4th Expedition. February – June 1966. Western Australia, Southwest and Midwest

C. J. O. Harrison

Personnel

From the British Museum (Natural History): Mr C. J. O. Harrison, leader, and Mr P. R. Colston, who had had previous experience with the 1st Expedition. Mr R. F. Dear, an Australian then working in England who had had extensive experience of Australian Wild-life and work in the bush. Caterer and cook: Mrs C. J. O. Harrison.

Diary of Expedition

5 – 16 *February*. Preparation at the CSIRO stores at Perth and the CSIRO property of Helena Valley, assisted by Mr John Warham, and by Dr Stephen Davies and his staff at Helena Valley. A day was spent accompanying the conservation Research Officer, Mr Riggert, to a lake near Regans Ford, northwest of Gingin, to investigate a rumour of a big mortality of water-fowl from suspected botulism. We found that deaths had occurred some months before but we were able to salvage a few skeletons of the Mountain Duck.

Leaving Perth we spent a night at Coolup where Mr Angus Robinson gave us advice and hospitality. Next day permits to collect in the forests of the southwest were obtained from the local officer of state forests at Margaret River.

17 *February* – 4 *March*. On property by the Homestead (Mr Cliff Kearney), **Scott River (11),** south bank. Camp was made in a small wooded paddock near the river crossing. At this point there was a narrow belt of riverine forest – Karri, Marree and Jarrah – bordered on slightly higher ground by heath with areas of *Banksia* scrub, Black Boys (Plate 1(a)) and some patches of paperbark in dry swamps. This showed that surface water must stand on this higher ground at some periods of the year. At this time the upper Scott River was a series of large pools, and did not flow. Collecting was mainly confined to the vicinity of the camp, but two visits were made to Swan Lake, a large coastal lagoon behind a high beach on the eastern side of the Hardy Inlet. The

Lagoon had a rushy marsh bordering the northern side, with peppermint trees and low scrub on the higher ground, and sparse growth on the dunes themselves. The endemic southwestern White-breasted Robin was found here. Visits were also made to Snake Spring and other areas to the south and southeast but the scrub on the inner dunes was difficult to traverse and some woodland areas were still recovering from fire. The weather was for the most part clear and mild, but there was a fine light rain at times. We were informed that the season was abnormal and that breeding had been up to two months late. We saw a number of species still feeding fledged young. There was also said to be more surface water present than occurred in more typical years. We had evidence of the movement of species along the coastal corridor in the presence of two Shrike-tits and some Magpie Larks. The specimens of the Port Lincoln Parrot collected showed a mixture of characters of the races *semitorquatus* and *zonarius*. The Red-winged Wren was found in the forest by the river.
4 – 19 *March*. Property of Mr J. R. Simpson at **Metricup (9)**, on the east side of Bussell Highway. The country was undulating, with cleared paddocks mixed with large stands of dry sclerophyll forest, much of it on ironstone. Visits were made south to the forests at **Margaret River (10)** where indigenous forest and mature pine plantations were present, and north to the Tuart and peppermint forests at **Wonerup (8)**, north of Busselton. The weather had become more autumnal, with considerable periods of overcast skies and strong southwest winds with intermittent light rain. The range of species encountered was similar to that at Scott River (11). The only two additional species were the Western Thornbill, which occurred in the forest tracts, and the Red-capped Parrot, another endemic southwestern species, obtained when individuals visited the homestead orchard to feed on apples.
20 – 30 *March*. Kiriwina (Mr Christopher Holden). Six miles east of **Mount Barker (12)**, on the north side of the Mount Barker to Porongorup road. The vegetation was open Jarrah forest with areas of heath, *Banksia* scrub and Karri. Much of it had been cleared for pasture but the paddocks were mostly small, with many trees along the edges, and small dams, the landscape having a superficial similarity to English farmland. The property had two small perennial swamps with rushes, paperbarks, and various small trees. The weather was variable, being very warm and dry at times, and some extensive bush fires occurred locally. The presence of a White-breasted Robin at this camp may have been due to a fire that was ravaging the Porongorups. We had been anxious to find the Grey Currawong, another southwestern species, and this was the first camp where we collected it. The birds were orchard-robbers and were difficult to approach. Some species such as the Regent and Elegant Parrots had apparently moved into this area following clearing and cultivation. Red-winged Wrens were found in the small swamps. A marked increase in subcutaneous fat was noted on a number of passerine species collected at this period.

Visits were made to the Porongorups and Stirling Range, both of which to European eyes were disappointing in their lack of altitudinal variation of fauna and flora. Albany was also visited and we were fortunate in being introduced to Mr Jack Higham of the Conservation Committee, to whom we were deeply indebted for introductions to the owners of the next two places which we visited.

30 *March* – 10 *April*. Beckon (Mr John Wimbush), **Green Range (13).** The country here was a mixture of heath and mallee with sparse Jarrah in places. The range itself was a low escarpment on the western border of a slight plateau. The land had been extensively cleared for pasture in this area but most of the settlement was very recent. Apart from one or two fine days the weather tended to be dull, cool, windy and often wet, especially at night; and we were grateful for the shearing shed which was placed at our disposal. One reason for visiting this area was to find the Purple-gaped Honeyeater, which has a restricted southwestern range in Western Australia and which we found fairly numerous locally. Two visits were made to the coast at Warriup, where there were high dunes with sparse scrub and several small lakes. Here we found small parties of the Red-eared Firetail. Cheyne Beach was also visited. Shore birds were very scarce, but the dune lakes at Warriup had a fairly rich avifauna.

10 – 21 *April*. Linwood (Mr George Orr), 12 miles east of **Newdegate (14),** just south of the Newdegate – Lake King road. The country had been extensively cleared but in this part still contained large tracts of mallee (Plate 1(b)), although some of these were due for clearance within a few years. Scattered areas of mallee were also present along the roads and around the homesteads. Changes in the water-table had produced large saline pools in low-lying areas, bordered by dead vegetation; in conjunction with the dams in the paddocks these appeared to attract a number of ducks and waders (Colston & Harrison, 1968). Visits were also made to a small area of heath just east of Newdegate (14). The weather was warmer and drier and winter rains were only just beginning towards the end of our stay. We were able to collect some typical mallee species, and since Lake Grace, 45 miles to the west, appeared to have been the eastern limit of some earlier ornithological studies, we were able to record a number of species further east than their previous noted ranges. The typical birds around the homestead were the wheatbelt form of the Port Lincoln Parrot, the Yellow-throated Miner and the White-eared Honeyeater, while the Southern Stone Curlew was heard at night. The Blue-breasted Wren was the local blue wren species.

21 *April* – 5 *May*. Helena Valley, **Perth (7).** Mainly overhauling vehicles and re-equipping. During this period the Hamersley region had heavy and unusually late rains, up to six inches in twenty-four hours on parts of the coast, and all roads were closed. Perth also had rain and two fine days were needed to dry tents and equipment. While at Perth I assisted Dr Stephen Davies in collecting samples of the White-tailed Cockatoo, from flocks in the Perth pine plantations, and we received from him some useful skeletal and preserved material of the species.

5 – 11 *May*. Perth (7) to Coolawanyah (4) via Wubin, Roy Hill and Wittenoon. We left just ahead of a series of thunderstorms, and spent the first night at a hotel in Wubin, but subsequently slept in the camping tops of the vehicles. The mallee, mulga and spinifex country through which we passed was said to have had its first rain for seven years, and it was obvious that much of the region, with rather poor surface drainage, had been flooded a short time before. There was surface water everywhere and the birds were distributed thinly rather than concentrated around waterholes. Crested Bellbirds, Crested Pigeons and Zebra Finches were very much in evidence; the last species busily nest-building in several localities. The region was rapidly drying out, but

the road surface was poor, there was water in places, and a number of large trucks bogged down near Mundiwindi. The creeks were flowing, and at Ethel Creek where we spent a night, vehicles had been delayed a few days previously. Here Budgerygahs were apparently nesting in the River Gums which were also inhabited by singing White-plumed Honeyeaters and small parties of Little Corellas. The effects of rain were evident in this region in the numerous pools of water in creeks, the greenness of the spinifex, and the extensive growth of short-lived herbaceous plants. In the long grasses of the Fortescue River crossing, north of Wittenoon, a flock of 27 Australian Bustards was seen on several occasions.

11 – 27 *May*. **Chichester Range (3)** and **Coolawanyah** Station **(4).** (Mr Roy Parsons, Owner; Mr Leslie Parsons, Manager). (Specimens collected at the Chichester Range camp and at Tambrey are grouped under the former name, while those collected further east at Coolawanyah, or on the Fortescue River or Hooley Creek are grouped under Coolawanyah). Camp was made at a windmill halfway between the Coolawanyah and Tambrey homesteads, by the Wittenoon – Roebourne Road. This was on the Chichester Range, a low sandstone ridge, broken into small hills and escarpments at intervals, covered mainly with gibber and spinifex, but with River Gums and sparse Mulga along the creeks. The rain had brought richer vegetation and in some places the gibber was partly concealed by a growth of young grasses which gave it a meadow-like appearance from a distance. The weather during our stay was consistently fine and relatively cool, with persistent northeast winds. Two young Emus came daily to bathe at a shallow pool by the camp, and a pair of Brown Hawks were always present. Raptors were both more numerous and more varied here and at Karratha (1) than at our other camps. Blue-and-white Wrens, about to breed or with newly-fledged young, were much in evidence, and a pair of Banded Plover were probably at the northern limit of their range here. There was evidence of other breeding probably associated with the recent rains. A pair of Australian Pipits, found feeding young, must have had eggs when we arrived; Rufous Songlarks were singing in trees along the larger creeks; a Red-plumed Pigeon had eggs; and collected specimens of the Singing Bushlark and Golden-backed Honeyeater showed very enlarged gonads.

A visit was made to the abandoned Tambrey homestead where the Blue-winged Kookaburra was encountered, and a party of the Variegated Wren were seen with newly fledged young. On a visit to the Hamersley Range Colston fortunately found, en route, the only group of Striated Grass-wrens encountered by us and was able to collect some. Birds of various species were collected at Hooley Creek where it crossed the Wittenoon road, and on the Fortescue River flats near Coolawanyah Homestead. These flats are mainly open grassy plains but near the river there is open gum forest with areas of bushes and long grasses bordering small patches of swamp. Here Budgerygahs were nesting in large numbers, and a Collared Sparrowhawk which we collected had been feeding on one. Parties of the Black-tailed Tree-creeper were also present and we were surprised to find parties of the Red Plumed Pigeon here as well as on the local sandstone outcrops.

27 *May*. Coolawanyah (4) to Karratha (1) via Roebourne. Colston saw distant parties of what appeared to be Flock Pigeons on the Pyramid Flats, and on the coastal flats by

the road between Roebourne and Karratha several Australian Bustards and Emus were seen.

27 *May* – 8 *June*. **Karratha** Station **(1)** on the Maitland river near the coast south of Roebourne. Camp was made at Marie's Pool on the river a little below the homestead, and two miles west of the coastal highway. At this point the river cuts through a small sandstone ridge in the flat coastal plain, and after passing over a resistant sill has scoured out a deep permanent pool in its bed. This is surrounded by trees, mostly River Gums, and has small reedheads and rushes in the shallower parts. In addition to this the river had water along most of its length at this time, although a number of flood channels just below the camp were dry. The sandstone ridge was covered with spinifex, but the coastal plain itself had benefited from the recent rains and was an almost continuous sea of grass. The weather was fine throughout the period of our stay. The homestead had a large flock of resident Little Corellas which regularly visited the pool; and Little Wood-swallows clustered to roost inside a shattered hollow treetrunk standing in reeds by the camp. Reed-warblers and parties of Star Finches inhabited the reedbeds and it was interesting to note that while the latter appeared from their enlarged gonads to be about to breed, the gonads were inactive in the Painted Finches of the sandstone ridge. Australian Pipits and Singing Bushlarks were seen carrying food to young in the grass of the plain, and a fledgling lark was seen. Rufous Songlarks were singing along the creek, and Pheasant Coucals were heard in the grass by the dry flood channels.

At its mouth the Maitland River divides into a number of channels dissecting the mud-flats and it was not possible to reach the sea in this area. At the upper limits of the tidal zone the channels were bordered by narrow strips of *Avicennia* mangroves, mostly about 4 – 8 feet high. Although sparse, these held small numbers of many mangrove species, and we visited this area to collect several times. We had glimpses of male Blue-and-white Wrens in low scrub adjacent to the mangroves, and these were very dark, almost blackish in colour.

We also visited **Peg's Beach (2)** at the head of Nickol Bay. The bay provided an extensive area of mud-flats at low tide and at the head of the bay was a good stand of mangrove, mostly *Avicennia* with a little *Rhizophora,* on fairly firm mud. The beach otherwise was shell sand with a thin line of mangroves at the water's edge in small sheltered bays and, adjoining the big mangrove stand, there was a long, wave-worn calcareous beach shelf which appeared to provide a resting place for waders at high tide. The species we found included eleven species of Palaearctic waders (Colston & Harrison, 1968), among them a flock of over 500 Bar-tailed Godwits. A pair of Beach Stone Curlew were also present, and a pair of White-breasted Sea Eagles were carrying twigs to a nest in the dunes. Bar-shouldered Doves were only found on the bushy growth bordering the landward side of the mangroves.

8 – 10 *June*. Karratha (1) to Carnarvon (5) by the coastal road. We called at Mardie and learnt that the Flock Pigeons, which in two previous years had come to Mardie Pool to drink, had not been seen for nearly a year. At Onslow camping site Little Crows were scavenging around the caravans, and a small flock of Star Finches fed among the weeds on the seaward edge. The west coast between Onslow and the Murchison River had not had rain and was dry and dusty with no water in the creeks. Few birds were seen.

10 – 22 *June*. **Bush Bay (6),** 24 miles south of Carnarvon (5), on Brick House Station. The camp site was just behind the beach, two miles off the coast road. Water was scarce. Most of the area was salt bush flats with low dune ridges covered with open scrub. A small samphire marsh behind the beach was the only area we could find with the ground sufficiently consolidated to hold a tent peg. On the seaward side of the beach below tide-level was a narrow zone of *Avicennia* mangroves.

There had been a little rain on the night we arrived followed by more on two occasions during the next ten days. This made it difficult for us to use tracks on clay soil. It may have stimulated breeding behaviour in some birds. Redthroats, Spotted Scrubwrens, Field-wrens and Wedgebills were in song on the dune ridges, the last species singing at night as well. Parties of Blue-and-white and Variegated Wrens were present, some of the latter being seen out in the mangroves. Samphire Thornbills were collected in the samphire marsh. Singing Bushlarks were seen near a bore.

With the kind help of Mr V. Lewis Jnr, who took us by boat, we were able to explore the mangroves of Oyster Creek just south of Carnarvon (5). A flock of White Egrets and several Red-backed Sea Eagles were present. White-breasted Whistlers were collected and Mangrove Herons were seen. On a later visit to this area Colston had a brief glimpse of a Mangrove Golden Whistler.

In the mangroves further south at Bush Bay (6) the Yellow Silvereye and the Dusky Warbler were still present, but there were only single sightings of a Mangrove Heron and a White-breasted Whistler. At Bush Bay the Yellow Silvereyes also occurred in small parties in the dune scrub up to a quarter of a mile inland.

5th Expedition. April – November 1968. Kimberleys, Western Australia and Arnhem Land, Northern Territory

B. D. MacDonald Booth

Introduction

The Expedition had been assigned the northwestern sector of Australia known as the Kimberleys, and tropical Northern Territory, a combined total area of some 220,000 square miles. To condense such a vast tract of country to practical proportions for our field work, it was first necessary to examine its climate and vegetational zones. Both regions have two well-defined seasons, the Wet lasting from December until about April, and an equally long Dry intervening period; factors which greatly influence the breeding and nomadism of these northern birds. Rainfall in the Kimberleys averages 600 – 1,000 mm annually and the whole region is one of woodland and grass savanna. A few widely dispersed collecting camps close to the gallery woodlands of the major rivers were all that was thought necessary to sample the avifauna of this region. The uncertainty of river crossings at the close of the Wet season could have been a problem.

In contrast, tropical Northern Territory presented a much wider variety of climatic and vegetational zones requiring a more specialised itinerary. We decided therefore to sample areas in which those species endemic to northern Australian occur, and selected as our principal field of work:

(a) A coastal strip of monsoon forest east of Darwin;
(b) The Arnhem Land massif, as the most striking and least examined feature of the region;
(c) The estuarine mangroves of the Roper River and the breakaway country of the McArthur River, as representative of the western section of the Gulf of Carpentaria.

Personnel

Major B. D. MacDonald Booth, leader, accompanied from England by Mr D. J. Freeman and Mr C. B. Frith, both from the British Museum (Natural History). Mr R. A. Whistler followed from England a week later. In Perth we were met by Mr W. H. Butler, a naturalist attached to the Western Australian Museum, and Mr A. Hiller a preparator from Sydney Zoo, making a team of six for the first part of the Expedition in the Kimberleys, reduced to four for the second part in Northern Territory.

Diary of Expedition

1 – 23 *April*. Preparations in Perth, aided in particular by Dr D. L. Serventy and the staff of the Division of Wildlife Research of CSIRO at Helena Valley, and by Dr David Ride, the Director of the Western Australian Museum, and his staff. On the 18th we left for Derby and made camp just outside.

1st Phase – Kimberleys, Western Australia

23 – 25 *April*. An overhaul programme on the vehicles was arranged with the Public Works Department workshops, Derby, enabling us to travel to our first base camp whilst the work was put in hand. This plan was fortunate as most roads were closed by rains shortly after we left. Before leaving however Major Sargent of the Army Survey Corps kindly took us to Koolan Island in an aircraft of the R.A.A.F., when we had unparalleled views of much of the southern Kimberley block over which we would later travel. The climate of the region is generally hot and arid. Cyclones are a regular feature of the Wet season when falls are often heavy and accompanied by high humidity near the coast. Wyndham in the northeast has the highest annual mean temperature in Australia. Rainfall exceeds 1,100 mm per annum in the extreme northwest of the Kimberley block, but decreases rapidly to the south and east.
25 *April* – 20 *May*. **Mount Anderson (28).** Our camp was established some 100 miles south of the main Kimberley block, close to the Mount Anderson homestead (Mr Adkins) in the Grant Range where the annual rainfall is about 550 mm during late April and early May the daily maxima average 93°F. dry bulb. These isolated hills rise abruptly in red and ochre tiers above the surrounding grasslands and woodland savanna. To the south the Fitzroy was still in heavy flood after recent rains, its course marked by the gallery woodlands that fringe the banks. Peaceful and Diamond Doves broke the morning silence with their incessant calling, and both Black-faced and Little Wood-swallows sailed close to the camp and about the rock face. It appears that April saw the nesting of the Brown Honeyeater, Yellow-throated Miner and several finches whose

breeding season was evidently just over; the seeds now ripening with the return of the dry weather.

Flowering *Melaleuca* attracted six or seven species of honeyeater and in just two days we examined over 150 birds in the mist nets. The tree steppe held Sacred and Red-backed Kingfishers, both insect eaters of this rather arid country, and Pallid and bronze Cuckoos were invariably heard calling, the former having recently bred. The taller stands of dead grass held parties of Variegated Wren and coveys of Brown Quail and Little Quail, the latter species with young. On the ill-drained soils where the grass layer was sparse the Singing Bushlark was common, and where the clumps of *Triodia* became thicker we found the scarce Spinifex-bird. One of the most noticeable features of this camp was the apparent lack of *Corvus,* though crows were found in some numbers further east up the Fitzroy River.

Owing to the unfordable conditions of the Fitzroy and Lennard Rivers we were delayed from moving north into the Kimberleys, and following a storm on the night of 21 May, when we received nearly 50 mm of rain, we moved site closer to Derby.

22 *May* – 1 *June*. **Derby (29).** Encouraged by our previous visit to this area we set up a collecting camp some 15 miles from the coast among pindan and tree savanna country, which featured the Baobab tree and termitaria on the yellow soils. Excursions were made to Yeeda Creek on the lower Fitzroy, where a specimen of the Black Falcon was obtained, and to a local marsh where the Swamphen was collected. A trip was also made to the tidal marsh and mangroves of **Point Torment (30),** a region notorious for mosquito and sandfly. In oppressive heat with humidity near saturation the sandfly were out in hoards. Mangrove Golden Whistler and the White-breasted Whistler were obtained along with the ubiquitous Mangrove Robin. The visit was however more than repaid by securing two specimens of the "Brown-tailed" Flycatcher, a bird not to be seen again during our field work in the northeast Kimberleys or indeed in the Northern Territory. During nine days visit to this coastal area we were able to add 17 species new to our collections.

1 – 6 *June*. **Geikie Gorge (31).** With the road into the Kimberleys closed by the height of the Lennard River we moved to Geikie Gorge by the junction of the Margaret and Fitzroy Rivers, some 130 miles inland. These rivers drain practically the entire southern range of the Kimberleys, and the Fitzroy was running 7 ft above fording depth just prior to our arrival. Camp was situated in the gallery woodland at the entrance to the overgrown gorge through which the river cuts a tortuous channel between the 200 ft walls of red sandstone. Our powered Avoncraft was the sole means of working the rough breakaway country up river, and as an aid to collecting it proved invaluable. Freshwater crocodile, sawfish and tortoise inhabit the river, whilst the heron, egret and bittern sit motionless in the over-hanging canopy or skulk among the matted stems of the riverside mangrove.

The surrounding ranges were believed to be an overlap zone of the white and brown-bellied forms of the Plumed Pigeon, and we were anxious to collect and study its behaviour and distribution. It was said to be absent in the Oscar Range in spite of seemingly favourable conditions there, so we turned our attention to the St George's Range away to the south.

7 – 10 *June*. **St George's Range (27).** Crossing the Fitzroy River to Gogo (Mr Timothy Emanuel) we followed a track for some 50 miles to the ranges. Travelling first through eucalypt savanna on the black and grey soil plains, where we shot our first wild boar, we moved out into the more arid acacia scrub, and finally to the barren plains of spinifex north of the sandhill country. There had been unconfirmed reports that the Night Parrot was seen here in recent years. In the foothills of the Range we found the white-bellied Plumed Pigeons in some numbers. The dark reddish-brown stone and pebble flats among the broken and tumbled rock were the most favoured by this pigeon, the warm hues of its plumage blending perfectly with the environment. Among a belt of flowering *Grevillea* the highly nomadic and rather scarce Black Honeyeater was obtained.

13 – 22 *June*. **Mount Bell (32).** By 12 June we had travelled north across the Lennard River to enter the main Kimberley block through Inglis Gap in the King Leopold Range, making camp at 1,800 ft near the base of Mount Bell (Plate 10(a)). Here beneath the eucalypt savanna the grass was rank and dense, whilst the adjacent quartzite ranges were covered with a thin layer of spinifex, some shrubs of *Acacia* and *Grevillea* and a scattering of the smaller eucalypts. We were now in a zone of 600 mm annual rainfall, comparable to our camp near Derby (29) barely a hundred miles to the west, yet a change to the more tropical elements of the Kimberleys was already noticeable. Along the course of a clear-water stream and above the canopy of *Pandanus* and flowering gums screamed Varied and Rainbow Lorikeets; Bar-breasted and Rufous-throated Honeyeaters crowded the flower clusters. Silver-crowned Friar-birds, Blue-faced Honeyeaters, and Olive-backed Orioles were all seen for the first time. An Azure Kingfisher frequented the stream in which lived fresh water tortoise. Rock wallaby were common and the Frilled Lizard *Chlamydosaurus kingii*, Possum *Petaurus breviceps* and Echidna *Tachyglossus hallucatus* were all found about our camp. The Grey Shrike-thrush was often heard among the dense foliage and perhaps the most interesting bird of this camp was the White-quilled Rock Pigeon which haunted the shade of the narrow rock ledges.

Mount Bell was a successful camp marred only by the constant droves of cattle fly that blacken anyone moving about in this hill country. We could not have wished for a better introduction to birds of the Kimberleys, a number of species being added to our collections that we had not come across in our camps outside the King Leopold Range.

Moving northeast we came to Mount House on 22 June, the basalt hills and latterly the homestead there being named after Dr House, the naturalist accompanying Brockman on his exploration of the Kimberleys in 1901. Mr and Mrs Richard Blythe kindly allowed us to use the station as a base for supplies and petrol sent up in preparation for the next four weeks field work in the central Kimberleys. We now moved east, first skirting the vast shale plains where a few small *Eucalyptus* and *Melaleuca* grow in the scattered pockets of soil, and then north into the savanna woodland that covers the greater part of the Kimberleys block receiving between 600 to 1,000 mm annual rainfall.

23 – 26 *June*. **Mount Caroline (33).** We made camp that evening close to the entrance of a spectacular gorge on the Hann River. Cascading down a series of rapids the river

winds its way through sheer walls of red quartzite that here and there encompass small but quite dense pockets of gallery woodland. Huge tumbled boulders lie strewn at the foot of the cliffs, and this was the haunt of the beautiful Brown-breasted Shrike-thrush, an extremely shy and furtive bird with a melodious song that is greatly amplified by the surrounding rock face. It was curious to find the Osprey so far inland, as we were later to do on the Manning Creek.

27 *June* – 6 *July*. **Joint Hill (34).** This highland plateau rising to 3,000 ft lies in a northwest-southeast line for some 300 miles between latitudes 16° and 18° S, dividing the two great watersheds of the Kimberleys. It was on these highlands that three collecting camps were set up to sample as much of the plateau as possible, and the first of these was established close by the Barnett escarpment some 18 miles south from where the Brockman Expedition discovered the Black Grass-wren *Amytornis housei* in 1901. Here three of us set off towards Joint Hill whilst leaving the three remaining members to collect locally. The going was extremely rough across boulder-strewn gullies which later gave way to spinifex flats. On 2 July, just four miles short of Brockman's camp, we found two parties of *Amytornis housei* on an outcrop of rock that rose up out of the surrounding woodland savanna. The outcrop was heavily eroded, a jumble of fallen blocks of dark red and black sandstone with matted spinifex concealing the deep clefts between the rock. The field party returned on 4 July to find that the others had also collected this grass-wren by the main escarpment. The re-discovery of this isolated species after 67 years well repaid the planning and trouble that had been taken to find it. Our first example of the lovely Purple-crowned Wren was collected in the riverside grass of the Manning Creek. Also a specimen of the Green-backed Warbler was obtained, a most unusual find for a species that was thought to be restricted to the mangroves of the coast, the nearest point of which was 150 miles distant.

6 – 15 *July*. **Mount Elizabeth (35).** Moving on north again we made a temporary camp in savanna woodland beside the upper Hann River from 6 – 11 July at 1,500 ft (Plate 10(b)). From here two parties were to examine the higher ranges bordering the Chapman River and the headwaters of the Charnley south of Mount Jameson. The Army Survey Corps, which had a small team operating in the area, reported that they had failed to locate the Chapman River track by helicopter, so we moved west towards the Charnley where it was hoped to set up a collecting camp. Birds collected from the upper Hann River have been listed with those from Mount Elizabeth.

This is one of the most remote cattle stations of the Kimberleys, and Mr Lacey's advice was most helpful in our search for a way through to the Charnley. After making 26 miles, steady rain set in which threatened to isolate us, and re-tracing our tracks we established another temporary camp at 1,950 ft where we could spend two useful days collecting by the headwaters of the Hann River. Here there were noticeably denser stands of woodland than we had seen elsewhere in the Kimberleys with two quite distinct tree layers and some shrub undergrowth. It was interesting to see the White-breasted Wood-swallow so far inland. The Red-backed Wren was common among the low shrubs and never far from water, and large numbers of the beautiful Gouldian Finch fed among the stony flats beside the river.

The Expedition moved south back to Mount House on 16 July to collect stores and fuel in preparation for our journey to the Durack Range on the eastern flank of the Kimberleys. After four days of incessant rain we left for Lake Gladstone some 40 miles to the southeast.

20 – 24 *July*. **Lake Gladstone (36).** This hill lake fringed with tall beds of rush was the home of the Grass and Water Whistling Ducks. Rafts of Green Pigmy Geese, the first we had seen in the Kimberleys, were astonishingly well concealed among the carpet of giant water lilies that hid much of the lake. Apart from the few Royal and Yellow-billed Spoonbills, Straw-necked Ibis and egrets, there were not the number of aquatic species we had hoped to find. Swathes were cut in the rush for our mist nets though no amount of walking up would induce Reed-warblers to trap themselves.

25 – 27 *July*. **Mount Remarkable (37).** After a few days at Lake Gladstone we moved east some 90 miles to Elgie Cliffs, an abandoned homestead 26 miles southwest of Mount Remarkable in the Durack Range. This range runs as a wall some 2,000 to 3,000 ft in height to encircle the Kimberleys to the east and south for nearly 200 miles, before joining up with the King Leopold Range which completes the crescent to the western seaboard. Camp was made at Elgie Cliffs, 26 miles west-southwest of the peak. The Black-backed Magpie became more frequent and less shy as we moved east, and its presence in the central Kimberleys is thought to be rather recent. Among the more arid tree steppe to the east of the range we obtained two specimens of the Ground Cuckoo-shrike, an astonishingly difficult bird to approach.

27 – 30 *July*. **Lissadell Hill (38).** Once off the Kimberley plateau we were free to turn north to Wyndham on the coast, stopping on the way at Bow River southwest of Lissadell Hill where the vegetation was similar to that of the main Kimberley block.

31 *July* – 11 *August*. **Wyndham (39).** Wyndham lies at the extremity of the northeast Kimberleys on the West Arm of the Cambridge Gulf, into which drain the major river systems of the eastern plateau. The aim of our collecting here was to sample the mangroves in the areas of the Forrest and King Rivers and for this purpose Mr Harris, the District Engineer, most kindly put the pilot boat at our disposal, without which we could not have coped with the terrific tidal race that occurs in the Gulf. To Mr Smith, the the coxswain, we were grateful for the patience he showed to those of us who explored the inundation flats and mangroves, which were shown on his 1894 chart as being an area "abounding in obnoxious flies". Due to the heavy silting off the mangroves we were obliged to use our 8 ft tender in getting ashore, and for travelling up the narrow and tortuous creeks it was invaluable. In the ten days spent on and off-shore some 60 skins were obtained of mangrove species including the whistlers *Pachycephala melanura* and *lanioides* and the Mangrove Robin, for comparison with those obtained at Derby over 500 miles around the coast to the southwest. Two specimens of the elusive Chestnut Rail were collected though we failed to collect the White-browed Crake, which was only seen for sure on one occasion. The Broad-billed Flycatcher was also new to our collections.

2nd Phase – Northern Territory

11 – 12 *August*. **Victoria River (40).** We left Wyndham on 11 August *en route* for

Arnhem Land. After 200 miles we came to the Victoria River where it flows through a gorge in the Stokes Range. Here we searched for the White-quilled Rock Pigeon *Petrophassa albipennis* in the hope of defining its eastern limit of range or of finding an overlap zone with *Petrophassa rufipennis* if one existed. We made camp and that evening managed to obtain a small series of this pigeon in the boulder-strewn country of the gorge. It was extremely interesting to note that the characteristic white wing-patch of this bird was absent or very nearly so. This appeared to be an isolated population, for the country through which we had travelled was mostly grass or woodland savanna.

12 *August* – 3 *September*. **Pine Creek (44).** The following day we passed through Pine Creek and taking a track east we entered quickly upon an area of woodland savanna that extended to Wandie Creek where we had arranged to set up camp alongside Mr and Mrs John Serventy, who were here extracting alluvial tin. Dr D. L. Serventy and his wife had flown to Darwin from Perth and had arrived just ahead of us. A ten day break had been planned before starting our field work in Arnhem Land. Whistler now had to return to England and Butler to Perth, thus leaving four members in two Land Rovers for the second phase of the Expedition.

Encouraged by what we had seen at Wandie it was decided to stage our first collecting camp there. The country had many features in common with parts of the Kimberleys, much the same savanna woodland but with much taller stands of rank grass. It was considerably warmer, with daily maxima averaging 99°F dry bulb during the last week of August and a very noticeable increase in humidity, and winds light to moderate easterly. The Dusky Honeyeater was a bird new to us as was the Golden-winged Parrot which is here said to nest in the termite mounds, some of which are 12 ft or more in height and are a useful indicator of the type of soil on which they are built. At night we obtained the Tawny Frogmouth, Spotted Nightjar and Owlet-nightjar, though this latter species was not nearly so common as it had been in parts of the Kimberleys.

4 – 27 *September*. **Darwin (41).** On 4 September we left Wandie and moved north towards Darwin (41) and into a rainfall zone of 1,500 mm. We made camp in tropical savanna woodland close to Noogoo Swamp just two miles from the mangroves of Hope Inlet. Coastal northern Territory was a welcome contrast to the monotonous savanna woodland of the Kimberleys in spite of the ever increasing humidity and mosquitoes.

An early search was made for the Streaked Grass-warbler *Cisticola juncidis leanyeri* in the Leanyer Swamp, though as this appeared to be dried out we turned our attention to Noogoo Swamp where we collected it alongside the more common *Cisticola exilis*.

The vegetational patterns of the coastal belt were remarkable and there was an astonishing transition among the species occupying these different zones. Our camp lay beneath tall stands of eucalyptus woodland some two miles from the shore, and certain tropical elements such as the Spangled Drongo and the Dollar-bird were already noticeable. Little more than a mile away the stands of trees quite suddenly became much taller and the canopy then thick and continuous as one entered a monsoon belt. Here among the heavy crowns of the trees and feeding on fruit were black-and-white Torres Strait Pigeons, Red-crowned Pigeons, Olive-backed Orioles and Yellow

Figbirds whilst in the under-storey were Varied Trillers and on the bare patches of burnt scrub at the foot of the trees were the beautiful and shy Green-winged Pigeons. This strip of forest (sometimes referred to as pseudo-rainforest) was barely more than 300 yards in width before being confined by a narrow band of inundation flats. Across the flats lay a tangle of dense monsoon thicket that covered the sand ridges fronting the shoreline. This belt of thicket, covered in vines and trailing thorn, was carpeted with sticks and dry leaves, much of the debris being used by the Scrub Fowl to build its gigantic nesting mounds.

The prominent species of this zone were the Melville Island Friar-bird and the Brown Whistler which was evidently breeding in August and also whilst we were there in September. Others included the rather scarce Cicada-bird and the extremely shy and furtive Rainbow Pitta which was only to be found in this dense thicket zone that fringed the shore.

Creeks had in many places cut deep into the land and were bordered by bright evergreen mangrove that often became so thick and inter-twined as to be almost impenetrable. The tree mangrove, standing on aerial roots and reaching 45 ft or more, formed a complete canopy that precluded much light even at mid-day. Here were Red-headed Honeyeaters, Shining Flycatchers, Little Shrike-thrushes, and the strident calls of the Black Butcher-bird which rang out across the mangrove and echoed through the adjacent woodland.

These four quite distinctive zones all held species that were rarely encountered in more than one of them, and yet this belt of coastal monsoon vegetation was barely two miles in depth. The birds of this camp were overwhelmingly tropical and some 23 species were new to our collections.

28 *September* – 18 *October*. **Oenpelli (43).** We were now ready to examine part of the Arnhem Land massif to the east, and on the 27th crossed the Adelaide, Mary and South Alligator Rivers. The following day we came to the East Alligator River that marked the western boundary of the 31,000 square mile Native Reserve of Arnhem Land. We were especially grateful to the Native Welfare Department of the Northern Territory for authority to enter and collect in this vast reserve, the interior of which is still mostly unexplored. The main massif is scored by deep canyons and valleys, the rock heavily eroded and often clothed in scrub or hidden by trees, and the narrow interlaced ravines are choked with thicket and vine. Between the ranges and the sea the open grass flood plains succeed one another and are only interrupted by lagoons and wooded outcrops of rock. This northern sector of Arnhem Land is said to receive between 1,575 – 1,650 mm (62 – 65 in) rainfall, and it was noticeable how the monsoon forest, or pseudo-rainforest, extended right back into the foremost valleys of the plateau itself some 45 miles from the coast.

Mr Cox of the Mission gave us every assistance and with his advice we made camp some two miles from the East Alligator River close by the main escarpment. Here buffalo graze everywhere on the adjacent plains and around the swamps and lagoons that held large assemblies of Magpie and pigmy geese, egret, ibis and spoonbill. Darters and Lotus Birds were common, and a dominant feature of the camp were the larger raptors.

Whistling Eagles were breeding everywhere, and some of their eyries held carapaces of the fresh water tortoises. The magnificent White-breasted Sea Eagle was often above our camp and we knew of one eyrie containing young back in the hills. Less numerous were the Black-breasted Buzzard, Red-backed Sea Eagle, Osprey, and the beautiful Black-shouldered Kite, a bird of the more lightly wooded country.

In spite of the wealth of bird life around us the attraction lay primarily in three species, all of them endemic to Arnhem Land. The first, the White-lined Honeyeater *Meliphaga albilineata* was found almost on arrival, and following the practice of McLennan who first collected this bird in 1915 we found we could attract it by remaining motionless and imitating its distinctive whistle. It was certainly more common than we had been led to expect, and later during an extremely arduous trek across the heavily ravined ranges we found its nest, which had not previously been described.

The Black-banded Pigeon was also found close to the main escarpment but only in the denser stands of woodland there or among the fruiting plums and *Ficus* of the gorges. It was remarkable how even small parties of this distinctive black-and-white pigeon could remain concealed, the disruptive pattern of the light hiding them in the sun-drenched canopy of the leaves. Until quite recently this pigeon was thought by some to be almost extinct.

The third species, the isolate White-throated Grass-wren *Amytornis woodwardi*, we found after several days searching among the spinifex clad sandstone ranges where it can so easily remain undetected.

On the 14th and 15th a journey was made to the **East Alligator River** mouth **(42)** to secure the Yellow Chat. The grass of the coastal flood plains was dry and withered and the ground, heavily pug-marked by buffalo, was baked hard. At the height of the dry season the bird's distribution must be patchy and it is then probably to be found along the fringes of the mangroves. Unfortunately we could not devote any further time to search for this species. Practically the only bird to be seen out on this barren plain was the Lemon-breasted Flycatcher which kept to the scattered line of bush along the river bank. Shortly after dark two Spotted Cat *Satanellus aculeatus* rummaged around our camp and a specimen was collected. Returning through the monsoon woodlands we came across small parties of Partridge Pigeon, the first we had seen, although we had been on the look out for it since leaving the Kimberleys. We were later to meet with it again in greater numbers through the eucalypt woodland to the west of the river.

A bird of special interest to us was the wren *Malurus dulcis* for we were keen to define the limits of its distribution to see if an overlap zone occurred with *Malurus lamberti*. The wren was found among the patches of spinifex and *Hibiscus* covered sandstone, where it was extremely adept at running along the shelves of rock and over the tumbled boulders.

During the three weeks we were camped near Oenpelli (43) the daily maximum temperatures were high, averaging 101.4°F. dry bulb and daily 72.1°F. With the onset of the rains not far away the nights were humid, and in the early mornings the mean relative humidity was 84.1%.

19 – 25 *October*. We travelled back to Darwin (41) and Pine Creek (44) to despatch material to Perth and to replenish stores, and thence continued south and east to the estuary of the Roper River, an overall journey of 540 miles around Arnhem Land, to make camp on the south side of the massif.

26 *October* – 3 *November*. **Mount Roper (45).** The Roper river skirts the southern ranges of Arnhem Land for some 200 miles before emptying into the Gulf of Carpentaria through a broad and ominous crocodile-infested estuary fringed by tall evergreen mangrove and inundation flats. The rainfall of this region is less than half that of the north face of Arnhem Land, being about 700 mm only. Our camp was beside a murky lagoon some 30 miles up-river, though we had to repair to St Vidgeons for supplies of fresh water. The lagoon provided us with the most un-prepossessing looking cat-fish up to about 15 lb, and was, as ever, infested with mosquito. Our attention was centered on the mangrove belt by the mouth of the river. These mangroves lay across a shimmering expanse of sand and baked salt flats, glaring in the sun and subject to mirage. Small parties of Australian Avocet, White-headed Stilt and Sharp-tailed Sandpiper were concentrated on the few remaining pools, around which ran Australian Pratincole.

Small tidal creeks ran beneath a canopy of mangrove and tall riverine woodland and held both Mangrove Golden Whistlers and White-breasted Whistlers, Rufous Fantails and the Shining Flycatcher. Breeding species included the Restless Flycatcher and the Northern Fantail. Everywhere the black mud seethes with life at low tide and among the protruding roots of the mangrove are the countless burrows of the cunning and bold mangrove crabs. It was not surprising therefore to find the Great-billed Heron.

Further up-river and among the tangled thicket we obtained the White-browed Robin which would have passed unnoticed were it not for its rather pleasing whistle. Two other species of interest were the Hooded Robin which was collected on the lightly wooded slopes around Mount Roper, and the undoubted sighting there of the Grey-headed Honeyeater a long way north of its known range.

4 – 6 *November*. Whilst travelling from the Roper to the McArthur River a census was made during two night runs of the species encountered on the woodland tracks. The Spotted Nightjar had replaced the Owlet-nightjar which was the predominant species recorded at night in the Kimberleys. Australian Bustard were also seen on these runs.

7 – 18 *November*. **Borroloola (46).** The first of two camps was made at Caranbirini waterhole close under the sandstone ranges 20 miles from the settlement near the McArthur. The river enters the sea some 120 miles southeast of the Roper and this was the eastern limit of our field work along the Gulf of Carpentaria. Having earlier obtained the grass wrens *Amytornis housei* and *woodwardi* we were now anxious to locate the third isolate, *Amytornis dorotheae,* which had only been recorded from this locality. The first week proved fruitless although we obtained the Brown-breasted Shrike-thrush *Colluricincla woodwardi,* and also managed to find a nest, with young, of this species near camp. This was the fifth disjunct population we had come across during the Expedition. On the 14th we moved camp to the south side of the river close to the junction of Clyde Creek. The McArthur River here cuts a way through the spinifex-clad ranges, with great stands of gallery forest growing from its high detritus banks. Its course at the end of the dry season is little more than a succession of pools

though the effects of flood damage could be seen high up among the trees, and the old catwalk bridge to the abandoned homestead hangs some 54 feet above the ford beneath.

Our camp was situated close to the river on the alluvial flats covered with coarse burr-laden grass and patches of naked thorn scrub. As darkness fell hosts of fruit bats gathered in the flowering gums above, squabbling and screaching, and sending a constant rain of debris upon us. The high silt banks of the flood channels were clothed with rank cane grass up to 10 ft in height and here the beautiful Purple-crowned Wren was common. For the next four days an extensive search was made for *Amytornis dorotheae* among the ranges, with a close watch being kept at dawn before the sandstone hills became unbearably hot. The colouring of these wrens so exactly matches that of the rock among which they live that they might easily escape detection, but it might be expected that our field experience with the other two isolate wrens would have enabled us to turn up *A. dorotheae*. Spinifex fires had blackened parts of the ranges and one simply does not know if these may have affected its distribution or to what extent the species may be endangered.

Among the lightly-wooded amphitheatres of the hills we collected the Grey-headed Honeyeater, which tends to confirm our sight record from the Roper. Other species of interest were the scarce Channel-billed Cuckoo, seen on two occasions, and the White-browed Robin, a most handsome bird that was obtained among the riverside thicket. Although we were constantly on the look out for the Painted Honeyeater we found no trace of it in spite of an unusual amount of the mistletoe growing beneath the hills fronting the sea.

The Wet season was almost upon us. During our ten days on the McArthur the temperature steadily increased, the daily maxima averaging 105.5°F. dry bulb. Humidity too was high at night. On the evening of the 17th a violent storm broke whilst two of the party were out on the ranges, and the following morning we all dried out and struck camp. We moved site to the unoccupied McArthur River mining camp where Mr Morris, the Manager, kindly gave us air-conditioned quarters for the night. This was something of a change after some seven months on trek. The following morning we set out to drive the 200 miles to Alice Springs and to Perth.

19 – 25 *November*. Our route lay through Newcastle Waters and Tennant Creek and south of Alice Springs we encountered storms and heavy flooding. South Australia was bitterly cold after our camp by the Gulf of Carpentaria. Hiller left us at Port Augusta and the three of us then continued west along the Nullarbor Plain, arriving in Perth on the morning of the 25th. We had been away over seven months and had completed a total of 16,500 miles.

South Australian Museum Expeditions. October – November 1963 and 1964

B. P. Hall

Two small expeditions from the South Australian Museum were partly financed by the Fund. The birds collected were divided between the British and South Australian Museums, some of the latter being sent on loan to London for examination. Only these specimens among the South Australian Museum share are included in this report.

The chief value of the specimens from these expeditions is that they were collected at a different season from those of the main expeditions, but unfortunately there are few data with them either of gonad condition, impermanent colours, habitat or behaviour. Because of this, generalisations in the discussion on these characters may not be applicable to these birds.

Collecting localities (all in South Australia)

1963. 3 October. **Kimba (78).**
3 – 9 October. **Nullarbor HS (69).**
13 – 14 October. **Sceale Bay (75).**
15 October. **Elliston (76).**
16 October. **Port Lincoln (77).**
17 October. **Whyalla (80).**
19 November – 1 December. **Angepena (84),** 45 miles west of Lake Frome, from which visits were made to Wertaloona, 13 miles west of Lake Frome.

1964. 23 – 30 October. **Nullarbor HS (69).**
1 November. **Yardea (72).**
1 November. **Minippa (73).**
2 – 3 November. **Lake Acraman (71).**

SYSTEMATIC LIST OF BIRD SPECIMENS

General remarks

The systematic section follows a standard format though the treatment of different families may reflect to some extent the special interests of the authors concerned. The name of the principal author (who should be treated as the sole author of the section when quoting from the Report) follows the heading for each division (Order or Family) with the names of assistants in brackets.

Information is divided under headings "Distribution & Abundance", "Habitat", "Behaviour", and so on, but where there is little or no information some paragraphs are combined under the general heading "Field Notes" or omitted, as appropriate. Data have been taken from the specimens and data sheets and may include facts which are already well known, but in the absence of a handbook on Australian birds it was felt that a useful purpose could be served by making this section more comprehensive than is usual in a systematic paper.

The sequence of the CSIRO Index (1969) and the scientific and common names used therein have been followed uncritically unless otherwise stated. Small amendments, chiefly in hyphenation and terminations, have been made without comment, in the interests of consistency or to conform to the International Code of Zoological Nomenclature (Article 27).

Specimens

Specimens are listed geographically under numbered collecting localities, with dates and individual weights. Unsexed specimens are listed as "ø". Young birds not fully grown are listed as "juv", and those fully grown as "imm", more precise terminology being used in the text where appropriate. Specimens made into rough skeletons are listed as "skel" and those preserved in fluid as "fl".

Specimens to other Museums

A number of specimens collected on the 1st Expedition were given to the National Museum of Victoria, Melbourne (NMV) and the Australian Museum, Sydney (AM) and some from the 2nd Expedition to the Queensland Museum, Brisbane (QM): none of these was available for examination when the report was compiled, so, although information from their data sheets is included, remarks on plumage condition and variation of series may not be applicable to these birds. A few specimens presented to the South Australian Museum, Adelaide (SAM) and the Western Australian Museum, Perth (WAM) were retained in London until the report was completed.

Distribution & Abundance

Incorporated into this section are sight records kept for all main camps on the 1st Expedition by P. R. Colston (PRC); by the 2nd Expedition for all species at Glendower (55) and for some species at other camps; by the 3rd and 4th Expeditions for all camps except Carnarvon (5); by the 5th Expedition for the Kimberleys generally (28 – 38) with more precise data on some species, and for all main camps in Northern Territory (41, 43 – 46). Except in northern Queensland and Carnarvon (5) and individual Kimberley camps, evidence is therefore provided of the birds which were absent as well as those which were present. Relative abundance was noted by the 3rd and 5th Expeditions, and elsewhere has been assessed as far as possible from the data sheets.

Habitat

The precision with which habitat could be described varied greatly from individual to individual (many collectors being on their first visit to Australia) and from area to area, according to the knowledge and helpfulness of local contacts (see Glossary of Botanical Terms). The 5th Expedition followed the terminology of Keast (1961). Authors have had to collate and interpret descriptions on the data sheets as concisely as possible, but a file of photographs retained in the Bird Room is available for students wanting more precise information.

Behaviour

Evidence on behaviour has been collated from data sheets. Observers' initials are added where detailed descriptions of behaviour or song are quoted (see p. 37).

Food

Evidence on food (including undigested stomach contents) has been collated from data sheets.

Annual Cycle

Data relevant to the time of breeding and the sequence of moult are included. Immaturity in birds has been assessed on their size, plumage, colour of mouth, eye, bill and legs, and (in part) on the state of pneumatisation of the skull (see p. 2). In the latter character too little is known on the length of time required for different species to acquire full pneumatisation to use this evidence independently.

Impermanent Colours

The facts are summarised as far as practicable from the data sheets. If differences can be related to age, sex, season or locality, this is done. The 2nd Expedition in an attempt to attain precision compared colours with *Stanley Gibbons colour guide for stamp collectors* No 3333. Initials of the colour names and numbers used in this guide were entered on the data sheets. Unfortunately use of the guides sometimes involved a delay in recording colours. Any time lapse before comparison has been entered on the data sheets (but not on the labels), and as far as possible has been taken into account in the report when evaluating variation in colour of parts such as the eye and bare skin which are liable to fade quickly.

Geographical Variation

Variation in colour, pattern or proportions exhibited by the specimens is noted, and the extent to which this agrees or disagrees with currently recognized subspecies. Other specimens in the British Museum are taken into account where relevant, but these are mostly old skins lacking data so that it has rarely been possible to discuss usefully variation in areas not covered by the expeditions.

Data available on data sheets and labels, but not included fully in report

(a) For each expedition the collector of each specimen can be identified by the letter(s) prefixing the collecting number as follows:

AH	A. Hiller	H	B. P. Hall
AM	A. R. McEvey	HB	W. H. Butler
B	R. Boswell	J	I. C. J. Galbraith
BB	B. D. McDonald Booth	L	(2nd) R. L. Pink
CF	C. B. Frith	L	(3rd) G. W. Lathbury
D	D. Goodwin	M	(1st) J. D. Macdonald
DF	D. J. Freeman	M	(3rd) A. Mathews
E	E. H. Galbraith	N	(2nd) I. McCallum
G	G. S. Cowles	N	(3rd) D. G. Nicholls

P	P. R. Colston	X	(2nd) J. Ashton
R	R. Lossin	X	(4th) R. Dear
RW	R. A. Whistler	Y	C. J. O. Harrison
S	(2nd) S. A. Parker	Z	L. Henry
S	(3rd) D. L. Serventy		
T	(2nd) T. Andersen		
T	(3rd) C. D. B. Thomas		

(b) Precise collecting locality, altitude, time of collection and weather conditions.
(c) Time lapse before weighing.
(d) Colour identification of impermanent colours according to Stanley Gibbons colour guide, and time lapse (if any) before comparison was made (applicable to 2nd Expedition specimens only).
(e) Full data on gonads, skull, habitat, etc.

DROMAIIDAE — Emus
D. Goodwin

Dromaius novaehollandiae Emu

Birds or tracks were seen widely in Western Australia (18-20, 28, 31) being fairly common at Mount Shenton (18) and Mount Anderson (28) – once one party of 20. In Northern Territory seen only at Oenpelli (43). In Queensland recorded at Glendower (55) and in the south (58, 60). In South Australia four at Clifton Hills (65), fairly common at Etadunna (66), but nowhere else in the centre (64, 67, 68, 70, 82). Seen at most camps in Victoria (88-90) and in New South Wales at Dunedoo (93).

SPHENISCIDAE — Penguins
D. Goodwin

Eudyptula minor Little Penguin

Specimens. WESTERN AUSTRALIA. Green Range (13) 8 Apr 66, 1 skel. Picked up dead high in the sand-dunes behind the beach at Warriup.

PROCELLARIIFORMES — Albatrosses & Petrels
D. Goodwin

Diomedea exulans Wandering Albatross

Four seen at Cheyne beach, Green Range (13).

Diomedea chlororhynchus Yellow-nosed Albatross

Six seen at Warriup, Green Range (13).

Macronectes giganteus Giant Petrel

Specimens. WESTERN AUSTRALIA. Green Range (13) 7 Apr 66, 1 skel. A dry carcase picked up on the beach at Warriup.

Puffinus carneipes Fleshy-footed Shearwater

About fifteen seen at Cheyne beach, Green Range (13).

PELECANIFORMES — Pelicans, Darters, Cormorants
D. Goodwin

Pelecanus conspicillatus Australian Pelican

About 500 were seen at Etadunna (66) and it was plentiful at Oenpelli (43). Elsewhere in small numbers from scattered localities (28, 31, 45, 58, 88).

Anhinga rufa Darter

Specimens. NORTHERN TERRITORY. Oenpelli (43) 7 Oct 68, 2 fl (1,373.2, 1,418.2 g). Pine Creek (44) 23 Aug 68, 1 fl (1,473.8 g). Mount Roper (45) 31 Oct 68, 1 fl. Borroloola (46) 15 Nov 68, 1 fl (1,478.2 g).

Distribution & Abundance. In Northern Territory fairly numerous on the lagoon at Oenpelli (43) but in small numbers elsewhere. Also seen in the Kimberleys, two at Geikie Gorge (31); in Queensland at Glenore (49), Thylungra (60) and Monkira (61); in South Australia four on Cooper's Creek at Innamincka (64) and several on Lake Killalpaninna, Etadunna (66); in New South Wales at Grafton (94) and Tenterfield (95).

Habitat. In Northern Territory on or near inland freshwater lagoons and at Pine Creek (44) on creek in savanna woodland. In South Australia on remaining rather lake-like areas of water in Cooper's Creek (64), where surrounded by rather dense riparian woodland, and on Lake Killalpaninna (66) in open sandy country.

Behaviour. Spend much time perched on some branch protruding from or near the edge of the water. When considerably disturbed will circle up high in the air for half an hour or more. Sometimes seen thus circling, like raptors on a thermal, when there had been no apparent disturbance (BB).

Food. Fish and vegetable matter (from vegetarian fish?).

Annual Cycle. No information.

Impermanent Colours. *Legs* straw yellow or off-white with yellowish buff or yellowish brown webs, yellowish black in one specimen from Oenpelli (43). *Iris* off-white to yellow, with greyish speckling in the Oenpelli (43) bird that had dark webs. *Bill* upper mandible olive brown, olive green or yellowish, darker and browner on culmen ridge: lower mandible yellowish to bright yellow. *Mouth* bright yellow, orange-yellow or pinkish. *Facial skin and gular pouch* yellow.

Parasites. Gizzards of two specimens heavily infested with parasitic worms in or near walls of gizzard.

Phalacrocorax carbo Black Cormorant

Seen at scattered localities in the east (55, 58, 61, 64, 88, 92) and possibly at Oenpelli (43).

Phalacrocorax sulcirostris Little Black Cormorant

Specimens. WESTERN AUSTRALIA. Perth (7) 14 – 16 Feb 66, 1 ♂ skel, 1 juv skel, 3 juv fl. NORTHERN TERRITORY. Pine Creek (44) 23 Aug 68, 1 fl (926.2 g). VICTORIA. Coleraine (89) 27 Jan 63, 1 ♂ skel.
Distribution & Abundance. In Western Australia at a lake near Nammen, approximately 35 miles N.W. of Gingin, Perth (7). In Northern Territory a few on the lagoon at Oenpelli (43) and at Pine Creek (44). In Queensland at Banoona (58) and Thylungra (60). In South Australia plentiful on Lake Killalpaninna, near Etadunna (66); also on Bool Lagoon (87). In Victoria at Nelson (88) and Coleraine (89). In New South Wales at Albury (92).
Habitat. Both inland and coastal lakes and lagoons. In Northern Territory (44) also on creek in savanna woodland.
Behaviour. At Pine Creek (44), a large mixed flock of this species, Little Pied Cormorants and Darters came each evening and, after circling overhead, landed to roost in the trees over the creek just before dusk.
Food. Small fish.
Annual Cycle. There were large numbers of nests with young in a breeding colony in partly submerged paperbark trees in lake near Nammen, near Perth (7) in mid-February 1966.
Impermanent Colours. *Legs* black. *Iris* light green to light greenish blue. *Bill* upper mandible black, light grey along cutting edge (male) or medium horn (female): lower mandible grey in male, horn in female. *Mouth* a light grey. *Facial skin and gular pouch* mauve. Juvenile: *Legs* greyish black. *Iris* brown. *Bill* pale grey. *Facial skin* light grey.

Phalacrocorax varius Pied Cormorant

Seen on the coast in Western Australia (1, 2, 16, 17) and at Glendower (55) in Queensland.

Phalacrocorax melanoleucos Little Pied Cormorant

Specimens. WESTERN AUSTRALIA. Mount Anderson (28) 15 May 68, 1 ♀. Joint Hill (34) 29 June – 4 July 68, 2 fl (648.0 g). Wyndham (39) 7 Aug 68, 1 fl. SOUTH AUSTRALIA. Innamincka (64) 27 Aug 65, 1 ♂ skel (857.4 g).
Field Notes. Widely distributed in Western Australia (1, 2, 7, 14, 31) and Northern Territory (40, 43, 45, 46) being especially plentiful on the lagoon at Oenpelli (43). Noted in the south of Queensland (58, 60), and the east of South Australia (64, 66, 87), and at Nelson (88), Victoria. On inland and coastal waters, from large lakes, lagoons and estuaries to wooded creeks.
Food. Freshwater crayfish and crabs.

Impermanent Colours. *Legs* black. *Iris* grey, greenish brown or dark brown. *Bill* upper mandible dark horn to black on top, yellow along cutting edge: lower mandible yellow, with blackish spot near tip in one specimen. *Mouth* bluish or grey, yellow near and along inside of bill edges.

Geographical Variation. All Australian specimens are referable to *P. m. melanoleucos.*

PODICIPITIDAE — Grebes
D. Goodwin

Podiceps novaehollandiae — Little Grebe

Specimens. Western Australia. Mount Anderson (28) 15 May 68, 1 ♀ (166.6 g). Lake Gladstone (36) 23 July 68, 1 ♀ (159.5 g). Northern Territory. Borroloola (46) 9 Nov 68, 1 fl (135.2 g). South Australia. Nonning (79) 1 Apr 65, 1 ♀ (200 g), 1 juv ♂ (208.8 g). New South Wales. Albury (92) 10 – 11 Mar 63, 1 juv ♀, 1 ♂ skel.

Field Notes. Also encountered in Western Australia at Scott River (11), and in Northern Territory at Oenpelli (43), where two pairs were seen on the lagoon. In Queensland fairly common at Glenore (49); at Thylungra (60) there were 15 together with a few Hoary-headed Grebes. In South Australia common at Etadunna (66) and Nonning (79). On inland waters: lagoons, pools, lakes and dams, usually with emergent vegetation.

Food. Stomachs contained beetles and other insects, one water snail and some vegetable matter. Seen both diving for food and picking flies off water surface.

Annual Cycle. Well-grown juveniles, still in company with their parents, in South Australia (79) in early April and in New South Wales (92) in mid-March.

Impermanent Colours. *Legs* greyish blue or grey with green or yellowish green on joints and at edges of scutes, sometimes predominantly yellowish green to dark olive green. *Iris* pale yellow to dark orange yellow, sometimes with dark outer ring: Immature, light brown. *Bill* upper mandible black or black with whitish tip (adults, breeding plumage); dark horn along ridge of culmen, orange at sides (adult, winter plumage); dark brown or blackish marked with pink or yellowish (immatures): lower mandible black to dull horn with whitish tip (adults); or as upper mandible but pale areas more extensive (immatures). *Mouth* white to bluish grey. *Bare skin* from base of lower mandible to eye yellowish green or yellow.

Geographical Variation. Specimens from the northern tropical regions average smaller but with proportionately larger bills (Mayr, 1943) but it seems best only to recognise one race within Australia (see Mayr, 1943, Keast, 1961).

Podiceps poliocephalus — Hoary-headed Grebe

Seen on the south coast (14, 88) and further north at Thylungra (60).

CICONIIFORMES — Herons, Egrets, Bitterns, Storks, Ibises & Spoonbills
D. Goodwin

Ardea sumatrana Great-billed Heron

One seen at Hope Inlet, Darwin (41) and two heard on the Roper river, Mount Roper (45).

Ardea pacifica White-necked Heron

Specimens. Western Australia. Mount Anderson (28) 18 May 68, 2 fl (1 ad, 1 imm).
Field Notes. In Western Australia seen also in the west at Karratha (1); in the northeast about six, mostly immature, were seen at Mount Anderson (28) and it was common at Geikie Gorge (31). In Northern Territory two seen at Pine Creek (44). In Queensland at Glendower (55), Banoona (58), a few at Thylungra (60) and ten at Monkira (61). In New South Wales several at Grafton (94). In freshwater creeks, lakes and pools, in savanna grassland and woodland.
Impermanent Colours. *Legs* black. *Iris* very pale yellow or yellow darkening to bronze (immature). *Bill* adult black; immature, upper mandible black: lower mandible black at tip and along cutting edge, base and central area yellow or white shading to yellow at end of light area. *Mouth* black or purplish. *Orbital skin* dark grey or black with a creamy-green or pale yellow streak running from eye to lower mandible.

Ardea novaehollandiae White-faced Heron

Specimens. Western Australia. Mount Anderson (28) 15 May 68, 1 imm ♀ (412.3 g). Mount Bell (32) 15 June 68, 1 ♂ skel (660.9 g). South Australia. Aroona Valley (82) 17 Apr 65, 1 ♂ (705.5 g), 1 ♀ skel (603.3 g). New South Wales. Grafton (94) 6 Apr 63, 1 ♂ skel.
Field Notes. In Western Australia seen also at Karratha (1), Coolawanyah (4), Perth (7), also at Metricup (9) where noted as frequent, Mount Barker (12), Green Range (13), Newdegate (14) and at Geikie Gorge (31), in northeastern Western Australia. Common at Mount Anderson (28), and Mount Bell (32). In Northern Territory at Pine Creek (44). In Queensland at Glendower (55), Banoona (58), Tyrone (59) and Thylungra (60). In South Australia fairly common at Innamincka (64) and about Lake Killalpaninna near Etadunna (66). One seen at Oodnadatta (67) and one at Nonning (79) and a few about Aroona Valley (82). In Victoria about 20 in the estuary of the Glenelg River (88). In New South Wales at Dunedoo (93) and Grafton (94). In vicinity of inland waters: creeks, pools and swamps.
Food. Beetles, insect larvae, grasshoppers, crabs, frogs.
Annual Cycle. The April male taken from Aroona Valley (82), was in moult. The young May bird from Mount Anderson (28) was still in juvenile plumage.
Impermanent Colours. *Legs* yellow; immature greenish yellow. *Iris* pale straw yellow to pale greenish yellow. *Bill* upper mandible black: lower mandible black with whitish base. *Facial skin* greenish yellow; immature, grey.

Ardea picata Pied Heron

Seen in fair numbers at Oenpelli (43) and in lesser numbers at Noogoo swamp, Darwin (41).

Butorides striatus Mangrove Heron

Specimens. Western Australia. Wyndham (39) 6 Aug 68, 2 fl (213.3, 224.0 g). Northern Territory. Darwin (41) 16 Sept 68, 1 fl (216.3 g). Queensland. Glenore (49) 17 Apr 64, 1 ♂ (209.9 g). Inkerman (50) 1 – 11 June 64, 2 ♀ (225.5, 234.6 g), 1 imm ♂ (262.3 g).

Field Notes. Seen also in Western Australia at Peg's Beach (2) and in New South Wales on the Clarence River at Grafton (94). At Glenore (49), in tropical savanna woodland along the Normanton River, but usually in or near mangroves (39, 41, 50) in which they quickly took cover. But individuals at Darwin (41) and Peg's Beach (2) were on open mud flats about 100 yards from cover.

Food. Crabs, mudskippers, grasshoppers and other insects. One stomach contained fragments of twigs, possibly from caddis larvae cases. One adult at Grafton (94) was observed catching small fish trapped inside some oyster trays at low tide.

Annual Cycle. A female taken at Glenore (49) on 17 April is in the final stages of a complete moult.

Impermanent Colours. *Legs* greenish yellow to dark olive green with soles of feet lemon yellow. *Iris* various shades of yellow. *Bill* upper mandible black, with yellow-green cutting edge in one adult: lower mandible yellowish green, light green or greyish green; one female had tip black and an unsexed specimen had basal half pea green and rest dull yellow. *Mouth* pale green or greyish green with pinkish throat. *Facial skin* yellowish green to pea green.

Geographical Variation. Mayr (1940a) takes Peters (1931) to task for synonymising *littleri* of northern Queensland, type locality Cooktown, with *macrorhynchus,* type locality New South Wales. He states that *littleri* is readily separable on its smaller size (wing 184 – 190 against 200 mm or over) and on its having more rufous brown colour below and on hind-neck and a bronze rather than a green gloss on its upperparts. In a later paper Mayr (1943) describes and discusses the races he recognizes, *rogersi, cinereus, stagnatilis* and *steini,* from western and northern Australia and from Timor.

The British Museum series of this species from Australia is not adequate for comparative purposes. The present specimens would appear to agree best with *macrorhynchus,* on the basis of both Mayr's description (see above) and comparison with two specimens from Cape Upstart, Queensland, and New South Wales. All, including the New South Wales specimen, have, however, wing lengths considerably under 200 mm. The rather dull sheen on their upperparts is predominantly greenish on new feathers but shows a dull bronzy purplish tinge on old feathers.

Bubulcus ibis Cattle Egret

Specimens. New South Wales. Grafton (94) 17 Dec 63, 1 juv fl.

This specimen, which was collected by Mr N. Burnett, was presented by him to the

Harold Hall Expedition. It was taken from a nest in trees at the edge of a swamp. The species was not recorded elsewhere.

Egretta alba White Egret

Specimens. WESTERN AUSTRALIA. Mount Anderson (28) 18 May 68, 1 fl (925.0 g). NEW SOUTH WALES. Grafton (94) 1 Apr 63, 1 ♂ skel.
Field Notes. In Western Australia about 100–150 birds were found, possibly a small breeding colony, in mangroves at Oyster Creek, Carnarvon (5): the Mount Anderson (28) bird was on an island lagoon in savanna woodland. Encountered singly or in twos and threes at various localities in Northern Territory (40, 45), Queensland (55, 58, 59), South Australia (64, 66), Victoria (88), and in New South Wales also at Tenterfield (95).
Food. Small fish.
Impermanent Colours. *Legs* black with a small amount of dirty yellow high up on inside of tibia. *Iris* golden (one) or straw-coloured (one). *Bill* yellow. *Mouth* yellow or pale violet. *Facial skin* olive yellow.

Egretta garzetta Little Egret

Specimens. NEW SOUTH WALES. Grafton (94) 17 Dec 60, 1 juv fl. This specimen, like the Cattle Egret, was collected and presented by Mr N. Burnett, and was also taken from a nest in a tree at the edge of a swamp.

Elsewhere about ten birds were seen on the Clarence River (94) in April.

Egretta intermedia Plumed Egret

Specimens. WESTERN AUSTRALIA. Lake Gladstone (36) 22 July 68, 1 ♂ fl (401.4 g). Wyndham (39) 4 Aug 68, 1 fl (397.6 g). NEW SOUTH WALES. Grafton (94) 5 Apr 63, 1 ♀.
Field Notes. At Wyndham (39) single birds and a party of about 60 were seen at an inland lagoon surrounded by savanna woodland. Very common at Grafton (94) in wet fields, swamps, and along creeks. Also seen at Mount Roper (45), at Nelson (88) and four birds on The Coorong (85).
Food. One bird (94) had eaten three frogs about 130 mm long, one cricket, one grasshopper, one larva and the egg case of a spider (?).
Annual Cycle. A female collected at Grafton (94), New South Wales, on 5 April, had very worn plumes. Some of its wing feathers appeared much less worn than others but it did not appear to be actually in moult. The largest oocyte in its ovary was about 4 mm in circumference.
Impermanent Colours. *Legs* olive green to black, tibia yellowish green, olive grene or mottled yellow and brownish. *Iris* pale straw yellow or cream. *Bill* yellow; in one specimen (immature?) upper mandible black, lower mandible dull yellow. *Mouth* pink or purplish. *Facial skin* greenish yellow to bright yellow.

Nycticorax caledonicus Nankeen Night Heron

Specimens. WESTERN AUSTRALIA. Mount Anderson (28) 18 May 68, 1 fl (562.3 g). Derby (29) 26 May 68, 1 ♀ imm (726.1 g). NORTHERN TERRITORY. Pine Creek (44)

1 Sept 68, 1 ad fl (715 g), 2 imm fl (519.1, 676.9 g). QUEENSLAND. Glendower (55) 28 Mar 64, 1 ♂ imm (576.5 g). Monkira (61) 1 June 63, 1 skel.

Distribution & Abundance. In Western Australia one seen at Perth (1), common at Derby (29) and found throughout the Kimberleys (30–36). In Northern Territory a few at Borroloola (46) but very common about the lagoons at Oenpelli (43) and at Pine Creek (44) where large flocks along the river contained many young birds. Found at various localities in Queensland (47, 50, 55, 61) and in small numbers in New South Wales (94).

Habitat & Behaviour. Usually in parties, sometimes singly, in creeks, lagoons and surrounding woodland (for roosting); at Derby (29) also in mangroves. At Inkerman (50) a Banyan tree was heavily streaked with their droppings. At Monkira (61) birds roosted close together in one or two favoured trees bordering the Diamantina River.

Food. Crayfish in one stomach.

Annual Cycle. The immature taken at Glendower (55) is in body moult from juvenile plumage. The new plumage is very similar in colour and pattern, although more mature in texture.

Impermanent Colours. *Legs* adult pale yellow, toes ribbed brown, under toes vivid yellow; immatures pale green or greenish yellow. *Iris* yellow, in one immature golden with black outer ring. *Bill* upper mandible dull black with pea green line along lower edge extending as far as nostril; lower mandible pea green with black tip; immatures similar but black on lower mandibles more extensive. *Mouth* pinkish to purple, roof of upper mandible pea green in adult, pale green in immatures. *Eyelids* yellow green or green. *Facial skin* pea green in adult, pale green or yellowish green in immatures.

Dupetor flavicollis — Black Bittern

Specimens. WESTERN AUSTRALIA. Derby (29) 26 May 68, 1 ♂ (317.2 g). Mount Caroline (33) 25 June 68, 1 imm fl. QUEENSLAND. Glenore (49) 23 Apr 64, 1 imm ♂ (317.0 g).

Field Notes. Also at Geikie Gorge (31), Pine Creek (44), Glenore (49) and Grafton (94). Encountered singly in reeds or on wooded river banks.

Impermanent Colours. *Legs* adult brown; immatures dark dull brown and olive yellow. *Iris* adult golden; immatures pale yellow. *Bill* upper mandible adult dark brown; immatures dark horn, and black with yellow-brown tomia: lower mandible adult light brown; immatures horn and olive yellow. *Mouth* pinkish. *Bare skin* in front of eye, adult pink; immatures yellowish.

Geographical Variation. This is discussed in detail by Mayr (1945). The British Museum material, although inadequate, tends to substantiate his conclusion that all Australian birds are best referred to *D. f. gouldi*.

Xenorhynchus asiaticus — Jabiru

Frequently encountered in the north (31, 40–46) and further south in Queensland (58) and New South Wales (94).

Threskiornis molucca White Ibis

Specimens. Western Australia. Mount Anderson (28) 15 May 68, 1 fl (1,244.0 g). Derby (29) 26 May 68, 1 fl (1,987.0 g). Northern Territory. Darwin (41) 16 Sept 68, 1 fl (1,387.9 g). Oenpelli (43) 12 Oct 68, 2 fl (1,302.0, 1,501.3 g).
Distribution & Abundance. In Western Australia recorded at Carnarvon (5), common at Wyndham (39). In Northern Territory fairly numerous at Noogoo Swamp, Darwin (42), plentiful at Oenpelli (43), one at Mount Roper (45). In Queensland about 100 Banoona (58), one Tyrone (59) and several at Thylungra (60). In South Australia, two Innamincka (64), one Etadunna (66) and about 30 Bool Lagoon (87). In New South Wales 10 at Albury (92 and about 150 Grafton (94).
Habitat. Tree fringed inland lagoon surrounded by grasslands and tree savanna (28). Elsewhere mudflats, marshes, swampy grassed areas, mangroves, creeks and edges of dams or rivers.
Impermanent Colours. *Legs* dark grey or black. *Iris* brown. *Bill* black. *Mouth* black. *Facial skin* blue-black.

Threskiornis spinicollis Straw-necked Ibis

Specimens. Queensland. Thylungra (60) 25 May 63, 1 ø skel.
Field Notes. Plentiful at Oenpelli (43), few at Mount Roper (45). Seen at Banoona (58), Thylungra (60), Monkira (61), about 150 Coorong (85), Furner (86), about 2,000 Bool Lagoon (87), Nelson (88). Usually on or near lagoons but on open grass at Thylungra (60).
Impermanent Colours. *Legs* black. *Iris* dark brown. *Bill* black.

Plegadis falcinellus Glossy Ibis

Specimens. Northern Territory. Oenpelli (43) 11 Oct 68, 1 fl (509.1 g).
Distribution & Abundance. Six seen at Oenpelli (43) and one at Bool Lagoon (87) feeding with about 2,000 Straw-necked Ibis.
Habitat. Inland freshwater lagoon, keeping mostly to the shallow edges on flooded grass (43); shallow brackish pool (87).
Behaviour. Found to be far less nervous than either the White or Straw-necked Ibis and were usually the last to rise when disturbed (43).

Platalea regia Royal Spoonbill

Specimens. Western Australia. Mount Anderson (28) 18 May 68, 1 fl (1,701.8 g). Derby (29) 26 May 68, 1 fl (1,684.3 g). Darwin (41) 18 Sept 68, 1 fl (1,298.7 g).
Field Notes. Seen on the edges of dams, creeks or lagoons at Banoona (58), Tyrone (59), Thylungra (60), Nelson (88), Grafton (94).
Impermanent Colours. *Legs* black. *Iris* crimson. *Bill* black. *Mouth* black (28, 29), or white (41). *Bare skin* yellow above eye; bright red on forehead just behind and between eyes; remainder around face and head black.
Parasites. See Appendix 1.

Platalea flavipes Yellow-billed Spoonbill

More concentrated in southern Queensland (59-61) with scattered records elsewhere (36, 43, 87).

ANSERIFORMES — Swans, Geese & Ducks
D. Goodwin

Anseranas semipalmata Magpie Goose

In the north a few seen at Mount Anderson (28) and Mount Roper (45) while present in estimated thousands near Darwin (41) and at Oenpelli (43).

Dendrocygna arcuata Water Whistling Duck

About 40 seen at Oenpelli (43) and a few in Western Australia (28, 36).

Dendrocygna eytoni Grass Whistling Duck

Specimens. Western Australia. Mount Anderson (28) 15 May 68, 1 imm ♂ (389.6 g). Lake Gladstone (36) 22 July 68, 1 ♂.
Field Notes. Abundant where seen in the Kimberleys (28, 36) and Northern Territory (43, 45). Three also seen at Thylungra (60), southern Queensland. Tending to form rather dense flocks of up to 300 birds (28) on freshwater lakes and lagoons.
Food. Vegetable matter. Mud and gravel also in stomachs.
Annual Cycle. The male collected on 15 May at Mount Anderson (28) was in body moult. The young male collected at Lake Gladstone (36) on 22 July was evidently just completing the acquisition of its first (juvenile) plumage; some feathers are coming in on the upper parts and the wing quills are not quite fully grown.
Impermanent Colours. *Legs* pink; immature light flesh. *Iris* orange; immature lemon. *Bill* upper mandible pink with black spots and blackish slate tip; immature dark reddish grey with black tip: lower mandible as upper but lighter. *Mouth* white, grey in immature. *Orbital ring* yellow in immature, not noted for adult.

Cygnus atratus Black Swan

Ten seen at Innamincka (64), fairly common at Etadunna (66) and recorded elsewhere from scattered southern localities (14, 58, 87, 88, 94).

Tadorna tadornoides Mountain Duck

Specimens. Western Australia. Perth (7) 14 Feb 66, 4 skel.
Field Notes. The birds collected were some of many found dead, probably from botulism, round a small slightly saline lake near Nammen, about 70 miles NNW Perth. Elsewhere five seen at Clifton Hills (65), a few at Etadunna (66) and in the south (14, 86, 88).

Tadorna radjah Burdekin Duck

Specimens. NORTHERN TERRITORY. Oenpelli (43) 14 Oct 68, 1 fl.
Field Notes. In fair numbers on the lagoon at Oenpelli (43). Also on the lagoon at Mount Roper (45) and twelve birds on Noogoo Swamp, Darwin (41).
Impermanent Colours. *Legs* off-white. *Iris* whitish. *Bill* whitish. *Mouth* whitish. *Orbital ring* yellow.

Anas superciliosa Black Duck

Specimens. NEW SOUTH WALES. Grafton (94) 30 Mar 63, 1 ♀ skel.
Field Notes. Seen also in Western Australia at Karratha (1), Metricup (9), Green Range (13), Newdegate (14), and Mount Anderson (28) where it was noted as fairly numerous. In Northern Territory a few at Oenpelli (43) and Mount Roper (45) and two at Borroloola (46). In Queensland at Glendower (55) Banoona (58) and twenty on lagoon at Thylungra (60). In South Australia small parties on The Coorong (85).
Impermanent Colours. *Legs* light brown. *Iris* dark brown. *Bill* olive, black at tip and around nostrils. *Mouth* pink.

Anas gibberifrons Grey Teal

Specimens. WESTERN AUSTRALIA. Mount Anderson (28) 15 May 68, 1 ♀. Lake Gladstone (36) 22 July 68, 1 ♂ (390.8 g). SOUTH AUSTRALIA. Angepena (84) 21 Nov 63, 1 ♂.
Distribution & Abundance. In Western Australia also at Karratha (1) and Newdegate (14). In Queensland small parties on lagoon at Banoona (58); also at Tyrone (59) and Thylungra (60). In South Australia two (possibly more as seen more than once) at Innamincka (64), flock of about 20 at Clifton Hills (65), common at Etadunna (66), especially on Lake Killalpaninna, five seen at Oodnadatta (67), six at Kingoonya (70) and about 20 at Aroona Valley (82). In the southeast recorded at several localities (86, 88, 92) and there were large numbers of either this species or *A. castanea* on The Coorong (85).
Habitat. Apparently almost any available fresh water: creeks, rivers, lakes or billabongs. The six at Kingoonya (70) had settled on a round galvanised water tank, perhaps about 12 or 15 feet in diameter. The water surface was about five feet above ground but below the level of the tank edge allowing the observer (DG) to get within a few feet of the birds.
Food. No identifiable stomach content, other than grit.
Annual Cycle. The female taken at Mount Anderson (28) in mid-May was in moult. Wing quills and about two thirds of body plumage new, most tail quills old and worn but tail moult begun. The female taken on 22 July (36) was in a less advanced state of body moult but had new wing quills. All its old tail and body feathers were very worn and bleached, those on underparts stained rusty orange, presumably from iron oxide. The male taken in South Australia (84) on 21 November showed some body moult but most of its plumage appeared fresh and new.
Impermanent Colours. *Legs* grey (male), greyish brown (female). *Iris* brown. *Bill* upper mandible bluish grey (male), black (female): lower mandible basal half of sides

bluish grey, anterior half dull yellow (male), black and brown (female). *Mouth* whitish with pinkish tongue.

Anas castanea Chestnut Teal

Specimens. WESTERN AUSTRALIA. Mount Barker (12) 23 Mar 66, 2 ♀ skel.
Field Notes. On a small dam in a paddock. Not recorded elsewhere except, possibly, on The Coorong (85) in South Australia.
Impermanent Colours. *Legs* grey with green tinge. *Iris* light reddish brown. *Bill* upper mandible very dark olive green: lower mandible as upper with pinkish orange in middle. *Mouth* orange pink. *Orbital skin* grey.

Anas rhynchotis Blue-winged Shoveler

Seen in Western Australia (14, 28) and in Queensland (58).

Malacorhynchus membranaceus Pink-eared Duck

Specimens. SOUTH AUSTRALIA. Oodnadatta (67) 1 June 65, 1 ♀.
Distribution & Abundance. In South Australia, where four individuals were seen at Innamincka (64), and at Clifton Hills (65) and at Oodnadatta (67); elsewhere only one adult was seen at Newdegate (14), Western Australia.
Habitat. Pools in creeks and sluggish rivers (in dry season). It was not seen (though looked for) among large numbers of water fowl on the large Lake Killalpaninna (66).
Behaviour. Encountered quietly swimming singly or in presumed pairs, or resting perched on large branches just above the water. Not so wary of humans as the Grey Teal.
Food. Vegetable matter.
Impermanent Colours. *Legs* light greyish blue. *Iris* not recorded. *Bill* upper mandible black: lower mandible black with pink tip. *Mouth* pink and grey.

Aythya australis White-eyed Duck

Specimens. WESTERN AUSTRALIA. Lake Gladstone (36) 23 July 68, 1 ♂ (929.0 g).
Field Notes. Usually on lakes. In Western Australia also at Perth (7) and Scott River (11). In Queensland at Glendower (53) and 20 on the lagoon at Banoona (58). In South Australia one seen in a pool on the river at Innamincka (64), and one on Lake Killalpaninna (66). A few at Bool Lagoon (87).
Annual Cycle. The specimen showed some body moult.
Impermanent Colours. *Legs* pale grey suffused black. *Iris* white tinged with blue. *Bill* anterior third pale grey, tip (nail) and base black. *Mouth* grey, tongue white.
Geographical Variation. No races, usually recognised. Material examined, though inadequate, confirms the view that *A. australis dampieri* Mathews of northwestern Australia, alleged to be paler and smaller, is not separable.

Chenonetta jubata Wood Duck

Specimens. QUEENSLAND. Tyrone (59) 10 – 12 May 63, 4 skel.

Distribution & Abundance. A few records from Western Australia (7, 11, 14). In Queensland about 24 birds were seen at Tyrone (59) and also recorded from Glendower (55), Banoona (58) and Monkira (61). A few were seen in northeastern South Australia (64-66) and a single female at Aroona Valley (82) was seen feeding and flying with a party of Grey Teal.
Habitat. Inland waters: creeks, rivers, dams and pools.
Annual Cycle. Specimens were either completing or had just completed moult.
Impermanent Colours. *Legs* olive. *Iris* dark brown. *Bill* olive. *Mouth* very pale yellow.

Nettapus pulchellus — Green Pigmy Goose

Specimens. WESTERN AUSTRALIA. Lake Gladstone (36) 22 – 23 July 68, 1 ♂, 2 ♀ (257.5, 300.5 g). Wyndham (39) 1 – 2 Aug 68, 1 ♂ (362.6 g), 1 fl (373.2 g). NORTHERN TERRITORY. Oenpelli (43) 1 – 11 Oct 68, 6 fl (284.6, 345.1, 354.7, 369.9, 408.9, 409.3 g).
Distribution & Abundance. Parties of 20 to 30 were seen at Lake Gladstone (36) and two parties of about 15 or 16 at Wyndham (39). In Northern Territory at Oenpelli (43) about 600 were seen on the main lagoon and smaller parties elsewhere, and small numbers were seen at Mount Roper (45) and Borroloola (46). In northern Queensland fairly common on the lakes at Glenore (49).
Habitat. Lakes and lagoons with aquatic vegetation.
Behaviour. Usually keep apart from other water birds and, where to some extent intermingled with them, probably fortuitiously so. When alarmed or suspicious congregate together on open water and when flushed fly low (not above about ten feet) uttering soft but high-pitched whistling calls. At Lake Gladstone (36), they did not leave the lake at night to feed elsewhere, as did the Water Whistling Ducks *Dendrocygna arcuata*. Never seen to dive.
Food. Seeds; grit and sand also in stomachs.
Annual Cycle. All the birds from Western Australia were in moult. The two males (36, 39) were in final stages of moult with a few new feathers coming in on head and neck but plumage otherwise fresh and new. The two females (36) were less advanced in the moult: their wing quills were new but tail and body were in full moult. One had cast all its old tail feathers and most of the new ones were at about the same stage of growth; the other had some old and some new tail feathers, the former showing extreme wear and bleaching.
Impermanent Colours. *Legs* dark olive green to black with a yellowish olive suffusion, webs black in some individuals but one female noted as having paler webs. *Iris* dark brown. *Bill* upper mandible black with pink or purplish tip: lower mandible pink or purplish pink, with black markings: pink on bills of some females paler and duller than in males. *Mouth* flesh pink or yellowish with pink tongue.

Oxyura australis — Blue-billed Duck

About five seen in Perth (7).

Biziura lobata Musk Duck

Seen at Warriup, Green Range (13) and at Nelson (88), where a film was made of the resident flock of about 50 birds showing the unusual escape behaviour. The birds paddled furiously and then broke into a fast scuttling across the surface. They seemed incapable of getting airborne. After 100 yards they tired and began giving up and took to diving to avoid being caught.

FALCONIFORMES — Kites, Eagles, Hawks, Harriers, Osprey, Falcons

D. Goodwin

Elanus notatus Black-shouldered Kite

Specimens. Western Australia. Newdegate (14) 21 Apr 66, 1 ♀ skel and flat skin, 1 fl.
Distribution & Abundance. In Western Australia seen also at Karratha (1), Peg's Beach (2), Chichester Range (3), Coolawanyah (4), and from the road about 30 miles east of Perth (7); also 20 individuals along the Great Northern Highway between Wubin and Mount Magnet. In Northern Territory three individuals seen near Oenpelli (43) and one over savanna grassland at Mount Roper (45). In Queensland one seen at Glendower (55), and several along the Brisbane-Dalby road 26 April and near Banoona (58) 14 June. In Victoria at Nelson (88) and in New South Wales at Grafton (94).
Habitat. Open savanna woodland, grassland or open cultivated country.
Behaviour. Seen hovering in a rather kestrel-like manner when looking for food, just as does the closely related *E. caeruleus* in Egypt. Favourite perch is a roadside telephone pole.
Annual Cycle. The adult female (14) was near the end of its wing moult, with two old outermost primaries in each wing.
Impermanent Colours. *Legs* lemon yellow to orange yellow with black nails. *Iris* deep red. *Bill* black. *Mouth* bluish grey to blackish.

Elanus scriptus Letter-winged Kite

Seen at Glendower (55).

Milvus migrans Fork-tailed Kite

Specimens. Queensland. Thylungra (60) 24 May 63, 1 ♂ skel. Monkira (61) 5 June 63, 1 ♀ skel.
Distribution, Abundance & Habitat. In Western Australia a single bird at Coolawanyah (4), and Derby (29) (see notes on *Falco subniger*); several at Mount Anderson (28) and very large numbers at Geikie Gorge (31). In Queensland very common at all western localities (47–50, 55, 60, 61). One seen in South Australia, near Cordillo Downs (63). In fairly open areas.
Behaviour. At Monkira (61) these kites habitually approached when a shot was fired, possibly through having become conditioned to expect food from carrion or injured birds.

Impermanent Colours. *Legs* yellow. *Iris* dark brown. *Bill* black at tip, base and cere yellow. *Mouth* yellowish grey.

Lophoictinia isura — Square-tailed Kite

Two seen at Mount Anderson (28) and one at Victoria River (40).

Hamirostra melanosterna — Black-breasted Buzzard

Seen in twos or threes in the north (28, 41, 43, 44) and singly further south (68, 79).

Haliastur indus — Red-backed Sea Eagle

Seen singly in Western Australia (1, 28) and in pairs in Northern Territory (41, 43, 45); at Mount Roper (45) nesting 30 ft up in mangroves at the end of October.

Haliastur sphenurus — Whistling Eagle

Specimens. WESTERN AUSTRALIA. Metricup (9) 16 Mar 66, 1 ♂ skel. Mount Barker (12) 21 Mar 66, 1 skel. Mount Anderson (28) 28 Apr – 15 May 68, 1 imm ø, 1 skel (857.1 g), 1 fl. Geikie Gorge (31) 4 June 68, 1 ♂. NORTHERN TERRITORY. Darwin (41) 21 Sept 68, 1 imm fl. SOUTH AUSTRALIA. Innamincka (64) 29 Apr 65, 1 skel (664.9 g). Oodnadatta (67) 1 June 65, 1 ♂ skel. VICTORIA. Harrow, 32 miles north of Coleraine (89) 24 Jan 63, 1 skel.
Distribution & Abundance. Widespread except in the centre, the 3rd Expedition only seeing a few at Warburton (23), Oodnadatta (67) and Kingoonya (70) though it was fairly common further east (64, 66). Elsewhere especially common at Mount Anderson (28), Darwin (41) and Oenpelli (43) where it was breeding. Numerous other sight records (7, 8-14, 44-46, 50, 55, 59-61, 88, 90, 92-95).
Habitat. Savanna woodland, open country and creeks. Always fairly near water and usually close to it.
Behaviour. Sometimes singly, often two or three in same area. Usually perched on high tree. Except for duller coloration general appearance very similar to that of the related Red-backed Sea Eagle *Haliastur indus*. At Mount Anderson (28) one was shot in flight three-quarters of an hour after sunset.
Food. Grasshoppers, fish, large ants, rabbits. Probably the latter were remains of rabbits that had been killed by other predators. At Innamincka (64) one was seen with a rabbit's leg, of which it was robbed by a Raven (see under *Corvus coronoides*).
Annual Cycle. In Northern Territory breeding in mid-August at Pine Creek (44) and in late September at Oenpelli (43). The immature taken in the Kimberleys (28) on 28 April is in body moult from, presumably, juvenile plumage and still retains its juvenile wing and tail quills. In southern Queensland (60) an adult was carrying a branch to the nest at the end of May.
Impermanent Colours. *Legs* white to pale yellow. *Iris* brownish yellow to dark brown. *Bill* upper mandible dark horn to steel grey: lower mandible horn to light steel grey: cere dark grey. *Mouth* pinkish or purplish pink. *Orbital skin* pink (one specimen).
Parasites. See Appendix 1.

Accipiter novaehollandiae Grey (White) Goshawk

One (almost certainly) seen at Albury (92).

Accipiter fasciatus Australian Goshawk

Specimens. WESTERN AUSTRALIA. Karratha (1) 4 June 66, 1 ♀ skel. Derby (29) 22 May 68, 1 imm ♀ (405 g). Geikie Gorge (31) 2 – 5 June 68, 2 imm ♂. QUEENSLAND. Glenore (49) 16 Apr 64, 1 imm ♀ (387.0 g). SOUTH AUSTRALIA. Angepena (84) 26 Nov 63, 1 imm ♀.

Distribution & Abundance. Elsewhere noted at Perth (7), Green Range (13), Glendower (55), Banoona (58) and Thylungra (60). In all probability much more widespread than these few certain records suggest.

Habitat. At Karratha (1) the specimen collected was perched on a rocky outcrop overlooking mangroves and surrounding estuary. Elsewhere in savanna woodland (29, 31, 49).

Food. Grasshoppers, crickets and locusts.

Annual Cycle. The November male from South Australia (84) is in fresh juvenile plumage. All other specimens show some (mostly rather little) body moult.

Impermanent Colours. *Legs* pale yellow in the one adult, immatures greenish yellow or dull yellow. *Iris* bright yellow. *Bill* black with base blue (adult) or grey (immatures) and extreme base of bill at gape sometimes (perhaps always) blackish: cere greenish yellow or dull yellow blotched brown (one immature). *Mouth* dark brownish or blackish, floor of mouth partly red in at least one specimen. *Orbital skin* yellow.

Geographical Variation. Brown & Amadon (1968) recognise two subspecies in Australia, nominate *fasciatus,* and *didimus* from the northern and northwestern coastal regions which tends to be smaller (but with some overlap of measurements) and paler in colour. Of the present specimens all seem best referable to *didimus* except the one from South Australia.

Accipiter cirrhocephalus Collared Sparrowhawk

Specimens. WESTERN AUSTRALIA. Coolawanya (4) 19 May 66, 1 ♀ skel. Derby (29) 25 May 68, 1 ♀ skel (346.5 g). Geikie Gorge (31) 2 June 68, 1 imm ♂ (124.5 g). Mount Elizabeth (35) 7 July 68, 1 ♂ (130.6 g). QUEENSLAND. Glenore (49) 17 Apr 64, 1 ♂ (188.8 g). Tully (54) 18 July 64, 1 ♂ (125.3 g). SOUTH AUSTRALIA. Kingoonya (70) 20 June 65, 1 ♀.

Distribution & Abundance. No more than one or two individuals seen where collected or elsewhere (10, 11, 55, 79, 82, 86) but the inconspicuous behaviour of all *Accipiter* species makes negative records over a short stay of little value.

Habitat. In Western Australia in gums on grassy river flats (4), thickets along river bank (29), grassland savanna (31) and riverside gums (35); in Queensland in tropical savanna woodland (49) and open forest (54). In South Australia in salt bush plain with strips and clumps of Mulga (70, 79) and among large gums and other vegetation near creek (82).

Behaviour. Food-seeking flight with alternate wing beating and long swift glides. Once seen soaring above trees.

A female at Derby (29) attacked rising Grey Teal which evaded her by diving. At Aroona Valley (82) another bird, thought to be a male from its apparent size, attacked from below three Magpie Larks, that were perched in a tall gum tree. It managed to separate one from its companions and at one point got well above and swooped down at it. This was repeated three or four times, but each time the Magpie Lark managed a split-second escape, apparently by dodging to one side, uttering a grating screech as it did so. Finally the hawk gave up the pursuit. (DG)

Food. Birds, grasshoppers.

Annual Cycle. The immature June male from the Kimberleys (31) was undergoing a complete moult from juvenile to adult plumage. About a third of the wing and body feathers were new (adult) but none of the tail feathers had been cast. The adult females taken on 17 April in Northern Queensland (49) and 10 June in South Australia (70) were both in final stages of complete moult.

Impermanent Colours. *Legs* males yellow or orange yellow, females yellowish green to yellow. *Iris* bright yellow. *Bill* dark brownish horn to black, bluish at base below cere: cere greenish yellow. *Mouth* bluish grey to blackish. *Orbital skin* bright yellow, greenish yellow in one female.

Geographical Variation. Brown & Amadon (1968) recognise, with some reservations, *A. cirrhocephalus quaesitandus* from north Australia and Cape York Peninsula as being paler above, less clearly barred below and paler than nominate *cirrhocephalus,* and averaging smaller in size. Peters (1931) treated *quaesitandus* as a synonym of *cirrhocephalus* but recognised *A. c. broomei* as the race from Northern Territory and Western Australia. This form (*broomei*) was alleged to differ by being darker in colour (Mathews, 1912a). Material in London, though inadequate, suggests that there may be a south-north cline in average size and colour. Contrary to Mathews' description, a female from Broome, Western Australia, is noticeably pale. If *quaesitandus* is to be maintained then the males from northeastern Western Australia (31) and (35) are certainly referable to that form; the male and female from Queensland (49, 54) are a little darker but seem best referred to *quaesitandus* and the South Australian specimen to nominate *cirrhocephalus.*

Erythrotriorchis radiatus — Red Goshawk

Seen at Thylungra (60).

Hieraaetus morphnoides — Australian Little Eagle

Specimens. WESTERN AUSTRALIA. Mount Anderson (28) 9 May 68, 1 ♀. NORTHERN TERRITORY. Pine Creek (44) 1 Sept 68, 1 fl.

Distribution, Abundance & Habitat. Seen in Western Australia also in the Darling Range, Perth (7), at Mount Shenton (18), and in the Kimberleys (28, 29, 31, 33-36) in savanna woodland. In southern Queensland at Banoona (58), Tyrone (59) and Thylungra (60). In South Australia at Innamincka (64) and Furner (86), and in Victoria at Nelson (88).

Behaviour & Annual Cycle. Encountered singly. The female collected was feeding on a monitor lizard, probably *Varanus geopalma,* which it was carrying when shot. A nest was found containing an almost fully fledged eaglet at Lake Cowan, 15 miles north of Norseman W.A. on 2 December 1962. It was a bulky structure about three feet across, built mainly of twigs and small branches with some fresh eucalypt sprays and situated 35 ft up in a eucalypt *Eucalyptus dundasi*. The eaglet flew off the nest, was caught, filmed and returned to the nest. Single adults twice circled high over the nest but did not call (PRC).

Food. Reptiles, including (perhaps only) *Varanus*. The nest (above) contained feathers of *Barnardius*.

Impermanent Colours. *Legs* creamy white or straw yellow with black claws. *Iris* brown (adult female) or dull yellow (unsexed specimen). *Bill* bluish or grey with black tip: cere greyish (adult female) or pale yellow (unsexed specimen). *Mouth* pale grey or pinkish. (See below for juvenile.)

Other Notes. The juvenile plumage was as follows: head rufous with grey lores and white chin; back blackish with cinnamon edges to the feathers; scapulars grey; primaries and secondaries black; upper wing-coverts dark grey forming a lighter forewing patch; upper breast rufous and rest of underparts light buff; tail short, dark grey with two narrow black bars and a half inch black subterminal bar with white tips to rectrices. The tarsi were feathered, the feet a very pale yellow with grey talons; the bill light grey with black tip, cere yellow; iris brownish grey.

Parasites. See Appendix 1.

Aquila audax — Wedge-tailed Eagle

Specimens. Western Australia. Carnarvon (5) late May 66, 1 skel (specimen found dead by road).

Distribution & Abundance. The many sight records of this species reflect its extreme conspicuousness, ease of identification and preference for open country. In Western Australia (3, 4, 8, 9, 10, 12–14, 17, 23, 25, 28) especially common at Mount Anderson (28). In Northern Territory two were seen at Pine Creek (44), one being immature. In Queensland noted in the north at Glendower (55) and further south at Tyrone (59), Thylungra (60) and Monkira (61). In South Australia common or fairly common at most main camps (64, 66, 70, 79, 82) but only one at Oodnadatta (67) and one or two at Granite Downs (68). In Victoria at Nelson (88) and Coleraine (89), and at all camps in New South Wales (92–96).

Food & Behaviour. In South Australia seen on carcasses of kangaroos and once two birds were feeding on a third that had apparently been shot (and beheaded) while at a dead kangaroo.

Haliaeetus leucogaster — White-breasted Sea Eagle

Five seen at Mount Roper (45), and at Oenpelli (43) in early October one pair had a single juvenile in a nest while a further pair with three fully fledged young were often seen. Two or three at Darwin (41) and seen elsewhere singly (1, 94).

Circus assimilis Spotted Harrier

Occasional scattered records (1, 3, 60, 82).

Circus approximans Swamp Harrier

Occasional scattered records (11, 58, 88, 89).

Pandion haliaetus Osprey

Specimens. WESTERN AUSTRALIA. Joint Hill (34) 2 July 68, 1 ♀ (1,447.7 g).
Field Notes. A few sight records from inland and coastal waters mainly of single birds (31, 43, 45, 56, 94, and from Albany, southwestern W.A. in March 1966).
Food. Fish.
Annual Cycle. A nest with young was found on a small island off Cape Gloucester (56) on 7 August. The Joint Hill (34) female was undergoing a complete moult.
Impermanent Colours. *Legs* light bluish green, claws black. *Iris* pale yellow. *Bill* black, pale bluish grey at base of lower mandible: cere dark bluish grey blotched with black. *Mouth* flesh pink.
Geographical Variation. Peters (1931) includes all Australian forms of the Osprey under *P. h. cristatus*. Brown & Amadon (1968) recognise, but with doubts, *P. h. cristatus* from southern Australia and Tasmania and *P. h. melvillensis* from tropical Australia which averages smaller in size, wings of females 435 – 464 mm as against 425 – 490 in *cristatus*. The wing of our female is 455 mm and thus it could fit either alleged race. Not enough specimens have been examined to judge the validity of *melvillensis* but it seems probable that Peters was wise in treating it as a synonym of *cristatus*.

Falco subniger Black Falcon

Specimens. WESTERN AUSTRALIA. Derby (29) 27 May 68, 1 ♀ (843.2 g).
Field Notes. This bird and one at both Karratha (1) and Glendower (55) were the only ones seen. The Derby (29) bird was one of a large association of raptors in an open savanna grassland plain sparsely vegetated with stunted *Acacia* near the Fitzroy River. Those present included two *Falco cenchroides*, two *F. longipennis*, several *F. berigora*, one *Milvus migrans* and an unidentified eagle, presumably either *Haliastur sphenurus* or *Hieraaetus morphnoides*. The specimen collected flew up from a fence post with rather slow wing beats, circling overhead to watch the collector, unlike the Karratha (1) bird which Colston observed trying unsuccessfully to catch a Magpie Lark, and which impressed him with its speed and alacrity.
Food. Bird, probably Galah, remains in stomach.
Annual Cycle. Some body moult. Second from outermost primary in each wing much newer than others near it. Tail quills all old and central two very worn. Ovary slightly enlarged with largest oocyte about 5 mm in diameter.
Impermanent Colours. *Legs* light greyish blue. *Iris* dark brown. *Bill* bluish grey with black tip: cere light greyish blue as legs. *Mouth* bluish grey with yellow gape flanges. *Orbital skin* greyish blue with brown marks.

Other Notes. In flight appeared a strongly-built, square-shouldered falcon and looked quite black, even when fairly close.
Parasites. See Appendix 1.

Falco peregrinus — Peregrine Falcon

Two seen at Pine Creek (44) and at Giles (25) with single records from scattered localities (16, 28, 41, 55, 86, 92, 93).

Falco longipennis — Little Falcon

Specimens. WESTERN AUSTRALIA. Mount Barker (12) 22 Mar 66, I imm fl. Derby (29) 22 – 27 May 68, 1 ♀, 1 imm ♀, 1 ♀skel (312.8 g). QUEENSLAND. Thylungra (60) 25 May 63, 1 ♂ skel.
Distribution & Abundance. In Western Australia also at Karratha (1), Newdegate (14), Warburton (23), Mount Anderson (28). In Northern Territory at Mount Roper (45). In South Australia at Nonning (79). Nowhere numerous, only one or a few individuals seen in any one area. (See notes on *F. subniger.*)
Habitat. In various types of open or fairly open country with trees, including grass paddocks (12), savanna woodland (29), edge of creek (60), and bluebush (79).
Behaviour. One adult near Warburton (23) appeared to make repeated attempts to catch Zebra Finches that were gathered in many hundreds round a drinking place, but did not succeed while watched. Presumably this was due to its being unable to concentrate on and single out an individual from the hundreds swirling all about it as it dashed among them. In flight and stance when perched this specimen seemed identical with the Hobby *Falco subbuteo*. At Thylungra (60) an adult was observed swooping at Budgerygahs and later at Black-faced Wood-swallows.
Food. Small birds; remains of *Mirafra* and probably a young *Coturnix*.
Annual Cycle. In all the specimens collected the gonads appeared inactive. The adult female from Derby (29) was just completing a full moult, while the immature female was in worn plumage but had just begun to moult into adult dress.
Impermanent Colours. *Legs* pale yellow, claws black or dark brown (one immature). *Iris* dark brown. *Bill* upper mandible bluish grey darkening to black at tip or entirely blackish (one immature): lower mandible bluish grey with blackish tip: cere grey, once pale yellow (adult female). *Mouth* grey, yellow or yellowish at gape. *Orbital skin* bluish grey or pale grey.
Parasties. The three specimens from Derby (29) all had parasitic worms in the lower parts of the body.

Falco hypoleucos — Grey Falcon

Specimens. WESTERN AUSTRALIA. Warburton (23) 9 Aug 65, 1 ♂ (335 g).
Distribution & Abundance. Appeared nowhere plentiful. Only seen at Chichester Range (3), Warburton (23), where two single birds were seen, Giles (25) and Mount Anderson (28) where a presumed pair were seen.

Habitat. Near Warburton (23) on open stony flat with scattered Mulga and bloodwoods. Near Giles (25) seen in flight above a rocky gorge in the Pass of the Abencerrages.

Behaviour. An adult on the Chichester Range (3) was observed in what was believed to be display flight, soaring about and then making a series of shallow dives at speed. Several times the wings were raised high above the level of the body. Frequently called with a series of sharp cries *Kitt-Kit*.

Food. Mammal fur and a small quartz stone in stomach.

Annual Cycle. The contour feathers of the specimen include some very worn and other newer feathers. No evidence of wing or tail moult.

Impermanent Colours. *Legs* yellow. *Iris* brown. *Bill* black at tip, yellow at base. *Mouth* pink. *Cere and orbital skin* orange-yellow.

Falco cenchroides — Nankeen Kestrel

Specimens. WESTERN AUSTRALIA. Karratha (1) 28 May 66, 1 fl. Mount Anderson (28) 18 May 68, 1 fl. Mount Remarkable (37) 25 July 68, 1 ♂ (165.0 g). QUEENSLAND. Thylungra (60) 26 May 63, 1 skel.

Distribution & Abundance. Recorded widely in Western Australia (1, 3, 4, 11-14, 16, 19, 21, 23-25, 28, 31, 37), Northern Territory (40, 44, 45), Queensland (47-50, 55, 59, 60), South Australia (63, 64, 66-70, 79, 84), Victoria (88, 89) and New South Wales (93, 95). Usually only a few individuals seen in any one area but "several" noted at Giles (25). (See notes on *F. subniger*.)

Habitat. Savanna woodland and open country with some trees.

Behaviour. Flight, hovering, and general appearance except for paler colouration, appeared identical to Kestrel *Falco tinnunculus*. A pair was watched in open woodland on 12 September 1965: the male brought prey to the female, who was perched in a gum tree. She greeted him with a call that sounded, so far as the observer could recall, identical to the begging call of female *tinnunculus* (DG). At Eyre (16) on 5 December a pair were very noisy and persistently tussled with a pair of Peregrines occupying an eyrie in limestone cliffs and were thought to be nesting in the vicinity. At Monkira (61) two or three roosting kestrels were several times disturbed at night from piles of loose stones and rubble, not more than 2 ft high, bordering a track in an almost treeless area.

Food. Grasshoppers, skinks. At Cocklebiddy (15) a single bird was seen catching large flying beetles (PRC).

Annual Cycle. The male from Mount Remarkable (37) was in mainly rather worn plumage but showed some newer body feathers. It had enlarged gonads.

Impermanent Colours. *Legs* yellow to pale orange with black claws. *Iris* dark brown or greyish brown (one male). *Bill* bluish grey with black tip: cere pale yellow to deep greenish yellow. *Mouth* pale greyish to purplish. *Orbital skin* pale greenish yellow to deep yellow.

Parasites. Mallophaga (sample M. 136) collected from specimen from Thylungra (60) Queensland.

Falco berigora Brown Hawk

Specimens. Western Australia. Cocklebiddy (15) 9 July 65, 1 ♀, 1 ♂ skel. Mount Shenton (18) 2 – 3 Aug 65, 2 ♂ (400 g), 1 ♀ (550 g). Derby (29) 27 May 68, 1 ♀ (570 g). Queensland. Moonlight Creek (47) 2 May 64, 1 ♂ (444.8 g). Glenore (49) 14 – 22 Apr 64, 3 ♂ (409.4, 456.3, 493.9 g), 3 ♀ (490.7, 630.6, 632.7 g), 1 imm ♀ (462.3 g). Tyrone (59) 14 May 63, 1 ♀ skel. South Australia. Etadunna (66) 18 May 65, 1 ♀. Granite Downs (68) 7 June 65, 1 ♂. Nullarbor HS (69) 31 Oct 64, 1 ♂.

Distribution & Abundance. Seen widely in Western Australia (1, 2, 4, 9, 11, 13, 14, 23, 24, 25, 28, 31). Always conspicuous where present and often noted as common (18, 23, 25, 31). In Northern Territory seen at most camps (40, 43-46) though rather scarce at Oenpelli (43). Similarly widespread in Queensland (also 55, 58-60 at least) and South Australia where it was noted as fairly common (66, 68) or common (70) and seen also at Bool Lagoon (87). Seen also in Victoria (91) and in New South Wales (92, 93). (See notes on *F. subniger*.)

Habitat. Various types of fairly open country with trees.

Behaviour. Usually encountered singly but three together at Tyrone (59) and a pair, apparently about to breed, on the old nest of a *Corvus sp.*, at Giles (25). Perches conspicuously on dead trees, fence posts and other look-out places. Usually showed very little fear of man; on one occasion, while investigating holes in the base of a tree into which some potential prey had probably disappeared, it was only temporarily distracted by the close approach of an observer. Also observed to chase and seize a wounded Port Lincoln Parrot, which escaped after a struggle: the Brown Hawk made no further attempt to catch the now badly hurt parrot but perched six feet away from it until disturbed by the collector.

Food. Grasshoppers, caterpillars, beetles, centipedes, scorpions (once), frogs, lizards and rabbits. In one case the rabbit remains were those of an adult animal and it seems highly likely that the mammal was dead, injured or held in a trap when taken.

Annual Cycle. Nesting activities (see above) were seen at Giles (25) in late August. Most of the specimens taken in Western Australia and in northern Queensland show some body moult, and female from Glenore (49), was completing wing and body moult and was in full tail moult. A male from Nullarbor HS (69), was moulting wing and body feathers but had not yet cast any of its tail feathers.

Impermanent Colours. *Legs* usually light bluish grey, sometimes whitish grey, pale blue or greenish grey; sage green in one immature female. *Iris* pale yellowish brown to dark brown. *Bill* pale greyish horn to greyish blue, usually with dark or blackish tip: cere yellow (one) or light bluish. *Mouth* purplish flesh to fleshy grey, sometimes very pale. *Orbital skin* most often light bluish grey, sometimes greenish yellow.

Geographical Variation. Brown & Amadon (1968) admit a dry-country race *centralia* as possibly distinct from *berigora* but suggest it might be better to regard the species "as monotypic and highly variable in Australia" – in contrast to Condon (1951b) who recognised six races. Our specimens and others examined show that in general southeastern birds average larger, those from coastal and more humid regions are darker and duller and those from arid regions paler and redder, but there is wide individual variation. All birds examined were measured and categorised as "light brown, brown, dark brown, light reddish, reddish or reddish brown." In our birds, skins from northern Queensland (47, 49) include three "brown", two "dark brown", one "light reddish" and one "reddish" bird: wings ♂s 325–347 (av. 334) mm, ♀s 335–361 (347). In Western Australia a Kimberley (29) female is "dark brown", wing 353 mm; from Mount Shenton (18) two males are "light reddish", wings 308, 314 mm, and the female is "brown", wing 354; the Cocklebiddy (15) female is "reddish", wing 355. In South Australia the northern (66, 68) male and female are both "light reddish", wing ♂ 317, ♀ 346, and the Nullarbor HS (69) male is "reddish brown", wing 312 mm.

GALLIFORMES — Fowl-like Birds
D. Goodwin

Megapodius freycinet — Scrub Fowl

Specimens. NORTHERN TERRITORY. Darwin (41) 12 Sept 68, 1 fl (1,090.7 g).
Distribution & Abundance. Fairly numerous at Darwin (41) and two seen at Oenpelli (43). In Queensland common at Ayton (52) and encountered at Tully (54).
Habitat. Pseudorainforest with thick ground vegetation (41).
Behaviour. Uusually seen in twos. Makes a lot of noise when on the ground in undergrowth. When flushed usually flies to the top of a nearby tree and perches, jerking head and neck up and down. Also seen to do this on the ground when hiding behind a rock (DJF).
Impermanent Colours. *Legs* pale orange-yellow. *Iris* pale brown. *Bill* dull orange. *Mouth* yellowish. *Orbital skin* reddish black.
Geographical Variation. On locality this specimen is referred to *M. freycinet tumulus.* See Mayr (1938) for discussion of subspecies and reasons for treating *M. reinwardt* as conspecific with *M. freycinet.*

Leipoa ocellata — Mallee Fowl

Seen at Newdegate (14).

Alectura lathami — Brush Turkey

Specimens. QUEENSLAND. Mount Dryander (57) 11 Aug 64, 1 ♀ (915.0 g).
Field Notes. Found in a rocky creekbed bordered by low dry rainforest and savanna woodland. Seen also at Tully (54). The bird collected was alone, scratching among the undergrowth, and showed little fear of man.
Food. Seeds. Soil, that looked as if it might be from earthworms, and grit also in stomach.
Annual Cycle. It was in early stage of wing moult.
Impermanent Colours. *Legs* blackish brown, olive on sides of shanks. *Iris* olivaceous. *Bill* blackish brown with rami pale olive. *Mouth* brownish, white in throat. *Skin of head and neck* pale red with small area adjacent to neck feathers dull lemon.
Geographical Variation. The specimen collected is referable to nominate *lathami.*

Coturnix pectoralis Stubble Quail

Specimens. WESTERN AUSTRALIA. Green Range (13) 4 Apr 66, 1 ♀. Newdegate (14) 19 Apr 66, 1 ♂, 2 ♀ skel, 1 ♀ fl. QUEENSLAND. Thylungra (60), 29 May 63, 1 ♀, 1 juv ♂.

Distribution & Abundance. Encountered also at Metricup (9), Cocklebiddy (15), Kingoonya (70) and Coleraine (89). At Cocklebiddy (15) and Kingoonya (70) only one individual was seen. An unidentified quail, thought to be of this species, was also seen at Karratha (1).

Habitat. In Western Australia in open stubble paddocks, short mallee scrub, heathland and coastal scrub. At Thylungra (60) in open grassland and at Kingoonya (70) in open bluebush country near to open woodland.

Behaviour. The Newdegate (14) birds were shot from a tractor ploughing, of which they showed no fear. Others were mostly single and were flushed at close quarters. A Brown Hawk *Falco berigora,* obviously interested in something under a small bush, drew attention to one (70) which was flushed from beneath the bush after the hawk flew off.

Food. Seeds, including wheat.

Annual Cycle. The young bird from Thylungra (60), 29 May, was about three weeks old, assuming the rate of growth to be similar to that of *Coturnix coturnix*.

Impermanent Colours. *Legs* pale pinkish buff, pale flesh or white; flesh pink in the immature. *Iris* light brown, golden brown or reddish brown. *Bill* pale horn with darker tips to mandibles, or grey (one); brown in the immature. *Mouth* whitish horn to pale pinkish brown.

Geographical Variation. Our specimens from Western Australia confirm Hartert's (1929) contention that *C. pectoralis praetermissa* is not separable from nominate *pectoralis*.

Other Notes. An adult female from Green Range (13) had only one wing, the other was completely missing from the humerus outward. Either the bird had been hatched with only one wing or had completely lost it at an early age. It was in good condition and plumage. This bird was twice flushed with a group of several birds. Each time it was the last to rise, kicking up a volley of dust, at the same time turning two 'cartwheels' two feet above the ground, before dropping down and scurrying off.

Synoicus ypsilophorus Brown Quail

Specimens. WESTERN AUSTRALIA. Derby (29) 24 May 68, 1 ♀ (90.8 g). Geikie Gorge (31) 2 June 68, 1 ♂ (74.5 g), 1 juv ♀ (62.7 g), 2 ♂ skel (80.0, 84.1 g). Joint Hill (34) 28 June 68, 1 ♀ (90.0 g). Mount Elizabeth (35) 8 July 68, 1 ♀. Mount Remarkable (37) 25 July 68, 1 ♀ (92.2 g). Lissadell Hill (38) 29 July 68, 1 ♂, 1 ♀. NORTHERN TERRITORY. Victoria River (40) 11 Aug 68, 1 fl. Darwin (41) 19 Sept 68, 2 fl, 1 ♀ skel (87.7 g).

Distribution & Abundance. Only positively identified where collected and at Moonlight Creek (47) and Inkerman (50). But almost certainly this species at Nelson (88) where common and identity later confirmed by resident ornithologists.

Habitat. Savanna woodland (29, 35, 37, 38, 41), savanna grassland (40) or grass swamps (34, 88), and at Darwin (41) in thin scrub between mangroves and light forest.

Behaviour. Encountered in small parties which rose only when the human was very close to them. The individuals of a party never rose all together. Flight low and straight but sometimes swerving to avoid trees; difficult to flush a second time. Call like the cheeping of a young domestic fowl sometimes heard from flushed birds as they rose.

Food. Seeds. Stomachs also contained sand and quartz grit.

Annual Cycle. The male of a possible pair taken at Lissadell Hill (38) on 29 July, had very large gonads. The female had a shelled egg in the oviduct and many enlarged oocytes in the ovary. Three other Kimberley females appeared to have developing ovaries but with the largest oocytes only 1 – 2 mm in diameter.

A young female taken at Geikie Gorge (31) on 2 June was probably about three to four weeks old, assuming growth rate to be similar to that of *Coturnix coturnix*.

Impermanent Colours. *Legs* dull yellow to yellowish brown; immature pale straw. *Iris* reddish brown, brown orange or deep red; immature dark brown. Possibly some seasonal change in iris colour as the deep red irides were recorded for the breeding pair (38) and from the two spirit specimens (40, 41) which may also have been breeding. *Bill* dark brown or blackish, blue-grey at base and sometimes along the edge of the culmen, lower mandible paler. *Mouth* off-white to pale flesh pink or purplish pink.

Geographical Variation. The present specimens are referable to *S. y. cervinus* if that race, allegedly paler and buffier, is recognised. Specimens examined suggest that geographical differences in colour and size are not clear-cut though birds from New South Wales and southern Queensland average larger than those from the northwest and Northern Territory.

GRUIFORMES — Bustard Quails, Cranes, Rails, Crakes, Bustards
D. Goodwin

Turnix maculosa — Red-backed Quail

Specimens. WESTERN AUSTRALIA. Mount Bell (32) 17 June 68, 1 ♀ (47.0 g). NORTHERN TERRITORY. Darwin (41) 7 – 21 Sept 68, 2 ♂ (31.5, 34.5 g), 1 ♀ (38.0 g). Oenpelli (43) 8 Oct 68, 1 ♂ (27.5 g). QUEENSLAND. Inkerman (50) 3 June 64, 1 ♀ (40.9 g).

Distribution & Abundance. Only identified where collected.

Habitat. Savanna woodland: also in swamp grassland and on a forest track in pseudo-rainforest (41) and on the mud and thicket surround of a lake (43).

Behaviour. Usually encountered singly but once five together at Darwin (41). Flush at close quarters. One seeking food walked quietly about among the grassy tussocks at the edge of a stream and did not appear to notice the observer (DJF) until it had actually touched his boot (32). Another was found at night squatting on a track in forest (41).

Food. Grass seeds of several species, insects, greenstuff. Grit and earth also in stomachs.

Annual Cycle. Specimens from Northern Territory in September and October appeared to have inactive gonads and to be in fresh plumage. The June female from Inkerman (50) was in full moult.

Geographical Variation. Our specimens from Westen Austrralia, supposedly *T. m. pseutes,* do not appear to be separable from those of Queensland and Northern Territory *T. m. maculosa.*

Turnix varia — Painted Quail

Specimens. WESTERN AUSTRALIA. Perth (7) 21 Nov 62, 1 ♂ (90 g). QUEENSLAND. Cape Gloucester (56) 7 Aug 64, 1 ♂ (81.6 g).
Field Notes. A few were found on laterite rock in savanna woodland in the Darling Range (7); a pair in Tuart forest at Wonerup (8); two birds in savanna woodland at Cape Gloucester (56) and several in woodland and dry sclerophyll forest in the southeast (90, 93, 94).
Food. Small seeds and insects. Grit also in stomachs.
Annual Cycle. The August bird (56) was in moult.
Impermanent Colours. *Legs* dull yellow or chrome yellow. *Iris* deep pink or orange-red. *Bill* grey or darkish horn, paler below. *Mouth* bluish grey.
Geographical Variation. These and other specimens examined confirm the view (Peters, 1934) that all Australian mainland specimens are best referred to *T. v. varia.*

Turnix castanota — Chestnut-backed Quail

A few coveys almost certainly seen at Mount Anderson (28).

Turnix pyrrhothorax — Red-chested Quail

Specimens. WESTERN AUSTRALIA. Joint Hill (34) 29 June 68, 1 ♀ (49.3 g). QUEENSLAND. Glenore (49) 13 – 20 Apr 64, 3 ♂ (33.1, 35.9, 35.8 g).
Field Notes. Only identified where collected: in grassland swamp at Joint Hill (34) and tall grass of tussock grassland on dried flood plain between areas of tropical savanna woodland. All males collected were with one or two others; the female was apparently alone.
Food. Small seeds.
Annual Cycle. April males (49) appeared to have active gonads, although one was in moult.
Impermanent Colours. *Legs* pinkish white to creamy buff. *Iris* cream-coloured or very pale lemon yellow. *Bill* grey or brown (once) white along edge of tomia; lower mandible paler. *Mouth* pale grey or pale mauvish. *Eyelids* transparent grey.
Geographical Variation. Our birds and the few others examined are insufficient to show whether the three subspecies recognised by Peters (1934) are valid but suggest they are not. A specimen from Boorlee or Boorloo (South Australia?), is as dull and greyish on the upper parts as Cape York and Glenore (49) specimens.

Turnix velox — Little Quail

Specimens. WESTERN AUSTRALIA. Cocklebiddy (15) 4 Dec 62, 1 ♂. Mount Anderson (28) 25 Apr – 12 May 68, 2 ♀ (49.6 g), 1 ♂ skel, 1 imm skel, 1 juv fl. Mount Bell (32) 19 June 68, 1 ♂ (37.0 g). QUEENSLAND. Moonlight Creek (47) 4 May 64, 1 ♂ (36.5 g). Glenore (49) 9 – 14 Apr 64, 3 ♂ (34.9, 36.5, 37.3 g), 1 ♀ (52.3 g).

Field Notes. Only identified where collected, in savanna grassland, savanna woodland, and spinifex. Always in either grassy areas or adjacent pebbly flats with some short vegetation.

Behaviour. Often only a single bird flushed; sometimes two to 30 or more in loose association. Large numbers found together only at Glenore (49).

Flush at close quarters and fly fast and straight for short distance. Very difficult to flush a second time.

Food. Seeds.

Annual Cycle. The December male from Cocklebiddy (15) had well-developed gonads. All the Kimberley adults (28, 32) in late April, May and June were undergoing complete moults and a juvenile was taken at Mount Anderson (28) on 25 April. Two April males from Glenore (49) were also moulting.

Impermanent Colours. *Legs* very pale pink or brownish. *Iris* opaque white or yellow. *Bill* horn-coloured, grey or brown usually with the lower mandible paler than the upper and the latter darkest along ridge. *Mouth* fleshy, pale grey or white.

Geographical Variation. Keast (1961) seems to imply that no subspecies are recognisable. Our birds and some from eastern and southern Australia from older collections suggest that this is so, although they are not sufficient to prove it. The three from the Kimberleys (28, 32) are all rather reddish in hue and thus agree with Mathews (1912a) description of *picturata*. An individual from Glenore (49) and one from South Australia are, however, equally rufous.

Parasties. See Appendix 1.

Grus rubicunda — Brolga

Specimens. WESTERN AUSTRALIA. Mount Anderson (28) 10 May 68, 1 ♂ skel.

Field Notes. Small parties of up to five birds seen throughout most of the Kimberleys (27-36) and at all camps in Northern Territory (40-46). Noted as plentiful at Noogoo Swamp, Darwin (41). In Queensland at Glendower (55), and in South Australia at least eight seen at Clifton Hills (65) and at Cadelga (62).

Usually associated with fresh water lagoons, in open grass or woodland and black soil plains.

Food. Gut contents listed as tubers and pebbles.

Annual Cycle. Gonads were small and the tail showed some moult. Adults with young were seen at Mount Anderson (28) in May and at Borroloola (46) in November.

Impermanent Colours. *Legs* black. *Iris* golden orange. *Bill* pale green base with brown tip. *Mouth* white. *Bare skin of head* orange-red.

Eulabeornis castaneoventris — Chestnut Rail

Specimens. WESTERN AUSTRALIA. Wyndham (39) 2 Aug 68, 2 ♂ (557.1, 626.7 g).

Field Notes. One other bird was seen at Wyndham (39) and two birds, almost certainly this species, in Northern Territory near Mount Roper (45). All were found in mangrove swamps but this rail was not encountered in the Queensland mangroves.

Food. Crabs.

Annual Cycle. Gonads appeared inactive.
Impermanent Colours. *Legs* yellow. *Iris* red. *Bill* greenish yellow. *Mouth* pale flesh, tinged bluish in places.

Porzana fluminea — Australian Spotted Crake

Seen at Nelson (88).

Porzana cinerea — White-browed Crake

A few seen at Noogoo Swamp, Darwin (41) and a single bird at Wyndham (39).

Tribonyx ventralis — Black-tailed Native Hen

Specimens. SOUTH AUSTRALIA. Etadunna (66) 18 May 65, 1 ♂ skel, 1 ♀ skel. Oodnadatta (67) 31 May – 1 June 65, 1 ♂ (385.3 g), 1 ♀. Lake Acraman (71) 2 Nov 64, 1 ♀.
Field Notes. Only encountered in South Australia, at or near edges of lakes, creeks and dams. One seen at Innamincka (64), about 10 at Clifton Hills (65), common at Etadunna (66) and three seen at Oodnadatta (67).
Annual Cycle. Some of the birds seen at Etadunna (66) on May 18 were not fully grown and must therefore, almost certainly have been reared locally. The male from Oodnadatta (67) on 31 May, had the wing feathers (primaries, secondaries and greater coverts) very worn, much more so than its body plumage: the female was in great contrast, having apparently fresh wing feathers. The female taken at Lake Acraman (71) on 2 November was in body moult but its wing feathers looked new.
Impermanent Colours. *Legs* dark red or vermilion (one female). *Iris* bright yellow to orange brown. *Bill* upper mandible, olive green or light green (one female), sometimes paler at tip: lower mandible basal two-thirds (approx.) orange to red, distal third light green to olive green. *Mouth* pale flesh pink.
Geographical Variation. The specimens in the British Museum substantiate Peters' (1934) doubts as to the validity of the races *territorii* and *whitei*.

Gallinula tenebrosa — Dusky Moorhen

Specimens. NEW SOUTH WALES. Grafton (94) 1 Apr 63, 1 ♀, 1 juv, 1 ♂ skel.
Distribution & Abundance. Seen in Western Australia near Perth (7); in South Australia at Clifton Hills (65), where two only seen, and Bool Lagoon (87): in New South Wales also at Dunedoo (93).
Habitat. Swamps, also ponds and river with adjacent reed or bush cover (65).
Behaviour. Gait and manners appeared identical with those of Moorhen *Gallinula chloropus*.
Food. Vegetable matter in stomachs.
Annual Cycle. Many chicks of different ages were seen, in company with their parents, at Grafton (94) on 1 April. The chick collected was about 12 to 14 days old. The female taken on same date was in body moult but had very worn primaries.
Impermanent Colours. *Legs* yellow, grey at ankle ("knee") and toe joints, and orange on tibia above ankle and on front of tarsus. *Iris* brown. *Bill* basal two-thirds deep bright

red, tip yellow: frontal shield orange or red. *Mouth* light greyish flesh. The chick had black legs. Its bill had the basal two-thirds pale dull flesh, then a narrow blackish band and a dull sandy-yellow tip.

Porphyrio porphyrio Swamphen

Specimens. WESTERN AUSTRALIA. Derby (29) 29 May 68, 1 ♂ (827.9 g), 1 ♀ (629.4 g).
Field Notes. In reed beds, included a small patch in a pond on a public green near Perth (7). Common at Bool Lagoon (87) and seen at a few other scattered localities (13, 45, 94).
Annual Cycle. Both birds collected were in body moult.
Impermanent Colours. *Legs* pale flesh pink, grey at joints, nails dark grey. *Iris* dark brown. *Bill* dark bright red, paler at tip: frontal shield dark bright red. *Mouth* whitish or purplish flesh.
Geographical Variation. The collected specimens are referable to *Porphyrio p. melanotus.*
Parasites. See Appendix 1.

Fulica atra Coot

Specimens. WESTERN AUSTRALIA. Lake Gladstone (36) 22 July 68, 1 ♂ (554.5 g), 1 imm ♀ (443.1 g), 1 ♂ fl (465.0 g), 1 imm fl (394.8 g).
Field Notes. A few seen on a large pond near Perth (7) and about 25 on Lake Gladstone (36) in Western Australia: not encountered in northern Queensland. In South Australia large numbers seen on Lake Killalpaninna (66). In Victoria a few at Nelson (88).

So far as seen its behaviour was identical with that of the species in Britain.
Food. Vegetable matter.
Annual Cycle. At Lake Gladstone (36) immatures present were well grown, the female collected was moulting into adult plumage, and the birds' sociability and inactive gonads showed they were not breeding.
Impermanent Colours. *Legs* olive grey. *Iris* blood red. *Bill* light grey: frontal shield, off-white or creamy white, looking pure white at a little distance. *Mouth* dirty white.

Eupodotis australis Australian Bustard

Specimens. WESTERN AUSTRALIA. Wyndham (39) 7 Aug 68, 1 fl.
Field Notes. In Western Australia seen at Karratha (1) and Peg's Beach (2): further north plentiful in the savanna grassland near St George's Range (27) and seen in the Kimberleys at Mount Anderson (28) and Geikie Gorge (31), in the centre at Van der Linden Lakes (24) while fresh tracks and a feather were found at Mungilli Claypan (20) and Hunt Oil Camp (22). In Northern Territory a few were seen in the savanna grassland and woodland surrounds of Pine Creek (44), Borroloola (46) and the Roper River (45) where it was also encountered out on the open mud flats.

In Queensland it was common on the Hann Highway near Glendower (55).
Impermanent Colours. *Legs* cream-yellow. *Iris* straw. *Bill* upper mandible cream with black culmen and tip: lower mandible cream. *Mouth* grey with white edges.

CHARADRIIFORMES — Waders, Gulls & Terns
D. Goodwin

Jacana gallinacea Lotus Bird

Specimens. Northern Territory. Oenpelli (43) 30 Sept – 11 Oct 68, 6 fl (62.9, 84.0, 111.6, 111.6, 113.9, 118.0 g), 2 imm fl (92.5, 111.4 g). New South Wales. Grafton (94) 1 Apr 63, 1 ♂ skel.
Specimens to other Museums. Nat. Hist. Mus., Reykjavik. 1 ♀ Oenpelli (43).
Field Notes. Abundant at Oenpelli (43) where it was usually found in loosely-associated parties of nine or ten birds, comprised of adults and immatures; once a lone individual. In inland marshy lagoons with floating vegetation, but no water-lilies. At Grafton (94) a pair with three young were on lily-grown pools of a riverside swamp. Not seen elsewhere.

At Oenpelli (43) the lone bird was wary and when approached went under the water for about 40 seconds, then rose and first ran over the vegetation and then flew.
Annual Cycle. Young birds were present in both April (94) and October (43), the April chicks were thought to be about five days old.
Impermanent Colours. The male with chicks (94) is assumed to be in breeding colouration. *Legs* breeding male light grey; others olive or yellowish green. *Iris* breeding male pale golden; others pale grey or light brown with white flecks (once). *Bill* breeding male bluish flesh with black tip; others light brown, lower mandible yellower with brown tip: frontal shield yellow to reddish buff. *Mouth* whitish to flesh pink. *Orbital skin* of breeding male was narrow and blue.

Rostratula benghalensis Painted Snipe

Specimens. Queensland. Glenore (49) 20 Apr 64, 1 ♀ (134.0 g).
Field Notes. The only bird seen was found at night on a track in tropical savanna woodland.
Impermanent Colours. *Legs* pale yellowish grey. *Iris* pale red-brown. *Bill* distal half pale orange, basal half greyish green.

Haematopus ostralegus Pied Oystercatcher

A few seen at Hope inlet, Darwin (41) and also recorded from scattered localities (1, 2, 13, 88).

Haematopus fuliginosus Sooty Oystercatcher

Seen in Western Australia (1, 2, 13) and at Port Augusta, South Australia.

Vanellus novaehollandiae Spur-winged Plover

Specimens. South Australia. Innamincka (64) 30 Apr 65, 1 ♀ (295.1 g), 1 ♂ skel. New South Wales. Dunedoo (93) 25 Mar 63, 1 ♂ skel.
Distribution & Abundance. In South Australia seen also inland at Blanchewater

29°32S; 139°26E), Clifton Hills (65), Etadunna (66) and Aroona Valley (82), appearing fairly common except at Clifton Hills (65). In the southeast in December and January about 400 were seen on The Coorong (85), about 300 on Bool Lagoon (87) and 50 at Nelson (88).

Habitat. Open ground, not necessarily extensive, but with short or scattered vegetation, near water. Thus at inland localities (64, 65, 82) in South Australia on banks of creeks and rivers, and at Dunedoo (93) on grassy field.

Food. Insects, small crustaceans, vegetable matter. Small stones also in stomachs.

Annual Cycle. Apparently not breeding.

Impermanent Colours. *Legs* purple or dark reddish purple, in two specimens (♂ & ♀) the scutes on tarsus and toes were noted as blackish. *Iris* pale yellow, darkening towards outer circumference (two), bright yellow (one). *Bill* lemon yellow, tip ivory white, tip of ridge of culmen dusky or brownish. *Mouth* fleshy. *Wattle* bright lemon yellow, deepening to golden yellow at lower ends.

Geographical Variation. The CSIRO (1969) list has been followed in treating *V. novaehollandiae* and *V. miles* as distinct species but we agree with van Tets *et al* (1967) that they are better regarded as conspecific: especially since our female from Innamincka (64), although referable to *novaehollandiae,* is to some extent intermediate in colour, in the amount of black on the sides of the neck and in wattle size.

Vanellus miles Masked Plover

Specimens. Western Australia. Derby (28) 25 May 68, 1 fl (262.8 g). Lake Gladstone (36) 22 – 23 July 68, 1 ♂, 1 ♀ (190.7 g). Queensland. Moonlight Creek (47) 2 May 64, 1 ♂ (233.8 g), 1 ♀ (251.9 g), 1 juv. Glenore (49) 10 – 23 Apr 64, 2 ♂ (250.8, 255.6 g), 5 ♀ (241.4, 247.8, 256.2, 286.0, 296.3 g). Glendower (55) 25 Mar 64, 1 ♀ (262.2 g).

Distribution & Abundance. Found also in Northern Territory in fair numbers at Darwin (41), Oenpelli (43) and Mount Roper (45), and a few near the homestead at Pine Creek (44). Usually in pairs or small parties on the ground.

Habitat. As for *V. novaehollandiae*. Thus in Western Australia in flooded fields (29) and on lakeside in savanna woodland (36); in northern Queensland in grassy swamps, at edges of waterholes, dams and rivers.

Food. Insects, including beetles; small snails, small crustaceans; grass seeds and other vegetable matter. Small stones in some stomachs.

Annual Cycle. The female taken at Lake Gladstone (36) on 23 July had an enlarged ovary, the biggest oocytes about 3 mm in diameter. In north Queensland a pair with a chick only a few days old were taken at Moonlight Creek (47) on 2 May; the adults were just beginning to moult. Most of the specimens taken at Glenore (49) and the female taken at Glendower (55) were also moulting.

Impermanent Colours. *Legs* mauvish pink to purple or carmine, with the scutes on tarsus and toes darker reddish, purplish or purplish brown: chick, medium grey. *Iris* pale whitish lemon to bright lemon yellow: chick, olive grey. *Bill* pale lemon yellow to deep yellow or greenish yellow, tipped cream or horn: lower mandible usually paler and with pale tip: chick, upper mandible greenish purple, lower mandible light grey. *Mouth* grey to greyish pink: chick, greyish white. *Wattle* lemon yellow to bright yellow: chick, rich buff. *Wing-spur* yellow. *Lower mandible* light grey. *Mouth* greyish white. *Wattles* rich buff.

Parasites. See Appendix 1.

Vanellus tricolor — Banded Plover

Specimens. Western Australia. Chichester Range (3) 13 May 66, 1 ♀ skel. Newdegate (14) 11 Apr 66, 1 fl. Queensland. Monkira (61) 2 – 6 June 63, 1 ♂, 1 ♀, 2 juv, 2 skel, 1 fl, 1 egg. South Australia. Lake Acraman (71) 2 Nov 64, 1 ♀. Nonning (79) 31 Mar 65, 1 ♂ skel. Angepena (84) 19 – 20 Nov 63, 1 ♂, 1 imm ♂.

Distribution & Abundance. In Western Australia seen also at Karratha (1), Metricup (9), Scott River (11), Mount Barker (12), Green Range (13), Cocklebiddy (15), and Mount Shenton (18). In Queensland a few at Thylungra (60) and locally common at Monkira (61). Usually in pairs.

Habitat. Open grassland, plains or mudflats with only short, scattered vegetation; often near water.

Behaviour. Voice, a shrill plaintive *er-chill-chat* repeated.

Food. Grasshoppers, beetles, other insects and seeds.

Annual Cycle. Breeding in Western Australia (3) in mid-May (female with large oocytes collected) and in south Queensland (61) in early June (egg and chicks collected, others seen), but in South Australia apparently later for the young November bird from Angepena (84) appears about six to eight weeks old: the female from Lake Acraman (71) was in moult.

Impermanent Colours. *Legs* purple to mauvish grey, sometimes with blackish sheen on front of tarsus. *Iris* yellow or greenish yellow. *Bill* yellow with blackish tip to culmen. *Orbital skin* yellow. *Wattle* red.

Geographical Variation. Western Australian specimens were separated as *V. t. gwendolenae* by Mathews (1912) as having, allegedly, a narrower pectoral band and differently-shaped wattle. The few Western Australian specimens in the British Museum do not show the former character and wattle-shape cannot be deduced with certainty from the skins. It seems likely that Peters (1934) was correct in placing *gwendolenae* as a synonym of *tricolor*.

Charadrius cinctus — Red-kneed Dotterel

Specimens. Western Australia. Wyndham (39) 4 Aug 68, 3 fl (51.1, 52.6, 53.4 g). Northern Territory. Borroloola (46) 15 Nov 68, 1 fl. South Australia. Innamincka (64) 30 Apr 65, 1 ♀.

Field Notes. Found on an open grass marsh in a loose mixed flock with *C. alexandrinus* at Wyndham (39). Only one bird was seen on the lagoon at Borroloola (46) and one in company with *C. melanops* on the bank of the Cooper at Innamincka (64). Not noted elsewhere.

Food. Insects.

Annual Cycle. The Innamincka (64) specimen was in moult.
Impermanent Colours. *Legs* brownish grey to purplish green, ankle joint ("knee") pinkish purple or red. *Iris* dark brown. *Bill* purple with black tip, and base and ridge of culmen. *Mouth* pinkish or purple with black tips to inner sides of mandibles.

Charadrius cucullatus Hooded Dotterel

Specimens. Western Australia. Newdegate (14) 19 Apr 66, 1 imm ♂ (and skel of trunk), 1 ♂ skel.
Field Notes. At Newdegate (14) six were feeding around edges of muddy islets of freshwater lake in loose association with eight Black-fronted Dotterels and five Greenshanks. Also eight birds were seen at Nelson (88).
Annual Cycle. The immature specimen had a few new body feathers growing but wing quills not moulting.
Impermanent Colours. *Legs* light flesh pink with blackish toes. *Iris* brown. *Bill* basal half (approx.) dark reddish pink, paler and less extensive in immature, tip black. *Orbital skin* adult scarlet, immature orange. *Mouth* dark flesh.

Charadrius alexandrinus Red-capped Dotterel

Specimens. Western Australia. Carnarvon (5) 14 June 66, 1 ♀ skel, 1 fl. Newdegate (14) 11 Apr 66, 1 ♂ skel. Wyndham (39) 3 – 4 Aug 68, 1 ♂ (31.0 g), 2 ♂ (28.4, 41.0 g), 2 fl (28.6, 32.1 g), 1 imm ♂ fl (28.9 g), 1 juv ♀ fl (27.6 g). South Australia. Clifton Hills (65) 10 May 65, 2 ♂, 1 ♂ skel.
Field Notes. In Western Australia common at Carnarvon (5): seen also at Karratha (1), Peg's Beach (2) and Perth (7). In Northern Territory plentiful on the shoreline at Darwin (41) and on salt lagoons at Mount Roper (45), where immatures were present. In South Australia fairly common near the Diamantina River at Clifton Hills (65) and Etadunna (66), and about 500 on The Coorong (85). In Victoria at Nelson (88) about 30 birds and a nest with four eggs found on beach. In New South Wales at Grafton (94).

Often in (presumed) pairs (5, 39, 65) on mudflats, sand or shingle beaches of sea or inland waters, muddy or sandy river edges. Occurred in same areas as *C. melanops* at Clifton Hills (65), and possibly elsewhere, but seemed in general to require more extensive mud, sand or shingle areas nearly bare of vegetation.
Food. Beetles; also unidentified vegetable matter and grit in stomachs.
Annual Cycle. In the southeast nests and eggs were found in December (85, 88). In contrast in the northwest (39) in early August immature and juvenile birds were seen and collected and two adult females had developing oocytes, in one the largest were 3 mm in diameter, the other contained a shelled egg in the oviduct. The first of these females had on the left side two new wing feathers, a scapular and a larger inner secondary, although the corresponding feathers on the other side appeared old and worn; the second appeared to be moulting scapulars and inner secondaries in a normal symmetrical manner. Both showed slight body moult and possibly in both moult may have been at least slowed down by their coming into breeding condition.

One male collected at Clifton Hills (65) in May was just completing its moult.
Impermanent Colours. *Legs* blue grey (once), dark grey to blackish; juvenile light grey. *Iris* dark brown. *Bill* black. *Mouth* fleshy grey to near black, tongue pinkish; juvenile light grey.
Geographical Variation. These and other specimens examined substantiate Peters' (1934) opinion that *C. a. tormenti* should be considered a synonym of *C. a. ruficapillus*.

Charadrius mongolus — Mongolian Sand-dotterel

A few seen at Hope Inlet, Darwin (41) and also recorded at Grafton (94).

Charadrius leschenaultii — Large Sand-dotterel

Specimens. NORTHERN TERRITORY. Darwin (41), 16 Sept 68, 1 ♂.
Field Notes. At a surprisingly late date, 10 – 14 June, small parties were seen at Peg's Beach (2), Bush Bay (6) and at Oyster Creek, Carnarvon (5). Seen on mudflats, appearing to prefer drier mud than other waders present did, although was often in shallow pools also.
Food. Tiny crabs. Shell and grit fragments also in stomach.
Annual Cycle. The specimen was in a complete moult, and had minute gonads. Some of the June birds seen were attaining summer plumage (PRC).
Impermanent Colours. *Legs* pale blue grey. *Iris* dark brown. *Bill* black. *Mouth* blue grey.

Charadrius melanops — Black-fronted Dotterel

Specimens. WESTERN AUSTRALIA. Newdegate (14) 19 Apr 66, 2 fl. Mount Anderson (28) 10 – 18 May 68, 2 ♂ (29.3 g), 2 ♀ (30.0, 33.3 g), 2 ad fl, 2 juv fl. QUEENSLAND. Thylungra (60) 26 May 63, 1 ♂ skel. SOUTH AUSTRALIA. Clifton Hills (65) 12 May 65, 1 ♂. Nonning (79) 31 Mar 65, 1 ♂ (28.0 g), 1 ♀ skel (33.2 g). Aroona Valley (82) 8 Apr 65, 1 skel (32.0 g). NEW SOUTH WALES. Albury (92) 13 Mar 63, 1 ♂ 1 ♀ skel, 1 fl.
Specimens to other Museums. NMV 1 ♂ skel Grafton (94).
Distribution & Abundance. In Western Australia breeding and common at Mount Anderson (28); also encountered at Karratha (1), Peg's Beach (2) and Geikie Gorge (31). In Northern Territory a few were seen at Oenpelli (43) and ten near a lagoon at Mount Roper (45). In Queensland at Tyrone (59). In South Australia it was common or fairly common at most main camps (64, 67, 79, 82). In New South Wales seen also at Grafton (94).
Habitat. Open, not necessarily extensive, areas near water. Muddy banks (only slightly sloping) of creeks, rivers, dams, ponds, etc. Once on muddy patch in swampy field (94).
Behaviour. Gait as other small plovers but more rounded wings and somewhat different, more "flapping" flight noticeable in the field.

At Clifton Hills (65) one bird was seen to fly aggressively at another. The latter turned and faced it, spreading its wings, then closed them and went into a display posture similar to the threat display of the Little Ringed Plover (Gillmor, in Sommons 1953:

197) with head pressed down on neck but with the purple feathers of the scapular region erected to a greater extent than the surrounding plumage, thus much increasing the apparent size of the purple areas (DG).

Calls heard were a scolding chatter, a tinkling trill and a loud *pwit*.

Food. Insects and crustaceans.

Annual Cycle. In May two chicks in down were taken at Mount Anderson (28). Their male parent, and two other adults taken at same time were in the latter stages of moult, as were a male from Nonning (79) and one from Albury (92) in March.

Impermanent Colours. *Legs* pale creamy flesh to orange yellow, claws dark. *Iris* dark brown. *Bill* pinkish orange to deep red with black tip. *Mouth* flesh pink to coral red, blackish on inside tips of mandibles. *Orbital skin* bright orange to dark crimson red. Chicks in down had light orange legs and brownish or orange-brown, black-tipped bills.

Peltohyas australis — Australian Dotterel

Specimens. WESTERN AUSTRALIA. Warburton (23) 14 Aug 65, 1 fl. QUEENSLAND. Monkira (61) 1 – 9 June 63, 1 juv, 4 fl. SOUTH AUSTRALIA. Nullarbor HS (69) 5 Oct 63, 29 Oct 64, 2 ♂. Kingoonya (70) 20 – 25 June 65, 3 ♂, 1 ♀, 1 ♂ skel, 1 fl.

Specimens to other Museums. NMV 1 fl Monkira (61). Zool. Inst. Acad. Sci. USSR 1 ad fl, 1 juv fl, Monkira (61).

Distribution & Abundance. Two were seen at Mount Shenton (18) and one at Warburton (23). Common in flocks of 10 to 30 at Kingoonya (70), and in family parties at Monkira (61). Also found breeding at Nullarbor HS (69).

Habitat. Open plains, paddocks or sandhill country. Usually stony or with only very short grass or scattered vegetation. The general appearance of some of the more barren areas where the species was found near Kingoonya (70) seemed similar in general character to wintering quarters of the Eurasian Dotterel *Charadrius morinellus* in Libya (DG).

Behaviour. Gait and head movements appeared almost if not quite identical to those of dotterels and ringed plovers *Charadrius* sp. and relatively, unlike those of coursers, *Cursorius* sp. Flight call *quiop,* frequently heard at night (PRC).

Food. Vegetable matter and seeds. Grit and sand also in stomachs.

Annual Cycle. Chicks about a week old were seen at Nullarbor HS (69) on 7 December 62 and some about three weeks old were taken at Monkira (61) in early June. Adults taken at Kingoonya (70), South Australia, in late June were in body moult. An adult male taken at Nullarbor HS (69) on 29 October was in wing moult, the four outermost primaries in each wing are old, the other primaries new. It did not appear to be (yet?) in body moult.

Impermanent Colours. *Legs* pale whitish yellow to light brown, feet usually darker. *Iris* brown. *Bill* base greenish or yellowish, rest dark brown or dark grey, blackish at tip. *Mouth* flesh pink to fleshy grey.

Geographical Variation. The few Western Australian specimens examined do not confirm the validity of the western race *whitlocki*.

Pluvialis dominica Eastern Golden Plover

Seen at Grafton (94).

Pluvialis squatarola Grey Plover

A few seen at both Hope Inlet, Darwin (41) and Perth (7).

Arenaria interpres Turnstone

Seen in Western Australia (1, 2) and at Hope Inlet, Darwin (41).

Numenius minutus Little Whimbrel

Several seen at Hope Inlet, Darwin (41) and at Oenpelli (43).

Numenius phaeopus Whimbrel

Fairly numerous at Hope Inlet, Darwin (41) and also recorded from Western Australia (1, 2, 39).

Numenius madagascariensis Eastern Curlew

Seen at Nelson (88).

Unidentified curlews were numerous at Hope Inlet, Darwin (41), 12 were seen at Wyndham, a few at Mount Roper (45) and some on the west coast (1, 2).

Tringa stagnatalis Little Greenshank

A few seen at Hope Inlet, Darwin (41) and also on the west coast (1, 2).

Tringa nebularia Greenshank

Twenty seen at Mount Roper (45) and a few at Hope Inlet, Darwin (41), with other widespread records (12, 14, 41, 79, 87, 88).

Tringa glareola Wood Sandpiper

Four seen near Albany, Mount Barker (12) on 25 March (Colston & Harrison, 1968).

Tringa hypoleucos Common Sandpiper

Specimens. Western Australia. Wyndham (39) 8 Aug 68, 1 fl.
Field Notes. In the north several were seen on tidal mud bordering the mangroves at Wyndham (39) and one or two elsewhere (40, 41, 45). During a week at Nelson (88) three only were seen along the edge of the Glenelg River.
Impermanent Colours. *Legs* olive grey. *Iris* brown. *Bill* dark grey. *Mouth* pale brown.

Tringa brevipes Grey-tailed Tattler

Specimens. Western Australia. Peg's Beach (2) 1 June 66, 1 ♂ skel, 1 ♀ skel, 2 fl.
Distribution & Abundance. At high tide assemblies of waders, some 20 of this

species were present at Peg's Beach (2) and 15 present for a week up to June 13 at Bush Bay (6). In New South Wales about 50 at mouth of Clarence River, Grafton (94) on 7 April and ten at Brunswick Heads, 70 miles to the north on 18 April. One seen at Darwin (41).

Habitat. Exposed mudflats or edges of muddy mangrove creeks. At high tide rest on rocks or tops of mangroves.

Annual Cycle. All birds collected were still in winter plumage with small gonads.

Impermanent Colours. *Legs* greenish yellow. *Iris* dark brown. *Bill* almost black with a green tinge; lower mandible paler and more greenish. *Mouth* greyish green.

Xenus cinereus — Terek Sandpiper

Fairly numerous at Hope Inlet, Darwin (41).

Calidris canutus — Knot

Eight seen at Hope Inlet, Darwin (41) and a few at Perth (7).

Calidris acuminata — Sharp-tailed Sandpiper

Specimens. NORTHERN TERRITORY. Mount Roper (45) 1 Nov 68, 5 fl (45.1, 47.9, 49.1, 49.5, 49.8 g).

Distribution & Abundance. In Western Australia about 30 birds were seen at Perth (7) and small numbers at Albany and Carnarvon (5). In Northern Territory scarce at Darwin (41) and about 20 at Mount Roper (45). In the southeast in December and early January about 2,000 were seen on The Coorong (85), about 100 at Nelson (88) and a large flock of some 3,000 at Altona Saltpans, Melbourne on 17 February 1963, many of which were attaining summer plumage.

Habitat. At Mount Roper (45) on a tidal salt lagoon surrounded by samphire about fourteen miles up-river from the sea.

Behaviour. In Northern Territory it was noted singly or in company with only one or two others, appearing to prefer not to associate with other waders present in the area. Tended to keep to edges of lagoon and to show less intense activity when feeding than other small waders. Not wary of man. Call note *Krip-Krip-Krip,* less reedy than *C. melanotos* and more musical (PRC).

Impermanent Colours. *Legs* olive green (four birds), or olive yellow "conspicuously different" (one bird). *Iris* dark brown. *Bill* sepia to black. *Mouth* purplish pink to flesh pink sometimes tinged grey.

Calidris ruficollis — Red-necked Stint

Specimens. WESTERN AUSTRALIA. Perth (7) 8 Feb 66, 2 ♂, 3 ♀ skel, 4 fl. Wyndham (39) 4 Aug 68, 1 ♂ (23.9 g), 2 fl (26.0, 26.5 g). SOUTH AUSTRALIA. Nullarbor HS(69) 9 Oct 63, 1 ♀ skin and trunk.

Distribution & Abundance. In Northern Territory plentiful at Darwin (41) and about fifty seen at Mount Roper (45). In the southeast in December and early January about 500 were seen on The Coorong (85), and 150 at Nelson (88).

Habitat. Edge of lagoon (7), freshwater grassland marsh (39), tideline of sea beach (69).
Food. Not recorded but one gizzard retained in spirit.
Annual Cycle. A male from Wyndham (39) has intermixed old and newer feathers on its upperparts. Its four outer primaries in each wing look newer than the rest but it does not show any growing feathers.
Impermanent Colours. *Legs* black. *Iris* dark brown. *Bill and mouth* black.
Other Notes. All the specimens collected at Perth (7) appeared to have been poisoned by botulism. Paralysis of legs, wings, limp head and some difficulty in breathing were outward signs. Other waders of different species in the vicinity appeared unaffected.

Calidris ferruginea — Curlew Sandpiper

Specimens. NORTHERN TERRITORY. Darwin (41) 16 Sept 68, 1 fl (59.1 g).
Distribution & Abundance. Encountered also in Western Australia at Peg's Beach (2), Perth (7); a few were seen on the banks of the Roper River (45), and at mouth of Moonlight Creek (47).
Habitat & Behaviour. At Darwin (41) a few individuals were seen on mud and sand flats between the mangroves and the sea in loose association with other small waders, tending to feed towards the sea as it receded.
Annual Cycle. The specimen was moulting into winter plumage.
Impermanent Colours. *Legs* blackish. *Iris* brown. *Bill* black. *Mouth* greyish white.

Calidris alba — Sanderling

One seen at The Coorong (85).

Limosa limosa — Black-tailed Godwit

About 30 seen at Mount Roper (45) and also recorded on the west coast (1, 2).

Limosa lapponica — Bar-tailed Godwit

Fairly numerous at Hope Inlet, Darwin (41) and also recorded on the west coast (1, 2).

Himantopus himantopus — White-headed Stilt

Specimens. WESTERN AUSTRALIA. Mount Anderson (28) 15 May 68, 1 imm fl. Wyndham (39) 4 Aug 68, 1 imm fl (134.3 g).
Field Notes. In Western Australia common at Wyndham (39) and with young. In Northern Territory four to six at Mount Roper (45). In Queensland, ten at Banoona (58) and one at Thylungra (60). In South Australia one at Clifton Hills (65), a pair at Etadunna (66) and Nonning (79), common at Bool Lagoon (87). In Victoria 20 or more pairs, several with young, at Altona saltpans near Melbourne in February.
Impermanent Colours. *Legs* salmon pink. *Iris* crimson. *Bill* black. *Mouth* pinkish white.

Cladorhynchus leucocephalus Banded Stilt

Specimens. South Australia. Etadunna (66) 17 May 65, 1 ♀ skel.
Field Notes A single bird at Etadunna (66) but about 1,500 on The Coorong (85) 18 – 19 December, and 2,000 at Lake George 70 miles further south on 13 January.
Food. Only some grit and weed in the stomach.
Annual Cycle. Ovary of female moderately enlarged.
Impermanent Colours. *Legs* pink. *Iris* dark brown. *Bill* black. *Mouth* pinkish and black towards tip.

Recurvirostra novaehollandiae Australian Avocet

Specimens. Northern Territory. Mount Roper (45) 30 Oct 68, 1 fl (241.3 g). South Australia. Clifton Hills (65) 10 – 11 May 65, 1 ♀, 1 ♂ skel, 1 fl.
Specimens to Other Museums. Nat. Hist. Mus. Reykjavik, 1 ♀ (236.8 g), Mount Roper (45).
Field Notes. In Western Australia seen at Perth (7) and Newdegate (14). Seven birds were present in Northern Territory on the Roper River (45). In South Australia fairly common in pairs along the Diamantina (65), and about 500 were seen on The Coorong (85).
Food. Insect larvae, seeds, vegetable matter; quartz grit also in some stomachs.
Annual Cycle. The May female from Clifton Hills (65), where birds were paired, appeared to be in early stages of moult.
Impermanent Colours. *Legs* light bluish grey. *Iris* dark brown, one with golden tinge. *Bill* dull black, sometimes paler at extreme base and at tip of culmen. *Mouth* whitish grey to dark grey.
Parasites. See Appendix 1.

Burhinus magnirostris Southern Stone Curlew

Specimens. Western Australia. Joint Hill (34) 2 July 68, 1 ♂ (719.1 g). Mount Elizabeth (35) 14 July 68, 1 ♀ (619.2 g), 1 ♂ fl (676.9 g). Northern Territory. Oenpelli (43) 29 Sept 68, 2 fl (535.8, 708.1 g).
Distribution & Abundance. In Western Australia heard at Chichester Range (3) and Coolawanyah (4) and seen also at Newdegate (14) and Mount Anderson (28). In Northern Territory a few at Darwin (41) and plentiful at Oenpelli (43). In the southeast a few at Coleraine (89) and a single bird at Grafton (94).
Habitat. In the Kimberleys in burnt savanna woodland on sandstone hill (34) and dry sand dunes by river in savanna woodland (35). In Northern Territory (43) in open plains near escarpment and forest edge. In New South Wales (94) edge of woodland.
Behaviour. Usually encountered in twos or threes, or singly (35, 94). Frequently utters a penetrating double whistle on a descending scale; often calls at night.
Food. Insects including beetles. Many of the insects very small for size of bird, the same sizes as those taken by *Malurus*.
Annual Cycle. The male from Mount Elizabeth (35) was in moult, the female showed some body moult.

Impermanent Colours. *Legs* pale yellowish green to olive yellow. *Iris* bright yellow with black speckling in one male. *Bill* black. *Mouth* flesh pink to purplish.
Geographical Variation. The specimens collected are referable to *B. magnirostris rufescens*.

Esacus magnirostris — Beach Stone Curlew

Two seen at Hope Inlet, Darwin (41) and also recorded on the west coast (1, 2).

Stiltia isabella — Australian Pratincole

Specimens. WESTERN AUSTRALIA. Wyndham (39) 4 – 7 Aug 68, 1 ♂ (55.7 g), 2 fl (62.6 g). NORTHERN TERRITORY. Mount Roper (45) 30 Oct – 1 Nov 68, 1 imm ø, 1 fl (49.4 g). QUEENSLAND. Monkira (61) 2 – 7 June 63, 1 ♀, 1 ø, 2 fl. SOUTH AUSTRALIA. Angepena (84) 1 imm ♂.
Specimens to other Museums. NMV 1 imm ♀, 1 fl, Monkira (61). Zool. Inst. Acad. Sci. USSR 1 fl, Monkira (61).
Distribution & Abundance. Noted only where collected except for a single bird on a lawn at Borroloola (46). Usually in small, scattered parties, but in large flocks at Monkira (61).
Habitat. Open plains, salt flats and open grassland. Usually at no great distance from water or in recently dried places.
Behaviour. Some individuals allowed a very close approach, others took wing at a considerable distance from the approaching human. A single bird at Wyndham (39) flushed at the collector's approach circled at about 70 ft up, then dropped to the ground and faced him with spread wings in an apparent threat display (WHB).
Food. Insects.
Annual Cycle. The breeding season is evidently variable since young birds in complete juvenile plumage (61, 84) or moulting from juvenile plumage (45) were taken in Queensland in June and in South Australia and Northern Territory in November. The August male from Wyndham (39) was in complete moult.
Impermanent Colours. *Legs* dark greyish brown to purplish black. *Iris* brown or dark brown. *Bill* adults basal half or third bright red, orange or reddish brown (one male), rest black: young birds in juvenile plumage brownish. *Mouth* whitish to pink or salmon pink.

Glareola maldivarum — Oriental Pratincole

One seen at Onslow near Karratha (1).

Stercorarius skua — Southern Skua

Two recorded in Western Australia (11, 13).

Larus pacificus — Pacific Gull

Three south coast records (13, 16, 74).

Larus novaehollandiae Silver Gull

Specimens. WESTERN AUSTRALIA. Karratha (1) 3 June 66, 1 ø skel. Bush Bay (6) 13 June 66, 2 ♂ skel, 1 fl, 1 imm fl.
Distribution & Abundance. Seen at Perth (7), Streaky Bay (74), The Coorong (85), Nelson (88).
Food. Small fish and marine worms.
Impermanent Colours. *Legs* red or red-orange; immature dark violet brown. *Iris* white. *Bill* dull red or dark red with black tip; immature dark horn. *Mouth* orange or red; immature yellow or creamy flesh.

Chlidonias hybrida Whiskered Tern

One seen at Mount Roper (45), and common on The Coorong (35).

Hydroprogne caspia Caspian Tern

Scattered single records (1, 2, 7, 41, 88).

Sterna nilotica Gull-billed Tern

About 30 seen at Wyndham (39), a few at Hope Inlet, Darwin (41) and also on the west and south coasts (1, 2, 86, 88).

Sterna nereis Fairy Tern

Seen and filmed on The Coorong (35).

Sterna bergii Crested Tern

Fairly numerous at Hope Inlet, Darwin (41) and also recorded on the west and east coasts (1, 2, 88).

Sterna bengalensis Lesser Crested Tern

A few seen at Hope Inlet, Darwin (41) and also on the west coast (1, 2).

COLUMBIDAE — Pigeons & Doves
D. Goodwin

Ptilinopus alligator Black-banded Pigeon

Specimens. NORTHERN TERRITORY. Oenpelli (43) 1 – 17 Oct 68, 2 ♂ (238.5, 255.7 g), 1 ♀ (233.6 g), 2 imm ♂ (161.0, 199.5 g), 1 imm ♀ (232.2 g).
Distribution & Abundance. Found only at Oenpelli (43) where specimens were collected 9 miles and 4 miles SSW of Oenpelli Mission.

Fairly numerous in suitable habitat in the restricted areas where it was observed.
Habitat. Monsoon forest in gorges and at the foot of sandstone escarpments (Plate 6(a)). Some were also seen at higher elevation, in trees on open rocky ground at about

300 ft. This latter habitat would appear similar to that in which the species was found by Austin *et al* (1963) who did not see it in true forest but only in trees on high, rocky ridges.

Behaviour. Usually encountered singly in thickly-foliaged trees. Not very frightened of man and when approached tended to fly or hop further off in the same tree rather than fly right out of it. Probably birds of prey are the only usual diurnal predators on adults of this species.

When perched in a tree the apparently conspicuous colour pattern is highly disruptive and the bird is often very hard to pick out. No call was heard.

Annual Cycle. The adult males had well-developed gonads and the female an enlarging ovary, the largest oocyte being about 5 mm in diameter.

The immature birds include a male in its first moult, still predominantly in juvenile plumage, probably no more than seven or eight weeks old. The other two are much older; the male had nearly completed its body moult but still retained many juvenile wing and tail quills, and the female had completed her moult into adult plumage except for one juvenile outermost primary (the corresponding feather missing from other wing) and two juvenile secondaries. She had a fully ossified skull.

This little series well illustrates how the yellow fringes of the juvenile feathers may fade and abrade before they are shed.

Impermanent Colours. *Feet* crimson. *Iris* red. *Bill* yellow, greenish or yellowish green at base. *Mouth* purplish. *Eye rims* males dull green to pale green, females pale green or pale bluish green.

The male in juvenile plumage had pale crimson feet: iris dull olive-yellow with a dark grey inner ring: bill like adults' but with brownish tip: eye rims, dull green.

Geographical Variation. No geographical variation is known nor, in view of the species' limited range, is it to be expected. The Banded Pigeon is sometimes treated (Cain, 1954) as a race of the Black-backed Fruit Dove *Ptilinopus cincta* of the Lesser Sunda Islands. I, think, however, that it is best to treat them, together with the Crimson-naped Fruit Dove *L. dohertyi* of Sumba as members of a superspecies (Goodwin, 1967a).

Other Notes. The juvenile plumage of this species has apparently not been described: based on the young male largely in its first plumage it is as follows:

The white parts of the head are suffused with grey; those of the neck and upper breast show fine pale grey barring or vermiculation except on the lower periphery of the white breast. The buff tinge on the lower parts of neck and breast is much less pronounced than in the adults. The median and greater wing coverts and scapulars that are grey edged with black in the adult are pinkish grey passing into dark pinkish bronze with dark olive green subterminal edgings and light yellow fringes. The lesser wing coverts are like the median ones but their visible parts are mainly dark olive green with yellow edges. The secondaries are similar but darker, with the central area mainly dark greenish. Primaries are much like the adults' but more tinged with green and with very narrow pale yellow fringes. Those primaries that have strongly indented ends in the adult lack this character. The juvenile belly and flank feathers are barred yellowish white and pale grey although in the present specimen most of the feathers in this area, are new adult ones of monotone grey. The tail feathers are narrower than the adults, their dark basal area is more tinged with green and they have narrow dark subterminal edges and pale yellowish fringes. The rump feathers and upper tail-coverts are a paler grey than the adults and have olive green subterminal edges and pale yellow fringes.

The general impression of the juvenile is a bird whose dark parts are predominantly bronzy green and yellow. Its colour pattern at once suggests its close relationship with the similarly-coloured juvenile of *P. cincta,* and with other and more typical fruit doves. It is of interest that the juvenile plumage of *P. alligator* is comparable with and similar to that of the blue pigeons *Alectroenas* sp. of Madagascar and nearby islands. In each case the adults show no green or pure yellow colours but the juveniles closely resemble those of the green *Ptilinopus* species.

This and closely allied species have been put in a separate genus *Leucotreron* in the CSIRO (1969) list but it seems preferable to treat the latter as a synonym of *Ptilinopus* (see Cain, 1954).

Ptilinopus regina — Red-crowned Pigeon

Specimens. Northern Territory. Darwin (41) 7 – 19 Sept 68, 2 ♂ (87.5, 92.2 g), 1 ♀ (73.9 g), 1 imm ♀ (73.5 g).

Distribution & Abundance. Found to be fairly common at Darwin (41) and one was seen at Cape Gloucester (56).

Habitat. Monsoon forest and tropical layered woodland.

Behaviour. When moving within a tree usually hops from branch to branch in usual fruit dove manner. When fleeing in alarm flight is slightly whirring, very fast and direct. Some other aspects of its behaviour have been recorded elsewhere (Frith, H. J., 1952, Wheeler, 1959) but not, apparently, a very low murmuring call that was heard from birds feeding in the canopy.

Food. Small green berries. Frith (1952) describes feeding habits of the species in eastern Australia.

Annual Cycle. One male had testes slightly enlarged, the other and the female had enlarged gonads; the immature female, probably about three to four months old, was in the latter stages of its post-juvenile moult.

Impermanent Colours. *Feet* grey-green or dark green. *Iris* golden-yellow, golden-orange or orange-brown. *Bill* green or yellowish green. *Eye rims* yellow, pea green in the male with enlarged testes.

Geographical Variations. Our specimens on locality should belong to *P. r. ewingii,* type locality Port Essington. As far as can be judged they match fairly well with two unsexed birds from Quail Island (south of Melville Island) – though these are bad and dirty skins – but differ from a male, a female and two unsexed specimens from Port Essington which are the only other representatives of *ewingii* in the British Museum. The Darwin (41) birds are a darker, brighter and less mauvish red on the cap and a darker and brighter mauve-pink on the lower breast, but if, as seems likely, they are clinally connected to those of Port Essington the differences do not seem enough to warrant subspecific recognition.

Ptilinopus superbus Purple-crowned Pigeon

Specimens. QUEENSLAND. Tully (54), 17 July 64, 1 ♀.
Field Notes. The only one seen. In dense tropical rainforest, calling from the top of a fig tree about 70 ft above the ground.
Food. Crop and stomach were empty.
Annual Cycle. Had developing oocytes in its ovary, the largest about 2 mm in diameter.
Impermanent Colours. *Feet* (legs feathered) dull red, soles buff. *Iris* lemon yellow. *Bill* dull greyish green, tinged with purple on tip and base of upper mandible. *Mouth* pinkish purple, yellowish at throat. *Lower eyelid* translucent grey.

Ptilinopus magnificus Wompoo Pigeon

Specimens. QUEENSLAND. Big Tableland (51) 1 July 64, 1 imm ♂ (340.8 g). Ayton (52) 22 June 64, 1 ♂ (380 g). Walter Hill Range (53) 23 July 64, 1 ♂ (317.3 g), Tully (54) 13 July 64, 1 ♂ (307.6 g).
Field Notes. Found at all localities of northern Queensland with rainforest (51–54, 57). Always in the canopy. Encountered in small parties feeding; once a pair together and once a bird alone and calling in a tree.
Food. Two of the specimens had their crops and stomachs full of purplish-black berries, one had eaten two quandong fruits (probably the Blue Quandong – ed.) and four smaller fruits.
Annual Cycle. All specimens had enlarged testes, including the immature. This specimen had a fully ossified skull. It was completing its first moult, the body plumage was all new but wings and tail were still in moult and it still retained some juvenile wing and tail quills.
Impermanent Colours. *Legs* light greenish yellow to dull yellowish green. *Iris* orange red or bright red (once). *Bill* light olive green to greenish yellow, basal half crimson or purplish red. *Mouth* pinkish purple, one mottled light orange and pink. *Orbital skin* translucent grey to greenish grey.
Geographical Variation. All the specimens are referable to *P. m. keri*. They have wing measurements of 197.5, 203, 205 and 205 mm.
Other Notes. For reasons for putting this and other *Megaloprepia* species in *Ptilinopus* see Goodwin (1959).

Ducula spilorrhoa Torres Strait Pigeon

Specimens. NORTHERN TERRITORY. Darwin (41) 12 – 20 Sept 68, 3 ♂ (443.2, 458.8, 467.3 g), 1 ♀ (396.6 g). Oenpelli (43) 1 ♀ skel (487.4 g).
Distribution, Abundance & Habitat. Fairly numerous at Darwin (41) in coastal pseudo-rainforest, plentiful at Oenpelli (43) in thick woodland between the foot of an escarpment and the lake shore. Not encountered in northern Queensland.
Behaviour. Seen singly and in couples in trees. One female was roosting 25 feet up in the top branches of a paperbark tree. Appeared to feed in early morning and evening and to shelter in thickly foliaged trees during the heat of the day.

Calls heard were described as a "resonant *whit-whoo,* the second note much louder than the first" and a resonant *whoop* repeated every 2 or 3 minutes for about 15 minutes. Possibly this is the same as the call described as *up-oooo* by Warham (1962).

Food. Green and blue berries with large stones were found in the crops and gizzards.

Annual Cycle. Two males had enlarged gonads. The other had one normal and somewhat enlarged testis, the other was blackish and only about one-third the size. Both females had ovaries slightly enlarged.

Impermanent Colours. *Legs* blue-green, blue-grey in one female. *Iris* very dark brown. *Bill* greenish yellow or yellow with greenish base. *Mouth* yellowish green, flesh pink in one female. *Eye-rims* dull blue.

Geographical Variation. All Australian specimens are referable to *D. s. melvillensis.* As Rand (1941) has pointed out, the usually greyish head is the best character of this race.

Other Notes. The genus *Myristicivora* is sometimes used for the four very closely related white and black fruit pigeons but seems best treated as a synonym of *Ducula* (Goodwin, 1960, 1967b).

Macropygia phasianella — Brown Pigeon

A pair seen on the coast near Tenterfield (95). Abundant in the Walter Hill Range (53).

Streptopelia chinensis — Spotted Turtle Dove

Distribution & Abundance. Abundant and conspicuous in and near Perth (7) and Nedlands. In late September most birds were in pairs but small parties and singletons (mate on nest?) also seen. Also seen in Melbourne.

Habitat. Gardens, parks and other open areas in or near towns and suburbs. Usually within a few minutes flight (at most) of some fairly tall and dense trees.

Behaviour. Sought food on ground that was bare or had only scattered or short vegetation. Perched and cooed freely from trees, buildings and telephone wires. Large numbers fed in the many open-topped pens in Perth Zoo.

Voice sounded deeper in tone than that of Indian and Ceylonese forms of the species. The advertising coo or "song" of Nedlands birds showed some individual variation and one bird might utter two variants. One individual's two variants were noted as *Coocoo, crroo-oo,* with the first two notes much softer and shorter than the others, and *Coo-oo, crroo-oo-ook.* This bird was heard (and seen) to give the first version about a dozen times, then it changed to the second and later it alternated to some extent, giving each only two or three times and than changing.

Display coo, given with the bowing display, a shorter *Coo! croo.*

Food. Seen to take bread and grain in Perth Zoo.

Geographical Variation. In appearance all birds seen resembled either the Chinese form, *S. c. chinensis,* the Malayan form *S. c. tigrina* or intergrades. No individuals resembling the smaller and more brightly spotted Indian race *S. c. suratensis* were seen. (See also Goodwin 1967b).

Streptopelia senegalensis Senegal Turtle Dove

Distribution & Abundance. Seen at Perth (7) and in the Nedlands district nearby. Also seen in Coolgardie, and, from moving vehicles, in most inhabited areas between there and Perth.

Habitat. Inhabited areas, especially in gardens and parks of towns and suburbs. Habitat near Perth and Nedlands shared with *S. chinensis* but present species more often seen well away from tall and dense trees.

Behaviour. Very similar to that of *S. chinensis*. See Goodwin (1956, 1967b) for description of displays. Seen seeking food on lawns and on bare or partly bare ground. Very large numbers, outnumbering *S. chinensis* by about 15 to 1, feeding in open-topped pens in Perth Zoo, where they were observed to give way to both *chinensis* and Feral Pigeons, *Columba livia*.

Food. Seen to take bread, grain, seeds or buds of lawn grass (probably *Poa annua*) and bits of pulp from fallen and squashed olives.

Geographical Variation. The main difference is between the duller Indian and brighter African forms (see Goodwin, 1967b). All Australian birds, of which a clear view was obtained, were referable to the latter.

Geopelia humeralis Bar-shouldered Dove

Specimens. Western Australia. Karratha (1) 29 May 66, 1 ♂ skel. Peg's Beach (2) 3 June 66, 1 ♂. Derby (29) 24 – 26 May 68, 2 ♀, 1 imm ♂. Wyndham (39) 6 Aug 68, 1 ♂. Queensland. Banoona (58) 30 Apr 63, 1 ♂, 1 ♀.

Specimens to other Museums. NMV 1 ♂ Banoona (58).

Distribution & Abundance. Seen also at most camps in the Kimberleys (28, 30-36) being very common at Geikie Gorge (31). Also common at most camps in Northern Territory (41, 43, 45) but only a few were seen at Borroloola (46).

In Queensland it was encountered also at Tyrone (59).

Habitat. In Western Australia collected in or near mangrove thickets (1, 2, 39) and in savanna woodland near water (29). In Queensland it was found in open woodland and low scrub.

Behaviour. Usually in small parties of three or four individuals, sometimes in couples and once singly. Encountered both perched in trees and seeking food on the ground.

Food. Seeds, in one case of spinifex grass.

Annual Cycle. A female taken in May (29) and a male taken in August (39) were both undergoing complete moults. The young male taken on 26 May is in juvenile plumage with the under wing-coverts not fully grown, the base of the bill unfeathered and some juvenile down still adhering to the tips of some cover feathers. It is probably between three and four weeks old.

Impermanent Colours. *Legs* deep pink to dark purplish red. *Iris* pale straw to golden yellow; orange and light brown in two males, grey in the juvenile. *Bill* light blue or bluish grey. *Mouth* greyish or pinkish. *Orbital skin* greyish blue, sometimes with a violet tinge under the eye.

Geographical Variation. Examination of these and all other specimens in the British Museum and the American Museum of Natural History (in 1962) confirms Mayr's (1951) account of geographical variation in this species.

If it is desirable to recognise a western race on the grounds of its paler upper parts and paler subterminal bars on the feathers of the occiput, then the specimen from Wyndham (39) is best referred to it as it agrees with those from further west in being rather paler than Queensland birds. This would imply recognition of *G. h. inexpectata*, of which *headlandi* would become a synonym.

Geopelia striata — Peaceful Dove

Specimens. WESTERN AUSTRALIA. Coolawanyah (4) 20 May 68, 1 ♀ skel. Joint Hill (34) 3 – 6 July, 2 ♀ (44.1, 44.7 g). QUEENSLAND. Moonlight Creek (47) 14 May 64, 1 ♀ skel. Banoona (58) 3 May 63, 1 ♂ imm. SOUTH AUSTRALIA. Innamincka (64) 26 – 27 Apr 65, 2 ♀ (52.5 g). Aroona Valley (82) 8 Apr 65, 1 fl. Angepena (84) 2 Dec 63, 2 ♀. NEW SOUTH WALES. Dunedoo (93) 18 Mar 63, 1 imm ♂, 1 ♂ skel, 2 fl.

Specimens to other Museums. NMV 1 fl Dunedoo (93).

Distribution & Abundance. In Western Australia the Peaceful Dove was seen at most Kimberley camps, and also at Karratha (1) and Peg's Beach (2). In Northern Territory plentiful at all camps (41-45) except Borroloola (46), where only a few were seen. Very common at Glendower (55) and other north Queensland localities. In South Australia abundant at Innamincka (64) and several small parties were seen at Aroona Valley (82) but was not found elsewhere by the 3rd Expedition (65-68, 70, 79). In New South Wales only at Dunedoo (93).

Habitat. Found in gum creeks (4, 64, 82, 84) and open savanna woodland (34, 47, 58, 82, 93). When in gum creeks tending to perch as often in lower trees as in big gums. A constant feature of the habitat was a near supply of water. In South Australia it was encountered far less generally than the Diamond Dove *G. cuneata* although where it did occur it was more abundant.

Behaviour. Usually in small parties of about half a dozen, sometimes in couples. Two that were together when collected at Angepena (84) proved both to be females on dissection.

Food. Small seeds and grit in crops and stomachs.

Annual Cycle. The male (skeleton) taken at Dunedoo (93) on 18 March had enlarged gonads. The two females from Joint Hill (34) were in moult; in one the moult was at an early stage the other was in full body moult and had three old primaries in each wing. The young male from Banoona (58) was moulting from juvenile to adult plumage. Nearly all the body feathers are new (adult) but most of the wing quills and all the tail quills are juvenile. The young male from Dunedoo (93) is in a similar condition but retains rather more juvenile contour feathers.

The two females from Angepena (84) were both in full moult. The two from Innamincka (64) were in fresh plumage with only one or two old or still growing new feathers.

Impermanent Colours. *Legs* pink or pinkish with the scutes dark purple, dark red or reddish brown, giving a spotted or mottled appearance at close quarters but a generally reddish appearance at a little distance. *Iris* pale yellow, light grey, light bluish grey, light blue or white: individuals of both sexes show these variations. *Bill* bluish grey, bluish horn or greenish blue, with the base brighter blue. *Orbital skin* light greenish blue or (less often) bluish green or powder blue. *Mouth* greyish or pinkish.

Geographical Variation. As Mayr (1951) states, *placida* and *tranquilla* represent at best two ends of a cline. Southern birds (*tranquilla*) tend to be darker than northern (*placida*) specimens. If these two forms are to be recognised then the Joint Hill (34) specimens are referable to *placida* and all others to *tranquilla*.

Mayr's statement that *tranquilla* (and *placida*) is pure grey above is not always correct. There is much difference between old and new feathers as well as apparently individual and micro-geographical variation. For example a specimen from Joint Hill (34) has the subterminal parts of the new feathers on wing coverts and scapulars of a warm pinkish fawn; the corresponding parts of its old feathers are very pale grey: one female from Angepena (84) has the subterminal parts of its back and mantle feathers dark slightly pinkish brown where the old feathers are light brownish grey, the other is similar but slightly paler and greyer. The two females from Innamincka (64) are appreciably greyer in colour than the Angepena (84) specimens but show comparable differences between old and new plumage.

Geopelia cuneata — Diamond Dove

Specimens. Western Australia. Chichester Range (3) 14 May 66, 1 fl. Giles (25) 20 Aug – 1 Sept 65, 1 ♀, 2 ♂ skel. Queensland. Moonlight Creek (47) 14 May 64, 1 ♂ skel (35.7 g). Thylungra (60) 14 – 23 May 64, 1 ♂ (35.7 g), 1 imm ♂, 3 ♀, 2 imm ♀, 1 juv ♂ skel, 1 fl. South Australia. Etadunna (66) 17 – 22 May 65, 2 ♂, 1 ♀, 1 fl. Aroona Valley (82) 10 – 16 Apr, 2 ♂ (33.2, 32.2 g), 1 ♂ skel (32.6 g), 1 ♀ (34 g).

Specimens to other Museums. NMV 1 ♂, 1 ♂ skel Thylungra (60).

Distribution & Abundance. In Western Australia found also at other camps in the west (1, 2, 4) and centre (20, 21) and at Mount Anderson (28): here it was common and at Giles (25) it was fairly common.

In Northern Territory it was seen only at Darwin (41), but was at all camps in northern Queensland except Glendower (55). In South Australia it was fairly common at Aroona Valley (82) and Etadunna (66) but two individuals only were seen at both Innamincka (64) and Granite Downs (68). At Oodnadatta (67) at least four birds were seen.

Habitat. Open woodland, savanna and gum creeks. Some trees and bushes, areas of bare ground and available surface water were constant features of its habitat. It was noticeable that although in South Australia it seemed more generally distributed than the Peaceful Dove, the reverse was the case in Northern Territory.

Behaviour. Usually in couples or small parties feeding or resting on the ground; sometimes perched in a tree, usually fairly low. At Thylungra (60) a large party were feeding on the ground together with Zebra Finches.

Food. Crops and stomachs contained small seeds, often of two or more species, together with (usually) tiny stones and other mineral matter. A specimen collected at Giles (25) had also eaten the comparatively large seeds of an Acacia. Other seeds identified were *Carrichtera annua, Echium plantagineum, Kochia* sp., *Bassia* sp. and grass seeds.

Annual Cycle. The only specimen certainly in breeding condition was a male taken at Aroona Valley (82) on 10 April. This must have been caring for nestlings as its crop was congested with milk. When shot it was in company with three others, having flushed with them from the ground and settled on a dead tree in a creek.

The immatures taken at Thylungra (60) were all in a late stage of the first moult with nearly all the contour feathers new but with some juvenile wing and tail quills not yet shed.

Impermanent Colours. *Legs* pale flesh pink, once "nearly white" (25) and once "pale salmon" (47). *Iris* reddish orange to bright scarlet, one light orange. None were noted as having the narrow silver inner ring which is common or usual in domesticated Diamond Doves in England. It is quite certain that this feature was not present in any of the South Australian specimens or in those from Giles (25). *Bill* black or blackish grey, powdery grey at base. *Mouth* fleshy grey. *Orbital skin* salmon red, orange red, or bright brick red. Two moulting males with small gonads, that were associating together at Etadunna (66) had, respectively, deep flesh pink and deep salmon pink orbital skin.

Geographical Variation. Mayr (1951) recognises no races and our specimens do not show any consistent geographical variation when allowance is made for sexual and individual differences. There appears to be very extensive and rather rapid colour changes to the feathers of the upper parts due to wear. Moulting birds have the new feathers on mantle and back much redder and darker than the old. This is very noticeable on moulting females from Thylungra (60) where the new feathers are deep reddish brown and the old pale greyish brown, giving an almost mottled effect.

The orbital skin of specimens obtained in South Australia appeared smoother and less extensive than is usually the case with domesticated Diamond Doves; this difference seemed less in the case of the Giles (25) specimens. It was, however, not possible to compare live or recently-killed individuals to test the validity of this impression.

Chalcophaps indica — Green-winged Pigeon

Specimens. NORTHERN TERRITORY. Darwin (41) 12 Sept 68, 3 ♀ (141.4, 146.1, 162.5 g). Oenpelli (43) 20 Oct 68, 1 ♂ (144.1 g).

Distribution, Abundance & Habitat. Fairly numerous in pseudo-rainforest at Hope Inlet (41) and in monsoonal forest in gorges in the Alligator Valley (43). Also at Ayton (52) in northern Queensland.

Behaviour. Singly and in small parties. All birds seen were initially on the ground, which often had been burnt, but some flew into trees when alarmed. From this species Hiller at Oenpelli (43) heard a single "rolling *rrrrooo* followed by *coo, coo* repeated" which was different from the "laughing cooing" he had heard from the species at Darwin (41). It seems more likely that these notes refer to different calls of the species rather than that there is such a marked regional variation of the same call.

Food. Seeds and small stones in crops and gizzards.

Annual Cycle. All specimens were in fairly fresh plumage. They appeared to have developing gonads, except for one female whose ovary was undeveloped.
Impermanent Colours. *Legs* male crimson-pink; females, dark dull pink. *Iris* dark brown. *Bill* male yellow, red at base; females pink or yellow (once) with purplish base. *Mouth* pinkish. *Orbital skin* male red; females grey, purplish grey, and purplish black.
Geographical Variation. Northern Territory birds are referable to *C. i. longirostris*. Individual and seasonal variation and foxing and fading of old skins account for most of the differences by which this race was originally separated from *C. i. chrysochlora* of eastern Australia and New Guinea but our specimens show that its recognition is justified on larger and longer bill (on average) and, more importantly, because the females have the shoulder patch white tinged with grey, nearly as conspicuous as that of the male. In this character the females resemble those of *C. i. timorensis* of the Lesser Sunda Islands and differ from those of *chrysochlora* and most other races in which the shoulder patch is greyish and inconspicuous. In size *timorensis* is intermediate, though there is overlap (Table 2).

Table 2
Size variation in *Chalcophaps indica*

		Wing (mm)	Bill (mm)
longirostris	11 ♂	157–165(161)	21–25(23.9)
timorensis	7 ♂	153–159(156)	22–23(22.6)
chrysochlora	18 ♂	150–162(154)	20–24(22.8)

Since *longirostris* is closer to *timorensis* in pattern (and marginally in size) than it is to *chrysochlora* it seems likely that the two forms occurring in Australia have been separated from each other longer than has *longirostris* from *timorensis*, or *chrysochlora* from the very similar *sandwichensis* of the New Hebrides and New Caledonia.

Thus *longirostris* may represent a second invasion of Australia by the species, and the case may be another to add to those which Mayr (1944) suggests have reached Australia from Timor. (See also Goodwin, 1967a.)

Petrophassa albipennis — White-quilled Rock Pigeon

Specimens. WESTERN AUSTRALIA. Mount Bell (32) 14 – 16 June 68, 3 ♂ (154.0, 163.1 g), 2 ♀ (123.2 g), 2 imm ♂ (123.0, 133.8 g), 1 ♂ skel (153.5 g), 1 ♀ skel (141.1 g), 2 fl. Joint Hill (34) 2 – 4 July 68, 3 ♀ (168.5 g), 2 fl (embryos), 2 eggs. Mount Remarkable (37) 25 July 68, 1 ♂ (120.4 g). NORTHERN TERRITORY. Victoria River (40) 11 Aug 68, 3 ♂ (103.6, 111.5, 129.0 g), 1 ♀ (106.8 g).
Distribution, Abundance & Habitat. Fairly numerous where collected. Always found in or in the immediate vicinity of rocky gorges, sandstone cliffs, or valley sides with broken rock and tumbled boulders (Plate 10(a)). Mr Shane Parker informs me (in litt.) that this species is restricted to sandstone areas.

Behaviour. Showed relatively little fear of man, allowing a fairly close approach and when flushed usually only flying a short distance. Flight by alternate clapping and gliding on somewhat decurved wings. When walking on a more or less horizontal rock ledge posture is crouching, the somewhat forward-bowed head bobs to and fro, the legs are hardly or not at all visible and the tail and ends of outer primaries are in contact with the substrate.

Encountered in pairs and small parties. Seen drinking or about to drink at mid-day. By day often resting in the shade of overhanging rocks. In the early morning some were flushed from crevices in a cliff face still in deep shadow where it is probable that they had been roosting. The only nest found (see Frith, C.B. 1970 for full description) was, however, on top of a flat-topped mushroom-shaped rock where it was fully exposed to the sun during much of the day.

Calls heard were a soft short *coo,* often repeated, from individuals watching the observer, probably a note of alarm or uneasiness; a "soft deep cooing"; a "sharp deep *coo*" and an "urgent triple *coo*". Possibly the last is merely a more intense version of the alarm *coo*. No displays were seen.

Food. Seeds and a small amount of sand and small stones were found in the crops and gizzards.

Annual Cycle. June specimens from Mount Bell (32) include a young bird in juvenile plumage that was only just growing the feathers around the base of the bill and was, therefore, only recently independent and probably not more than four or five weeks old, and an older (and heavier) bird still partly in juvenile plumage. All the adults were in moult, involving wing and tail quills as well as body plumage, except one female which appears to be in fresh wing and body plumage but with some worn tail feathers still unshed; and another which had just started its wing moult and all of whose tail feathers are old. One of the males had testes about 5 x 3 mm, the others' were slightly smaller. The degree of body moult was not, so far as one can judge from the skins "heavy".

The male from Mount Remarkable (37) was in new plumage including wing and tail quills. Its testes were small.

At Joint Hill (34) in July a female that was in almost completely new plumage but still moulting its tail was incubating two eggs that were within a few days of hatching. The two other females were also in new plumage with one of them likewise still in tail moult.

The specimens from Victoria River (40) were all in new plumage or nearing the completion of a full moult. One (only) of the males had enlarged gonads. The female had slightly developed oocytes in its ovary.

Impermanent Colours. *Legs* black or very dark grey, often with purplish tinge; possibly browner in some females (32–37): dark grey with purple tinge (40). *Iris* dark brown. *Bill* black. *Mouth* fleshy pink or purplish white with edges and tips of mandibles dark grey or black. *Orbital skin* black or dark grey (32–37): pale grey or grey (40).

The youngest bird had legs grey with brownish black scutes; iris brown; bill, dark olive green; orbital skin dark grey.

Geographical Variation. The Victoria River (40) specimens represent *P. a. boothi* (Goodwin, 1969) which is characterised by having little or no white on the primaries. It is also slightly smaller and more reddish in colour.

The series from Mount Bell (32), nominate *albipennis,* indicates that my former suggestion (Goodwin, 1967a) that males of this species might prove to average redder in colour than females was invalid. There is some individual variation and the two females are, respectively, the reddest and the greyest. The male from Mount Remarkable (37) is slightly redder in colour than any of the other specimens except the Victoria River (40) birds.

The three females from Joint Hill (34) are the coldest and greyest in colour of those collected and agree well with the type and topotypical female of *P. a. alisteri* in New York. I previously (Goodwin, 1967a) doubted the validity of *alisteri* because the only male examined was much redder in colour, but I have since discovered that this male was collected a considerable distance inland (Forrest River) from the type-locality (Napier Broome Bay). It is apparent that in northwestern Australia there is some local variation in colour, due probably to intense selection for coloration matching the local rock substrate, since this pigeon nests on exposed rock faces. Whether greyer and redder birds are sufficiently segregated to allow subspecific recognition requires further study, and the taxonomic position is further complicated by the discovery that the type of *albipennis* came from an off-shore island at the mouth of King Sound, not from Wyndham (Parker, 1968b).

Petrophassa rufipennis Chestnut-quilled Rock Pigeon

Specimens. NORTHERN TERRITORY. Oenpelli (43) 30 Sept – 10 Oct 68, 5 ♂ (157.7, 158.8, 162.5, 163.8, 177.0 g), 3 ♀ (126.9, 143.3, 153.4 g), 1 imm ♀ (116.8 g), 1 fl.

Distribution, Abundance & Habitat. Only met with 9 and 11 miles SSW of Oenpelli Mission where it seemed fairly numerous. In eroded sandstone range and sandstone/granite escarpments and gorges.

Behaviour. In small parties of from two to ten individuals. Some birds were encountered singly but in all such cases others were seen or heard (flying) in near vicinity and had probably been in association with the single bird.

Most birds discovered during the middle hours of the day were sheltering under rock caves, near the entrance. Others when (presumably) on look-out or when flushed, perched on rocky outcrops or ledges, often in full sunlight. A party that were disturbed when drinking at a small rock spring beneath thickly foliaged trees, flew about 30 feet upwards onto a platform of rock and later dispersed further and higher up the escarpment. One of these birds was collected, and was found to have a large quantity of water in its crop together with food and grit so that it had evidently fed before going to drink. None was ever seen to perch in trees.

When only slightly alarmed may walk away with slight bobbing head movements but never seen to run for an appreciable distance as the Plumed Pigeon does. Takes wing with a loud characteristic clapping. Flight consists of alternate flapping, and gliding on slightly decurved wings. Can gain sufficient momentum to glide upwards for short distances as Austin (in Goodwin, 1967b) has also observed the Squatter Pigeon *Geophaps scripta* do.

Food. Seeds of three species, and grit, were found in some crops: there were seed remains and grit in all gizzards.

Annual Cycle. All the adults were in some stage of a complete moult, most of them nearing its end. One male had two growing primaries in one wing; the third from outermost, which is nearly fully grown, and the outermost which is just showing. In its other wing the outermost primary is old and worn and would probably have very shortly been cast. The tail moult appears to be asymmetrical; several tail feathers are evidently shed at about the same time. Another bird, for example, has six adjacent tail feathers at one side of its tail all in same stage of growth.

The immature female is in its first moult and retains much juvenile plumage, especially on the wings. If this specimen is typical, the juvenile plumage is much like that of the adult but the fringes of most of the covert feathers are more rufous in tone, the primary coverts are entirely chestnut instead of chestnut with sepia brown tips, and there are no concealed iridescent wing spots.

Impermanent Colours. *Legs* dark grey or brown tinged with purple which is often strongest on upper part of tarsus: soles of feet light grey. *Iris* dark brown. *Bill* black. *Mouth* pale flesh, purplish pink or whitish pink; tongue often, perhaps always, whiter than rest of mouth. *Orbital skin* very dark grey, dark greyish purple, purplish black or bluish black.

Geographical Variation. Our birds do not differ from two from an older collection taken 10 miles east of South Alligator River, when allowance is made for the older skins having faded slightly.

Other Notes. What was probably a nest of this species was found in a niche in a rock wall (Frith, C.B. 1970).

Phaps chalcoptera — Common Bronzewing

Specimens. WESTERN AUSTRALIA. Metricup (9) 17 Mar 66, 1 ♀. Green Range (13) 6 Apr 66, 1 ♂. Mount Elizabeth (35) 8 July 68, 1 ♀ QUEENSLAND. Banoona (58) 30 Apr 63, 1 ♂. SOUTH AUSTRALIA. Granite Downs (68) 7 – 10 June 65, 2 ♂, 1 ♀ skel. Kingoonya (70) 18 June 65, 1 ♀. NEW SOUTH WALES. Dunedoo (93) 20 Mar 63, 1 fl.

Distribution & Abundance. In Western Australia this species was also encountered at other camps of the southwest (11, 12, 14) and centre (19, 20, 22, 23, 25). In Northern Territory it was seen near Borroloola (46), and in Queensland and New South Wales only recorded where collected. In South Australia it was seen at all camps on the 3rd Expedition except Clifton Hills (65) and Etadunna (66).

As no particular search was made for this widespread species, except on the 3rd Expedition, and it is secretive by day, it is most unlikely that all these negative records necessarily imply its complete absence from the respective areas. It was very noticeable that at Granite Downs (68) where many dozens, possibly hundreds, were coming to water at nightfall one seldom saw more than one or two pairs during the course of a day in the bush, even when a particular look-out was kept for them.

Habitat. This species appears able to adapt to a relatively wide variety of habitats. It was met with in open scrub woodland, in both open and rather thick savanna woodland, in Mulga scrub, and in wooded rocky areas at Aroona Valley (82). It was also seen among the tall trees in the thickly wooded parts of the Adelaide National Park, and flying over open wheatfields (at dusk, so probably it was on its way to drink) near Perth.
Behaviour. By day it was usually seen in pairs or, less often, in small parties of six to ten individuals. Most often the birds were in the shade on the ground under a shrub, sometimes perched well up in a tree. Once a number of this species were flushed, together with Crested Pigeons, from an open area among the native pines in Aroona Valley (82).

The crepuscular drinking habits of this bird are well known and have been described elsewhere (e.g., Goodwin, 1967b). Birds seen in flight always flew through the woodland or scrub, usually low, and never above the canopy. One perched about 15 feet up in a tree dropped to a height of about five feet before flying off.
Food. Crops and stomachs of specimens collected at Granite Downs (68) contained largely (one entirely) seeds of *Acacia* but also seeds of *Cassia, Indigofera,* a few Proteaceae (probably *Hakea*), a few Malvaceae (probably *Hibiscus*) and a few succulent leaves. A female taken at Kingoonya (70) had eaten Chenopodiaceae (probably *Rhagodia*).
Annual Cycle. All specimens collected were in some stage of moult. The June female from Kingoonya (70) had practically completed hers, only one outer primary being not fully grown.

The March female from Metricup (9) although in full moult, had an oocyte about 5 mm long in its ovary. This bird shows particularly well how even the iridescent parts of the wing feathers alter with wear and bleaching.
Impermanent Colours. *Legs* dark purplish red. *Iris* medium brown to very dark brown. *Bill* black or blackish brown, powdery grey on cere. *Mouth* fleshy. *Orbital skin* fleshy grey to dark grey.
Geographical Variation. This species shows little geographical variation and what there is appears to be largely clinal (Keast, 1961; Goodwin, 1967a). If the three races recognised by Peters (1937) are to be maintained then specimens from South Australia seem best included with *murchisoni* as they are as reddish in tone as any Western Australian specimen. The specimen from Mount Elizabeth (35) is rather pale and is best referred to *consobrina.* Only the Banoona (58) specimen is referable to nominate *chalcoptera.*

Different degrees of wear in old and new feathers can produce a considerable difference in appearance between specimens in old and others in new plumage from the same locality.

Phaps elegans — Brush Bronzewing

Seen at Nelson (88).

Ocyphaps lophotes — Crested Pigeon

Specimens. WESTERN AUSTRALIA. Giles (25) 30 Aug 65, 1 ♂ (200 g). Mount Anderson

(28) 19 May 68, 1 fl. Mount Remarkable (37) 25 July 68, 1 ♀ (175 g). QUEENSLAND. Glendower (55) 30 Mar 64, 1 fl (180.5 g). Banoona (58) 3 – 6 May 63, 1 ♂ skel, 1 fl. Monkira (61) 7 June 63, 1 ♂, 1 ♀. SOUTH AUSTRALIA. Innamincka (64) 27 Apr 65, 1 ♂, 1 ♀. Clifton Hills (65) 11 May 65, 1 ♂ (204.8 g). Oodnadatta (67) 31 May 65, 1 ♀. Granite Downs (68) 7 June 65, 1 ♂. Kingoonya (70) 25 June 65, 1 ♀. Nonning (79) 31 Mar 65, 1 ♀ (205.8 g). Wirrealpa (83) 15 Apr 65, 1 ♂ (230.3 g), 1 ♂ skel (215 g). NEW SOUTH WALES. Dunedoo (93) 22 Mar 63, 1 ♂. Grafton (94) 3 Apr 63, 1 fl (180.5 g).

Specimens to other Museums. NMV 1 fl Monkira (61), 1 ♀ skel Streaky Bay (74).

Distribution & Abundance. Seen in most areas, although not at every camp, in western and southern Western Australia (but not at Cocklebiddy (15) or Eucla (17)), northern Queensland, extreme southeastern South Australia and southern New South Wales. In Northern Territory only at Mount Roper (45) and Borroloola (46). Fairly common or common wherever seen.

Habitat. Arid areas with mulga scrub and some larger trees, open eucalyptus woodland, creek beds, cleared land with some trees or scrub and, at Grafton (94), open swamp and woodland.

Behaviour. Usually in pairs or small parties of from four to 30 individuals. At Innamincka (64) where large numbers, probably in the aggregate several thousands, were present in and near the riparian woodland along Cooper's Creek, they still usually moved about in small parties. Hundreds, however, took wing at the same time when frightened by birds of prey.

Fond of perching on bare trees that command a good view around but spend much time on the ground, even when not seeking food. In the early morning, before going to drink, often spend some time loafing on the ground in some sunny spot. In the heat of the day seek shade and then often rest on the ground under some tree or shrub. Drink about an hour after sunrise, filling their crops with water. Some, but probably not the same individuals, drink in the evening, but where its drinking habits were repeatedly observed (64, 68) very few came to drink in the evening, compared with the morning numbers.

At Innamincka (64) many roosted in (presumed) pairs in thickly-foliaged trees that were scattered about the flood plain. These trees, which in size and appearance suggested a very thickly-leaved but non-prickly Hawthorn, were clearly selected for roosting for they were much less numerous than other trees offering less dense cover which were not used. At Clifton Hills (65) where there were no similarly thick trees some birds roosted in isolated clumps of small trees on the gibber plain but others in trees in dry creek beds.

Feeds on the ground, often in paddocks where stock has been kept or on the dusty dung-strewn ground around about dams and drinking troughs.

The advertising coo or "song" is a soft musical *Oo-oo* or *Oo-roo* neither note very loud but the first *oo* a little longer and more emphasised than the second. It is very similar to the advertising coo of the Stock Dove *Columba oenas* but, I think, less loud and a little higher pitched. The display coo, which accompanies the rapid bowing, fanning and raising of the tail and frontal presentation of the slightly spread wings, is a hurried grunting *Oo(wa), oo(wa), oo(wa)* the syllable in brackets being an apparent inspiration and not audible except at close quarters. The distress call is a musical repeated *oo*. It is often uttered repeatedly in conflict situations, as when birds coming to drink perched in sight of water but seemed afraid to fly down to it (until some bold or very thirsty individual took the initiative).

Food. Crops and stomachs of specimens collected contained only vegetable food, chiefly seeds, except that surprisingly a male and a female (61) had eaten, respectively, three maggots (in addition to many seeds), and grasshoppers. Vegetable foods identified were:—seeds of *Trianthema* sp., *Acacia* sp., *Glossogyne* sp., *Sida* sp. and *Sida corrugata, Enchylaena tomentosa, Calotis multicaulis, Eucalyptus* sp.; bulbs of *Anguillaria dioica,* flowerbuds of *Stenopetalum* sp., flowerbuds and young fruits of *Lepidium oxytrichum* and small pieces of *Eucalyptus* "manna".

At Nonning (79) many Crested Pigeons fed daily on an area of dusty ground where the soil contained great numbers of seeds of *Medicago,* probably *M. denticula,* on which they were almost certainly feeding.

Annual Cycle. Specimens from Giles (25), Mount Remarkable (37), Monkira (61) and Dunedoo (93) have enlarged or developing gonads. Apart from sometimes roosting and associating by day in presumed pairs, no sexual activity was seen except at Etadunna (66) (15 – 23 May). Here the Crested Pigeons appeared to be beginning to breed. In the early morning especially, both the advertising coo and the display coo were frequently heard and birds were watched displaying (bowing display and display flight) on many occasions.

The only juvenile birds seen were two at Carnegie (19) on 16 September. These appeared to be about three and half to four weeks old and were near an adult, probably their father.

The majority of specimens collected were in some stage of moult; most taken in late May to August were nearing completion of the moult. Two of the moulting males from Innamincka (64) and Clifton Hills (65) were moulting the wings asymmetrically. A suggestion of such asymmetrical wing and tail moult is shown by two (only) of the four other moulting males but by none of the four moulting females, although one of these was also collected at Innamincka (64).

Impermanent Colours. *Legs* pinkish red with a slight purplish tinge, to purplish red. *Iris* usually with a yellow or orange inner ring and a darker outer ring of orange, orange-brown, or brownish red. Sometimes, however, this division into an inner and outer ring was not noticeable. *Bill* dark grey, dark olive, or blackish; powdery grey on cere. *Mouth* pinkish to fleshy grey. *Orbital skin* purplish pink to deep purplish red, sometimes red or pink with very little purplish tinge.

Geographical Variation. Mayr (1951) recognised the race *whitlocki* of Western Australia stating that it "differs strikingly" from nominate eastern specimens in having reduced white tips to the tail feathers and narrower black bars on the scapulars and upper wing coverts. The present specimens and others examined show only individual differences in the latter character; indeed the Giles (25) male has wider black bars than the majority of eastern birds. The two specimens from Western Australia (25, 37) do however have smaller white tail tips and on this character *whitlocki* can be accepted.

Lophophaps plumifera Plumed Pigeon

Specimens. WESTERN AUSTRALIA. Giles (25) 20–26 Aug 65, 2 ♂, 3 ♀. St George's Range (27) 8 June 68, 3 ♂, 1 imm ♂, 1 imm ♀ 7 fl (67.9, 76.2, 83.1, 93.1, 94.7, 96.1 g). Mount Remarkable (37) 25 July 68, 1 ♂ (89.9 g), 1 ♀ (76.0 g). Lissadell Hill (38) 28 July 68, 1 imm ♂ (68.5 g). NORTHERN TERRITORY. Borroloola (46) 11 Nov 68, 1 ♂ (95.6 g). QUEENSLAND. Monkira (61) 2 June 63, 2 ♀, 1 ♂ skel. 1 ♀ skel, 1 fl.

Distribution, Abundance & Habitat. Seen also near the roadside west of Borroloola (46) but otherwise in Western Australia and Northern Territory only where collected, appearing always fairly plentiful. Always associated with rocky outcrops, hill ranges or rocky creek beds, though it might be found in relatively open country near-by. In the St George's Range (27) it was on the edges but not in the deeper gorges and more precipitous rock faces further in.

Behaviour. Usually in small parties of from three to 20 individuals. Near Giles (25) where it was starting to breed, it was also in pairs on the gently sloping hill tops of the Rawlinson Range adjacent to the Pass of the Abencerrages. These pairs appeared to be spaced out although often two or more pairs were in sight of one another. When they came down to feed in the beds of both the main and subsidiary creeks they usually joined into small, loose knit parties, within which the pair was still the obvious social unit.

Although "pure pigeon" in their displays (Goodwin, 1967c) Plumed Pigeons show an astonishing convergent resemblance in gait and flight, and to some extent in plumage, to partridges of the genus *Alectoris* and the American quails of *Lophortyx* and allied genera.

Food. Many specimens, most of which were collected in the morning, had empty or nearly empty crops but contained remains of seeds, and grit, in their gizzards. Some had seeds and small amounts of earthy matter, and one also a small succulent leaf, in their crops. Seeds identified from crops (25) were:—grass, probably *Setaria* sp., *Hibiscus* sp. cf. *H. sturtii*, a few *Sida* and a few *Eucalyptus*.

Annual Cycle. All three females taken near Giles (25) had well-developed ovaries. One specimen, contained two large oocytes, 6 and 10 mm in diameter. The mate of this bird had testes about 5 mm in length; the other male, which was also seen displaying sexually before it was collected, had slightly smaller gonads. A second female, also found in company with a displaying male and also with a large oocyte (4.5 mm), appears to be in the very last stages of a full moult, the two outermost primaries being about three-quarters grown.

In the Kimberleys a juvenile in its first moult was taken at Lissadell Hill (38) in July. It has about two-thirds of the body plumage new (adult) and had begun its wing moult. Two other juveniles, taken in June (27) were, respectively, a little further and a little less advanced in the moult from juvenile to adult plumage.

Impermanent Colours. *Legs* very dark greyish purple, appearing quite black. *Iris* (25-46) bright deep golden yellow, noted in one female only as slightly paler; at Monkira (61) male orange, females lemon yellow. *Bill including cere,* very dark horn, dark grey or black, some paler at the extreme tip. *Mouth* fleshy grey to blackish grey.

Orbital skin appearing wholly red but at Giles (25) detailed notes showed there were some grey areas immediately above, below and behind the eye, and the post-ocular area was a more orange, less brick red than the skin between the bill and the eye.

Geographical Variation. Mayr (1951) treats *ferruginea* and *plumifera* as conspecific and recognises three subspecies of the white-breasted form: nominate *plumifera* from the east Kimberleys, Northern Territory and Queensland (Cloncurry and Vindex); the larger *leucogaster* from central Australia; and *proxima* from the western Kimberleys at Hall's Creek Road, Middle Fitzroy River and Margaret River. This latter he described as being more reddish in colour on the upperparts and brown parts of the upper breast, resembling the brown-breasted form *mungi* except for its white breast-band and belly.

Earlier (Goodwin, 1967a) I found little difference in colour between specimens of *leucogaster* and *proxima*, but of our new series birds from St George's Range (27) are more reddish than others, especially on the bases of the feathers of the mantle and lower hind neck (most marked in two of the three adult males). Since birds from Mount Remarkable (37) and Mount Bell (32) agree best with nominate *plumifera, proxima* must have a very circumscribed range in the western Kimberleys. It was thought that some of the individual variation among the series from St George's Range (27) might indicate introgression between the white-breasted *plumifera* and the brown-breasted *L. f. mungi*, in particular one female shows only a trace of white on the breast-band. However, this proves to be a young bird in the latter stages of its first moult with white-tipped feathers coming in; it might well have acquired a normal white breast-band. No brown-breasted birds were seen within the range of *L. plumifera*.

Specimens from Giles (25) representing *L. p. leucogaster,* as well as being larger also have on average darker grey foreheads, but the differences between all three races are slight and overshadowed by the considerable differences between old and newly-grown feathers (Goodwin, 1967a).

Lophophaps ferruginea — Red Plumed Pigeon

Specimens. WESTERN AUSTRALIA. Chichester Range (3), 24 May 66, 1 fl, 2 eggs.

Field Notes. Seen in the Chichester Range (3), in the Hamersleys and on flats of the Fortescue River (4): in spinifex country with scattered mulga and small sandstone outcrops: once (4) in a flock among grass clumps in a belt of savanna woodland bordering the river. Apparently similar in behaviour to *L. plumifera.*

Annual Cycle. The bird was incubating the two eggs.

Impermanent Colours. *Legs* black. *Iris* red. *Bill* blackish. *Orbital skin* vivid scarlet.

Geophaps scripta — Squatter Pigeon

Specimens. QUEENSLAND. Glenore (49) 11 Apr 64, 1 ♂, 1 ♀. Inkerman (50) 26 May 64, 1 ♀ (199.2 g). Mount Dryander (57) 14 Aug 64, 3 ♀ (213.9, 214.0, 225.0 g), 2 imm ♀ (167.8, 182.0 g).

Field Notes. Fairly plentiful where collected, in tropical savanna woodland. Always on the ground when first encountered. When disturbed often flew up and perched in trees, usually on fairly exposed branches. One pair were found at 17.30 hrs near a billabong where they had presumably come to drink. Usually in couples, once a flock of about 30 individuals.

Food. Seeds, especially small grass seeds.

Annual Cycle. The two specimens taken at Glenore (49) in April were in moult, the male having four and the female five outer primaries still to shed; all their rectrices were old. The female from Inkerman (50) had a few new contour feathers growing and well-developed oocytes in its ovary, the largest 5 mm in diameter. The three females from Mount Dryander (57) were in final stages of a complete moult. The two immature females were in a complete moult from juvenile to adult plumage. Most body feathers are new but they were about half way through moult of wing quills and the tail moult had just begun in one and not yet in the other. Like the adults they had fully pneumatised skulls.

Impermanent Colours. *Legs* purplish grey to dull purple. *Iris* dull brown to blackish brown with very large pupils. *Bill* black. *Mouth* greyish or nearly black, with throat and palate showing pale pinkish or whitish areas. *Orbital skin* rose pink mottled with yellowish, purplish red, or reddish orange at front and (except in one bird) back; bluish white to pale grey around the eye itself.

Geographical Variation. Birds from northern Queensland were separated as *G. s. peninsulae,* type locality Coen, Cape York, as being smaller (wing 140 as against 148 mm in nominate *scripta*), paler in colour, and with the iridescent wing patches purple instead of green.

I have not seen specimens of *peninsulae* from the type locality. The present specimens do not appear separable from nominate *scripta*. It is possible that comparison of larger series would reveal a size cline from south to north. This is not apparent in our birds and two from southern Queensland in the British Museum. The latter have wings of 140 and 142 mm, three from Mount Dryander (57) measure 139, 141 and 144 mm, one from Inkerman (50) 138.5 and one from Glenore (49) 139 mm.

All of the 19 specimens I have examined have the "speculum" partly green *and* partly purple, when viewed with the light behind the observer, but it can look entirely green at other angles to the light.

Geophaps smithii Partridge Pigeon

Specimens. Northern Territory. Oenpelli (43) 27 Sept – 19 Oct 68, 2 ♂ (195.7, 225.0 g), 3 ♀ (171.2, 180.3, 196.3 g), 6 fl.

Distribution & Abundance. Only found 10 miles southwest, nine miles southsouthwest and 16 miles northwest of Oenpelli Mission (43) appearing fairly common in all three areas.

Habitat. Savanna woodland and medium eucalyptus forest, in or near areas with some grass on ground and near to water.

Behaviour. Found seeking food on the ground in parties of from three to ten individuals. In one instance a scattered "flock" of about 60 birds was encountered, which seemed to consist of several smaller parties in rather loose association. When approached birds first walk away and if pressed take flight, very often flying up onto a conspicuously open perch in a tree, whence they watch the human with relatively little fear, until fired at. Thus their reactions are similar to those of woodland-living game birds (Phasianidae and Tetraonidae) to mammalian predators (including man in areas where they are not shot), a further of the many examples of convergence between the largely terrestial bronzewings and the gamebirds.

Their flight, as is well-known, also consists of a game-bird like alternation of rapid flapping and gliding on set wings.

On the ground among grass, leaves, and dead branches they are difficult to see if they remain still but when they move the conspicuous white shoulder patch, formed by covert feathers which overlap the carpal area of the wings in life, often gives them away. So to a lesser extent does the conspicuously-coloured face. It is, however, unlikely that either would be readily visible to a bird of prey looking down from above.

Food. Seeds of grasses and other plants, and some sand and grit, were found in crops and gizzards of specimens collected.

Annual Cycle. All the specimens retained as skins and one (probably all) of the spirit specimens were in process of a complete moult when collected. This is furthest advanced in a female collected on 19 October, in which all the wing quills are new.

None of the specimens had gonads in breeding condition.

Impermanent Colours. *Legs* pinkish purple, purplish grey or grey tinged with purple. *Iris* off-white to pale cream, sometimes speckled with blackish, and with sepia inner ring. *Bill* black or greyish black. *Mouth* pale flesh. *Orbital skin* vermilion or scarlet.

Geographical Variation. All the birds collected are nominate *G. s. smithii*. The western form *G. s. blaauwi,* which differs strikingly in having the orbital skin ochre-yellow instead of red (Mees, 1968), was not obtained or seen by the expedition.

Histriophaps histrionica Flock Pigeon

Large flocks seen near Karratha (1) and one male at Clifton Hills (65).

PSITTACIDAE — Parrots & Cockatoos
B. P. Hall (C. B. Frith & F. Carline)

Trichoglossus haematodus Rainbow Lorikeet

Specimens. Western Australia. Derby (29) 22 – 24 May 68, 1 ♀ (114.9 g), 1 fl. Mount Bell (32) 16 – 19 June 68, 2 ♂, 2 ♀ (112.7 g), 1 ♀ skel, 1 fl (101.9 g). Lissadell Hill (38) 28 July 68, 1 fl (118.4 g). Queensland. Moonlight Creek (47) 1 – 11 May 64, 4 ♂ (134.8, 136.3, 137.9, 139.1 g), 1 ♀ (117.4 g), 1 imm ♂ (109.0 g). Glenore (49) 9 – 24 Apr 64, 3 ♂ (114.8, 116.5, 128.2 g), 1 ♀ (112.1 g), 1 imm ♂ (106.0 g). Inkerman (50) 2 – 3 June 64, 7 ♂ (119.2, 122.2, 123.1, 126.0, 133.2, 133.6, 134.6 g), 7 ♀ (106.6, 107.0, 108.7, 108.8, 109.7, 110.0, 119.5 g). Ayton (52) 21 June 64, 1 ø (115.8 g). Tully (53) 23 July 64, 1 ♂ (147.0 g). Cape Gloucester (56) 7 – 8 Aug 64, 4 ♂ (129.2, 143.3, 143.8, 147.7 g), 3 ♀ (130.9, 131.3, 137.6 g). Mount Dryander (57) 13 Aug 64, 2 ♂ (150.7, 156.7 g), 4 ♀ (120.0, 136.9, 142.6, 143.0 g).

Distribution & Abundance. Seen also in Northern Territory, a few at Oenpelli (43) and Mount Roper (45), plentiful at Darwin (41) and very common in flocks at Pine Creek (44). Also at Grafton (94) and Tenterfield (95) in New South Wales.
Habitat. Usually in savanna woodland (Plate 10(a)) but also in tropical rainforest (52, 53) and open gum forest (56, 57).
Behaviour. Paired, or in small parties or flocks of up to 30 birds. Usually found feeding on flowering eucalypts or tea-trees. At Moonlight Creek (47) the voice of *T. h. rubritorquis* (see below) was noted as being slightly higher pitched than that of eastern birds (IM).
Food. Stomachs and crops contained grubs, some soft seeds, blossom and nectar.
Annual Cycle. Birds were mainly in post-breeding condition, replacing wing and tail feathers. The two immatures were identified as such on their rather duller plumage and lower weights, though their skulls were full pneumatised and the bills not notably duller than the adults'.
Impermanent Colours. *Legs* grey. *Iris* males orange-red to red, sometimes with white inner ring: females, paler orange. *Bill* orange-red or red with a yellower tip. *Mouth* deep yellow to orange with grey interior. *Cere and orbital skin* black or dark grey.
Geographical Variation. The birds from Western Australia and from Moonlight Creek (47) belong to the distinctive western form *T. h. rubritorquis,* with the dark green and red collar. On the east a cline of increasing size is recognised from Cape York southwards. This is reflected in the weights of our specimens. If this cline is recognised nomenclatorially the northern birds (49, 50, 52) are *T. h. septentrionalis,* while those from further south (53, 56, 57) are intermediates between *septentrionalis* and *moluccanus* of New South Wales.

Trichoglossus chlorolepidotus — Scaly-breasted Lorikeet

Specimens. QUEENSLAND. Mount Dryander (57) 14 Aug 64, 1 ♂ (82.3 g), 3 ♀ (70.5, 76.0, 83.4 g). NEW SOUTH WALES. Grafton (94) 30 Mar 63, 1 ♂, 1 ♀, 1 ♀ skel.
Specimens to other Museums. NMV 1 fl Grafton (94).
Field Notes. Found in open gum forest (57) and "medium timbered forest" (94), usually feeding on blossom quite high in trees, in small flocks or pairs. Not encountered elsewhere.
Food. Stomachs contained only a yellowish nectar paste.
Annual Cycle. Mount Dryander (57) birds were in moult; those from Grafton (94) were in very fresh plumage with minute gonads.
Impermanent Colours. *Legs* grey to greenish-brown. *Iris* yellow to pale red. *Bill* bright orange or red, with light orange tip. *Mouth* greyish-yellow. *Cere and orbital skin* grey-brown.
Geographical Variation. The New South Wales birds are slightly larger than those from Queensland but we agree with most authors in thinking the difference is not great enough to warrant recognition of two races.

Psitteuteles versicolor — Varied Lorikeet

Specimens. WESTERN AUSTRALIA. Derby (29) 22 May 68, 3 ♂ (54.8, 57.9, 58.9 g), 1 ♀

(52.6 g). Mount Bell (32) 14 – 19 June 68, 1 ♂ (59.1 g), 1 ♀, 2 fl (60.2 g). Mount Caroline (33) 25 June 68, 1 ♂ skel. Joint Hill (34) 3 July 68, 1 ♂ (58.4 g). QUEENSLAND. Glenore (49) 10 Apr 64, 1 ♀ (57.1 g). Glendower (55) 26 Mar 64, 1 ♂ (62.2 g), 2 ♀ (50.8, 61.7 g), 1 imm ♂ (48.2 g).

Field Notes. Seen also at Pine Creek (44), where it was plentiful, but not elsewhere. Found in savanna woodland (Plate 10(a)) and open forest in small flocks feeding in flowering eucalypt with screeches and chattering, or dashing with a very fast and direct flight over the tree tops.

Food. Only nectar was found in stomachs.

Annual Cycle. The Western Australian birds and the female from Glenore (49) were in worn plumage or moult, with small gonads. The three March adults from Glendower (55) (see below) were in fresh or very fresh plumage, with the gonads somewhat enlarged. The immature male resembled the two females in its rather duller plumage but appears to be moulting on the throat into brighter plumage.

Impermanent Colours. *Legs* blue grey. *Iris* (when noted immediately after death) yellow sometimes with black inner ring, but apparently turning quickly to brown (49, 55). *Bill* orange with grey tip and cutting edges. *Mouth* orange, grey in throat. *Cere and orbital skin* whitish.

Geographical Variation. There is considerable individual variation in the amount of pink on the breast of both males and females which appear otherwise to be adult. In this respect, of the specimens collected only one female from Queensland matches a western bird, the two other Queensland females having less pink. However other Queensland birds from older collections are as rich in colour as the Derby (29) and Kimberley (32, 34) birds, suggesting there is a sub-adult plumage.

Glossopsitta concinna — Musk Lorikeet

Specimens. VICTORIA. Victoria Range (90) 31 Jan 63, 1 fl. NEW SOUTH WALES. Dunedoo (93) 23 Mar 63, 1 ♂, 1 ♀, 1 skel, 3 fl.

Specimens to other Museums. NMV 1 fl Victoria Range (90), 1 ♀ skel Dunedoo (93).

Field Notes. Small flocks of up to 15 birds were encountered at both localities, but it was not seen elsewhere. Always found in tree tops, whether feeding or resting, in lightly timbered woodland on the lower slopes of the Victoria Range (96) and in eucalypt woodland at 1,600 ft at Dunedoo (93).

Annual Cycle. The two Dunedoo (93) skins examined were just completing moult. All gonads were small.

Impermanent Colours. *Legs* yellowish brown. *Iris* orange. *Bill* black and orange. *Mouth* grey.

Glossopsitta porphyrocephala — Purple-crowned Lorikeet

Specimens. WESTERN AUSTRALIA. Cocklebiddy (15) 10 July 65, 1 ♀.

Field Notes. Seen but not collected at all main camps in the southwest (9, 11, 12, 13, 14) and in the east at Furner (86), Victoria Range (90) and Little Desert (91). At Cocklebiddy (15) small parties of up to 16 birds were found in mallee scrub, though the female collected was one of a pair. Birds apparently looking for flowering trees fly fast, circling and half-circling just over and between tree-tops, suddenly alighting. During flight they utter a loud, harsh very incisive trilling screech, repeated constantly, but ceasing abruptly as they alight (DG).

Food. Only nectar was found in the stomach.

Annual Cycle. The female collected was in fairly worn plumage with a small ovary.

Impermanent Colours. *Legs* light olive-grey. *Iris* dull medium brown. *Bill* greyish-black. *Mouth* fleshy-grey. *Orbital skin* dull grey. *Cere* fleshy brown.

Glossopsitta pusilla — Little Lorikeet

Specimens. New South Wales. Grafton (94) 5 Apr 63, 1 ♀.

Field Notes. Seen also at Furner (86) and Tenterfield (95). The only ones seen in Queensland were a pair collected (♂ 40 g, ♀ 34.2 g), but not retained, in open forest on the Gregory River, Mount Dryander (57). The Grafton (94) female was one of a flock feeding on blossom high in the canopy of sclerophyll forest.

Annual Cycle. The Grafton (94) female is in fresh plumage with a small ovary. The gonads of the Queensland August birds were slightly enlarged.

Impermanent Colours. *Legs* greenish-brown or grey. *Iris* light brown (94) or yellow (57). *Bill* upper mandible black, lower mandible dark brown. *Mouth* grey. *Cere* lilac; *orbital skin* pale blue.

Opopsitta diophthalma — Fig Parrot

Specimens. Queensland. Big Tableland (51) 2 July 64, 1 ♂ (42.9 g). Ayton (52) 30 June 64, 1 ♂ (39.1 g).

Field Notes. Not seen elsewhere. At Big Tableland (51) four birds were found feeding in the top of a tall Johnstone Hardwood in the rainforest at 2,200 ft. At Ayton (52) the two birds seen were not in the rainforest proper, which was confined to the gullies, but in open forest at 250 ft above the gullies. The call was a soft lorikeet-like shriek (IM).

Food. Yellow seeds and blossom remains were found in stomachs.

Annual Cycle. Both birds in very fresh plumage with small gonads.

Impermanent Colours. *Legs* greenish-grey to light greenish-blue. *Iris* brownish-black. *Bill* light grey with dark slate tip. *Mouth* white with blackish patches. *Orbital skin and cere* black.

Geographical Variation. These specimens have the distinctive facial pattern of *O. d. macleayana* and come from the northern limit of the range.

Calyptorhynchus baudini — White-tailed Black Cockatoo

Seen at all main camps in the southwest (9, 11, 12, 13), in gallery forest and *Banksia* scrub. Dr Stephen Davies of CSIRO, Perth, presented three skins, four skeletons, four skulls and 11 fluid specimens to the British Museum from the series collected at Perth for his own work.

Calyptorhynchus funereus Yellow-tailed Black Cockatoo

Specimens. NEW SOUTH WALES. Tenterfield (95) 13 Apr 63, 1 ♂ skel.
Field Notes. Found at Tenterfield (95) in eucalypt forests at 3,800 ft. The bird collected was perched on a dead tree screaming, while eight others fed in the eucalypts below. Seen also in the Victoria Range (90) and in fairly large flocks in the pine plantations east of Nelson (88). Not seen in Queensland.
Annual Cycle. It was noted as in good plumage, with small testes.
Impermanent Colours. *Legs* grey. *Iris* dark brown. *Bill* blackish relieved with horn colour. *Mouth* pinkish-grey. *Eyelids* reddish.

Calyptorhynchus lathami Glossy Black Cockatoo

Specimens. NEW SOUTH WALES. Tenterfield (95) 10 Apr 63, 1 ♂, 1 ♀.
Field Notes. The pair collected were feeding together on *Casuarina* in open forest at 3,000 ft. Not seen elsewhere.
Annual Cycle. Both birds were in full moult but the gonads were enlarged, the female with eggs forming in the ovary.
Impermanent Colours. *Legs* grey. *Iris* brown. *Bill* grey. *Mouth* grey. *Orbital skin* black.

Calyptorhynchus banksi Red-tailed Black Cockatoo

Specimens. WESTERN AUSTRALIA. Mount Anderson (28) 10 – 18 May 68, 1 ♀, 1 ♂ skel, 2 fl. QUEENSLAND. Glenore (49) 14 Apr 64, 1 fl (763.0 g).
Distribution & Abundance. Common in flocks of up to 20 along Snake Creek at Mount Anderson (28), fairly numerous at Darwin (41), and Pine Creek (44), less common at Lake Gladstone (36), Oenpelli (43), Mount Roper (45) and Borroloola (46). In Queensland recorded also at Glendower (55).
Habitat. Savanna woodland.
Behaviour. Mostly seen in direct flight over tree tops, calling loudly one call at a time, evenly spaced. Often in small feeding flocks in tree-tops, very noisy with much flapping about.
Food. Grubs and seeds, possibly also some grains of rice.
Annual Cycle. The ovary of the May female from Mount Anderson (28) contained an 18 mm oocyte. The bird was in moult.
Impermanent Colours. *Legs* light grey to black. *Iris* brown to dark brown. *Bill* horn grey to white. *Mouth* yellowish to greyish-brown. *Orbital skin and cere* grey to black.
Geographical Variation. On locality all birds belong to *C. b. macrorhynchus.*

Callocephalon fimbriatum Gang Gang

Seen only on the Victoria Range (90).

Cacatua galerita Sulphur-crested Cockatoo

In Western Australia recorded from Geikie Gorge (31). In Northern Territory one bird was seen at Oenpelli (43), two at Pine Creek (44) and a few at Mount Roper (45).

In Queensland, noted at Glendower (55), Banoona (58) and Tyrone (59). In New South Wales at Albury (92) and Dunedoo (93).

Cacatua leadbeateri — Major Mitchell

Specimens. WESTERN AUSTRALIA. Warburton (23) 16 Aug 65, 1 ♂, 1 ♀. SOUTH AUSTRALIA. Between Nullarbor HS (69) and Sceale Bay (75) 10 Oct 63, 1 ♂. Kingoonya (70) 25 June 65, 1 ♀ skel. Lake Acraman (71) 2 Nov 64, 1 ♂. Minippa (73) 1 Nov 64, 1 ♂. Nonning (79) 29 Mar 65, 1 ♀ (394.6 g).

Field Notes. At Kingoonya (70) a flock of about 20 birds, including two Galahs, was found feeding in the grass. Collected birds were mostly from small flocks but south of Cocklebiddy (15) about 300 birds were in eucalypt scrub between the cliffs and dunes. Less common where seen further north, at Mungilli Claypan (20) and Warburton (23), in open mulga: at Granite Downs (68) only one was seen among a flock of Galahs. Not recorded at other camps by the 3rd Expedition.

Food. Stomachs contained seeds, including *Eucalyptus, Bassia* sp. *Citrullus lanatus* (decorticated) (a water melon found growing on dusty ground) and unidentified yellow cotyledons.

Annual Cycle. The gonads examined (23, 70, 79) were not enlarged. The August Warburton (23) birds were in worn plumage while the South Australian October birds were mostly very worn but with some moult on the neck, and the November birds were in mainly fresh body-plumage, moulting wings and tail. The March bird was more worn on the body and had nearly completed moult of wings and tail.

Impermanent Colours. *Legs* light grey to grey. *Iris* light brown (3 ♀) to very dark brown (1 ♂). *Bill* whitish. *Mouth* grey. *Eyelids* grey.

Geographical Variation. On the evidence of our limited series there seem grounds for recognising Mathews' three southern races, *C. l. superflua* of South Australia being smaller than *C. l. leadbeateri* of New South Wales and Victoria (wings ♂ ♀ 255-261 against 258-275), and with less yellow in the crest, while the yellow is even further reduced in the two Warburton (23) birds, referable to *C. l. mollis*.

Cacatua sanguinea — Little Corella

Specimens. WESTERN AUSTRALIA. Karratha (1) May 66, 1 skull only. Mount Anderson (28) 16 May 68, 1 ♂, 1 ♂ skel, 1 fl. Derby (29) 26 May 68, 1 ♂ skel (605 g). QUEENSLAND. Glendower (55) 29 Mar 64, 2 fl (382.7, 448.0 g). SOUTH AUSTRALIA. Aroona Valley (82) 8 Apr 65, 1 ♀ skin (430.0 g).

Distribution & Abundance. Common in the Kimberleys (28-36) mainly in flocks of up to 500: fairly common Darwin (41) and Oenpelli (43). The 1st Expedition saw it only in southwest Queensland (60, 61), and the 3rd Expedition only in eastern South Australia where it was common at Aroona Valley (82); uncommon, Innamincka (64); large flocks of up to 1,000 birds at Clifton Hills (65), Etadunna (66) and Lake Harry (29°26S: 138°13′E); two birds only, Oodnadatta (67). The 4th Expedition encountered flocks in river gums on the creeks of the Hamersley region wherever there were waterholes.

Habitat. Savanna woodland (28, 29), open forest (55), creeks (1, 82), always near water.
Behaviour. At Aroona Valley (82) a large flock was feeding in Red River Gums, *Eucalyptus camaldulensis;* others were feeding and fighting over the little shrub *Rumex vesicarius* (Polygonaceae). They roosted in the river gums with Galahs, the Corellas on alighting always taking the highest perches. Often in large flocks on the ground, at Lake Harry among scattered palms near the bore. Very noisy in flight, the alarm call being a harsh shriek.
Food. Stomachs contained seeds, including oats *Avena avena* and melon seeds.
Annual Cycle. Of the Mount Anderson (28) birds the skinned male was in fresh plumage; neither male had enlarged testes but the female's ovary contained enlarged oocytes (17 x 6 mm). The Aroona Valley (82) female was in full moult and had a slightly enlarged ovary.
Impermanent Colours. *Legs* grey. *Iris* dark brown. *Bill* whitish, greyer near base. *Mouth* grey. *Orbital skin* blue; *skin under chin* pale purplish grey.

Cacatua tenuirostris — Long-billed Corella

Two flocks each of about 100 birds seen at Bool Lagoon (87).

Cacatua roseicapilla — Galah

Specimens. Western Australia. Mount Shenton (18) 2 – 3 Aug 65, 1 ♂, 1 ♀ (300 g), 1 ♀ skel. Mungilli Claypan (20) 13 Sept 65, 1 ♂. Warburton (23) 10 Aug 65, 2 ♂. Mount Anderson (28) 14 – 18 May 68, 1 ♂ fl (316.0 g), 2 fl (299.0 g). Queensland. Glendower (55) 31 Mar 64, 1 ♂ (327.8 g), 1 ♀ (303.3 g). Banoona (58) 3 May 63, 2 ♂. South Australia. Clifton Hills (65) 10 May 65, 2 ♂, 1 ♀. Whyalla (80) 17 Oct 63, 1 ♂. Aroona Valley (82) 8 Apr 65, 1 ♂ (332 g), 1 ♀ (312 g). New South Wales. Albury (92) 9 – 10 Mar 63, 1 ♂ skel, 1 ♀ skel. Dunedoo (93) 20 Mar 63, 1 ♂, 1 ♀.
Distribution & Abundance. Noted as common at most camps where sight records were kept but absent from the southwest (7-17), the extreme north (41, 43) and extreme east (94, 95).
Habitat. Open forest (55), savanna woodland, mulga scrub, sheep paddocks.
Behaviour. Goodwin made the following observations on behaviour:—

"Display. Two displays were frequently seen from birds at Aroona Valley (82) which were still showing interest in their nest sites. In what I termed the 'Screech display', the bird typically stands upright, 'leans back' a little and raises its wings, often one wing more than the other. Then it suddenly swings its head and the forepart of its body forward and downward, raising its wings fully above it and uttering a harsh, hissing screech as it does so. As it utters this sound the mouth is opened wide but the cheek feathers still completely cover the lower mandible when viewed from the side. This display usually incorporates a rocking side-to-side movement and is not very stereotyped. The screech may be uttered as the bird leans back and raises its wings and the forward movement may be omitted.

This display was usually directed towards the observer but on a few occasions a pair displayed in this way together when at a distance from and not apparently reacting to his presence. It appears likely to be a threat display and, like many threat displays, shown when there is conflict between fear and aggression. The movements shown appear to be derived from those used when taking flight.

The other display I termed the 'nodding display'. In this the crest is fully erected, showing red bases to its feathers at back, and the bird bows quickly forward, uttering a sharp *ink*! This display was often seen from one or both birds of pairs perched in roosting trees or, more especially, near a nest site. Sometimes it alternated with the screech display. Although the birds were probably motivated partly by fear one had the impression that the nodding, like that of pigeons, indicated an intention of holding ground – 'Here I stay' – rather than of fleeing.

Feeding. When feeding in gum trees often hangs upside down very much as the Wood Pigeon *Columba palumbus* does and, like that species, usually drops off onto a lower branch instead of regaining the perch it is hanging from."

Food. Seeds, including *Dicrastylis* (Chloanthaceae), *Boerhavia* (Nyctaginaceae), *Kochia* sp., and some unidentified red angular corrugated seeds. The ground at Aroona Valley (82) on which some birds were feeding daily with Crested Pigeons held seeds of an introduced fodder plant, *Mendicago* sp., probably *M. denticulata,* and at Clifton Hills (65) the feeding ground was covered with the adventive weed *Centaurea melitansis* (Compositae).

Annual Cycle. None of the gonads indicated breeding condition though in birds from Mount Shenton (18) and the Mungilli Claypan (20) they were somewhat enlarged; these birds were in worn plumage. (Gonads not noted in the October specimen from the Eyre Peninsula.) All other specimens in mixed plumage (though not all had feathers actually growing) except one from Aroona Valley (82), one from Banoona (58), and one from Dunedoo (93) which had just completed moult.

Impermanent Colours. *Legs* grey. *Iris* male dark brown, female light brown or pink. *Bill* whitish. *Mouth* usually whitish but once yellow (♂ 20), once yellowish white (♀ 55). *Orbital skin* Mount Shenton (18) grey; Mount Anderson (28) purple red (♂) or deep pink; Clifton Hills (65), pinky brown (1 ♂) or dark pink (1 ♂, 1 ♀); Glendower (55) deep pink (♂ ♀); Aroona Valley (82) pinky brown (1 ♀); not recorded elsewhere.

Geographical Variation. Forshaw (1969: 77-78) recognises only two races, *C. r. roseicapillus* in which the orbital skin is pink, and *C. r. assimilis* of central Western Australia with the orbital skin grey and the crest deeper pink. He believes that Kimberley birds may prove to be *C. r. roseicapillus* and this is confirmed by our series (see above), though *C. r. assimilis* was observed at Derby (29). Although the Gibson Desert (20, 23) specimens did not have the orbital skin noted they have a pink crest similar to the Mount Shenton (18) birds which are nearly topotypical *assimilis*. Other variation is difficult to assess because of the seasonal variation in plumage, but four skins in the British Museum from southern New South Wales and Victoria are darker both above and below than others and there is a cline of decreasing pigmentation northwards and westwards, which can be recognised by referring our South Australian (65, 80, 82) birds to *C. r. kuhli* of Northern Territory.

Nymphicus hollandicus — Cockatiel

Specimens. WESTERN AUSTRALIA. Warburton (23) 16 Aug 65, 1 ♂ (83.0 g). Derby (29) 22 May 68, 1 ♂ (84.3 g), 1 ♀ (88.9 g). QUEENSLAND. Glendower (55) 26 Mar 64, 1 fl (83.6 g). Tyrone (59) 12 May 63, 1 ♂ skel. Thylungra (60) 26 May 63, 2 ♀.

Field Notes. Seen also at Karratha (1), Chichester Range (3), Giles (25), Mount Anderson (28), Lake Gladstone (36), Victoria River (40), Pine Creek (44), Glendower (57) and Banoona (58), but nowhere common except at Glendower (57).

Seen in small flocks in mulga woodland (23), savanna woodland (29), River Gums (55) and in trees and scrub on the edge of sand-dunes (60). Noted as perching in topmost branches.

Food. Seeds, at Warburton (23) including *Danthonia* and *Aristida* sp., a few other grass seeds and a few unidentified horned seeds.

Annual Cycle. May birds just starting moult at Derby (29) and in mid-moult or just finishing at Thylungra (60). Just finishing moult at Warburton (23) in August. All gonads small.

Impermanent Colours. *Legs* grey or black. *Iris* dark brown. *Bill* grey paler below. *Mouth* grey. *Cere* grey.

Polytelis anthopeplus — Regent Parrot

Specimens. WESTERN AUSTRALIA. Mount Barker (12) 21 – 23 Mar 66, 3 ♂, 1 ♀, 2 ♀ skel, 2 fl.

Field Notes. All were collected from small flocks in the tops of Red Gum and Jarrah trees on the edge of cleared paddocks. Others were seen at Newdegate (14) but not elsewhere.

Food. Gum seeds.

Annual Cycle. All birds just completing moult: gonads small.

Impermanent Colours. *Legs* grey. *Iris* light brown to orange, or "brown with narrow orange outer ring". *Bill* pink with yellowish tip to lower mandible. *Mouth* fleshy-pink. *Orbital skin* blue-grey.

Aprosmictus erythropterus — Red-winged Parrot

Specimens. WESTERN AUSTRALIA. Mount Anderson (28) 13 May 68, 1 ♂ (157 g), 1 ♀ (141.2 g), 1 imm ♂ skel (141.7 g), 3 ♀ skel (148.0, 157.7, 159.0 g), 2 ♂ fl (145.7, 156.7 g), 1 ♀ fl (149.7 g). QUEENSLAND. Banoona (58) 5 May 63, 1 ♂, 1 ♀. Tyrone (59) 10 May 63, 1 ♂ skel, 1 ♀ fl. SOUTH AUSTRALIA. Cordillo Downs (63) 6 May 65, 1 ♀. NEW SOUTH WALES. Dunedoo (93) 21 Mar 63, 1 ♀.

Specimens to other Museums. NMV 1 ♂ skel Banoona (58), 1 ♂ Dunedoo (93).

Distribution & Abundance. Recorded at all camps in the Kimberleys (29-36), and in Northern Territory at Darwin (41), a few at Pine Creek (44), while at Borroloola (46) it was fairly common with fully fledged young. Very common at Glendower (55): seen at all camps in southern Queensland (58-60) except Monkira (61): also at Innamincka (64) and Tenterfield (95).

Habitat. Savanna woodland with eucalypts and acacia: in isolated eucalypts in sheep paddocks (93). At Mount Anderson (28) a party of seven were feeding in a *Crotolaria* thicket.
Behaviour. Collected birds were either in small flocks of six to seven birds (28, 63) or paired (58, 59, 93) except for one "seedy-looking bird" at Banoona (58). The alarm was noted as "*Tchik-tchik-tchik* – very like *Barnardius*" (DG). A group in a distant tree produce flock-babble which sounds distractingly like nearby small passerines (ICJG).
Food. Seeds – at Cordillo Downs (63) identified as *Amyema preissii* (Loranthaceae), with four seeds of a Labiate and one of *Gahnia* (perhaps *G. aspera)* (Cyperaceae). At Banoona (58) seeds of Pepper Tree, at Dunedoo (93) mistletoe berries, and at Mount Anderson (28) seen feeding on seeds of a creeper in *Crotolaria.*
Annual Cycle. The fully fledged young seen at Borroloola (46) at the beginning of November indicate winter breeding, but no specimens showed significant gonad activity and all were in moult except the pair collected at Banoona (58), both of which were in fresh plumage, though the single "seedy bird" (NMV) was in full moult.
Impermanent Colours. *Legs* grey or dark brown (28). *Iris* orange, brownish-orange (♀ 63). *Mouth* red or yellow (28), yellow with black tongue (58), fleshy grey, inside of upper mandible yellow (63). *Cere* dull brown, yellow round nostrils – noted only in one female (63). The duller colours of the Cordillo Downs (63) female suggests that when in moult colours may be variable, for there is no indication of immaturity in the plumage.
Geographical Variation. The Kimberley birds (28) belong to the smaller northern race, *A. e. coccineopterus* and the remainder to *A. e. erythropterus.*

Aprosmictus scapularis King Parrot

Two birds were seen in tropical rainforest on the Walter Hill Range (53) on 21 July 64, of which one female was collected but not kept. Weight 198.0 g. Ovary slightly enlarged, c.10 mm. Stomach contents – tiny yellow seeds.

Platycercus elegans Crimson Rosella

Specimens. SOUTH AUSTRALIA. Furner (86) 8 – 9 Jan 63, 1 imm ♀, 1 fl. NEW SOUTH WALES. Tenterfield (95) 10 – 12 Apr 63, 1 ♂, 1 ♂ skel, 1 ♀ skel.
Specimens to other Museums. NMV 1 ♀ skel Furner (86), 1 ♂ Tenterfield (95).
Field Notes. Found commonly in parties of three to four in open woodland at Furner (86). At Tenterfield (95) seen feeding in casuarinas and on grassy ground. Seen also at Victoria Range (90) and Albury (92).
Food. Large quantities of medium-sized seeds, possibly grain.
Annual Cycle. No gonad activity. The Tenterfield (95) male is in fine fresh plumage, and the immature plumage of the Furner (86) bird is also fresh, indicating that it is probably not more than a month or two old.
Impermanent Colours. *Legs* grey or black. *Iris* brown. *Bill* adult whitish or grey; immature whitish with pink tinge. *Mouth* pink to grey.
Geographical Variation. On locality the Tenterfield (95) birds belong to the typical race, and the Furner (86) ones to *P. e. victoriae,* if this race is valid (Condon, 1941: 135).

Platycerous eximius Eastern Rosella

Specimens. SOUTH AUSTRALIA. Furner (86) 9 Jan 63, 1 ø skel. NEW SOUTH WALES. Albury (92) 9 – 12 Mar 63, 1 ♀, 2 imm ♀, 1 ♂ skel. Dunedoo (93) 18 – 20 Mar 63, 2 ♂, 1 imm ø, 1 fl.

Specimens to other Museums. NMV 1 ♂ Albury (92), 1 ♂ skel Dunedoo (93).

Field Notes. Found in woodland, from open agricultural savanna to light forest. In small flocks at Furner (86). In New South Wales seen also at Grafton (94), and Tenterfield (95), usually in groups of four, occasionally singly.

Food. Seeds. Three stomachs also contained grit.

Annual Cycle. All gonads small. The adult male from Albury (92) was completing moult of wings and head; the young birds were in different stages of body moult from immature to adult plumage. The males from Dunedoo (93) had just completed total moult and the third bird, though sexed as a male, appears to have been moulting from immature to adult female plumage.

Impermanent Colours. *Legs* light to dark grey. *Iris* brown to dark brown. *Bill* light blue-grey to grey, lower mandible with a darker tip. *Mouth* flesh to pinkish-grey.

Geographical Variation. The meeting place between the northern *P. e. ceciliae,* in which the mantle feathers are edged with deep yellow, and *P. e. eximius* in which they are greenish-yellow, is known to be in New South Wales. The Dunedoo (93) males are particularly interesting since one is typical of each race. The immature bird has the mantle even greener, and resembles the female from Albury (92) in this respect and is also similar in size. The few sexed specimens in the British Museum suggest that greenness may be a sexual character though it is not mentioned by Forshaw (1969: 191).

Platycercus icterotis Western Rosella

Specimens. WESTERN AUSTRALIA. Metricup (9) 8 – 10 Mar 66, 3 ♂, 2 fl. Scott River (11) 19 Feb – 1 Mar 66, 2 imm ♂, 3 imm ♀ (see below), 2 ♂ skel, 1 ♀ skel, 1 fl. Green Range (13) 31 Mar 66, 1 imm ♀.

Field Notes. In sclerophyll forest on ironstone (9); at Swan Lake, Scott River (11) in Jarrah, *Banksia* and low tea-tree thickets, perched low in the trees or feeding on the ground in near-by heathland. Also seen at Mount Barker (12) and Newdegate (14). Found in pairs or groups of up to ten birds. One small party consisted of immature birds only (11).

Food. Stomachs contained seeds including grass once. Birds were observed feeding on "gum-nuts" in mallee (13).

Annual Cycle. Young birds were present at all localities. The immature females are identified as young on comparison with Serventy & Whittell (1967: Plate 2) no adult females being available. They have varying amounts of red on the underparts and, at most, a tinge of yellow on the cheeks. All adults were in moult and had small testes.

Impermanent Colours. *Legs* grey to dark grey. *Iris* brown. *Bill* adults blue-grey with darker tip to lower mandible; immatures horn to grey. *Mouth* fleshy-grey.

Geographical Variation. On locality all belong to the coastal, green-backed race, *P. i. icterotis.*

Platycercus adscitus — Pale-headed Rosella

Specimens. QUEENSLAND. Banoona (58) 28 Apr–4 May 63, 1 ♂, 1 ♀, 1 ♂ skel, 1 ♀ skel.
Specimens to other Museums. NMV 1 ♂, 1 fl Banoona (58). AM 1 ♂, 1 ♀ Banoona (58).
Field Notes. All collected birds were found feeding on ground in open woodland, including iron-bark and firs. In parties of two to 12 birds. Seen also at Glendower (55), where it was common, and at Tyrone (59).
Food. Small seeds.
Annual Cycle. Both specimens were in moult, the male in mainly fresh plumage and the female in mainly worn. The ovary of one female (skel) was somewhat enlarged but all other gonads were very small.
Impermanent Colours. *Legs* grey. *Iris* brown. *Bill* upper mandible pale bluish grey with white tip, lower mandible whitish.
Geographical Variation. Both birds examined have the rump pale blue and feathers of the mantle edged with golden yellow, the characters of *P. a. palliceps*. The female has a number of red-tipped feathers on the forehead, cheeks and nape, some of which are fresh and some growing, so these are not necessarily a character of immaturity.

Platycercus venustus — Northern Rosella

Specimens. WESTERN AUSTRALIA. Mount Bell (32) 18–20 June 68, 2 ♂ (92.0, 111.9 g), 2 ♀ (88.0, 91.6 g), 1 imm ♀ (81.4 g), 1 ♂ skel (97.0 g), 1 fl. Mount Elizabeth (35) 6 July 68, 1 ♂ (96.8 g). NORTHERN TERRITORY. Darwin (41) 7 Sept 68, 2 juv ♂ (77.8, 86.4 g).
Distribution & Abundance. Found at Derby (29) on the mainland and on Koolan Island; common at Mount Bell (32) and found at all camps in Western Australia further east except Lake Gladstone (36) and Wyndham (39). At Wyndham (39) work was mostly coastal and suitable habitat was hardly encountered.

In Northern Territory found at Darwin (41), Pine Creek (44), and Borroloola (46).

Alone or in pairs except for a family party of four birds at Mount Bell (32).
Habitat. In the Kimberleys (32, 35) between 1,500–1,800 ft in open savanna woodland with some local available water. At Darwin (41) in quite dense woodland close to the coastal strip of monsoon forest.
Behaviour. Most often found feeding in outer foliage of flowering eucalypt. Silent when approached, remaining quite still and sometimes perching lengthways along branches.
Food. Seeds and nectar, grit in some gizzards.
Annual Cycle. Kimberley (32, 35) birds in post-breeding moult with inactive gonads. The immature female did not have the skull fully pneumatised and moult is less advanced than in adults. The two juveniles from Darwin (41) in September indicate that breeding there was later than in the Kimberleys.
Impermanent Colours. *Legs* grey. *Iris* adults brown; juveniles brown-black. *Bill* adults upper mandible light grey, lower mandible darker; juveniles white with some yellow on sides. *Mouth* adults whitish to grey: juveniles cream. *Cere* adults dark grey: juveniles yellow.

Geographical Variation. In our series the colour of the underparts varies from pale yellow ochre (1 ♂, 1 imm ♀) to pale lemon yellow (1 ♂, 1 ♀) and pale lime green (2 juveniles) but no bird has the heavy black edges to the feathers which are present in three old British Museum skins from Port Essington and Melville Island, and the Kimberley adults have more blue-green on the lower throat. The juveniles are only slightly duller than the adults and are greener above as well as below. No subspecies are currently recognised, and better series are required to judge the extent of individual variation.

Barnardius zonarius Port Lincoln Parrot

Specimens. WESTERN AUSTRALIA. Chichester Range (3) 14 May 66, 1 ♂, 1 ♀. Coolawanyah (4) 21 May 66, 1 ♂. Wonerup (8) 12 Mar 66, 1 imm ♀, 1 ♀ skel. Metricup (9) 9 Mar 66, 1 imm ♀. Margaret River (10) 14 Mar 66, 2 ♂, 1 ♀. Scott River (11) 18 – 28 Feb 66, 1 ♂, 3 ♀, 1 imm ♂, 1 imm ♀. Newdegate (14) 11 Apr 66, 1 ♂, 2 ♀, 1 fl. Cocklebiddy (15) 10 July 65, 1 ♂, 1 ♀. Mount Shenton (18) 30 July 65, 3 ♂, 1 fl. Mungilli Claypan (20) 12 Sept 65, 1 ♀. Warburton (23) 10 Aug 65, 1 ♀. Giles (25) 23 Aug – 2 Sept 65, 1 ♂ (165 g), 1 ♀ (130 g), 1 imm ♀. SOUTH AUSTRALIA. Oodnadatta (67) 31 May 65, 3 ♀ (121, 136 g), 2 fl. Granite Downs (68) 7 – 9 June 65, 1 ♀, 1 ♂ skel. Kingoonya (70) 25 June 65, 1 ♀, 1 ♂ skel. Streaky Bay (74) 9 Dec 62, 1 ♂ skel. Nonning (79) 29 – 30 Mar 65, 1 ♂ (180 g !), 2 ♀ (123.6, 136.5 g).

Specimens to other Museums. SAM 1 ♂ Yardea (72) 1 Nov 64.

Distribution & Abundance. In Western Australia found also at other camps along the Gunbarrel Highway (19, 21, 24, but not 22). One bird only was seen at Eucla (17). In South Australia no *Barnardius* was seen at Clifton Hills (65) or Etadunna (66).

Habitat. In the southwest (8-11) in open forest – Tuart, Jarrah, Marri, Karri and a pine plantation: elsewhere in mallee, mulga and often in gums along creek beds. At Giles (25) the commonest bird in a large stand of Desert Oaks *Casuarina decaisneana* on the Gunbarrel Highway 45 miles to the west.

Behaviour. Single birds and pairs seen bathing from dead boughs sticking up from or hanging into muddy water at Oodnadatta (67): often going belly deep along the bough, one actually jumping into the water and floating a second before flying out. Birds often bent head down and under the lifted wing to one side, at the same time ducking body forward and tilting the other side more into the water (DG). (See notes on *B. barnardi* for differences in voice.)

Food. Seeds of *Lysiana exocarpi,* buds of *Eremophila* sp., a few *Solanum* sp., mistletoe seeds. Wood-boring grubs and caterpillars. Grit and charcoal in some stomachs.

Annual Cycle. Birds from Chichester Range (3) and Coolawanyah (4) were just starting moult, the female had the ovary somewhat enlarged. All adults from the southwest were in post-breeding moult with small gonads: the three adjudged to be immature were not moulting wings and tail but were moulting body from a duller plumage to adult. The Newdegate (14) April birds were nearer completion of moult. All the birds from Cocklebiddy (15) and the centre of Western Australia (18-25) July to September, were in worn plumage, most just starting moult: gonads slightly enlarged.

In South Australia birds from Yardea (72) in November and Nonning (79) in March were finishing moult; all others, May – June, are in indeterminate plumage, with no significant gonad activity.

Impermanent Colours. *Legs* grey. *Iris* dark brown. *Bill* pale blue-grey with darker tip to lower mandible. *Mouth* pinkish, perhaps yellower in males. *Cere* dark grey.

Geographical Variation. Our series shows the clear-cut distinction between the large, dark birds of the southwest (8-11) with red on the forehead and green on the abdomen *B. z. semitorquatus* and the remaining populations. The three from Newdegate (14) are intermediate, having the red greatly reduced or absent and having more yellow on the abdomen. The differences between the other populations are slight. Birds from the Chichester Range (3) and Coolawanyah (4) have paler blue cheeks, paler wing coverts and rump, and can be referred to *B. z. occidentalis.*

Birds of central Australia, *B. z. myrtae,* are alleged to be a lighter, yellower green on the back than *B. z. zonarius,* which ranges from the interior of Western Australia to the Eyre Peninsula (Condon, 1941: 127). This character is very variable in our series, the birds with most yellow-green being two from Cocklebiddy (15) and the bluest being one from Mount Shenton (18). A topotypical *zonarius* from Port Lincoln is intermediate between them. It seems doubtful whether *myrtae* should be recognised.

The Yardea (72) male differs from all others in having some orange on the yellow feathers of the collar and the upper abdomen, as are found in some of the hybrids *zonarius* x *barnardi* from the Aroona Valley (82) but otherwise in the colour and pattern of the head it is typical *zonarius.*

Barnardius barnardi — Ringneck Parrot

Specimens. QUEENSLAND. Tyrone (59) 10 May 63, 1 ♀. SOUTH AUSTRALIA. Innamincka (64) 26 Apr 65, 1 ♂, 1 ♀. NEW SOUTH WALES. Dunedoo (93) 19 Mar 63, 1 ♂.

Specimens to other Museums. NMV 1 ♂ Tyrone (59).

Distribution & Abundance. Not seen at other camps in southern Queensland (58, 60, 61) or New South Wales (92, 94, 95): only a few at Innamincka (64): always in small parties or pairs.

Habitat. Open woodland, light timber (59, 93). Open woodland near creek (64).

Behaviour. At Innamincka (64) birds after fighting indulged in a great display of lateral tail-wagging with rapidly repeated excited calls. One displayed in an upright posture with tail spread and wings thrust forward to show the blue shoulders and orange on lower breast. A series of calls *chuck-chuck-chuck, pree-pree-pree* were noted as similar to those of the hybrid *barnardi* x *zonarius* at Aroona Valley (82) but softer and less ringing in tone. Calls of *B. zonarius* at Oodnadatta (67) were similar but with different "accent", the *chuck* being replaced by a throatier sound (DG).

Food. Small seeds. Seeds of *Eucalyptus* sp. with a few galls containing flies.

Annual Cycle. All birds in fresh plumage, one (♀ 64) just completing moult on head. This female had the ovary slightly enlarged but the others all had small gonads.

Impermanent Colours. *Legs* grey. *Iris* light brown to dark greyish-brown. *Bill* upper mandible pale horn-grey, lower mandible slightly paler. *Mouth* pale greyish. *Orbital skin* dark grey. *Cere* medium grey.

Geographical Variation. All birds differ. The male from Dunedoo (93) has the crown bright yellow-green, the cheeks mixed green and blue, the mantle grey-blue and the underparts wholly green. The female from Tyrone (59) is similar but duller, especially on the mantle, and has a patch of clear yellow on the upper abdomen. The male from Innamincka (64) has the crown paler than the other male, the cheeks wholly green, the mantle blue-green and a small orange patch on the upper abdomen: the female has the crown yellow, the cheeks yellow-green, the mantle green, the breast yellow-green and a more extensive orange patch on the abdomen. We cannot judge whether these differences fall within the range of individual variation expected in *B. b. barnardi.*

Barnardius barnardi x zonarius — Port Lincoln x Ringneck Parrot

Specimens. South Australia. Aroona Valley (82) 9 – 17 Apr 65, 1 ♂ (142.3 g), 1 ♀ (127.0 g), 3 imm ♂ (125.8, 134.0, 149.5 g), 1 ♂ skel (142.4 g). Angepena (84) 1 Dec 63, 1 ♀.

Distribution & Abundance. Very common at Aroona Valley (82) in pairs and small parties of up to four or five birds. The bird from Angepena (85) was one of a pair collected at Wertaloona, 32 miles to the east.

Habitat. Low scrub and large gum trees along creek beds (82).

Behaviour. Many birds came to an artificial water seepage near an old homestead, where masses of feathers were scattered, presumably from moulting birds. The alarm call was a sharp *chink*! *chink*! *chink*! (DG). (See also notes on *B. barnardi.*)

Food. Seeds and berries, including *Eucalyptus camaldulensis* and *Lysiana exocarpi* (a mistletoe).

Annual Cycle. All at Aroona Valley (82) had small gonads. One male is clearly fully adult with black primaries; it has the wings and tail fresh and was finishing moult on the head. The other three males have been adjudged immature because they have brown primaries, but it is not easy to follow the plumage sequence from them. The one lightest in weight is the brightest in plumage, with bright blue cheeks and green on the crown: the abdomen has a considerable amount of orange in the yellow; some body feathers and the central rectrices were growing while the remainder of the tail is fairly worn. The other two look duller, perhaps representing a more worn stage of immature plumage, while the central rectrices were fully grown, appearing fresher than the outside rectrices; they show some body moult.

The December female from Angepena (84) is fairly worn, and, from the dullness of the red on the forehead in comparison with the Aroona Valley (82) female, might be immature. No note was made of the state of the gonads.

Impermanent Colours. *Legs* grey. *Iris* dark brown. *Bill* pale blue-grey, slightly darker at lower mandible. *Mouth* white to greyish flesh.

Geographical Variation. These hybrids are variable, as would be expected, but closer to *barnardi* in having red on the forehead. They are sometimes treated as a race of *B. barnardi, B. b. whytei.* In the Aroona Valley (82) series, adult males have the head mainly green with a brown tinge: one immature (the blue-cheeked bird discussed above) has a narrow black band between the red and the green on the forehead: the other immatures have the head browner, while the female has the crown yellow green: all have a broad band of yellow on the upper abdomen with varying amounts of orange in the centre of the feathers, greatest in the blue-cheeked male and the female: the fully adult male has the mantle bluer while the female is yellower on the breast and under tail-coverts. The Angepena (84) female has the red on the forehead dull, the crown brownish, washed with blue, the breast and mantle blue-green and the abdomen clear yellow.

Purpureicephalus spurius Red-capped Parrot

Specimens. WESTERN AUSTRALIA. Wonerup (8) 12 Mar 66, 1 imm ♂, 1 fl. Metricup (9) 8 – 16 Mar 66, 1 imm ø, 2 ♀ skel, 3 fl. Margaret River (10) 9 Mar 66, 1 ♂. Mount Barker (12) 23 Mar 66, 1 imm ø.
Field Notes. In small parties in Tuart and peppermint forest: seen feeding near the tops of Tuarts (8), on ripe fruit in low trees in apple orchard (9) and on York Gums at the edge of a swamp (12).
Food. See above.
Annual Cycle. The adult male (10) had small gonads and was in mainly fresh plumage, moulting body and tail. The immature birds were all moulting on the body or head with the wings and tail partly worn. The one sexed as a male differs from the other two in being larger, especially the bill, and having the breast greyish tinged with faint red barring, instead of a mere spot: all three have the under tail-coverts mixed red and yellow. It seems likely therefore that the other two are young females; they are slightly duller than two unsexed birds in the British Museum believed to be adult females.
Impermanent Colours. *Legs* grey. *Iris* dark brown. *Bill* blue-grey with dark grey tip. *Mouth* greyish-pink.

Psephotus haematogaster Blue Bonnet

Specimens. QUEENSLAND. Thylungra (60) 26 May 63, 1 ♀. SOUTH AUSTRALIA. Etadunna (66) 18 May 65, 1 ♀. Kingoonya (70) 22 – 25 June 65, 4 ♂, 1 ♀, 1 ♀ skel, 1 ø skel. Yardea (72) 1 Nov 64, 1 ♂, 1 ♀. Whyalla (80) 17 Oct 63, 1 ♀.
Distribution & Abundance. Fairly common at Etadunna (66): common at Kingoonya (70), once in a party of about 30 birds. One bird seen at Nonning (79), and others at Banoona (58).
Habitat. Usually mulga or other low trees and scrub with open sandy ground.
Behaviour. Feeds both on the ground and in trees. On ground has a toddling walk but takes long hops when hurried. Flight and call (*chuck-chuck-chuck,* but with many variants) very like *Barnardius barnardi;* also has a less loud and higher-pitched *cheenk,* possibly a contact call (DG). The flock at Kingoonya (70) were feeding on open ground covered with a prickly creeper near the edge of Lake Labyrinth (dry). All birds at Etadunna (66) were paired, and also those from Yardea (72) and Whyalla (80). Others were in small parties.

Food. Seeds, including at Etadunna (66), Loranthaceous seed (perhaps *Amyema preissii*), *Kochia* sp. (Chenopodiaceae), a few other Chenopodiaceae: at Kingoonya (70), in one crop two species of grass, a few *Bassia* sp. and a few *Kochia* sp.; in another mostly *Bassia* sp. with some grasses and one *Erodium* seed, perhaps *E. cygnorum*.
Annual Cycle. Some of the Kingoonya (70) birds had somewhat enlarged gonads but in others and those from Thylungra (60) and Etadunna (66) they were small. The season of moult is either irregular or varies in different parts of the country, for the October bird from the Eyre Peninsula (80) and the May bird from southern Queensland (60) are in similar fresh plumage, just completing moult, in contrast to the very worn November birds from Yardea (72) which were just starting moult; birds from Etadunna (66) and Kingoonya (70) in May and June are also in moult but the old plumage is less abraded than in the Yardea (72) pair.
Impermanent Colours. *Legs* grey. *Iris* brown. *Bill* uniform bluish grey, or greenish grey. *Mouth* yellowish grey, paler at the sides.
Geographical Variation. The plumage condition of this series makes comparison unsatisfactory but the Etadunna (66) female is noticeably paler and can be referred to the race of the Lake Eyre basin, *P. h. pallescens*. The Thylungra (60) female has some red on the wing coverts, showing an approach to *haematorrhous* in this character, but has clear yellow under tail-coverts like all the series. This bird and the Whyalla (80) male are slightly darker on the breast than the remainder (70, 72), but this may be due to fresh plumage. Following Cain (1955: 467) all populations with yellow under tail-coverts are regarded as *P. h. haematogaster* except for the paler *pallescens*.

Psephotus haematonotus — Red-rumped Parrot

Specimens. SOUTH AUSTRALIA. Innamincka (64) 27 Apr – 1 May 65, 2 ♂, 1 ♀ (52.5 g). Aroona Valley (82) 8 – 15 Apr 65, 1 ♂ (68.4 g), 4 ♀ (63.4, 65.0, 65.2 g), 1 imm ♂ (70.3 g), 1 ♀ skel (53.9 g). Furner (86) 7 – 9 Jan 63, 1 ♀, 1 ♂ skel. NEW SOUTH WALES. Albury (92) 9 – 13 Mar 63, 1 imm ♂, 1 ♀ skel, 1 fl. Dunedoo (93) 18 – 25 Mar 63, 1 ♂, 1 ♀ skel.
Specimens to other Museums. NMV 1 ♂ skel Dunedoo (93).
Distribution & Abundance. Very common at Innamincka (64) and Aroona Valley (82). Seen also at Little Desert (91), Tenterfield (95) and Banoona (58).
Habitat. Mostly open woodland. At Aroona Valley (82) some birds were collected in arid scrub at the Brachina Homestead at the western foot of the Flinders Range, others in native pine or River Gums at 1,600 ft. At Dunedoo (93) in Yellow Box.
Behaviour. Single or in small parties of up to twelve birds, sometimes in company with *Barnardius* (82).
Food. Seeds: at Aroona Valley (82) one crop contained only *Eucalyptus* sp: another, grass seeds *Aveneae*, some *Echium plantagineum* (nearly all decorticated) and a few *Boerhaavia* (decorticated): a third some unidentified seeds including separated yellowish cotyledons, a few broken fragments of grass seed and one *Kochia* seed. Some stomachs contained grit.
Annual Cycle. No gonad activity. At Aroona Valley (82) and Innamincka (64) birds

were in full moult, the immature male moulting body into adult plumage. The Furner (86) female is in fresh plumage: the immature male from Albury (92) in fairly fresh plumage and the male from Dunedoo (93) rather more worn.

Impermanent Colours. *Legs* grey. *Iris* yellowish to dark brown. *Bill* adult male dark grey or dark brown, blackish at tip; females and immatures dark horn or light grey with darker tip, lower mandible sometimes paler. *Mouth* white, cream or pink. *Cere* dark grey: *orbital skin* blue-grey.

Geographical Variation. The Innamincka (64) birds are typotypical of *P. h. caeruleus* Condon, 1941 and show the colour characters of that race (paler and more bluish) even in worn plumage; but they do not seem to be smaller, though measurements are inexact because of moult (wing 2 ♂ 120-126, 1 ♀ 122 against 4 ♂ 124-125, 5 ♀ 117-121 for other birds): one male is considerably greener on the breast than the other. A male and female in the British Museum from Mount Hutton (25.50S: 148.47E), southern Queensland are closer to *caeruleus* than to *P. h. haematonotus,* though slightly greener on the rump. Our females from Aroona Valley (82) vary in the intensity of olive wash on the breast and green on the rump and may include some immature birds: all are paler than the female from Furner (86).

Psephotus varius Mulga Parrot

Specimens. Western Australia. Mount Shenton (18) 30 July 65, 1 ♂. Giles (25) 23 – 31 Aug 65, 2 ♂, 1 ♀. South Australia. Oodnadatta (67) 31 May 65, 2 ♂ (56.0, 60.0 g), 1 ♀ (59.0 g). Granite Downs (68) 4 – 10 June 65, 1 ♂, 1 ♀. Kingoonya (70) 17 – 25 June 65, 1 ♂, 1 ♀, 1 ♀ skel, 1 ♂ fl. Minippa (73) 1 Nov 64, 2 imm ♂, 1 imm ♀. Streaky Bay (74) 9 Dec 62, 1 ♀ skel. Elliston (76) 15 Oct 63, 1 ♂. Nonning (79) 31 Mar 65, 1 ♂, 1 ♀. Angepena (84) 27 – 29 Nov 63, 1 ♂, 1 ♀.

Distribution & Abundance. In Western Australia fairly common where collected and at Warburton (23). Seen also along the Gunbarrel Highway as far west as Mungilli Claypan (20) but not at Carnegie (19): a few at Eucla (17). Common or fairly common where collected by the 3rd Expedition in South Australia (67, 68, 70, 79) but not seen in the eastern camps (64-66, 82).

Habitat. In gums or mulga, often in quite small trees or scrub.

Behaviour. Usually pairs (25, 70, 76) or small parties, one with *Barnardius zonarius* (67). At Granite Downs (68) a bird came to water well after sunrise, unlike Bourke Parrots.

Food. Seeds, including *Chenopodium* sp. perhaps *C. melanocarpum, Kochia* sp., *Bassia* sp., *Enchylaena tomentosa, Erodium* sp., near *E. cygnorum,* a few *Lysiana exocarpi,* some grasses. Also once some stem fragments possibly *Trianthema* sp. (18), and once some small greenish-yellow grubs (79). Several stomachs included grit.

Annual Cycle. The three November immatures from Minippa (73) are in fresh plumage: the female and one of the two males are much browner on the breast and mantle than the other but both males have only a little red on the abdomen. (All three birds are dirty, perhaps from being on burnt ground, and difficult to compare.) Of other spring birds the October male from Elliston (76) is very worn and the late November

pair from Angepena (84) are very fresh, one just completing moult. None of these have the gonads noted. Elsewhere all males had small testes but a female from Giles (25) in August and one from Kingoonya (70) in June had somewhat enlarged ovaries. Except for the two from Kingoonya (70) which were in fresh plumage, all birds taken in May (67), June (68), July (18) were in moult, the July male only just starting.
Impermanent Colours. *Legs* grey. *Iris* dark brown. *Bill* blue-grey with darker tip. *Cere* grey. (No data for immature birds.)
Geographical Variation. Among the males variation is found chiefly in the amount of red-brown on the hind crown and orange on the abdomen, but the two from Oodnadatta (67) differ widely in these characters, suggesting they are individual rather than geographical. The amount of wear accounts for variation in the colour of the green plumage. Of the five females the one from Oodnadatta (67) stands out as being emerald green rather than olive green on the breast: three have the frontal band yellow (67, 68, 79) and two have it red and yellow (70, 84): all have a conspicuous red shoulder-patch, darkest in the Oodnadatta (67) bird. All specimens come within the range of *P. v. varius* (Condon, 1941: 143).

Psephotus chrysopterygius — Golden-winged Parrot

Specimens. NORTHERN TERRITORY. Pine Creek (44) 30 Aug 68, 1 ♂ (56.1 g), 1 ♀ (54.2 g).
Distribution & Abundance. Between 16 – 30 August parties of two (collected), four, six and 12 birds were seen on four occasions on the track between Pine Creek (44) and Wandie Creek 20 miles ESE. Not seen elsewhere by 5th Expedition, or in northern Queensland, though looked for especially.
Habitat. Savanna woodland. Rocky area with long dead grass, plenty of termite mounds in the vicinity.
Behaviour. Flight low and undulating when flushed, usually from ground. Call described as a soft twittering (DJF).
Food. Seeds.
Annual Cycle. No gonad activity. Wings in moult, in the male all the primaries were new except the two outermost: the next four were fully grown and the rest part grown. The female was just a little more advanced.
Impermanent Colours. *Legs* grey. *Iris* brown. *Bill* light blue-grey, dark tip. *Mouth* very pale pink. *Cere* grey.
Geographical Variation. These belong to the western form *P. c. dissimilis,* sometimes regarded as a species.

Neopsephotus bourki — Bourke Parrot

Specimens. WESTERN AUSTRALIA. Mount Shenton (18) 30 July – 3 Aug 65, 1 ♀, 1 ♂ skel. SOUTH AUSTRALIA. Granite Downs (68) 5 – 7 June 65, 2 ♂, 3 ♀ (48.9 g), 1 ♂ skel (49.4 g), 2 fl.
Distribution & Abundance. Abundant at Granite Downs (68) (see below): common at Mount Shenton (18) and Warburton (23): a few were seen at Carnegie (19) and Mungilli Claypan (20), one at Giles (25), and two at Thylungra (60).

Habitat & Behaviour. Mulga woodland and scrub. At Granite Downs (68) among red sand hills near Larry's Well and Sailor's Well, where birds usually came to drink just before dawn or after dusk; large numbers (up to 1,000 birds) coming in pairs or small parties to drink at large puddles or water troughs in company with many *Phaps chalcoptera*. During the day flocks of about 30 birds were common, perched in or flying among Mulga trees. Pairs were seen inspecting holes in dead Mulga stumps. Flight fast and low, going through the trees rather than over, with alternating bursts of wing beats and glides on decurved wings; somewhat suggestive of that of *Ocyphaps lophotes* and producing a very similar sound (DG). Calls very like Budgerygahs' but softer; in flight a mellow, musical *chu-wee* and whistling warblings *chirrup* and *ch'up;* alarm or protest call *krik-krik*. No calls heard resembled the *tit-tit-tit* of *Neophema elegans* or *chrysostoma* (DG).

Food. Small seeds including a few *Bassia* sp. and unidentified flattened ovate decorticated seeds were in one crop examined (18). Also, once, some red pulp. Most stomachs contained grit or small stones.

Annual Cycle. No significant gonad activity though some ovaries were slightly enlarged. June birds in fresh plumage or just completing moult (68), July birds fairly fresh (18).

Impermanent Colours. *Legs* grey or dark brown (once). *Iris* dark brown. *Bill* dark grey paler at base. *Mouth* grey or whitish. *Cere* dark grey.

Geographical Variation. None recognised. All females are slightly paler than the males, both above and below, and the Mount Shenton (18) female is paler than those from Granite Downs (68), probably due to less fresh plumage.

Neophema elegans — Elegant Parrot

Specimens. Western Australia. Mount Barker (12) 21 – 24 Mar 66, 1 ♂, 1 imm ♂, 2 imm ♀, 1 imm ø, 1 ♂ skel, 1 ♀ skel, 2 fl. Green Range (13) 2 Apr 66, 1 imm ♂, 1 fl. Newdegate (14) 21 Apr 66, 1 ♀ skel. South Australia. Aroona Valley (82) 12 Apr 65, 1 ♀ (44.0 g).

Distribution & Abundance. In parties of up to 15 birds at Mount Barker (12). Not common in the Flinders Range (82) where the bird collected was one of a party of four. Not seen elsewhere.

Habitat. Tussock grass with scattered trees near tea-tree swamp (12): grassy areas with clumps of scrub, low bushes and mallee (13): in native pine and eucalypts at 1,600 ft on the Flinders Range (82).

Behaviour. Most birds collected in Western Australia were feeding in grass. They were shy and difficult to approach. At Aroona Valley (82) the call was a thin, high-pitched *tit-tit-tit* like that heard later from *Neophema chrysostoma* (DG).

Food. Grass seeds. (One stomach examined.)

Annual Cycle. The Aroona Valley (82) female had the ovary slightly enlarged and is in fresh plumage. Western birds (12-14) had small gonads, the adult male was completing moult in wings and tail while the immatures were fairly worn or starting moult into adult plumage, showing a few blue feathers of the frontal band.

Impermanent Colours. *Legs* grey. *Iris* brown. *Bill* upper mandible dark grey, lower mandible paler. *Mouth* white or flesh. *Cere* grey.
Geographical Variation. Mathews differentiated western birds (*N. e. carteri*) as being darker above, greener below, with a small amount of yellow on the tail feathers, and having an almost black frontal line. Cain (1955: 472) found the only constant character of *carteri* was the darker blue frontal band. This is certainly not evident in the Mount Barker (12) male, as compared with the Aroona Valley (82) female, but the latter is much yellower below, and slightly yellower above.

Neophema chrysostoma — Blue-winged Parrot

Specimens. South Australia. Innamincka (64) 29 Apr 65, 1 imm ♂. Victoria. Nelson (88) 4 Jan 63, 1 ♀.
Field Notes. Abundant at Nelson (88) in grasslands and light woodlands on the edge of coastal dunes: birds when flushed flew off fast and climbed steeply. At Innamincka (64) only a single bird was seen coming to drink at Cooper's Creek, while the bird collected was among sandhills covered with small groups of gum trees in the flood-plain (dry) of the creek, 96 miles southwest down the Strezlecki track. Call similar to that of *N. elegans* (DG).
Food. Grass seeds.
Annual Cycle. A pair were seen feeding young at Nelson (88) where the female collected was in moult. The April immature bird is fairly worn, rather duller than the adult with only a faint blue band on the forehead.
Impermanent Colours. *Legs* grey. *Iris* dark brown. *Bill* adult dark horn above, paler below; immature grey above, paler below. *Mouth* adult whitish.

Neophema pulchella — Turquoise Parrot

Seen at Tenterfield (95).

Melopsittacus undulatus — Budgerygah

Specimens. Western Australia. Chichester Range (3) 13 – 15 May 66, 1 ♀ skel, 1 fl. Coolawanyah (4) 15 May 66, 2 ♂ skel, 1 ♀ skel. Carnegie (19) 16 Sept 65, 1 imm ♂. Warburton (23) 9 Aug 65, 1 ♂, 1 ♀, 1 imm ♀. Queensland. Thylungra (60) 22 May 63, 1 fl. South Australia. Lake Acraman (71) 3 Nov 64, 1 ♂, 1 ♀. Kimba (78) 3 Oct 63, 1 ♂, 1 ♀.
Distribution & Abundance. In Western Australia seen also at Mount Anderson (28) (common), Karratha (1), Lake Gruszka (21), Giles (25) (several small flocks). In Queensland, common at Glendower (55) and seen also at Tyrone (59) and Monkira (61). The 3rd Expedition did not meet budgies in South Australia except for a small flock at Clifton Hills (65).
Habitat. Mostly in open country, grass or spinifex, with scattered trees, Mulga (23) or River Gums (4).
Behaviour. Mostly seen in small flocks, but at Warburton (23) up to 300 birds. Also some single individuals which had evidently lost contact with their companions.

Goodwin notes that "These flew high on an erratic course with frequent changes of direction, uttering the disyllabic contact call *cherroop* very loudly and frequently. This behaviour had often been seen from escaped and lost domestic budgies in England and it was interesting to have proof that it was a normal pattern of the species".
Food. Tiny seeds, including some believed to be spinifex.
Annual Cycle. Birds were seen nesting at Coolawanyah (4) in May, and those collected there and on the Chichester Range (3) showed gonad activity. The immature taken at Warburton (23) in August was in fairly fresh, dull plumage, while the adults had small gonads and were completing moult in the wings. The Carnegie (19) male a month later was in full moult; since it had a brown eye, and some barring on the forehead, this is presumed to be from immature to adult. No note was made of the gonads for South Australian birds but October birds from Kimba (78) were in the same stage of moult as the Warburton (23) adults, and the November birds (71) were fairly worn.
Impermanent Colours. *Legs* pale grey. *Iris* adults white (60, 71), yellowish-white (4, 23), or yellow (3): immature male dark brown, female white. *Bill* adults and immature female yellowish-olive, with tip and edge of culmen paler: immature male yellowish horn. *Mouth* whitish to grey. *Cere* adult and immature males bright blue; adult female (23) dark brown enclosing light green patches; immature female, blue.

Pezoporus wallicus — Ground Parrot

Six seen in an area of swampy heathland near the mouth of the Clarence River at Grafton (94) on 4 April 63.

CUCULIDAE — Cuckoos

D. Goodwin

Cuculus pallidus — Pallid Cuckoo

Specimens. WESTERN AUSTRALIA. Peg's Beach (2) 1 June 66, 1 imm fl. Chichester Range (3) 13 May 66, 1 ♀. Cocklebiddy (15) 9 July 65, 1 ♀, 1 ♀ skel. Mount Anderson (28) 27 Apr – 5 May 68, 1 ♂, 1 imm ♂, 1 ♂ skel. Derby (29) 26 May 68, 1 ♀ fl (91.1 g). Mount Elizabeth (35) 7 July 68, 1 ♀. QUEENSLAND. Moonlight Creek (47) 2 May 64, 1 ♂ (82.3 g). Glenore (49) 9 Apr 64, 1 ♂ (81.0 g), 1 ♀ (88.5 g), 1 imm ♀ (67.1 g). Thylungra (60) 23 – 28 May 63, 1 ♂, 1 fl. SOUTH AUSTRALIA. Innamincka (64) 28 Apr 65, 2 ♀ (82.4 g). Etadunna (66) 15 – 20 May 65, 1 ♂, 3 ♀, 1 ♂ skel. Oodnadatta (67) 31 May – 1 June 65, 1 ♂, 1 ♂ juv skel, 1 fl.
Distribution & Abundance. Common at Cocklebiddy (15): where collected elsewhere and at Mount Shenton (18) other individuals, but not large numbers, were seen.
Habitat. Found in many types of fairly open country with some trees and shrubs: including spinifex plain and scrub on coastal dunes (2, 3), savanna woodland and riparian vegetation (28-49), in dry creeks with gum trees, Gidgee or box on edge of gibber desert, or in arid areas with scattered trees or bushes (64-67).

Behaviour. Usually seen singly but a male and female were together at Mount Anderson (28) on 27 April and a male and female (collected) and a third individual (sex not noted) were seen in company at Derby (29).

Very often encountered perched on dead trees. Behaviour when seeking food exactly like that of the European Cuckoo, the bird watching and then flying to take an insect from the foliage or the ground. One was seen to hop about on the ground, taking rather short small hops, again precisely like *C. canorus* when on the ground.

Gould (1865) describes the flight of this species as being "straight and rapid and not unlike that of *Cuculus canorus*". This type of flight was also seen once (DJF), from a female flying in the rain at Mount Anderson (28), but at other times the flight was with periodic wing-closure and consequent woodpecker-like undulations, very unlike the flight of *C. canorus*. A call, transcribed as *kew-kew-kew-kew,* heard from a female in flight was not like any call of *C. canorus*.

An immature at Mount Anderson (28) was perched in a dead tree. It uttered a two-syllabled call, the first note higher-pitched than the second. This call appeared to attract the attention of many birds including Willie Wagtails, Black-faced Cuckoo Shrikes, Singing and Golden-backed Honeyeaters, Long-tailed Finches and Zebra Finches. Zebra Finches were seen to perch on the young Pallid Cuckoo and appeared to feed it in response to its begging (CBF).

The specialised method of regurgitation and complementary behaviour of the young (see Nicolai, 1964) make it very doubtful whether an estrildine, such as the Zebra Finch, would or could feed a non-estrildine species. The stomach contents of this young Cuckoo (see below) do not suggest that it had actually been given any food by Zebra Finches. It seems likely that the Zebra Finches were attracted by the young Cuckoo and responded to its food-begging (which if directed towards a Zebra Finch perched on top of it might supply in supra-normal form some of the stimuli that would be given from a begging young Zebra Finch in typical estrildine begging posture) by lowering their bills to its gape, but as it did not then respond in the "correct" way they did not actually regurgitate food.

In view of the fact that this young Cuckoo was already in full body moult the possibility that its apparent food-begging may have been a defensive response to mobbing cannot be excluded. Gaping in defensive threat occurs in many bird species, including *Cuculus canorus*.

Food. The stomachs of 11 specimens contained hairy caterpillars; two others had a mixture of remains of hairy caterpillars and other insects, in one case grasshoppers. Two other stomachs contained "caterpillars", two others unidentified insect remains, and one wolf-spiders and grasshoppers.

A young bird which, from its behaviour (q.v.) may still have been fed by its fosterers or only recently become independent, contained grasshoppers (including one with a rather wasp-like banded body), small and large caterpillars, and three seeds (mainly intact, and too large to have been fed to it by a Zebra Finch).

Annual Cycle. Young birds were collected in April (28, 49) and June (2, 67); the two skinned (28, 49) were largely in juvenile plumage. Gonads of most adults collected appeared to be in non-breeding condition, a female collected on 5 May at Mount Anderson (28) had the largest oocytes in its ovary, about 2 mm in diameter.

Impermanent Colours. *Legs* olive, usually paler between the scutes: immatures; one male light grey, two females very dark olive green and dull olive brown. *Iris* medium brown to very dark brown, often with a whitish or yellowish outer ring. *Bill* dull black, lower mandible paler with dark tip, gape yellow or orange: immatures with light part of lower mandible dark yellow. *Mouth* bright deep yellow to orange-red. *Orbital skin* lemon yellow to bright yellow; dark yellow and pale olive green in two of the immatures.

Geographical Variation. Peters (1940) queries the validity of the western Australian race *C. pallidus occidentalis*. I can see no difference at all in colour or markings, other than those due to wear or individual variation, between our specimens from different parts of Australia. Those from South Australia and New South Wales average larger than Western Australian specimens but there is a considerable overlap in measurements: wings being:—Western Australia, 6 ♂ 180–196 (187.8 mm), 7 ♀ 179–194 (185.8); South Australia, 2 ♂ 191–193, 5 ♀ 183–185 (184.6 mm); New South Wales, 4 ♂ 188–198 (194.8), 4 ♀ 186–194 (188.5).

If the British Museum specimens are typical it would seem best to consider *occidentalis* as a synonym of nominate *pallidus*.

Cacomantis variolosus — Brush Cuckoo

Specimens. Western Australia. Geikie Gorge (31) 3 June 68, 1 ♀ (36.4 g). Mount Elizabeth (35) 10 July 68, 1 ♀ (32.8 g). Lake Gladstone (36) 21 July 68, 1 imm ♀. Northern Territory. Oenpelli (43) 13 Oct 68, 1 ♂ (36.1 g). Mount Roper (45) 28 Oct 68, 1 ♂ (33.7 g). Queensland. Moonlight Creek (47) 14 May 64, 1 imm ♂ (28.0 g). Glenore (49) 16 – 17 Apr 64, 1 ♂ (37.5 g), 1 juv (28.8 g). Inkerman (50) 5 June 64, 1 ♀ (30.0 g).

Distribution & Abundance. Single birds were also seen or heard in Northern Territory at Pine Creek (44), and Borroloola (46).

Habitat. Usually savanna woodland; often near water. At Oenpelli (43) found in trees at water's edge in a grassland plain and at Mount Roper (45) in mangroves beneath riverine forest.

Behaviour. All birds appeared to be alone. The call from the Oenpelli (43) male was described as a "plaintive monotonous rising whistle, suddenly stopping and often repeated".

Food. Insect remains. Caterpillars were identified in four cases, and dragonfly remains once.

Annual Cycle. All the specimens taken in Western Australia (June and July) were undergoing complete moults. The adult taken at Glenore (49) in mid-April was also in full moult and the juvenile had the wing and tail quills not fully grown and was probably not more than three weeks old.

Impermanent Colours. *Legs* males orange yellow to olive, sometimes with darker scutes; females and immatures greenish grey. *Iris* males chestnut to dark brown, once with paler outer ring; females and immatures brown or olive. *Bill* adults black or dark brown, the lower mandible olive or dull yellow with a darker tip; immatures mainly horn orange to creamy buff. *Mouth* males orange or vermilion; females and immatures yellow or light orange.

Cacomantis castaneiventris — Chestnut-breasted Cuckoo

Specimen. QUEENSLAND. Ayton (52) 1 July 64, 1 ♂ (31.9 g).
Field Notes. Collected in dense tropical rainforest. A Whipbird *Psophodes olivaceus* was seen nearby and had possibly been watching or mobbing the cuckoo. Not seen elsewhere.
Food. Insects, including ticks, beetles and grasshoppers.
Annual Cycle. This specimen had nearly completed a full moult, but most of the tail feathers were still old.
Impermanent Colours. *Legs* light yellow, claws black. *Iris* yellowish brown. *Bill* upper mandible, black; lower mandible, black at tip, rest brownish black. *Mouth* greyish black with palate a salmony orange-red. *Orbital ring* lemon.

Cacomantis pyrrophanus — Fan-tailed Cuckoo

Specimens. WESTERN AUSTRALIA. Metricup (9) 7 Mar 66, 1 imm fl. Scott River (11) 27 Feb 66, 1 imm ♂ skel. Green Range (13) 6 Apr 66, 1 ♀. QUEENSLAND. Mount Dryander (57) 3 Aug 64, 1 juv (37.8 g). Tyrone (59) 12 May 63, 1 fl. NEW SOUTH WALES. Dunedoo (93) 20 – 25 Mar 63, 1 imm ♂, 1 ♀, 1 ♂ skel, 1 ♀ skel, 1 imm fl.
Specimens to other Museums. NMV 1 ♀ skel Grafton (94) 5 Apr 63.
Field Notes. In Queensland usually collected in fairly open woodland but also in low, dry rainforest (57) and tropical layered woodland (50) and in Western Australia also in sparse vegetation on dunes (13). Sometimes on the edge of woodland or in second growth at edge of rainforest.

Also encountered at other Queensland localities (50-52, 54). The species was probably fairly numerous where found but always seen singly.
Food. Furry caterpillars, sometimes also beetles and other insects.
Annual Cycle. The juvenile taken at Mount Dryander (57) on 3 August had the wing and tail quills not fully grown and was probably only about three to three and a half weeks old. The female taken at Green Ranges (13) on 6 April was in full moult. The female taken at Dunedoo (93) on 23 March was nearing the completion of a full moult; and the immature male taken at Dunedoo (93) on 20 March was rather more than half way through the moult from immature to adult plumage, this moult involving wing and tail quills.
Impermanent Colours. *Legs* adults olive brown to yellowish buff, sometimes with the scales rather darker and usually with the soles of the toes yellow, sometimes feet paler or more yellow than legs: juvenile dull darkish pink with bright yellow feet. *Iris* adults yellow brown to dark brown, at least once with the peripheral areas paler than the central ring; juvenile brownish grey. *Bill* adults black or brownish black, lower mandible brown, yellowish or fleshy at base; juvenile blackish brown with a buff base to the lower mandible. *Mouth* adults yellow-orange to red-orange, juvenile dull yellow.

Orbital skin pale bright yellow to deep yellow; juvenile greenish yellow.
Geographical Variation. Specimens examined confirm that *C. p. albani* of Western Australia and *C. p. athertoni* of northern Queensland are best regarded as synonyms of *C. p. prionurus* of New South Wales (Peters, 1940). Western birds match *prionurus* in colour and while those from northern Queensland may be a trifle darker on average the differences are much less than those between old and new plumages. Typical *prionurus* average slightly larger but there is some overlap in wing length: *prionurus* 8 ♂ 143–147 (144.5), *albani* 4 ♂ 139–144 (142.2), *athertoni* 2 ♂ 114 mm.

Chrysococcyx osculans — Black-eared Cuckoo

Specimens. WESTERN AUSTRALIA. Carnarvon (5) 11 June 66, 1 ♂ skel (and flat skin). QUEENSLAND. Moonlight Creek (47) 6 May 64, 1 ♀. Glenore (49) 10 – 25 Apr 64, 2 ♂, 3 ♀.
Field Notes. Most often seen singly; never more than three together. At Glenore (49) and Moonlight Creek (47) found in tropical savanna woodland. One, perched in a dead tea-tree, uttered a soft *cheep cheep*. The Carnarvon (5) specimen was picked up dead in an area which, except for the riverside trees, was open grassland. It appeared to have killed itself by striking a wire in flight.
Food. Three stomachs contained remains of insects; beetles, one of them with orange wing-cases, being present in all. One contained a grey unidentifiable paste and some yellow seeds.
Annual Cycle. All specimens were in moult.
Impermanent Colours. *Legs* dark slate grey to nearly black. *Iris* males yellowish brown (once) and rich brown (once); females dark brown (twice) medium brown (once) and dull olive brown (once). *Bill* black. *Mouth* black with throat pink or mottled with grey (47, 49) or scarlet with only inside of bill black (5). *Orbital skin* slate grey; *lower eyelid* translucent grey.
Geographical Variation. The Carnarvon (5) specimen and two in the British Museum from South Australia and New South Wales do not appear separable from Queensland birds.

Chrysococcyx basalis — Horsfield Bronze Cuckoo

Specimens. WESTERN AUSTRALIA. Chichester Range (3) 12 – 20 May 66, 1 ♂, 1 ♀ skel, 1 fl. Coolawanyah (4) 15 – 20 May 66, 1 ♂ skel, 1 fl. Mount Shenton (18) 3 Aug 65, 1 ♂. Mount Anderson (28) 28 Apr – 13 May 68, 1 ♂, 1 ♀, 1 juv ♂ (21.0 g), 1 ♂ skel, 3 ♀ skel (18.7, 20.2 g), 1 fl. Derby (29) 26 May 68, 1 fl (22.0 g). QUEENSLAND. Moonlight Creek (47) 2 May 64, 1 imm ♀ (22.2 g). Glenore (49) 13 – 21 Apr 64, 1 ♂ (18.8 g), 1 ♀ (22.2 g), 1 imm (21.1 g), 1 juv (8.1 g). SOUTH AUSTRALIA. Oodnadatta (67) 31 May 65, 1 ♀ (27.5 g). Nullarbor HS (69) 6 Oct 63, 1 imm ♀: 29 Oct 64, 1 imm ♀. VICTORIA. Nelson (88) 30 Dec 62, 1 ♀. Coleraine (89) 19 Jan 63, 1 imm. Little Desert (91) 26 Jan 63, 1 imm ♀.
Distribution & Abundance. Seen at all Kimberley camps (27-36) and at Karratha (1) and Cocklebiddy (15), and heard at Newdegate (14). At Mount Shenton (18) it was noted as common but elsewhere only one or two individuals were seen.

Habitat. Found in open savanna woodland, open bushy plains with scattered trees, shrub steppe, among low bushes and spinifex, and in coastal scrub.
Behaviour. Usually single. A juvenile at Glenore (49) in company with three *Malurus melanocephalus* was heard apparently to "imitate" their call. It would be interesting to know if this species, like the parasitic viduines, really does mimic the calls of its fosterers or whether its innate calls, or some of them, are closely similar. Two other juveniles were, respectively, in company with Buff-rumped and Brown Thornbills (89, 91). The latter were undoubtedly and the former probably fostering the cuckoo in question.
Food. One stomach contained 14 geometrid caterpillars and one red bug; another contained small insects including a wasp; three others contained only hairy caterpillars; the remainder contained mainly caterpillars and once what appeared to be the remains of purplish berries. A juvenile (28) that was probably still being fed by its unknown fosterers, had a stomach-full of small grasshoppers.
Annual Cycle. In May the male from the Chichester Range (3) had testes about 44 mm in length and was completing its moult, while the female was near to laying with the largest oocyte in its ovary about 10 mm in diameter. At Mount Anderson (28) the young bird cannot long have been out of its foster parents' nest as its (juvenile) wing and tail quills are not fully grown: the two adults were in full moult.

The immature from Moonlight Creek (47) on 2 May was beginning its moult from juvenile to adult plumage. Both adults from Glenore (49) were in full moult; both of the immatures were in unworn juvenile plumage, in one of them the wing and tail quills were not fully grown.

Most other adults were in some stage of moult.
Impermanent Colours. *Legs* dark slate grey to black, often with an olive tinge; soles of feet dirty white to yellowish. *Iris* males deep pink or golden brown with whitish outer ring, or – in two birds in full moult – pale red brown or dark brown: females and young birds, pale grey or whitish sometimes with darker mottling or yellow tinge. *Bill* males dark bluish grey to black, usually paler at base; females dark brown to black, often with the upper mandible brownish horn, the lower mandible paler, in one yellow-brown. *Mouth* black, sometimes with some yellow, especially at gape in females; young juveniles bright yellow, older juveniles creamy or yellow mottled with grey. *Eye rim* noted for one female as grey, for one juvenile as yellowish grey.

Chrysococcyx plagosus Golden Bronze Cuckoo

Specimens. WESTERN AUSTRALIA. Scott River (11) 19 Feb – 1 Mar 66, 1 imm ♂ skel, 1 ♀ skel, 1 fl. QUEENSLAND. Glenore (49) 22 Apr 64, 1 ♂ (18.9 g). Glendower (55) 1 Apr 64, 1 ♂ (22.8 g). NEW SOUTH WALES. Albury (92) 11 Mar 63, 1 imm ♂. Grafton (94) 3 Apr 63, 1 ♂ (18 g).
Specimens to other Museums. NMV 1 ♂ Grafton (94).
Field Notes. In Western Australia collected in or near peppermint thickets in open bushland; also seen at Metricup (9) in Queensland in tropical savanna woodland and riverine forest, and in New South Wales in open forest along a creek and in mixed woodland. All birds were single except for two seen together at Grafton (94), and at

Scott River (11) one young bird in company with Scarlet Robins and another with Yellow-rumped Thornbills.

Food. Two stomachs contained caterpillars, another unidentified insect remains.

Annual Cycle. The three adults had inactive gonads and were in moult, with the majority of the contour feathers new. The March immature (92) is in fresh plumage, its wing and tail quills are fully grown but show very little sign of wear.

Impermanent Colours. *Legs* males grey or (once) bronze green; female greenish brown; immatures greenish grey. *Iris* usually brown but one male (49) orange and one immature pale grey. *Bill* dark brown to black, sometimes greenish yellow or fleshy at base. *Mouth* males fleshy (94), orange (92) or greyish black with pinkish throat (49, 55 examined 5–10 minutes after death), females and immatures yellow.

Parasites. See Appendix 1.

Chrysococcyx minutillus Little Bronze Cuckoo

Specimens. WESTERN AUSTRALIA. Point Torment (30) 30 May 68, 1 imm (16.5 g). QUEENSLAND. Moonlight Creek (47) 15 May 64, 1 ♀ (18.7 g). Glenore (49) 21 – 23 Apr 64, 1 ♂ (17.9 g), 1 ♂ imm (16.5 g). Inkerman (50) 27 May – 8 June 64, 2 ♂ (16.9, 18.3 g), 1 ♀ (18.6 g). Ayton (52) 23 – 30 June 64, 2 ♂ (18.8, 20.2 g).

Distribution & Abundance. Besides where collected, once believed seen at Mount Roper (45).

Habitat. Dense mangroves (30, 47); riverine tropical savanna woodland (49); tall thick timber with matted undergrowth (50) and woodland near the edge of tropical rainforest (52).

Behaviour. All birds appeared to be alone except for the immature at Glenore (49) which was being fed by its foster parent, a *Gerygone olivacea.*

Food. Hairy caterpillars, and some unidentified insects.

Annual Cycle. The adult male from Glenore (49) was just completing a moult from immature to adult plumage: the contour feathers appeared new but there were a few juvenile wing and tail quills not yet cast. A male and female from Inkerman (50) were also completing wing moults, the male's tail feathers appear not to have started moulting but the female was also moulting its tail.

The young bird from Glenore (49) was still being fed by its foster-parent.

Impermanent Colours. *Legs* dark grey to black, one male from Ayton (52) dark greyish purple and the other blackish green; immatures slate grey. *Iris* males orange: females dark brown. *Bill* black, sometimes with paler base to lower mandible; immature dark brown, paler below. *Mouth* black grading to pink in throat: immatures whitish or pale yellow. *Orbital skin* males bright reddish-orange; females, yellowish white or translucent grey; immature (49) pale yellow.

Geographical Variation. Our series has been compared with the types of *minutillus* and *russatus* and with British Museum skins of the latter race from Cape York. The female from Moonlight Creek (47) matches fairly closely with the type of *minutillus* and the young bird from Point Torment (30) must also be regarded as that race. Of the remaining Queensland birds those from Ayton (52) and most (but not all) of the Cape

York series match the type of *russatus*. They have a considerable amount of rufous on the undersides of the wings, on the tail and sides of the breast, a rufous suffusion on the upper parts and heavy barring below. The specimens from Inkerman (50) and Glenore (49) and two from Cape York are less rufous and less heavily barred below and appear to be intergrades between *minutillus* and *russatus*.

Eudynamis scolopacea — Koel

Specimen. QUEENSLAND. Glenore (49) 16 Apr 64, 1 juv ø.
Field Notes. The juvenile collected was in tropical savanna woodland consisting mostly of stunted paperbark trees. A *Philemon citreogularis,* presumably its foster parent, was seen to feed it. Elsewhere two birds were seen at Oenpelli (43) and one at Grafton (94).
Annual Cycle. This bird is in juvenile plumage with its wing and tail quills not completely grown. It seems unlikely that it was more than, at most, about three and a half weeks old.
Impermanent Colours. *Legs* bluish grey. *Iris* dark olive brown. *Bill* deep buff. *Mouth* bright red-orange. *Lower eyelid* olive yellow.

Scythrops novaehollandiae — Channel-billed Cuckoo

Specimens. QUEENSLAND. Glendower (55) 13 – 23 Mar 64, 3 ♀ (559.7, 585.9, 632.3 g), 1 imm ♂ (565.8 g).
Field Notes. Encountered alone or in small groups, either perched high in the trees or in low vegetation in the riverine forest along the Flinders River (55). Elsewhere two presumed pairs were seen at Borroloola (46).
Food. Two stomachs contained a mass of tiny yellow seeds embedded in a fibrous vegetable paste (remains of wild figs?) and the wing cases of a few small beetles.
Annual Cycle. One of the three adults was in moult; the wings were moulting asymmetrically, while all the tail feathers appeared old. The other two appeared to be in mixed plumage but with no actual signs of moult. All had somewhat enlarged ovaries. The young bird is in juvenile plumage that was already showing signs of wear.
Impermanent Colours. *Legs* slate grey (two) or dark sage green (one). *Iris* bright scarlet. *Bill* pale greyish brown with white tip, once with the lower mandible purer grey than the upper. *Mouth* pinkish white. *Facial skin* deep pink to scarlet. The immature male had pale greyish green legs; olive brown iris; a very pale greyish horn bill, and greyish brown facial skin.

Centropus phasianinus — Pheasant Coucal

Specimens. WESTERN AUSTRALIA. Mount Bell (32) 19 June 68, 1 ♂. Joint Hill (34) 28 June – 2 July 68, 1 ♂ (330.0 g), 1 ♀, 1 ♀ skel (494.8 g). NORTHERN TERRITORY. Borroloola (46) 14 – 18 Nov 68, 1 ♂ skel (343.0 g), 2 fl (322.0, 340.4 g).
Distribution & Abundance. Apart from where collected, where it appeared fairly common in suitable habitat, was seen also on the west coast (1, 2), a few at Darwin (41) and Pine Creek (44) and it was plentiful in suitable habitats at Oenpelli (43) and Mount Roper (45).

Habitat. Found in savanna woodland with grass or other ground cover, also in tall grass (1) or reeds and bushes (43) by river and in *Pandanus* swamp in sandstone gorge near Joint Hill (34).

Behaviour. Usually encountered on the ground, but when flushed sometimes flew into a tree (once only after being flushed twice from ground) and then hopped up from branch to branch until fairly high. At other times took refuge in long grass and could not be flushed from it.

Food. Grasshoppers.

Annual Cycle. The male from Mount Bell (32) was in body moult but had not yet started moulting wing or tail. The male and female from Joint Hill (34) were both in complete moult although the female had five apparently enlarging oocytes in her ovary, four of which were about 2 mm in diameter. The skeletal and spirit specimens were also in moult.

Impermanent Colours. *Legs* bluish or greenish grey. *Iris* in two partially melanic birds examined immediately after death, male bright red, female pinkish straw: in normal birds examined 2-16 minutes after death, males chestnut or yellowish with brown speckles; females grey brown. *Bill* males horn or grey with darker, sometimes black, culmen; females probably paler and browner. *Mouth* whitish to dark flesh. *Orbital skin* (noted in only one female) grey.

Geographical Variation. On locality these are referred to *C. p. macrourus*. The small number of adequately sexed and localised skins in the British Museum suggests that *macrourus* hardly warrants separation from nominate *phasianinus*.

STRIGIFORMES — Owls

D. Goodwin

Ninox connivens — Barking Owl

Specimens. NORTHERN TERRITORY. Darwin (41) 19 Sept 68, 1 ♂ (571.6 g). Oenpelli (43) 10 – 14 Oct 68, 1 ♀ (415.6 g), 1 fl.

Field Notes. The two collected at Darwin (41) were in monsoon forest near the coast. Two others were seen at Oenpelli (43) in paperbarks near water along the edge of the escarpment.

All were alone, perched in thick tree or (once) bush cover.

Food. Beetles, small mammal (unidentified).

Annual Cycle. The Darwin (41) male was undergoing a complete moult, the moult of the wing quills being still at an early stage.

Impermanent Colours. *Feet* yellow or yellowish grey. *Iris* yellow. *Bill* greenish grey, bluish grey or greyish horn, blackish at tips of mandibles and cutting edges. *Mouth* purplish pink to flesh pink. *Cere* greenish yellow.

Geographical Variation. Mees (1964b) has discussed geographical variation in this species. Our specimens are referable to *N. c. occidentalis*. They are darker than most of the few other specimens of this race in the British Museum and, especially on the upper parts, show some approach to the Cape York form *N. c. peninsularis,* but this may be partly due to their fresher condition.

Ninox novaeseelandiae Boobook Owl

Specimens. WESTERN AUSTRALIA. Chichester Range (3) 16 May 66, 1 ♀ skel and flat skin. Derby (29) 20 May 66, 2 skel. Geikie Gorge (31) 1 June 68, 1 ♂ (237.5 g). Mount Bell (32) 14 June 68, 3 ♂ (224.2 g). NORTHERN TERRITORY. Pine Creek (44) 25 Aug 68, 1 ♂ (187.0 g). QUEENSLAND. Moonlight Creek (47) 11 – 13 May 64, 2 ♀ (270.3, 281.8 g). Glenore (49) 17 – 20 Apr 64, 1 ♂ (254.2 g), 2 ♀ (278.8, 289.9 g). Glendower (55) 29 Mar 64, 1 ♀ (311.9 g). SOUTH AUSTRALIA. Innamincka (64) 2 May 65, 1 ♀ (239 g). Etadunna (66) 22 May 65, 1 ♂. NEW SOUTH WALES. Albury (92) 9 – 10 Mar 63, 1 ♂, 1 ♀.

Field Notes. Recorded frequently in the west and southwest (1-4, 7-9) and common at Geikie Gorge (31). In Northern Territory also at Borroloola (46). In the southeast seen also at Furner (86) and Dunedoo (93). Found in savanna woodland (most localities), eucalyptus woodland in rocky gorge (32), and big gums near rivers (31, 55).

Food. Grasshoppers, beetles. One bird had also eaten a rock gecko; another spiders, a praying mantis and a mouse.

Annual Cycle. One April female from Glenore (49) was moulting heavily on throat and upper breast. Its plumage otherwise looks new so this may be the last stage of a full moult. A male taken near Albury (92) on 10 March is almost certainly in final stages of moult. Some new contour feathers are still growing and the two central tail feathers are not quite fully grown.

Impermanent Colours. *Feet* pale grey with yellowish soles. *Iris* males pale greenish yellow, yellowish grey or fawn; also dark brown (31) and "stone brown with dark outer ring" (44): females pale buff to golden buff, one greenish yellow. The differences may be part individual, part sexual and part due to age. Fleay (1926) found that in several captive Boobook Owls the males had irides with "a greyish tint" whereas those of the females were "quite green", while Mees (1964b) states that a dark iris is characteristic of immature birds. The two males with brown irides (31, 44) had fully pneumatised skulls, but could have been younger than pale-eyed specimens. *Bill* pale grey, or bluish grey, tips and cutting edges blackish. *Mouth* bluish, sometimes (or always) with two black spots on palate.

Geographical Variation. This has been discussed by Mees (1964b), who recognised only four mainland Australian subspecies. These are *N. n. boobook* from Victoria, New South Wales, southern Queensland and eastern South Australia, a large form which shows some variation in coloration; *N. n. ocellata,* from much of Queensland, Northern Territory, South Australia and Western Australia, which is also very variable in colour but generally paler, and often very much paler, than any other Australian subspecies; *N. n. rufigaster,* from southwestern Australia, a very dark race but more rufous below than *boobook;* and *N. n. lurida,* which is the darkest race and has a very restricted range in the forests of northeastern Queensland.

Our specimens fit quite well into this arrangement. The bird from Chichester Range (3) is referable to *rufigaster* and those from northwestern Australia, Northern Territory,

and Moonlight Creek (47) are referable to *ocellata*. Mees includes the Glenore (49) area within the range of *ocellata,* but our three birds from there are markedly larger (wings 2 ♀ 245 mm) and darker than those from Moonlight Creek (47) (wings 2 ♀ 234, 238 mm). The specimen from Glendower (55) is like those from Glenore (49) but more markedly spotted, and the four birds (49, 55) seem best referred to *N. n. boobook* together with those from South Australia (64, 66) and New South Wales (92).
Parasites. See Appendix 1.

Tyto alba — Barn Owl

Specimens. Western Australia. Mount Caroline (33) 30 June 68, 1 ♂. Queensland. Inkerman (50) 3 – 5 June 64, 2 ♂ (393.3, 416.1 g), 2 ♀ (386.2, 450.2 g).
Distribution & Abundance. Also encountered at Borroloola (46) and Banoona (58).
Habitat. Baobab trees in savanna woodland (33); at Inkerman (50), tropical layered woodland by day, hunting in nearby open grassland at night.
Behaviour. Singly or (once) in pairs. Hunts over grassland with rather low, erratic-looking flight; two birds were collected when they landed in the grass.
Food. Small mammals, including rats and mice.
Annual Cycle. All Inkerman (50) birds appeared to have active gonads. The largest oocytes in the two females were about 3 mm and 15 mm in diameter.
Impermanent Colours. *Legs* whitish, one with dirty brown scales. *Iris* brown. *Bill* white or (one) very pale greyish pink; bare or nearly bare skin at base of both mandibles pink or flesh pink. *Mouth* white to pale purplish pink.
Other Notes. One of the males from Inkerman (50) is very grey on the upperparts with only a slight trace of buff. This is clearly an individual not a geographical variation. No other Australian specimens in the British Museum have as little buff although one from Moreton Island, Queensland and one without locality, approach it fairly closely.

Tyto novaehollandiae — Masked Owl

Seen at Newdegate (14).

CAPRIMULGIFORMES — Frogmouths & Nightjars

D. Goodwin

Podargus strigoides — Tawny Frogmouth

Specimens. Western Australia. Mount Anderson (28) 2 – 17 May 68, 1 ♀, 2 ♂ skel (256.2 g), 1 fl. Joint Hill (34) 29 June 68, 1 ♀ (209.4 g). Mount Elizabeth (35) 14 July 68, 1 fl. Wyndham (39) 6 Aug 68, 1 ♂ (204.9 g). Northern Territory. Oenpelli (43) 6 Oct 68, 1 fl (228.4 g). Pine Creek (44) 22 Aug 68, 1 fl (240.4 g). Mount Roper (45) 30 Oct 68, 1 ♀ imm (159.9 g). Queensland. Glenore (49) 15 Apr 64, 1 ♂ (261.6 g), 1 ♀ (226.2 g). Inkerman (50) 2 June 64, 1 ♀ (260.8 g). Tully (54) 17 July 64, 1 ♀ (235.1 g). Glendower (55) 17 – 29 Mar 64, 2 ♂ (339.8, 363.9 g), 1 ♀ (290.1 g), 1 ø (359.0 g). Cape Gloucester (56) 5 Aug 64, 1 ♂ (298.2 g). New South Wales. Dunedoo (93) 18 Mar 63, 1 ♂ skel.

Distribution & Abundance. In Western Australia also encountered at Metricup (9), Mount Barker (12), Geikie Gorge (31) and Mount Remarkable (37), where it was common. In Northern Territory a few individuals were seen at most camps (41, 43-45) and one at Borroloola (46). In Queensland noted also at Tyrone (59). In South Australia the only bird seen was at Innamincka (64). In New South Wales recorded also at Grafton (94) and Tenterfield (95).

Habitat. Fairly open woodland. Thus in savanna woodland in Western Australia, Queensland and Northern Territory, in thin scrub with mangroves and rocks on the coast at Wyndham (39) in mixed woodland in New South Wales and in open forest in northern Queensland. It appeared to be absent from dense riverine forest. Its feeding habits appear to require more or less bare or open areas of ground that can be overlooked from convenient perches.

Behaviour. Usually found singly; once two birds together and once a pair near their well-grown young, one of which was collected, at Mount Roper (45). Birds active after sunset (seen by torchlight) did not adopt the cryptic pose when perching.

Food. Grasshoppers, beetles and large centipedes.

Annual Cycle. Specimens taken in Western Australia in May, June and August were all in moult, the August bird in a later stage. Two specimens taken in northern Queensland in April (49) and one in early June (50) were also undergoing complete moults. The immature taken at Mount Roper (45) on 30 October is moulting its body plumage, apparently before the first set of outer primaries was completely grown.

Impermanent Colours. *Legs* pale greenish yellow or greenish grey to dark greenish brown, sometimes with paler areas between the scutes. *Iris* yellow or orange with a darker outer ring. *Bill* greenish grey or brown sometimes slightly paler on lower mandible. *Mouth* pale lemon yellow or cream (once pink) usually with the back of the mouth pinkish and with conspicuous purple capillaries.

Geographical Variation. The collected specimens fall within races recognised by Peters (1940). Those from Western Australia and Northern Territory are referable to *P. s. phalaenoides;* those from Glenore (49) and Inkerman (50) in northwestern Queensland to *P. s. gouldi,* which is rather larger and, perhaps, averages a little darker in colour. Eastern Australian birds show a north–south cline in increasing size and increased pigmentation. If the smaller and paler northern populations are separated as *P. s. cornwalli* then our birds from Tully (54) and Cape Gloucester (56) are that race, while those from Glendower (55), which average darker and greyer, show more approach to *P. s. strigoides.*

Parasites. See Appendix 1.

Podargus papuensis — Papuan Frogmouth

Specimens. Queensland. Inkerman (50) 1 ♂ 29 May 64 (599.2 g).

Field Notes. Found alone in mangroves along creek about a mile from sea. The only record.

Food. Unidentified insects.

Annual Cycle. The single specimen was in complete moult, including wing and tail.

Impermanent Colours. *Legs* buffy-green. *Iris* orange. *Bill* brownish grey. *Mouth* lemon with dark tips to inner side of mandibles, throat flesh pink.

Aegotheles cristatus Owlet-nightjar

Specimens. WESTERN AUSTRALIA. St George's Range (27) 9 June 68, 1 ♀ (51.0 g). Mount Anderson (28) 7 – 17 May 68, 1 ♂ (46.3 g), 1 ♀. Derby (29) 23 – 26 May 68, 2 ♂ (44.8 g), 1 ♀. Joint Hill (34) 29 June 68, 1 ♀ (54.2 g). Mount Elizabeth (35) 6 – 9 July 68, 1 ♂ (59.7 g), 1 ♀ (53.6 g). NORTHERN TERRITORY. Pine Creek (44) 25 Aug 68, 1 ♂ (48.0 g), 1 fl (42.3 g). Mount Roper (45) 31 Oct 68, 1 ♀ (47.1 g). QUEENSLAND. Moonlight Creek (47) 5 – 11 May 64, 2 ♂ (48.9, 57.4 g). Glenore (49) 24 Apr 64, 1 ♀ (43.8 g). Inkerman (50) 3 June 64, 1 ♀ (45.4 g). Glendower (55) 11 Mar 64, 1 ♀. SOUTH AUSTRALIA. Etadunna (66) 22 May 65, 1 ♀. Oodnadatta (67) 31 May 65, 1 ♀ (49.0 g). Angepena (84) 24 Nov 63, 1 ♀.

Distribution & Abundance. Seen also at Mount Bell (32) in Western Australia and one at Borroloola (46) in Northern Territory. Few individuals were seen at any one locality, no doubt owing to the bird's retiring habits.

Habitat. Fairly open dry woodland seemed the most usual habitat: also in mangroves (47), in gums along a creekbed (67) and in open country (84).

Behaviour. All were alone although at Etadunna (66) two were seen in the same area within a short time. Often found at night looking out of the openings of holes in trees or on the ground in tracks or roads where it appeared to feed. The Etadunna (66) female flew with a dipping flight very like that of the Peaceful Dove. When it alighted it bobbed its head and up and down in a manner very like that of owls of the genus *Athene*. When mobbed by Yellow-throated Miners and White-plumed Honeyeaters it fled from them and hid in a hole in a small eucalypt.

Food. Grasshoppers, beetles, ants and other unidentified insects.

Annual Cycle. Some of the specimens taken in Western Australia in May and June show some body moult, as does a female taken in northern Queensland (55) on 11 March. A male taken on 11 May in northern Queensland (47) was undergoing a complete moult. A male taken in Northern Territory on 25 August (44) and a female taken on 31 October (45) had enlarged gonads.

Impermanent Colours. *Legs* pale yellowish olive, pale flesh, pale buff or creamy grey. *Iris* brown. *Bill* upper mandible grey, dark horn or blackish: lower mandible paler grey, yellowish horn or flesh with black or dark tip. *Mouth* pale pink, sometimes nearly white.

Geographical Variation. This has been discussed by Serventy & Whittell (1967) and Mees (1961a). The former authors recognise, *inter alia, A. c. rufus* (type locality Derby) which they describe as being "very rufous . . . the ground colour rufous barred with grey". Mees treats *rufus* as a synonym of *leucogaster* and recognises only two races (a) *leucogaster* from Northern Territory and Kimberley Division which he describes as being "sand-coloured (brownish) above, freckles below confined to upper breast. Usually with some rufous coloration": (b) nominate *cristatus* "greyish to blackish above, many freckles below", from the south extending perhaps into northern Queensland.

Our specimens suggest that rufous and more or less rufous-tinged individuals occur in most or all areas and confirm Mees' (1961a) opinion that "there is not such a thing as a rufous race". The three nominate *cristatus* from South Australia (66, 67, 84) (two grey and one reddish brown) are darker below than all others. The remainder are all paler below, greyish individuals being extensively white on the belly, rufous individuals having the belly tinged pinkish buff. The specimens from the northwest (27-35) are paler in colour (whether reddish or grey) than those from Northern Territory and northern Queensland, with the exception of one very pale female from Inkerman (50). Whether it is worth maintaining *rufus* distinct from *leucogaster* because of this slightly paler coloration seems questionable.

Eurostopodus guttatus — Spotted Nightjar

Specimens. WESTERN AUSTRALIA. Mount Anderson (28) 17 May 68, 1 fl. Derby (29) 12 May 68, 1 ♂ skel. NORTHERN TERRITORY. Pine Creek (44) 25 – 31 Aug 68, 1 ♀ (80.4 g), 1 fl (73.6 g). Borroloola (46) 12 Nov 68, 1 ♀. QUEENSLAND. Glenore (49) 13 – 20 Apr 64, 1 ♂ (84.2 g), 2 ♀ (81.3, 96.7 g), 2 imm ♂ (86.4, 88.5 g), 1 imm ♀ (85.5 g). Inkerman (50) 29 May – 5 June 64, 2 ♂ (95.7, 104.1 g), 2 ♀ (89.0, 99.7 g), 1 imm ♂ (98.2 g). Glendower (55) 1 imm ♀ (75.6 g).

Distribution & Abundance. A few other individuals, besides those collected, were seen at Mount Anderson (28) and Pine Creek (44). Also recorded at Mount Bell (32) and Monkira (61).

Habitat. Savanna woodland; also in open grassland and at the edge of dry, bushy mangroves at Inkerman (50).

Behaviour. Usually alone, but three on track at Pine Creek (44). By day flushed from ground, often from some open stony area among trees or from side of track. One at Inkerman (50) perched three feet high on limb of fallen tree. Was noted (AH) as active shortly after dusk but none seen two to three hours after full darkness. Heard in June calling at Monkira (61) on several nights, from about 21.00 hours until midnight – a peculiar repeated *Caw-caw-caw, gobble-gobble-gobble.*

Food. Grasshoppers, moths, beetles, shield-bugs.

Annual Cycle. A male and two females from Inkerman (50) in early June were in late stages of a complete moult. Young birds from Glenore (49) in April, from Inkerman (50) in May and from Glendower (55) in March were moulting their contour feathers from juvenile to adult plumage.

Impermanent Colours. *Legs* pale pinkish brown to greyish black. *Iris* medium to very dark brown; opaque grey in immatures. *Bill* pale pinkish brown to dark brown or black; paler at base of lower mandible. *Mouth* pink or pinkish purple with red veins, changing to mottled yellow, purple and pink soon after death. *Eye rims* buffish or grey (one).

Geographical Variation. Matthews (1912a) separated Northern Territory birds *E. g. harterti* as being darker in colour than Victorian specimens. Deignan (1950) described a new race *E. guttatus gilberti* from those areas of the Northern Territory with an annual rainfall of more than 40 inches, as being darker both above and below than *harterti.*

The female from Pine Creek (44) is a darker grey, especially on the head, than any of the others. It is not, however, darker on the buff parts, either above or below. Allowing for wear, individual variation and age differences, there do not seem to be any significant differences among our birds, and other British Museum specimens, except that three from Laverton, Western Australia, are rather paler than average.

Caprimulgus macrurus — White-tailed Nightjar

Specimens. QUEENSLAND. Inkerman (50) 29 May 64, 1 ♀ (60.2 g). Tully (54) 15 – 22 July 64, 2 ♂ (73.7, 76.4 g), 1 ♀ (72.7 g).
Field Notes. Only found in northern Queensland, in open forest, rainforest and mangroves; at forest edge or along roads or openings; not in dense forest. Encountered singly on the ground or on low perches. At night its call, like a resonant piece of wood struck with a hammer, was unmistakable, and was heard also at Mount Dryander (57).
Food. Moths.
Annual Cycle. The female from Inkerman (50) was in early stages of wing moult; one male from Tully (54) was moulting its tail, presumably the final stages of a complete moult.
Impermanent Colours (of ♀ from Inkerman (50) only). *Legs* blackish brown. *Iris* dark brown. *Bill* yellow brown with dark tip. *Mouth* purple, creamy peripherally. *Orbital skin* ochre.
Geographical Variation. All Australian and New Guinea examples of this species are currently referred to *C. m. yorki*. The Inkerman (50) female is much smaller and paler in colour than the Tully (54) specimens although seemingly adult from plumage and skull. It has a wing measurement of 173 mm as against 185 for the Tully (54) female, 177 and 180 for the only other Australian specimens (one from Cape Grenville, one with no locality) in the British Museum.

APODIDAE — Swifts
D. Goodwin

Collocalia spodiopygia — Grey Swiftlet

Specimens. QUEENSLAND. Tully (54) 10 – 16 July 64, 9 ♂ (10.6, 10.7, 11, 11, 11.1, 11.5, 11.6, 11.8, 11.9 g), 4 ♀ (10.9, 11.1, 11.9, 12.5 g), 2ø (10.5, 10.9 g).
Field Notes. Flocks of from 10 to over 50 birds were seen feeding in flight, often soaring, over grassy clearings in rainforest and nearby areas of open woodland. They had presumably come from their breeding site on Dunk Island.
Food. Flying ants, small beetles, unidentified insects.
Annual Cycle. One specimen had some new feathers growing on back. Otherwise its plumage looked fresh, as did that of others. One male with a fully pneumatised skull had much larger gonads, about 4 mm in length, than any other specimen's. Only three of all the specimens collected appeared to have fully pneumatised skulls.

Impermanent Colours. *Legs* pale purplish brown sometimes with outside of leg grey, claws black. *Iris* medium to blackish brown. *Bill* black. *Mouth* dull pink with whitish grey palate and blackish tips to insides of mandibles. *Orbital skin* black, apparently fading to grey soon after death.

Hirundapus caudacutus — Spine-tailed Swift

Scattered southeastern records (86, 88, 90, 92–94).

Apus pacificus — Fork-tailed Swift

About 100 seen at Perth (7) and two (probably) at Oenpelli (43).

CORACIIFORMES — Kingfishers, Bee-eaters, Rollers

D. Goodwin

Alcyone azurea — Azure Kingfisher

Specimens. Western Australia. Mount Bell (32) 15 June 68, 1 ♂ (46.6 g). Mount Caroline (33) 25 June 68, 1 ♂ (29.2 g). Joint Hill (34) 1 July 68, 1 ♂ (28.1 g). Mount Elizabeth (35) 8 – 10 July 68, 1 ♂ (30.5 g), 1 ♀ (31.1 g), 1 fl (29.3 g).
Field Notes. Invariably in immediate vicinity of fresh water, whether stream, river or lagoon; in savanna woodland, freshwater mangroves, *Pandanus* and eucalypts. Seen also at Geikie Gorge (31), and in small numbers at all main camps in Northern Territory (41, 43-46). Call, a sharp high-pitched repeated whistle.
Food. Fish, and (once) water beetles.
Annual Cycle. The two specimens from Mount Elizabeth (35) were in moult; the male from Mount Bell (32) appears to be in fresh plumage.
Impermanent Colours. *Legs* orange. *Iris* dark brown. *Bill* black, sometimes with brownish tip. *Mouth* orange, blackish along inner edge of mandibles.
Geographical Variation. This has been discussed by Keast (1957b). Our birds are *A. a. pulchra.*

Alcyone pusilla — Little Kingfisher

Specimens. Queensland. Tully (54) 14 July 64, 1 ♂ (15.0 g).
Field Notes. Only seen where collected: at edge of small creek in tropical rainforest.
Food. Shrimps, one small fish and insects.
Annual Cycle. Plumage apparently fairly fresh.
Impermanent Colours. *Legs* medium brown. *Iris* blackish brown. *Bill* black. *Mouth* blackish brown peripherally, pinkish in throat, remainder greyish.
Geographical Variation. This has been discussed by Keast (1957b). The present specimen seems, as would be expected, referable to *A. p. halli.* It agrees with the three other specimens of this form in the British Museum except that it has only a very faint trace of greenish on its forehead. This character is, however, variable (Keast) and is, incidentally, shown by some New Guinea specimens of *A. p. pusilla.*

Dacelo gigas Laughing Kookaburra

Specimens. WESTERN AUSTRALIA. Scott River (11) 24 Feb – 3 Mar 66, 1 ♂, 1 ♀ skel, 2 fl. QUEENSLAND. Ayton (52) 21 June 64, 1 ♂ (272.8 g). Glendower (55) 24 Mar 64, 1 ♂ (289.5 g). Mount Dryander (57) 3 Aug 64, 2 ♂ (303.5, 305.6 g), 4 ♀ (290.8, 328.8, 344.6, 353.4 g). NEW SOUTH WALES. Dunedoo (93) 19 Mar 63, 1 ♂ skel.
Distribution & Abundance. Appeared fairly plentiful where collected and seen widely in the southeast (86, 88, 90, 92, 93, 95). It was also abundant near Adelaide and, as an introduced species, near Perth.
Habitat. In Western Australia found in open forest and parkland. In Queensland in open savanna woodland (57), and in dense riverine forest near more open woodland (52, 55). In New South Wales, South Australia and Victoria in open woodland. Some large trees seem essential for this species.
Behaviour. Occasionally encountered singly but usually three or more individuals either in company or within a small area.
Annual Cycle. The March male from Glendower (55) was in the final stages of complete moult. A female taken at Mount Dryander (57) on 3 August was in fresh new plumage. The largest oocyte in its ovary was about 2 mm across, suggesting that the bird was beginning to come into breeding condition.
Impermanent Colours. *Legs* dull green. *Iris* yellowish brown to dark brown; the differences appear to be individual rather than sexual. *Bill* upper mandible, dark horn to black, sometimes paler at tip or near base: lower mandible, creamy white to yellowish horn, often fuscous at sides of basal area. *Mouth* whitish to pale pink, purplish pink in throat, tongue whitish.
Geographical Variation. Keast (1957b) discusses this and recognises *D. g. minor* (Cooktown north through Cape York) on its smaller size, although the figures he gives show some overlap. He found there was a north-south increase in size but an apparently rather abrupt drop between the Cairns and Cape York populations, the former having wing lengths of 209-220 mm (average 215), the latter of 192-210 (average 202). A single specimen from Cooktown, the type locality of *minor,* had a wing of 200 mm.

The six specimens from Mount Dryander (57) have wings of 203-216 mm (average 208.8); the specimen from Ayton (52) has a wing measurement of 203. Fourteen specimens in the British Museum collection from Victoria, South Australia and southern Queensland have wing measurements of 207-225 mm (average 216.9).

Dacelo leachii Blue-winged Kookaburra

Specimens. WESTERN AUSTRALIA. Karratha (1) 28 May 66, 1 fl. Chichester Range (3) 17 May 66, 1 ♀ skel. Mount Anderson (28) 14 May 68, 1 ♀ (280 g). Mount Bell (32) 18 June 68, 1 ♂ (286.2 g). Mount Caroline (33) 26 June 68, 1 ♂ (276.7 g). Mount Elizabeth (35) 7 – 13 July 68, 1 ♂, 1 ♀, 1 fl. QUEENSLAND. Inkerman (50) 28 May – 5 June 64, 2 ♂ (318.5, 338.4 g). Glendower (55) 2 Apr 64, 1 imm ♂ (284.9 g), 1 ♀ (390.8 g).
Distribution & Abundance. In Western Australia found also at Coolawanyah (4). In Northern Territory at Mount Roper (45) and Borroloola (46), and noted as very plentiful at Darwin (41), Oenpelli (43), and Pine Creek (44).

Habitat. Savanna woodland, riverine forest and in gums along creeks. Very often, but by no means always, near to water.

Behaviour. Sometimes singly, sometimes in couples or small parties. Perches on branches that give a clear view around. Noted as being more afraid of man than *Dacelo gigas* normally is.

Food. Grasshoppers and centipedes. One stomach contained eggshell fragments. One specimen caught itself in a mist net through attacking a Peaceful Dove which was already entangled.

Annual Cycle. The immature male taken at Glendower (55) on 2 April shows some body moult. It is notable that the juvenile primaries hardly show any sign of wear in contrast to its tail and inner secondaries. Adults in final stages of complete moult were taken in April (55) and May (1, 28).

Impermanent Colours. *Legs* males dull yellow to light olive brown, females dull yellow or pinkish buff; immature male dull olive green. *Iris* males white to straw-coloured, sometimes brown at outer edge; females pale golden brown (one), fawn (one) or very light grey with brownish outer ring (one). *Bill* upper mandible, dark brown to bluish black, sometimes paler at tip and pinkish buff along commissure; lower mandible, white to pale grey or pinkish buff. *Mouth* pale fleshy pink to purplish, sometimes nearly pure white.

Geographical Variation. See Keast (1957b). All our birds seem best referred to the nominate form, although the adult male from Inkerman (50) is rather dark and approaches *D. l. kempi* from the northern half of Cape York Peninsula. Probably the birds from this (50) area are intergrades. Both the present series and others examined confirm Keast's remarks about the degree of individual variation in this species. Plumage wear can also produce considerable differences in appearance between birds in old and new dress.

Halcyon macleayii — Forest Kingfisher

Specimens. NORTHERN TERRITORY. Darwin (41) 5 – 21 Sept 68, 2 ♂ (36.7, 38.5 g), 1 fl. NEW SOUTH WALES. Grafton (94) 5 Apr 63, 1 ♀.

Distribution & Abundance. Plentiful at Darwin (41) and Oenpelli (43). Specimens were also collected in northern Queensland (50, 54) but were not kept.

Habitat. At Darwin (41) in pseudo-rainforest and at the edge of swamps and grassy flats fringed by *Pandanus* palms and savanna woodland. In northern Queensland in open forest at Tully (54) and in tall timber near a creek at Inkerman (50). At Grafton (94) found on the bank of a swamp.

Behaviour. Usually solitary but once a pair (41), and twice parties of three individuals (41, 94), in one of which a male was seen displaying with fanned tail. Calls heard were harsh chattering and a high-pitched repeated whistle.

Food. Insects, including dragon-flies.

Annual Cycle. The two males from Darwin (41) are in fresh-looking plumage although one, which was in a burned-over area, already has its plumage much stained and slightly worn. The female from Grafton (94) is in the final stages of a complete moult.

Impermanent Colours. *Legs* slate grey to black, dark brown in one unsexed (fluid) specimen. *Iris* dark brown. *Bill* black, basal half of lower mandible whitish except along cutting edge. *Mouth* purplish pink, inside of mandibles black near tips.

Geographical Variation. Our specimens from Northern Territory (41) and others from older collections have noticeably bluer rumps and scapulars than all birds in comparable plumage from New South Wales and most from Queensland, though one or two from Queensland and one of two from Port Essington are intermediate. This supports Mathews (1912a) who separated Northern Territory birds as *H. m. distinguendus* from birds of New South Wales which he believed to represent *H. m. macleayii*. Mayr (1937b) recognised only one race in Australia, *H. m. macleayii,* and pointed out that since its type locality had been restricted to Port Essington this name must apply to Northern Territory birds, and eastern birds if they proved separable would have to take the name *incinctus*. Our series is not large enough to judge whether the differences mentioned are great enough and consistent enough to warrant recognising the two races.

Halcyon pyrrhopygia — Red-backed Kingfisher

Specimens. WESTERN AUSTRALIA. Coolawanyah (4) 15 May 66, 1 imm ♀ skel. Warburton (23) 8 Sept 65, 1 ♂. Mount Anderson (28) 3 – 13 May 68, 2 ♂ (52.3 g), 1 ♀, 1 ♂ skel, 2 ♀ skel (52.4 g), 1 fl (49.0 g). Derby (29) 21 May 68, 1 ♂. Mount Elizabeth (35) 13 July 68, 1 ♂ skel (46.9 g). QUEENSLAND. Glenore (49) 16 Apr 64, 1 ♀ (57.2 g). Thylungra (60) 23 May 63, 1 ♂ skel, 1 fl. Monkira (61) 4 June 63, 1 ♀, 1 ♂ skel. SOUTH AUSTRALIA. Nullarbor HS (69) 8 Oct 63, 1 ♂. Lake Acraman (71) 3 Nov 64, 1 ♀.

Specimens to other Museums. NVM 1 ♂ skel Monkira (61).

Distribution & Abundance. Seen at most Kimberley camps and also at Mount Roper (45), and collected by the South Australian Museum also at Angepena (84).

Habitat. Woodland, scrub or grassland with trees. The single bird from Warburton (23) was found by the road in an area of sandhills covered with scrub in the Gibson Desert 45 miles west of the Mission.

Behaviour. Usually encountered singly but quite often in pairs. Commonly perches on a dead bough that, presumably, gives it a good look-out. One bird was on the ground when first seen.

Food. Stomachs contained mostly insects and grasshoppers mixed with pieces of grass. Frog and centipede remains recorded once.

Annual Cycle. Specimens in final stages of moult were collected in May in Western Australia (28, 29), in April in northern Queensland (49) and in June in southern Queensland (64). Others from the same localities were in fresh plumage while those taken in October and November in South Australia were in worn plumage (69, 71).

Impermanent Colours. *Legs* usually grey but apparently browner in eastern and southern birds (60 (1 ♂), 61, 71): immature, brownish grey. *Iris* brown, once dark grey (60). *Bill* black, with basal half of lower mandible white, pale horn or pinkish grey except along commissure which is black. *Mouth* whitish to purplish pink, blackish towards tips of inner sides of mandibles.

Geographical Variation. Our birds and others in the British Museum confirm Keast's (1957b) view that *H. p. obscura* of northwestern Australia, Northern Territory and northern Queensland is not separable from nominate *H. p. pyrrhopygia* when birds in comparable plumage state are compared.

Halcyon sancta Sacred Kingfisher

Specimens. WESTERN AUSTRALIA. Peg's Beach (2) 3 June 66, 1 fl. Scott River (11) 21 – 26 Feb 66, 1 ♂, 1 ♂ skel. Mount Anderson (28) 30 Apr – 19 May 68, 2 imm ø, 5 fl. Derby (29) 26–27 May 68, 2 ♂ (39.1, 39.2 g), 1 ♀ (45.5 g). Point Torment (30) 30 May 68, 2 fl (42.8, 45.3 g). Geikie Gorge (31) 3 June 68, 1 fl (45.2 g). Wyndham (39) 6 Aug 68, 1 ♀ (42.9 g), 1 fl (41.9 g). SOUTH AUSTRALIA. Furner (86) 9 Jan 63, 1 ♀. VICTORIA. Coleraine (89) 22 Jan 63, 1 fl. NEW SOUTH WALES. Dunedoo (93) 23 Mar 63, 1 skel, 1 fl.
Specimens to other Museums. NMV 1 skel Dunedoo (93).
Distribution & Abundance. Widespread in Kimberleys, Northern Territory and northern Queensland with many sight records (27-36, 41, 43-46, 47, 50, 55), but only a few individuals at each place except Glendower (55) where it was common. One seen at Perth (7) and one on Cooper's Creek, Innamincka (64).
Habitat. Open forest of Karri, Marri and Jarrah (11); river gums and savanna woodland, not always near water, in the Kimberleys; mangroves (30, 39, 47, 50); open forest (86, 89) and tall trees along a dry creek bed (93).
Behaviour. Usually found singly but at Mount Anderson (28) three individuals, one at least an immature, were roosting in the same tree and at Dunedoo (93) a presumed family party were together.

Two individuals were seen taking prey from the water or the water's edge and another catching grasshoppers. Usually silent both during the swift straight flight and when perched, but a rattling call sometimes heard.
Food. Three stomachs contained grasshoppers; one a large grub and one, from a bird taken in mangroves, small crabs and a black beetle.
Annual Cycle. Two immatures taken at Mount Anderson (28) on 30 April and 1 May were in body moult. Two adult males taken at Derby (29) on 26 May were in final stages of a complete moult and had very small gonads. The January female (86) is in worn plumage but has one fresh central tail feather, perhaps due to adventitious loss.
Impermanent Colours. *Legs* usually bluish grey to black with whitish soles, twice dark brown. *Iris* dark brown. *Bill* black with basal part of under mandible whitish except along edge of commissure where it is dark. *Mouth* pinkish, blackish at tips of insides of mandibles.
Geographical Variation. Peters (1943) queries the validity of *H. sancta westralasiana* which was described as being bluer above and having white instead of buff underparts. Our series and older specimens in the British Museum confirm Keast's (1961) opinion that geographic variation in this species is negligible within Australia. In worn plumage the green and greenish blue tones become paler and bluer and the buff of the collar and underparts may be nearly pure white. In comparable plumage all adult males are a deeper and warmer buff on the underparts and nuchal collar than are females.

Halcyon chloris Mangrove Kingfisher

Two seen at Peg's Beach (2).

Merops ornatus Rainbow Bee-eater

Specimens. Western Australia. Chichester Range (3) 16 May 66, 1 fl. Mount Anderson (28) 27 Apr 68, 1 ♂. South Australia. Lake Acraman (71) 3 Nov 64, 1 ♀. Minippa (73) 1 Nov 64, 1 ♀. New South Wales. Albury (92) 9 Mar 63, 1 ♂ skel, 1 ♀ skel, 1 fl.

Distribution & Abundance. Common or fairly common at camps in the north (28, 31, 36, 43, 44, 55) and noted elsewhere in Northern Territory (41, 45, 46) and northern Queensland. Also at Thylungra (60), one at Innamincka (64), at Dunedoo (93), and on the west coast (1, 4, 7).

Habitat. Gum creeks (3, 71), savanna woodland (28, 92), eucalyptus forest (73), and in trees of the flood plain of the Cooper (64).

Behaviour. Most often seen in small parties, sometimes in couples and twice single individuals. Behaviour like that of European Bee-eater, hawking at flying insects from a look-out perch or soaring about over the tree tops and making sallies at prey.

Annual Cycle. No evidence of breeding from specimens seen or collected.

Impermanent Colours. *Legs* dark grey to black, one male noted as having pink colour on upper part of leg. *Iris* usually red, brown in birds from Albury (92). *Bill* black. *Mouth* pinkish flesh.

Food. Insects, including wasps.

Eurystomus orientalis Dollar-bird

Specimens. Northern Territory. Oenpelli (43) 19 Oct 68, 1 ♀ (186.0 g). Borroloola (46) 7 Nov 68, 1 fl (133.8 g). Queensland. Glendower (55) 24 Mar 64, 1 imm fl.

Field Notes. Found in savanna woodland or open riverine forest at all major camps in Northern Territory but only single or in pairs except at Borroloola (46) where a number of birds were seen each day.

Food. Beetles and other insects.

Annual Cycle. The Oenpelli (43) female had a shelled egg in its oviduct and four large oocytes in its ovary.

Impermanent Colours. *Legs* vermilion; immature pale red. *Iris* dark brown. *Bill* vermilion; immature black with orange base to lower mandible. *Mouth* pale yellow; immature bright yellow. *Orbital skin* dark reddish brown; immature translucent grey.

PITTIDAE — Pittas
C. B. Frith

Pitta versicolor Noisy Pitta

Specimens. Queensland. Tully (54) 25 July 64, 1 imm ♀ (105.9 g).

Field Notes. This bird, the only one seen, was perched on a horizontal stick one foot

from the ground in a thicket in rainforest.
Food. Medium-sized insects.
Annual Cycle. The skull was not fully pneumatised, and the black of the throat was mixed with grey, showing that the bird was not fully adult.
Impermanent Colours (20 mins after death). *Legs* pale purplish-flesh. *Iris* blackish brown. *Bill* black with pinkish gape. *Mouth* whitish purple. *Eye rim* black; *orbital skin* translucent white.

Pitta iris Rainbow Pitta

Specimens. NORTHERN TERRITORY. Darwin (41) 14 Sept 68, 2 ♀ (64.9, 65.5 g). Oenpelli (43) 16 – 17 Oct 68, 2 ♀ (57.8, 59.5 g).
Distribution & Abundance. The only birds seen were the two at Hope Inlet, north-east of Darwin (41) and three in two separate localities at Oenpelli (43).
Habitat. Coastal monsoon forest (41) and non-coastal forest in sandstone gorges (43). Dense vegetation, fallen logs and leaves. Hard dry ground.
Behaviour. Moved over ground and onto logs in short hops, covering about 10 feet at a time before pausing and re-starting: came to a whistled imitation of *Pitta versicolor*'s call *Walk-to-work*. The bird seen but not collected was perched in a tree 15-20 feet up, singing. At both places the two females collected were possibly in some sort of association as the second one to be collected was subsequently found in the same spot as the first.
Food. Insects, including grasshoppers, larvae of wood-boring beetles, a moth and ants: also small skinks.
Annual Cycle. All birds had the ovary slightly enlarged and their plumage shows some wear.
Impermanent Colours. *Legs* light flesh to reddish-grey, feet dark brown. *Iris* dark brown. *Bill* black. *Mouth* whitish.

MENURIDAE — Lyrebirds
D. J. Freeman

Menura alberti Albert Lyrebird

At Tenterfield (95) one bird was seen in rainforest and three or four more were heard singing locally.

ALAUDIDAE — Larks
D. J. Freeman

Mirafra javanica Singing Bushlark

Specimens. WESTERN AUSTRALIA. Karratha (1) 30 May 66, 1 imm ♀. Chichester Range (3) 12 – 14 May 66, 5 ♂, 2 ♀ skel, 1 fl. Mount Anderson (28) 29 Apr – 10 May 68, 1 ♂ (21.7 g), 2 ♀ (24.1 g), 1 imm ♀, 4 ♂ skel, 1 ⌀ skel (24.1 g), 1 imm ♂ skel, 1 fl. Mount

Bell (32) 19 June 68, 1 ♂ (21.0 g), 2 ♀ (22.2, 22.7 g). Joint Hill (34) 28 June 68, 2 ♂ (20.7, 23.7 g), 1 ø (20.5 g). Mount Elizabeth (35) 7–9 July 68, 2 ♀ (21.7, 21.7 g). Mount Remarkable (37) 25 July 68, 2 ♀ (22.3, 22.5 g). Wyndham (39) 7 Aug 68, 1 ♀. NORTHERN TERRITORY. Mount Roper (45), 1 ♂ (20.9 g). QUEENSLAND. Moonlight Creek (47) 2–5 May 64, 1 ♂ (22.9 g), 2 ø (22.5, 23.0 g). Glenore (49) 20 Apr 64, 4 ♂ (23.6, 24.1, 24.4, 24.7 g), 1 ♀ (23.0 g), 1 imm ♀, 2 imm ø (23.1, 23.3 g). Inkerman (50) 25 May – 3 June 64, 1 ♂ (24.0 g), 2 ♀ (22.8, 24.1 g), 1 ø (23.8 g). Glendower (55) 18 Mar – 1 Apr 64, 7 ♂ (22.6, 23.5, 23.7, 23.8, 23.9, 24.3, 25.8 g), 1 ♀ (24.8 g), 3 ø (21.4, 21.6, 23.2 g). Thylungra (60) 28–29 May 63, 4 ♂, 3 ♀, 2 ø, 1 imm ♀, 1 fl. SOUTH AUSTRALIA. Bool Lagoon (87) 14 Jan 63, 3 ♂, 4 ♀ 2 ø, 1 imm ♂, 1 ø skel. NEW SOUTH WALES. Tenterfield (95) 15 Apr 63, 1 ♂, 1 ø.

Distribution & Abundance. Locally common where collected. In Western Australia seen at Coolawanyah (4) and throughout most of the Kimberleys (27, 29, 30, 31, 33, 36). At Mount Bell (32) population estimates varied from 15 birds in half an acre to 50 birds in forty acres. Usually flushed singly or in pairs, maximum eight, and considered fairly abundant, though well spaced out throughout habitat.

In Northern Territory a few birds seen near Borroloola (46). In northern Queensland common at Glendower (55) but not encountered on the east coast (51-54, 56-57) nor in the dry interior of South and Western Australia.

Habitat. Generally open areas of short grass or spinifex with isolated trees and bushes (1, 3, 28, 47, 50, 60). At Mount Anderson (28) among savanna grassland with scattered acacia: "they very much favour grasslands where small mud-pans and pebble-flats occur.........they must pick up a good deal of food from such situations; easy to pick up wind-blown seeds which tend to gather round the margins of these clearings" (BB). At Wyndham (39) where they were living on the coast they occupied thinly grassed mud-flats that were probably affected by very high tides.

In northern Queensland (47, 49) found in tropical savanna woodland with tussock-grass clearings, and further south (60) in open black soil plains with mixed grasses up to three feet high. Here, too, they were in savanna and showed preference for pockets of dry, rank herbage – a typical habitat of non-breeding birds.

In South Australia (87) found in cultivated paddocks planted with *Phalaris* grass for cattle. Along the Roper river (45) it was noted (BB) that where Bushlarks were occupying the short grass flats beside the river they were replaced in the more barren ground by the pipit *Anthus novaeseelandiae* (BB): the same was true in northern Queensland.

Behaviour. Normally found on the ground. When flushed flew low with fluttering of wings and jerky movements for rarely more than 50 yards before diving down for cover again, sometimes uttering short twitterings every few seconds. Occasionally settled on post or lower branches of dead tree. Full song could be heard from dawn until well after sunset: sometimes uttered from a low fence post, sometimes when flying high in the air, almost hovering. In the Chichester Range (3) birds were mostly paired and evidently breeding. Birds at Bool Lagoon (87), South Australia, were apparently paired or single but not performing territorial displays.

Food. Stomachs contained seeds from grasses and spinifex, insects, especially ants, grasshoppers and caterpillars.

Annual Cycle. Specimens from Chichester Range (3) were the only ones whose behaviour was typical of breeding birds (see above). They all showed enlarged or enlarging gonads and one female had a large egg yolk in the oviduct. They have either completed or were in the final stages of moult, plumage being generally fresh. The female with the egg yolk has some new and some old feathers and shows no visible sign of pin feathers. This may indicate a case of 'arrested' moult, suggesting that possibly, in the drier areas at least, Bushlarks cease moult with the onset of breeding conditions that depend on local rainfall and resume this activity upon completion of the reproductive cycle. Further north in the Kimberleys (28, 32, 34, 35, 37, 39) birds showed no gonad development and are either in fairly fresh plumage or some stage of moult. The bird from Mount Roper (45) is in fresh plumage while those from Queensland (47, 49, 50, 55, 60) are in rather worn plumage, with moult, and showed no significant gonad activity. Adults from Bool Lagoon (87) showed some gonad activity and are in rather worn plumage with slight indications of moult on head, back and tail.

Immatures, with a more 'scaly' plumage, especially about the head and back, were collected at Karratha (1) and Mount Anderson (28) in May, Glenore (49) in April, Bool Lagoon (87) in January and Thylungra (60) in May. All but one undetermined (1) have the skull recorded as not fully pneumatised. Others in comparable plumage but with skulls recorded as fully pneumatised are probably sub-adult and were collected at Chichester Range (3) in May (breeding), Mount Bell (32) in June, Mount Elizabeth (35) in July and Glendower (55) in March.

Impermanent Colours. *Legs* flesh-pink to pale grey or occasionally brownish-grey. *Iris* yellow brown to dark brown. *Bill* upper mandible black, dark brown, dark horn (grey), tomium white or yellowish: lower mandible pale, usually with darker tip. *Mouth* bright yellow, yellow, orange, rosy or white.

Geographical Variation. The races of *Mirafra javanica* in Australia have been reviewed by Mayr & McEvey (1960) and for the most part our specimens show the same trends as they described and their names have been used. In size there is little variation, either geographical or sexual (mean wings 30 ♂♂ 74.1 mm, 21 ♀♀ 73.6). In colour there is a general tendency for western birds to be more rufous than eastern but there is individual variation in this as in the degree of throat spotting and the size of dark centres to feathers of the back.

Seven birds from Chichester Range (3) can be ascribed to the race *woodwardi,* though three of them show tendencies towards Kimberley birds. These are paler at Mount Anderson (28), darker at Mount Bell (32) and then progressively less rufous eastwards (34, 35, 37) to Wyndham (39). The central Kimberley (32–35) birds represent typical *subrufescens* while western birds (28) show resemblances to the paler *halli* of the Roebuck Bay Plains area, and the eastern (37) birds are close to the dark grey *forresti* of which the Wyndham (39) female is typical. Mayr & McEvey extend the range of *subrufescens* from the Kimberleys to the Gulf of Carpentaria, including the Roper River (45). Our single specimen from this locality is very close to *forresti,* differing only in having a slight rufous wash below. Further specimens eastwards from Wyndham (39) would help define the range of *subrufescens* more clearly.

Many Queensland specimens are too worn for satisfactory comparison. Inkerman (50) birds are the darkest and represent an extension of range northwards. They are closest to *normantoni*. Thylungra (60) birds are pale, rather worn and best referred to *rufescens*, though some of the new feathers of these birds are as dark as those on Glendower (55) birds which come within the range of *queenslandica*, though throat spotting is variable. Further south, Tenterfield (95) birds agree closely with *horsfieldi* and birds from Bool Lagoon (87) belong to the distinct *keasti*.
Other notes. Soil samples were retained for specimens from Mount Anderson (28), Mount Bell (32), Mount Remarkable (37), Wyndham (39) and Mount Roper (45).

Alauda arvensis — Skylark

Specimens. SOUTH AUSTRALIA. Bool Lagoon (87) 14 Jan 63, 1 imm ♂, 1 imm ø.
Field Notes. Fairly plentiful in open grass paddocks at Furner (86), Nelson (88), and Albury (92). At Bool Lagoon (87) in association with *Mirafra javanica* and *Cinclorhamphus cruralis*.
Annual Cycle. Both specimens are in fresh immature plumage. One, which had a yellow gape flange, is paler than the other. Neither of the skulls was fully pneumatised and gonads showed no activity. Adults were found feeding young about four days old in nest at Nelson (88) in early January.
Impermanent Colours. *Legs* yellow. *Iris* brown. *Bill* dark horn. *Mouth* yellow. *Gape flange* (once) yellow.

HIRUNDINIDAE — Swallows & Martins

D. J. Freeman

Cheramoeca leucosternum — White-backed Swallow

Specimens. QUEENSLAND. Thylungra (60) 23 May 63, 1 ♂. SOUTH AUSTRALIA. Innamincka (64) 29 Apr 65, 2 ♀ (11.0, 12.6 g), 1 ♀ skel. Clifton Hills (65) 11 May 65, 1 ♀ (15.8 g).
Field Notes. Always either singly or in small parties hawking fairly low over country of sparse vegetation – spinifex, Bluebush, saltbush with some scattered trees. Seen also at other localities in the south and east (58, 60, 61, 66, 82, 94) and in the drier regions of Western Australia (19, 23, 25).
Food. Insects, including beetles.
Annual Cycle. No specimens showed any gonad activity and all skulls investigated were recorded as fully pneumatised. Birds from South Australia (64, 65) show some moult of primaries and rectrices.
Impermanent Colours. *Legs* grey, grey-brown or reddish-grey. *Iris* medium brown to black. *Bill* brownish black or black. *Mouth* fleshy pink, light pinkish-yellow or whitish, once purple on palate.
Geographical Variation. There are no recognised races. The female from Clifton Hills (65) was heavier and is also larger than the two from Innamincka (64), wing 107 against 96.5 and 98 mm.

Hirundo neoxena — Welcome Swallow

Specimens. WESTERN AUSTRALIA. Cocklebiddy (15) 8 – 9 July 65, 1 ♂, 1 ø skel. SOUTH AUSTRALIA. Kingoonya (70) 25 June 65, 1 ♀. NEW SOUTH WALES. Dunedoo (93) 18 Mar 63, 1 ♂ skel.

Distribution & Abundance. In Western Australia scattered along the south coast (15-17) becoming more concentrated in the southwest corner (9, 10, 13, 14). Further north recorded at camps 1-4. A few birds seen at Mount Anderson (28) during May were possibly late wintering *H. rustica*. In South Australia fairly common between Nonning (79) and Clifton Hills (65). Seen on the Nullarbor Plain and further east along the coast (74, 76, 88) to New South Wales at Albury (92) and Dunedoo (93).

In northern Queensland seen at Glendower (55) and at Proserpine near Mount Dryander (57).

Habitat. Most types of open country including agricultural woodland savanna (93), open woodland of mixed gum and acacia (15), saltbush (69, 70), Bluebush plain with occasional Mulga (70).

Behaviour. Some were flying in and out of potholes on the Nullarbor Plain inland from Eucla (17). One of these potholes, about ten feet across and six feet deep, was found to contain empty nests. Flight, habits and nests appeared identical with those of *H. rustica* (DG).

Food. Insects, including beetles and ants.

Annual Cycle. No specimen showed any gonad activity and neither of the birds skinned was in moult.

Impermanent Colours. *Legs* black. *Iris* dark brown to red-brown. *Bill* black. *Mouth* yellow.

Geographical Variation. Eastern birds *H. n. carterae* in the British Museum are similar in size to western *H. n. neoxena* but differ in having the rufous crown darker and slightly more extensive. The rufous throat and upper breast are also darker and more sharply demarcated from the lower breast. The steel-blue gloss on the back is variable and possibly attributable to plumage condition. The bird collected at Cocklebiddy (15), which is in the intermediate area between the two populations, shows the demarcation of eastern birds but lacks the colour intensity.

Petrochelidon nigricans — Tree-martin

Specimens. WESTERN AUSTRALIA. Coolawanyah (4) 20 May 66, 1 fl. Mount Barker (12) 21 – 27 Mar 66, 2 ♀, 1 ♀ skel, 3 fl. Mount Anderson (28) 10 – 18 May 68, 1 ♂ (15.3 g), 1 imm ♂ (12.8 g), 2 ♂ skel (14.8, 15.4 g), 2 ♀ skel (14.8, 16.1 g), 3 ø skel (14.2 g), 1 fl. SOUTH AUSTRALIA. Innamincka (64) 27 Apr 65, 1 ♂ (15.5 g), 1 ♀ (17.4 g), 1 ø. Etadunna (66) 17 May 65, 2 ♂. Oodnadatta (67) 31 May 65, 1 ♀, 1 ø. Aroona Valley (82) 8 Apr 65, 1 ♂ (14.9 g), 1 ♂ skel (16.5 g). Furner (86) 9 Jan 63, 1 ♀ skel.

Distribution & Abundance. Fairly common throughout most of South (64-68, 70, 82, 86) and Western Australia (1-4, 7, 14, 18, 23, 28). Seen either singly or in flocks of up to 20 birds. Once (12) about 150 were seen together. Also in Victoria (88) and New South Wales (93, 95).

In northern Queensland specimens were collected but not retained. Details as follows: Moonlight Creek (47) 4 May 64, 1 ♂ (14.1 g), 1 ♀ (13.9 g). Inkerman (50) 2 June 64, 1 ♂ possibly imm (12.4 g). Tully (54) 25 July 64, 2 ♂ (13.8, 14.2 g). Not recorded in Northern Territory.

Habitat. Usually open water, swampland or grassland in savanna woodland. Not always associated with water. In northern Queensland (50) seen flying over tropical woodland, grassland and clearings near rainforest (54) and around low offshore mangrove belt (47). In South Australia (66) flying over open sand-dune country with low herbage. On the Flinders Range (82) seen among eucalyptus at about 1,600 ft in Brachina Gorge.

Behaviour. Commonly seen in groups flying out low over water or grassland from open branches of dead trees up to 40 ft high. Noted (BB) to catch insects above water level (28) and never from the surface. Other birds recorded with this species were *P. ariel* (28), *Artamus minor* (4) and *Collocalia spodiopygia* (54). At Mount Barker (12) Tree-martins roosted on dead branches of the taller trees in a small swamp. A migrant came on board a ship passing through the Torres Straits during June 1963 (PRC).

Food. Small insects.

Annual Cycle. Specimens collected during March and May in Western Australia (12, 28) and one during January in South Australia (86) show body moult and had either completed, or were completing, the replacement of primaries and rectrices. Other specimens from South Australia (64, 66, 67, 82) are in fresh plumage except for one that was completing moult on rump and crown.

Of all specimens examined only one, a female from South Australia (67) showed any gonad activity.

The immature male from Mount Anderson (28) had a yellow gape and the skull recorded as not fully pneumatised. It was in moult and had a layer of subcutaneous fat. Secondaries on this bird and four others are edged white. They are also slightly duller than the remaining specimens. These may be indications of immaturity, although of three skulls examined one was partly pneumatised and two were fully so. The edging on the secondaries may well be retained until birds are fully adult for the amount of white varies individually among birds in different states of plumage.

Impermanent Colours. *Legs* pinkish-grey (imm?), brown and dark grey to black. *Iris* red-brown, dark brown to blackish-brown, once olive-grey. *Bill* black, once greyish-black. *Mouth* white to yellow and yellowy-orange, pink and a mottled grey, pink and yellow.

Geographical Variation. The race *neglecta* was described by Mathews as differing from nominate *nigricans* in its smaller size and duller coloration. Our specimens indicate that the dullness of western birds may be attributable to either plumage condition or immaturity. Eastern birds showing more gloss are in fresher plumage. Mees (1961a) discounted the colour differences but was still prepared to recognise *neglecta* on size alone. Our collection shows that the size difference is slight, eastern birds having wings ♂ 102–108 mm (105.7), ♀ 102–106 (103.9) against western ♂ 100–105 (102.4), ♀ 101–106 (102.6). Since sight records from central districts of South Australia (68) and western Australia (19, 23) suggest that the two populations are not segregated and the species is known to be partly nomadic it does not seem useful to recognise two subspecies.

Petrochelidon ariel Fairy Martin

Specimens. WESTERN AUSTRALIA. Mount Anderson (28) 10 – 15 May 68, 1 ♀ (10.5 g), 1 ø skel. Mount Caroline (33) 21 June 68, 1 ♂. Mount Elizabeth (35) 13 July 68, 1 ♂ (10.0 g), 1 ♀. QUEENSLAND. Tully (54) 25 July 64, 3 ♂ (9.8, 9.8, 10.5 g), 1 ♀ (9.9 g), 1 ø (9.1 g). Monkira (61) 5 June 63, 1 ♀.
Distribution & Abundance. In Western Australia common only at Mount Anderson (28). A few seen in central Kimberleys (33, 35) and on the south coast at Cocklebiddy (15) and Eyre (16). In Northern Territory small numbers recorded at Darwin (41), Roper River (45) and Borroloola (46). Seen also in Queensland at Glendower (55) and Thylungra (60), and in coastal South Australia at Furner (86) and inland at Nonning (79).
Habitat. Open grass and woodland, usually near a creek or lagoon. At Tully (54) over grassland between rainforest and open tea-tree forest.
Behaviour. Generally perching in trees and flying out over surrounding water or grassland to catch insects on the wing. In association with *P. nigricans* throughout the north, and at Tully (54) both species were in association with *Collocalia spodiopygia.*
Food. Insects.
Annual Cycle. A nest with eggs was found at Borroloola (46) in November. One male from Tully (54) with testes slightly enlarged was the only specimen which showed any gonad activity. Of two birds collected together at Mount Anderson (28) one is in fairly fresh plumage whereas the other was completing moult of primaries and rectrices. Central Kimberley birds (33, 35) are in fresh plumage while those from Tully (54) show signs of moult on the crown and secondaries. The bird from Monkira (61) is in fresh plumage.
Impermanent Colours. *Legs* dull purple, slate, dark grey, dark brown or black. *Iris* dark grey-brown, brown, or once, black. *Bill* dark grey or black. *Mouth* cream, light yellow or lemon but also various combinations – mottled grey, pink and yellow; pinkish and yellowish grey.
Geographical Variation. The specimen from southwest Queensland (61) differs from the rest in having the crown a deeper russet colour. This may be due to fresher plumage or evasion of the harsher bleaching effects of the sun in lower latitudes. There are no recognised races in Australia.

MOTACILLIDAE — Pipits

B. P. Hall

Anthus novaeseelandiae Australian Pipit

Specimens. WESTERN AUSTRALIA. Chichester Range (3) 13 May 66, 1 ♂, 1 ♀, 1 ♂ skel, 1 fl. Scott River (11) 1 Mar 66, 1 ♂, 1 imm ♀, 1 ♀ skel, 1 fl. Cocklebiddy (15) 8 – 9 July 65, 1 ♂, 1 fl. Eucla (17) 2 July 65, 3 ♂, 1 ♀ skel. Mount Shenton (18) 2 – 3 Aug 65, 1 ♂, 1 ♀ skel. Carnegie (19) 17 Sept 65, 1 ♀. Warburton (23) 12 Aug 65, 1 ♀. Giles (25) 23 – 31 Aug 66, 2 ♀. St George's Range (27) 8 June 68, 1 imm ø. Wyndham (39) 6 Aug

68, 1 ♀, 1 fl. NORTHERN TERRITORY. Darwin (41) 3 – 5 Sept 68, 1 imm ♂, 1 fl. QUEENSLAND. Moonlight Creek (51) 4 – 16 May 64, 2 ♂ (24.5, 26.3 g), 2 ♀ (21.7, 23.7 g), 1 ø (19.3 g), 1 imm ♂ (22.4 g), 1 imm ø (20.2 g). Inkerman (50) 27 May – 6 June 64, 4 ♀ (21.5, 22.2, 23.2, 25.0 g), 1 imm ♀ (20.3 g). Glendower (55) 24 Mar 64, 2 ♂ (25.9, 28.7 g). Banoona (58) 29 Apr 63, 1 ♂, 1 ♀. Thylungra (60) 24 May 63. Monkira (61) 4 – 7 June 63, 1 ♂, 1 ♀. SOUTH AUSTRALIA. Cadelga (62) 7 May 65, 1 ♀. Clifton Hills (65) 11 May 65, 1 ♂. Etadunna (66) 18 – 20 May 65, 1 ♀, 1 ♀ skel. 1 fl. Oodnadatta (67) 31 May 65, 1 ♂. Granite Downs (68) 7 June 65, 1 ♂. Nullarbor HS (69) 4 – 8 Oct 63, 4 ♀: 25 – 29 Oct 64, 2 ♂, 1 ♀, 1 ø. Kingoonya (70) 20 – 24 June 65, 2 ♀, 1 ø. Aroona Valley (82) 8 Apr 65, 1 ♀ (24.5 g). NEW SOUTH WALES. Albury (92) 9 – 13 Mar 63, 2 ♂, 2 imm ø. Dunedoo (93) 24 Mar 63, 1 imm ♂. Tenterfield (95) 15 Apr 63, 3 ♂, 1 ♀.

Specimens to other Museums. NMV 1 ♂ Thylungra (60), 1 ♂ Monkira (61), 1 ♂ Albury (92), 1 ♂ Tenterfield (95). SAM 1 ♂ Nullarbor HS (69).

Distribution & Abundance. Found in all main camps in the southern half of the continent except Metricup (9) and Mount Barker (12). Less common in the centre (19-25 and 65-68) than further south or at Innamincka (64). Not common in the Kimberleys (27-38) nor in Northern Territory though recorded (43, 45): in northern Queensland encountered only where collected.

Habitat. All types of open country including bare gibber (65), spinifex (25), mulga (18), Bluebush and saltbush (15), grass near coastal dunes (4), saltpans (47), cultivated land (95). At Mount Roper (45) and Glendower (55) found near to *Mirafra javanica* but on barer ground. A similar ecological distinction was found elsewhere in Queensland and New South Wales where both species occurred (54, 60, 95) but in Western Australia near Coolawanyah (4) and at Karratha (1) both birds were together in new grass and spinifex.

Behaviour. Normally found on the ground, but perches readily in bushes or on fence posts. Usually single or in pairs, but small parties or flocks recorded at various camps (15, 17, 59, 60, 61).

Food. Insects, including grasshoppers, caterpillars and beetles, form the staple diet, but many stomachs also contained grit and seeds of various plants, including the introduced *Echium plantagineum* (Boraginaceae).

Annual Cycle. At Coolawanyah (4) a nest of half-grown young was found, and at Karratha (1) adults were carrying food to young (Harrison & Colston 1969). The only other positive indication of breeding seasons is given by immatures from southwestern Australian (11) in March, northwestern Queensland (51, 54) in May and early June, and in New South Wales (92, 93) in March. All adults from the same places were in moult or fresh plumage. Birds from eastern and central Western Australia (10-25) in July, August and September had the gonads enlarged or slightly enlarged and are in worn plumage or starting moult (18, 25). Four out of five from southern Queensland (60, 61) in May and early June had very large gonads and all five were in full moult.

Most others were in some stage of moult with no gonad activity.

Impermanent Colours. *Legs* pale sandy flesh to yellow-brown. *Iris* dark brown. *Bill* upper mandible dark horn or brown: lower mandible fleshy-pink or whitish with dark tip. *Mouth* yellow or pinkish.

Geographical Variation. Little variation in size, the wings of the whole series falling between ♂ 83-90 and ♀ 77-85 mm, but with southern birds averaging larger with longer bills. Birds of the northeast and centre, on the other hand, tend to have longer hind claws.

The darkest and greyest birds are those of the north (39, 41, 51, 54) and the extreme southwest (11), while all from New South Wales are also dark but browner. The remainder are on the whole paler and more sandy.

Our series thus falls into the subspecies recognised by Mayr (in Peters' 1960) as follows: *A. n. rogersi* northern Queensland (51, 54), Northern Territory (41) and extreme northeastern Western Australia (39) dark, grey averaging small with a long hind claw. *A. n. bilbali* southwestern Australia (11): dark grey, averaging larger than *rogersi* with a shorter hind claw. *A. n. australis* New South Wales (92-95): dark, rich brown, averaging large. *A. n. subaustralis* everywhere else (3, 15-27, 55-82): paler and more sandy.
Parasites. See Appendix 1.

CAMPEPHAGIDAE — Cuckoo-shrikes
D. J. Freeman

Pteropodocys maxima Ground Cuckoo-shrike

Specimens. Western Australia. Mount Remarkable (37) 26 – 27 July 68, 2 ♀ (121.5, 121.5 g).
Distribution & Abundance. Usually single or in small parties of up to five birds. In Western Australia seen at Chichester Range (3), Coolawanyah (4) and at Mount Shenton (18); five together at Mount Elizabeth (35); three at Gibb River homestead, between Joint Hill (34) and Mount Elizabeth (35); also seen between Mount Remarkable (37) and Lissadell Hill (38). In South Australia fairly common at Nonning (79). In Queensland a few noted at Thylungra (60) and Monkira (61). In New South Wales six on the west side of the range at Tenterfield (95).
Habitat. Dry open country. Savanna grassland with isolated trees and rocky out-crops (37); large open plains with areas of pebble and bare earth flats (37).
Behaviour. Shy and alert. Usually first seen on ground and when flushed rise steeply to a little below tree-top level, or will settle momentarily in the topmost branches before making off again. Flight call, *queel.*
Food. Grasshoppers, pieces of grass and small black, chitinous remains also found in stomachs.
Annual Cycle. Neither bird had the ovary enlarged and both skulls were recorded as fully pneumatised. Both show some body moult, especially on the upper parts. Tails are very worn and one bird was replacing central rectrices. Primaries show some wear and secondaries were being replaced, with old and worn feathers conspicuous.
Impermanent Colours. *Legs* grey and grey-black. *Iris* mid-brown and whitish-grey. *Bill* black. *Mouth* white. The considerable difference between iris colouring suggests that one bird might be immature but there is no other evidence to support this.

Geographical Variation. There is insufficient material in the British Museum for subspecific determination, Peters (1960) recognises two races, according to which our specimens would be grouped with *P. m. pallida* from the north. Keast (1958h), however, suggests that geographical variation is virtually absent though birds from hotter areas may be slightly paler.

Coracina novaehollandiae Black-faced Cuckoo-shrike

Specimens. Western Australia. Karratha (1) 6 June 66, 1 ♀. Chichester Range (3) 12 – 15 May 66, 2 imm ♂. Coolawanyah (4) 19 – 21 May 66, 1 imm ♂, 2 imm ♀. Perth (7) 12 Feb 66, 1 imm ♂ skel. Scott River (11) 12 – 26 Feb 66, 1 ♀, 1 imm ♂. Mount Shenton (18) 30 July 65, 1 ♂. Carnegie (19) 16 – 17 Sept 65, 2 ♂. Mount Anderson (28) 1 – 18 May 68, 1 ø, 1 imm ♀, 1 imm ø, 1 ø skel, 3 fl. Derby (29) 21 May 68, 1 ø skel. Geikie Gorge (31) 2 – 5 June 68, 1 ♂, 1 imm ♂ (101.1 g), 1 imm ♀. Mount Caroline (33) 24 June 68, 1 ♂ (115.3 g). Mount Elizabeth (35) 15 July 68, 1 fl (119.0 g). Mount Remarkable (37) 26 July 68, 1 ♀. Lissadell Hill (38) 28 July 68, 2 fl (117.5, 118.4 g). Wyndham (39) 6 Aug 68, 1 fl (115.1 g). Northern Territory. Borroloola (46) 8 – 12 Nov 68, 1 ♂ skel (105.2 g), 1 fl (116.7 g). Queensland. Thylungra (60) 26 May 63, 1 ♂. Monkira (61) 7 June 63, 1 imm ♀. South Australia. Etadunna (66) 20 May 66, 1 ♂. Angepena HS (84) 22 Nov 63, 1 imm ♂. Albury (92) 11 – 13 Mar 63, 1 ♂, 1 imm ♂, 1 imm ♀, 1 imm ø, 1 fl. New South Wales. Dunedoo (93) 18 Mar 63, 2 ♂.

Specimens to other Museums. NMV 1 ♂ skel Albury (92). SAM 1 ♂ Nullarbor HS (69) 30 Oct 64.

Distribution & Abundance. Common or fairly common throughout its range, whether singly, in small parties or large groups. In Western Australia seen also at camps 2, 8-14, 21, 25 and 27-39. Present at all camps in Northern Territory except Oenpelli (43); and also at all in New South Wales and in Queensland where especially common along the Diamantina River (61). In Victoria at Victoria Range (90) only. Widespread throughout South Australia (65-67, 70, 82, 86, 88).

Habitat. Almost any type of country with trees, from Mulga (18, 19) to thick mangroves.

Behaviour. Shows preference for "look out" spots in tall trees and when disturbed makes off fairly high with characteristic undulating and gliding flight. High flying birds will often drop steeply to settle in upper foliage. Calls were varied and heard in flight or from tree tops: they included a loud, but not deep or harsh *cree-eu-cree-eu* uttered by one bird after its mate was shot, and various shrieks or croaks.

Occasionally with babblers (18) and once (28) in loose association with *Lalage sueurii*, *Chrysococcyx* sp., *Artamus leucorhynchus*, *A. cinereus*, *Geopelia cuneata*, *G. striata*, and *Climacteris melanura*, in a party some fifty in number, moving through trees together and keeping fairly high (WHB). The mangrove birds (39) were making their way through the taller trees, settling momentarily on the topmost branches. Feeding activities twice observed; one bird was seen banging a locust on its perch before eating it and another was seen flying down from tree to seize large insects. Rain-bathing recorded (CJOH) near Perth (7).

Food. Insects including grasshoppers and beetles, vegetable matter and seeds. One crop (7) was packed full with a compact nutmeg-sized lump of leaf-like material and some black chitinous fragments.

Annual Cycle. Four birds moulting from juvenile to first-year plumage were from Angepena (84) in November, Albury (92) in March and Mount Anderson (28) in May. They are well advanced in this moult showing only scattered brown juvenile feathers mainly on the head, back and secondaries; primaries and rectrices, not to be replaced in this moult, are fairly fresh. None showed any gonad activity and all but one had the skull recorded as fully pneumatised. One first-year bird collected in May (28) shows no body moult: four in March (92), May (3, 4) and June (61) show some body moult: three in May (3, 4) and June (3) are more advanced with the black throat of the adult plumage visible: two in May (4) and June (31) show the same characters but are also replacing primaries: one in February (11) is in almost complete adult plumage with a few scattered light feathers persisting on the throat, only the outer primaries to be changed and the rectrices, unlike those of any other young bird, being replaced.

Adults with inactive gonads, little or no body moult and plumage showing wear, mainly on quills, were from Western Australia in May (28), June (31, 33) and July (18, 37) and from Queensland and South Australia in May (60, 66). The New South Wales (92) March bird showed no gonad activity and is in fresh plumage with the outer two primaries half grown, the remainder fully grown. Three birds that showed slight gonad development and heavy body moult were from Western Australia in February (11) and June (1), and from New South Wales (93) in March. Two September males from Carnegie (19) had enlarged testes and are rather worn, just beginning body moult, perhaps illustrating the ability of desert birds to moult and breed simultaneously (Immelmann, 1963).

Impermanent Colours. *Legs* black, dark grey; immature pale grey. *Iris* dark brown. *Bill* black, often grey (once reddish) towards base of lower mandible; immature pale slate-grey. *Mouth* white, to pinkish-grey.

Geographical Variation. Keast (1958h) reviewed the species in Australia and separated four subspecies mainly to represent clinal extremes of wing and bill size. Mees (1961a), however, suggests that Keast's extremes are exaggerated and do not allow sufficiently for individual variation, and on the mainland he recognises only *C. n. melanops* and the distinct race *subpallida* of northwest Western Australia.

On the whole our specimens support Mees. The Karratha (1) female (wing 198: bill 32 mm) is the palest and represents typical *subpallida*. Nearly as pale are two males from Carnegie (19) (wing 198-200; bill 29.5-32.5 mm) and non-breeding birds from the Kimberleys (28-37) (wings ♂ 190-195, 188 mm; bill ♂ ♀ 30-31), all of which could be regarded as intermediates between *subpallida* and *melanops*. Further south the Mount Shenton (18) male is darker and longer-winged (wing 207; bill 30 mm) and the Scott River (11) female (wing 196; bill 29.5 mm) darker again. Eastern birds (60, 61, 66, 92, 93) are generally darker and larger than western (wings ♂ 195-203 mm, ♀ 188-197 mm; bill ♂ ♀ 28-33 mm).

Thus there seem to be two populations within Australia, the sedentary *subpallida* occupying the central and western arid areas, and the migratory *novaehollandiae* from the eastern, northern and southern wetter areas, showing some clinal variation in wing and bill size.

Coracina papuensis (including *robusta*) Little Cuckoo-shrike

Specimens. WESTERN AUSTRALIA. Geikie Gorge (31) 3 June 68, 1 ♀ (60.0 g), 1 fl. Mount Bell (32) 15 – 17 June 68, 2 ♂ (66.7 g), 3 ♀ (67.9 g), 1 imm ♀ (56.7 g). Mount Caroline (33) 24 June 68, 2 ♂ (67.6, 68.6 g), 1 ♀ (60.5 g), 1 fl. Mount Elizabeth (35) 10 July 68, 1 fl (67.8 g). Lissadell Hill (38) 28 July 68, 2 fl (55.9, 63.7 g). NORTHERN TERRITORY. Darwin (41) 7 – 11 Sept 68, 1 ♂ (62.2 g), 2 ♀ (60.7, 63.8 g). Borroloola (46) 12 Nov 68, 1 ♀ skel (60.8 g). QUEENSLAND. Moonlight Creek (47) 1 – 15 May 64, 2 ♂ (61.7 g), 3 ♀ (58.6, 66.9 g), 4 imm ♀ (55.4, 57.7, 58.0, 58.1 g), 1 imm ø (58.1 g). Glenore (49) 23 Apr 64, 1 imm ♀ (55.9 g). Inkerman (50) 27 May – 9 June 64, 3 ♂ (70.6, 74.7, 75.8 g), 3 ♀ (64.5, 73.3, 73.5 g), 2 imm ♀ (56.9, 65.1 g). Ayton (52) 23 – 24 June 64, 1 ♂ (73.1 g), 1 imm ♀ (63.0 g). Tully (54) 14 – 20 July 64, 4 ♂ (64.9, 67.0, 69.0, 72.4 g), 1 ♀ (69.2 g). Glendower (55) 13 – 17 Mar 64, 1 imm ♀? (56.4 g), 2 imm ø (53.5, 58.8 g). Cape Gloucester (56) 6 – 8 Aug 64, 1 ♂ (63.9 g), 2 imm ♂ (53.0, 59.1 g). Mount Dryander (57) 11 – 14 Aug 64, 5 ♂ (54.3, 55.4, 58.6, 62.3, 63.5 g), 2 imm ♂ (52.0, 54.1 g). NEW SOUTH WALES. Grafton (94) 30 Mar – 1 Apr 63, 1 ♂ (70.0 g), 1 ♀, 1 imm ø (60.0 g).

Distribution & Abundance. Generally singly, in pairs or small parties of up to five birds. In Western Australia a few seen at Mount Anderson (28) and where collected in central Kimberleys (31-35). In Northern Territory very common at Darwin (41), fairly numerous at Pine Creek (44) and a few elsewhere (43, 45, 66). Seen at Glendower (55) but not at Wernadinga (48) or in southern Queensland and New South Wales, except at Grafton (94).

Habitat. Most types of woodland throughout range. In Western Australia (31, 32) eucalypt woodland by river and in rocky gorge country. More open grassland further east (35). In northern Queensland found in thicker vegetation, tall trees with thick layers of shrubs or flowering tea-trees. At Inkerman (50) in a variety of habitats including dense river mangrove adjacent to open grassland, densely timbered "island" surrounded by mud flats, and homestead garden with shrubs and layered vegetation. Also found in riverine forest (55). In New South Wales (94) in open forest country.

Behaviour. Voice recorded as short single, squeaky whistle declining at end of note; an undulating shriek; a high-pitched *keer-serk* uttered at minute intervals; also a soft churring. On two occasions the calls of injured birds attracted others from nearby. In northern Queensland never seen far from the much more numerous *C. novaehollandiae*. Once seen (50) chasing *Geopelia striata* (SAP). Birds were seen banging grasshoppers on a perch to kill them, catching insects on the wing and feeding on insects in the foliage.

Food. Stomachs contained remains of grasshoppers, ants, beetles, a two inch long dragonfly, a wasp and a green mantis. Also seeds and berry stems.

Annual Cycle. Of ten young birds retaining brown juvenile feathers one in March (94) had them persisting on the head, neck, back and upper breast. The remainder in

April (49), May (47), June (32, 33) and August (56) had virtually completed this moult into first year plumage and retained only one or two of such feathers on the rump. All these birds had juvenile quills, three had fully pneumatised and three not fully pneumatised skulls. Gonads were either too small for determination or showed no activity. Four birds in immature (first year) plumage in May (47), June (50) and August (57) showed no moult and are extensively worn on primaries and rectrices. Five immatures in May (50), June (52), and March (55) show varying degrees of moult to adult plumage. Either body moult was advanced or quills were being replaced. Three have skulls recorded as fully pneumatised.

Of nine adults from the Kimberleys (31-33) seven showed no gonad activity and were in fairly fresh plumage with slight body moult. The two others (of five from Mount Bell (32)), showed some gonad activity and are more worn, without moult. All but one (31) had skulls recorded as fully pneumatised. Northern Territory (41) adults showed some gonad development and were not as advanced in moult as Kimberley birds though collected later in the season. Adults from Inkerman (50) and Ayton (52) showed slight gonad activity and are in fairly fresh plumage. Body moult was well advanced and primaries and rectrices either new or being replaced. Moonlight Creek (47) and Tully (54) birds are in comparable plumage and gonad condition, though individuals show more body wear. One male (56) showed some body moult and had slightly enlarged gonads. Of six birds from Mount Dryander (57) two showed gonad activity: all had some body moult and wings and tail rather worn in appearance.

Two adults were seen feeding young on the wing at Grafton (94) during late March and early April (PRC).

Impermanent Colours. *Legs* black, dark grey with grey or ochraceous soles: in juveniles appearing brown in the skin (see Galbraith, 1969). *Iris* brown. *Bill* adults black; immatures very dark brown or black with a brown base to the lower mandible. *Mouth* adults black with some pink or grey in the throat; immatures, mainly whitish (see Galbraith, 1969).

Geographical Variation. Galbraith (1969) has shown that the dark-breasted *robusta* of eastern Australia is best regarded as conspecific with the northern pale-breasted *C. papuensis*. The two birds had previously been separated specifically on the melanism amongst immature *robusta* and a supposed overlap in ranges. Galbraith demonstrates that the melanism is variable and not necessarily confined to a particular plumage and that there is no evidence to support any overlap of breeding ranges.

C. p. hypoleuca, supposedly characterised by a pure white breast (Keast, 1958h), occupies northern Western Australia, Northern Territory and northwestern Queensland. Our specimens from within this range show local, and some individual, variation. Kimberley birds, including the immature, have distinct white throats with some grey on the breast: at Darwin (41) birds are darker above and on the breast: those from Moonlight Creek (47), including the young birds, are the whitest on the breast. There is an abrupt transition across the Carpentaria Gap: adults have some pale grey on the breast and are larger and heavier than those from Moonlight Creek (47) while young birds (49, 50) have even darker breasts with some barring, though one is rather paler than the other two. The amount of grey on the breast in adults as well as young increases southwards. Thus birds from Tully (54) are slightly darker than those from the Gulf (49, 50) and Ayton (52), but all can be assigned to *C. p. stalkeri*. Further south (56, 57) the darkness is more apparent with only reduced amounts of white remaining on the chin and in some specimens faint barring on the breast is noticeable. The New South Wales (94) bird is darker still with the white on the throat almost negligible. Southern birds (56, 57, 94) are *C. p. robusta*.

Other notes. ICJG adds: Dr G. M. Storr has pointed out to me that the abnormal melanic female recorded as from Bowen at 20° S (Galbraith, 1969) is actually from Port Bowen, now Port Clinton, at 22°29′S. This correction still places it far to the north of any other melanic specimen, but removes it from the Burdekin area so that interaction with *stalkeri* cannot be involved.

Parasites. See Appendix 1.

Coracina lineata — Barred Cuckoo-shrike

Field Notes. Encountered twice only. One bird collected but not retained, Big Tableland (51) 2 July 64, 1 ♂ (100.0 g). This bird was one of two at the edge of dense rainforest, 60 feet up in tall tree, calling amongst foliage. Voice a repetitive *chirp*. Several birds also seen on 29 July in fruiting tree in municipal caravan park at Tully (54) well away from forest.

Food. Stomach contained two large fruits, small berries and four orange and black winged insects.

Annual Cycle. Gonads not enlarged, skull recorded as fully pneumatised.

Impermanent Colours. *Legs* black, soles ochraceous-brown. *Iris* whitish-lemon. *Bill* black. *Mouth* dull pink, distally black.

Edoliisoma tenuirostre — Cicada-bird

Specimens. NORTHERN TERRITORY. Darwin (41) 11 – 20 Sept 68, 1 ♂ (66.5 g), 1 ♀ (65.6 g).

Field Notes. In Northern Territory seen only at Hope Inlet (41) in tall, thick mangroves and monsoonal forest adjacent to mangroves.

In northern Queensland single birds were seen in the main rainforest block (51–54).

Food. Stomachs contained insect pulp and a few seeds.

Annual Cycle. Both birds had inactive gonads. The male had the skull recorded as fully pneumatised, the female only partly so. The female is in fresh plumage with slight body moult and the tail nearing completion; the male less advanced, has some body moult and the rectrices rather worn. At Tenterfield (95) a local observer pointed out a nest in a eucalypt in dry sclerophyll forest at about 2,200 ft. Apparently it had contained young during late March – early April.

Impermanent Colours. *Legs* black with buff or yellow soles. *Iris* dark brown. *Bill* black or blackish-brown sometimes paler or streaked white on lower mandible. *Mouth* blackish, dark blue-grey and blackish brown.

Geographical Variation. Our birds are referable to *E. t. melvillensis*.

Lalage sueurii White-winged Triller

Specimens. WESTERN AUSTRALIA. Chichester Range (3) 17 May 66, 1 fl. Coolawanyah (4) 15 May 66, 2 ♀ skel. Mount Shenton (18) 31 July 65, 1 ♂. Carnegie (19) 17 Sept 65, 1 ♀, 1 ø, 1 ♀ skel. Lake Gruszka (21) 10 Sept 65, 1 ♂. Warburton (23) 12 Aug 65, 2 ♂. Van der Linden Lakes (24) 6 Sept 65, 1 ♀. Giles (25) 31 Aug 65, 1 ♀. Mount Anderson (28) 28 Apr – 14 May 68, 4 ♂ (24.1 g), 2 ø, 1 imm ø, 1 ♂ skel, 2 ø skel. Joint Hill (34) 27 June – 3 July 68, 1 ♀ (22.9 g), 1 ♂ skel. Mount Elizabeth (35) 14 July 68, 2 ♂ (25.3, 26.2 g), 2 fl (19.3, 23.8 g). Lissadell Hill (38) 29 July 68, 1 ♂ (23.2 g). NORTHERN TERRITORY. Oenpelli (43) 1 Oct 68, 1 fl (21.6 g). QUEENSLAND. Glenore (49) 11 Apr 64, 1 imm ø (22.8 g). Inkerman (50) 3 June 64, 1 ♂ (24.5 g). Glendower (55) 18 Mar 64, 1 imm ø (22.9 g). Tyrone (59) 15 May 63, 1 ø. Thylungra (60) 24 May 63, 1 fl. Monkira (61) 2 June 63, 1 ♂. SOUTH AUSTRALIA. Nullarbor HS (69) 8 – 27 Oct 64, 1 ♂, 1 imm ♀. Kingoonya (70) 19 June 65, 1 ♂. Angepena (84) 20 Nov 63, 1 ♀. Coleraine (89) 22 Jan 63, 1 fl.

Distribution & Abundance. Recorded also in Western Australia at Perth (7), Mungilli Claypan (20), Lake Gruszka (21) and in the Kimberleys (21-37). In Northern Territory one seen at Mount Roper (45) and at Borroloola (46). In Queensland fairly common at Glenore (49), Inkerman (50) and Glendower (55).

Constantly seen throughout most of its range, though in small numbers and with individuals or pairs well spaced out. Only at Mount Anderson (28) and Tyrone (59) was it present in any numbers with some flocks of up to 12 or 15 birds.

Habitat. In northern Queensland found only in wetter regions, such as tropical savanna woodland and riverine forest. In Northern Territory (43) in monsoon forest edge to savanna grassland and black soil plain.

Throughout the remainder of its range found in fairly dry, open country usually near water. In low gums and scrub along creek running through spinifex (3); in creek bed with green vegetation in sparse Mulga woodland (18); bloodwoods and low bushes (19); on high dead branches of trees and shrubs in sandhills (24); savanna grassland with flowing creek (28); Mulga woodland with low sandalwood bushes and long grass (59); open Bluebush plain with scattered clumps of trees (70) and low scrub with open forest (84, 89).

Behaviour. Often found feeding on the ground or perched low down in bushes or trees. If disturbed will move into the higher reaches of trees before reluctantly taking to wing. Flight undulating and direct. In northern Queensland (49, 50), where undergrowth possibly too thick to permit satisfactory ground feeding, seen feeding amongst tea-tree blossom and upon flies in upper foliage. At Mount Elizabeth (35) in loose association with *Artamus cinereus, A. minor, Pachycephala rufiventris* and various honeyeaters, especially *Conopophila rufogularis* and *Lichmera indistincta.* At Mount Shenton (18) one pair amongst a party of *Petroica cucullata,* a *Chrysococcyx* sp., *Aphelocephala* sp.; and *Pachycephala rufiventris.* Voice a very sweet resonant *che'uu, che'uu, che'uu,* sometimes prefaced by a low piping note almost certainly made by this species though a bronze cuckoo was nearby (DG). Comparisons between this triller and cuckoos were made on several occasions. Cuckoo-like flight and single *peep* noted (WHB) at Mount

Anderson (28), and at Kingoonya (70) "hopping on ground seeking food, appearing rather short-legged and hence cuckoo-like" (DG). In mid-November in the Darling Range (7) a male was seen chasing a female through the trees singing all the time. The song was a repetitive trill *teu-teu-teu* followed by a *stit-stit-stit* and the courtship flight was buoyant and floating (PRC). Subsequently a hen was discovered incubating eggs on 25 November.

Food. Stomachs contained insects, grasshoppers, caterpillars, grubs, bees, leaf beetles, small wing cases of *Coleoptera,* vegetable matter and seeds.

Annual Cycle. Nesting near Perth (7) in November, but the breeding season probably protracted and varying slightly in different parts of the country. In the Kimberleys a young bird undergoing first moult was collected in April (28), and one was moulting from immature to first year plumage in May (28). Males were moulting into eclipse dress (see Mayr, 1940b) in early May (28) and moulting again into breeding dress in mid-July (35), while at the end of July (38) this moult was complete and gonads were enlarging. In Queensland the cycle was apparently similar, with a young bird undergoing first moult in March (55) and a male in fresh eclipse in early June (61). The male from Inkerman (50), also early June, is possibly sub-adult; it is thinly feathered on the back and head, the feathers being dull black suffused or edged with brown: pin feathers are black with grey bases, suggesting that breeding dress is being assumed: primaries, however, are fairly fresh with bold white edges, whereas those of adults moulting out of eclipse dress show more wear: the tail is more worn than the wings.

In South Australia the June male from Kingoonya (70) shows equal numbers of black and brown feathers on the head and back but no moult is apparent: from Nullarbor HS (69) in October the male is in almost complete breeding plumage and the young female in spotted juvenile plumage: the November female (84) shows slight body moult with the wings and tail worn. In central Western Australia the moult from eclipse was later than in the Kimberleys, starting in the end of July (18) or August (23) in three males, while one taken on 11 September (21) is in almost complete breeding dress with gonads enlarging.

Impermanent Colours. *Legs* black, dark grey, dark brown and once flesh (pink). *Iris* dark brown. *Bill* males black (in breeding dress); black or dark brown with grey or yellow base to lower mandibles; females black or brown with yellow base to lower mandible: immatures brown. *Mouth* pink, yellow or black.

Lalage leucomela — Varied Triller

Specimens. NORTHERN TERRITORY. Darwin (41) 7 – 20 Sept 68, 3 ♂ (32.3, 32.3, 33.3 g), 2 ♀ (31.8, 32.9 g), 1 imm ♀ (32.6 g), 4 fl (30.2, 33.0, 33.7, 37.5 g). QUEENSLAND. Inkerman (50) 29 May 64, 1 ♀ (32.8 g).

Distribution & Abundance. In Northern Territory seen only near Hope Inlet (41). Fairly plentiful though not more than three birds seen together. In northern Queensland seen also at Mount Dryander (57).

Habitat. Coastal (41) or riverine (50) mangroves and adjacent monsoonal forest; also the canopy of dry rainforest (57).

Behaviour. Keeps under cover and approaches to collectors' squeakings with a certain furtiveness. At Darwin (41) usually in fairly thick secondary growth, or sitting quietly, often in the sun, in open situations at the forest edge. Once recorded flying high from tree to tree. At Inkerman (50) a single bird was with a flock of *Zosterops lutea.*

Food. Small insects, grasshoppers, caterpillars, vegetable matter, seeds and berries.

Annual Cycle. All adults were in moult, those from Darwin (41) showing new and old body feathers and having wings and tail rather worn. The Inkerman (50) bird was in body moult, with the central rectrices in sheath and several fresh primaries. Gonads of Darwin (41) birds were inactive while the ovary of the Queensland (50) bird was slightly enlarged.

The immature female (41) has white tips to feathers of the head, nape, scapulars and back, and the skull not fully pneumatised. It was apparently moulting into adult plumage.

Impermanent Colours. *Legs* black, dark grey or brown. *Iris* blackish-brown to brown. *Bill* black. *Mouth* adult, various suffusions of pink, grey, purple and black; immature, orange.

Geographical Variation. The Darwin (41) specimens are referable to the race *L. l. rufiventer*. The Inkerman (50) female falls, geographically, between the two eastern subspecies *L. l. leucomela* and *L. l. yorki* in which the females are similar: it could represent an extension of range for either of them.

TURDIDAE — Thrushes
C. J. O. Harrison

Drymodes brunneopygia — Southern Scrub-robin

Specimens. Western Australia. Newdegate (14) 14 – 18 Apr 66, 1 ♂, 2 ♀, 1 skel (and trunk of ♂), 1 fl.

Distribution, Abundance & Habitat. Not seen elsewhere. Encountered singly in low shrubby cover, in mallee on the edge of salt pans or on the edge of open heath.

Behaviour. Very skulking, keeping to cover and moving quietly on the ground under bushes. In response to squeaking noises approached gradually but cautiously, keeping behind branches. Perched very upright. Called with a loud, intense and melodious single note.

Food. Small black beetles.

Annual Cycle. Specimens showed moult of rectrices and some body feathers. Gonads were small.

Impermanent Colours. *Legs* male black, female dark grey or greyish-brown. *Iris* dark brown. *Bill* black. *Mouth* grey. One female with dark grey bill and flesh-coloured mouth may have been sub-adult.

Geographical Variation. Referable to *D. b. pallidus,* the race of Western Australia (Keast, 1961).

Zoothera dauma Australian Ground-thrush

Specimens. NEW SOUTH WALES. Tenterfield (95) 12 Apr 63, 1 ♀.
Field Notes. A solitary bird in thick undergrowth in rainforest. Not seen elsewhere.
Annual Cycle. In apparently fresh plumage. Ovary minute.
Impermanent Colours. *Legs* grey. *Iris* dark brown. *Bill* black. *Mouth* yellow.
Geographical Variation. On locality this specimen is referable to *Z. d. heinei*.

TIMALIIDAE — Quail-thrushes & Babblers
G. S. Cowles

Orthonyx spaldingi Northern Chowchilla

Specimens. QUEENSLAND. Ayton (52) 1 July 64, 1 ♂ (173.6 g), 1 ♀ (126 g). Walter Hill Range (53) 21 July 64, 2 ♂ (150.0, 165.8 g). Tully (54) 24 – 25 July 64, 2 ♂ (156.6, 180.3 g), 1 ♀ (143.9 g), 2 ♂ skel (208.8, 213 g).
Distribution & Abundance. Only seen where collected. Usually two birds together, once four.
Habitat. Undergrowth of dense rainforest (Plate 9(a)) and thickets in the clearings.
Behaviour. Seeks food on the ground by scratching like a domestic hen amongst the fallen leaves. Moves about noisily, calling occasionally with a hen-like clucking and a subdued *chowchilla*. A squawk like that of a domestic hen was given when alarmed.
Food. Insects.
Annual Cycle. The gonads of some birds were enlarged and coming into breeding condition at all localities during July. The plumage shows no moult and is in unworn condition, except for the tails, which are all very much abraded at the tip.
Impermanent Colours. *Legs* black. *Iris* dull brown to blackish-brown. *Bill* black. *Mouth* at tip black with the soft parts whitish to dark grey. *Orbital ring* dull white with a bluish tinge.
Parasites. Each bird had a heavy infestation of mallophaga eggs on the head. These could not be identified but it is interesting to note that similar eggs were also present on the heads of *O. spaldingi* collected as skins over 90 years ago.

Cinclosoma punctatum Spotted Quail-thrush

Specimens. NEW SOUTH WALES. Tenterfield (95) 12 Apr 63, 1 ♂.
Field Notes. Only found once, in low scrub on the Great Dividing Range. One of two birds seen running along the ground.
Food. Hard seeds and insects.
Annual Cycle. The testes were minute, the tail and rump area are in moult, the back is worn but the underparts appear to be fresh.
Impermanent Colours. *Legs* light brown. *Iris* light brown. *Bill* black. *Mouth* grey.
Geographical Variation. This specimen has been referred to the nominate race *punctatum*.

Cinclosoma castanotum Chestnut Quail-thrush

Specimens. WESTERN AUSTRALIA. Coolgardie 1 Dec 62, 1 ♀. Giles (25) 23 Aug 65, 1 ♀.

Distribution, Abundance & Habitat. Seen only where collected. On main Perth – Adelaide road near Coolgardie in mixture of low saltbush and eucalypts. Near Giles (25) at least two birds (probably three) of either sex were together in low Mulga scrub on red sand, adjacent to dry creek bed with eucalypts and spinifex; also seen in stony area nearby.

Behaviour. A call, probably of alarm, was a high-pitched thin *see-see,* not quite so long drawn out as that of *C. cinnamomeum,* heard at Clifton Hills (65). After the female was shot a sharp *kweeoo, kwit-kwit,* was uttered repeatedly, presumably by the male (DG).

Food. Stomachs contained Mulga seeds and grit.

Annual Cycle. The Coolgardie specimen is in the early stages of moult; all the old plumage shows considerable wear. The Giles (25) specimen is also considerably worn, with some feathers of the upperparts much less faded and worn (hence presumably newer) than others but with no positive evidence of moult. Its ovary was small.

Impermanent Colours. *Legs* dark grey to blackish grey. *Iris* brown or dark brown. *Bill* black or blackish grey. *Mouth* fleshy grey, tongue (noted once) yellowish flesh.

Geographical Variation. The female from Coolgardie has the upperparts predominantly dark brownish grey, the central part of its rump is chestnut. It seems referable to *C. castanotum dundasi.*

The female from Giles (25) differs in having the mantle predominantly and the back and rump entirely chestnut. As Macdonald (1968b) has pointed out, its upper plumage almost exactly matches those of females of *C. cinnamomeum (marginatum)* taken in the same general area.

From Condon's (1962b) descriptions of the races of this and other *Cinclosoma* species the present specimen is evidently referable to *C. castanotum clarum.*

Cinclosoma cinnamomeum (including *alisteri*) Cinnamon Quail-thrush

Specimens. WESTERN AUSTRALIA. Mount Shenton (18) 2 Aug 65, 1 ♀. Giles (25) 23 Aug – 2 Sept 65, 2 ♀. QUEENSLAND. Monkira (61) 2 – 8 June 63, 1 ♂, 1 ♀, 1 ♂ skel, 1 ♂, 1 fl. SOUTH AUSTRALIA. Cadelga (62) 7 May 65, 1 ♂. Clifton Hills (65) 10 – 12 May 63, 1 ♂, 4 ♀, 1 fl. Etadunna (66) 22 May 65, 1 ♂, 1 ø. Nullarbor HS (69) 10 Oct 63, 1 ♂. Kingoonya (70) 22 – 25 June 65, 2 ♂, 1 ♀. Angepena (84) 25 Nov 63, 1 ♂.

Specimens to other Museums. NMV 1 ♀, 1 ♂ fl. Monkira.

Distribution & Abundance. Two seen at Carnegie (19) and five at Mount Shenton (18). Fairly common at Giles (25), Clifton Hills (65) and Etadunna (66). One seen at Granite Downs (68). Common at Kingoonya (70) and five seen at Aroona Valley (82).

Habitat. Open sandy or gibber plains (61, 62, 66), sometimes with low Mulga scrub (18, 25, 61, 65), Bluebush (69, 70), or saltbush (66), or light timber (84).

Behaviour. Found singly, in pairs or small parties of up to six birds, always on the ground, usually feeding near bushes. When disturbed will either hide near the base of a bush or run quickly or fly low over the ground. When seeking food, digs with the bill in a side to side movement (DG). Calls heard were a shrill, high-pitched *see-see, see-see,* also *sisisis.*

Food. Stomachs contained mainly seeds, including *Calandrinia* and *Enchylaena tomentosa* (65), together with beetle and ant remains and grit.

Annual Cycle. Birds with the gonads in full breeding condition were found in early June in Queensland (61), and in Western Australia a female with an egg in the oviduct was collected at Giles (25) in early September. In South Australia (70, 65) in May and June the gonads were slightly enlarged; elsewhere they were small.

Most birds were in fairly fresh plumage, except at Giles (25) where they were in worn condition. In South Australia a June bird (70) was just completing moult and one in November (84) had the body, secondaries and tertiary feathers in moult.

Impermanent Colours. *Legs* dark grey, greenish-grey and black. *Iris* dark brown, reddish-brown. *Bill* blackish-grey, black. *Mouth* yellow, flesh grey, flesh pink, black.

Geographical Variation. Most South Australian specimens seem referable to nominate *cinnamomeum.* The three from Kingoonya (70) are, however, noticeably deeper and redder on the upperparts and on the sides of the breast and flanks. They would appear to represent *C. c. samueli,* from Condon's (1962b) description of that race.

The specimens from Giles (25) and Mount Shenton (18) are considerably darker and rather longer-tailed. They are referable to *C. c. marginatum.*

The Nullarbor HS (69) specimen is referable to the very distinct subspecies, *alisteri,* which is sometimes given specific rank.

Parasites. See Appendix 1.

Pomatostomus temporalis including *alisteri* Grey-crowned Babbler

Specimens. Western Australia. Mount Anderson (28) 1 – 15 May 68, 1 ♂ (54.2 g), 2 ♀ (54.1, 57.7 g), 1 ø, 1 ♂ skel, 6 ♀ skel (59.4 g), 2 ø fl. Joint Hill (34) 3 July 68, 2 ♂, 1 ♀ skel. Lissadell Hill (38) 29 July 68, 1 ♂ (64.7 g), 2 ♀ (55.3, 66.0 g). Queensland. Moonlight Creek (47) 1 – 12 May 64, 7 ♂ (58.6, 60.6, 60.7, 60.8, 61.2, 64.0, 66.3 g), 3 ♀ (57.1, 59.2, 61.7 g). Wernadinga (48) 20 May 64, 2 ♂ (71.4, 74.9 g), 3 ♀ (67.2, 68.9, 72.5 g). Glenore (49) 8 – 14 Apr 64, 2 ♂ (62.8, 68.0 g), 9 ♀ (51.9, 57.3, 59.0, 61.7, 62.0, 62.6, 62.5, 64.2 g), 1 imm ♂ (67.4 g). Inkerman (50) 26 May – 2 June 64, 4 ♂ (62.9, 64.2, 65.8, 66.5 g), 3 ♀ (59.6, 60.0, 64.1 g). Glendower (55) 19 Mar – 1 Apr 64, 3 ♂ (66.6, 71.6, 75.3 g), 5 ♀ (68.4, 69.0, 70.2, 72.1 g), 3 ø (62.7, 63.1, 65.3 g), 1 imm ♂ (66.2 g). Banoona (58) 3 May 63, 1 ♀, 1 ø skel, 2 fl. Tyrone (59) 10 May 63, 2 ♂. New South Wales. Albury (92) 12 – 13 Mar 63, 2 ♂, 1 ♀ skel. Dunedoo (93) 23 Mar 63, 1 ♀, 1 ♂ skel. Grafton (94) 3 Apr 63, 1 ♂, 2 fl.

Specimens to other Museums. NMV 1 ♂ skel Grafton (94). AM 1 ♀ Banoona (58), 1 ♂ Tyrone (59).

Distribution & Abundance. Seen also in the Kimberleys (27, 31, 32, 33, 35) and in the eastern Hamersleys on the road to Chichester Range (3).

Habitat. Savanna woodland, usually close to water or in the trees near dry creeks. Tropical savanna and open forest.

Behaviour. Feeds on the ground amongst dead leaves and bark and in low bushes, usually in parties of 4-12 individuals. Hops heavily from branch to branch, is active and noisy. Calls notes described as harsh, bubbling, squeaking, rapid scolding, chattering and cat mewing.
Food. Small insects, and once, a grasshopper.
Annual Cycle. Birds were in full breeding condition at the end of July in Western Australia (38). In Queensland (55) fledglings or juveniles were collected 1 April. Birds with worn plumage were found in May (28, 47), June (50) and July (34, 38). Moult was in progress during March and May (92, 47, 48, 58) (tail, secondaries and body), while during the same months some birds were completing moult (59, 92).
Impermanent Colours. *Legs* light grey to black. *Iris* varying from light yellow to dark brown in both sexes without apparent correlation with age or locality. *Bill* black, ridge of culmen and lower edge of mandible grey. *Mouth* bright yellow, grey-white flesh.
Geographical Variation. This series shows considerable variation in both colour and size, mostly of a clinal nature. Birds from the extreme northwest (28) are the darkest and most rufous. Eastwards the underparts become less rufous with an increase of white on the throat, and the upper-parts become greyer, the greyest birds being found in the southeast. Among eastern birds size increases from southern Queensland (58, 59) through New South Wales (92-94).

In these circumstances it does not seem useful to retain more than three Australian subspecies out of the ten recognised by Deignan (in Peters', 1964). The dark, rufous birds from Mount Anderson (28) agree with *P. t. nigrescens,* type locality Carnarvon, and other Kimberley birds (34, 38) are near to *nigrescens.* The oldest name available for the larger southern birds (58, 59, 92-94) is *trivirgatus,* leaving the smaller and greyer birds of northern Queensland (47-55) as nominate *temporalis.*
Parasites. See Appendix 1.

Pomatostomus ruficeps Chestnut-crowned Babbler

Specimens. QUEENSLAND. Thylungra (60) 22 May 63, 1 ♀, 1 juv ♂, 1 ø juv. SOUTH AUSTRALIA. Cordillo Downs (63) 6 May 65, 1 ♀, 1 fl. Innamincka (64) 3 May 65, 2 ♂ (50.1, 51.4 g).
Specimens to other Museums. NMV 1 ♀ Thylungra (60).
Distribution & Abundance. A few parties were also seen at Etadunna (66). It was fairly common at Innamincka (64). The Cordillo Downs (63) birds were collected at the side of the track 73 miles north of the homestead.
Habitat. Open arid plains with small scattered trees near dry creek beds.
Behaviour. Always seen in parties of 12 to 15 birds, adults and juveniles together. Feeds both on the ground and in the trees and is highly sociable. Feeding parties hop rapidly or fly onwards in loose order after those that initiate movement in a definite direction, but there is no synchronised movement of the party (DG). The parties appear to move faster than the White-browed Babbler *P. superciliosus* and are more wary. The calls heard were sharper and harsher than those of *P. superciliosus,* a hard,

rapid *tchat-a-chat,* in a very scolding note, like that of the Song Thrush *Turdus philomelos,* but harsher (DG).

Food. Insects and seeds.

Annual Cycle. Birds appeared to have finished breeding in May in both Queensland and South Australia; the young birds collected at Thylungra (60) appeared only recently fledged while the adults collected (63, 64) were in or had recently completed moult, with gonads inactive.

Impermanent Colours. *Legs* grey or dull black. *Iris* brown or black. *Bill* dark grey, lower mandible blue-grey at base. *Mouth* pale flesh or yellow, juveniles with a pink gape.

Other Notes. Juveniles are a darker chestnut on the head than the adults and the upper plumage is dark brown not grey: they have deep chestnut patches at the sides of the lower breast where the brown meets the white.

Pomatostomus superciliosus White-browed Babbler

Specimens. Western Australia. Newdegate (14) 11 – 14 Apr 66, 1 ♂, 3 ♀. Eyre (16) 8 July 65, 2 ♂. Eucla (17) 1 July 65, 1 ♂. Mount Shenton (18) 31 July – 3 Aug 65, 1 ♂, 3 ♀. Carnegie (19) 16 – 17 Sept 65, 1 ♀, 1 ø. Hunt Oil Camp (22) 9 Sept 65, 1 ♀. Warburton (23) 13 – 14 Aug 65, 2 ♂, 1 ♀, 1 ø skel. Giles (25) 21 – 24 Aug 65, 1 ♂, 3 ♀, 1 ♂ skel. South Australia. Granite Downs (68) 4 – 10 June 65, 1 ♂, 3 ♀, 1 ø. Nullarbor HS (69) 5 – 7 Oct 63, 1 ♂, 1 juv ø. Kingoonya (70) 17 – 27 June 65, 1 ♂, 1 ♀, 1 ø, 2 ø skel, 2 fl. Nonning (79) 29 Mar – 5 Apr 65, 5 ♂, 1 ♂ skel (43.1 g), 1 fl (39.0 g). Aroona Valley (82) 8 – 16 Apr 65, 3 ♀ (40.2, 41.0, 43.0 g). Angepena (84) 18 – 26 Nov 65, 2 ♂. Victoria. Coleraine (89) 22 Jan 63, 2 ♀. Little Desert (91) 26 Jan 63, 1 fl.

Specimens to other Museums. SAM 1 ♀ Nullarbor HS, 1 ♀ Minnipa (73) 1 Nov 69.

Distribution & Abundance. Especially common at Giles (25), Granite Downs (68), Nonning (79) and Kingoonya (70), and found at several camps where not collected (20, 60, 74, 90, 91, 94).

Habitat. Open acacia and eucalypt woodland, low scrub and low bush, coastal scrub (16).

Behaviour. Seen always in parties of 4-15 individuals feeding on the ground or preening in the low branches of trees or flying low through the trees. Very inquisitive and noisy, readily attracted by squeaking. A nasal sounding call, like that of a Zebra Finch amplified, seems to indicate alarm or to be a contact call: a loud *shack-shack* was given by a bird when separated from the main flock (DG). Mutual preening was observed.

Food. Beetles, caterpillars, ants and seeds.

Annual Cycle. Two eggs were collected 29 June 65, 12 miles north of the Nullarbor Homestead (69). Two years previously a juvenile had been collected just south of Nullarbor HS on 7 October 63.

Except at Granite Downs (68), gonads were slightly enlarged or nearing breeding condition (16, 18, 19, 22). Birds at Nonning (79) in South Australia were in heavy moult at the end of March, and in Western Australia (14) one female had the wings and tail in moult in mid-April.

Impermanent Colours. *Legs* dark grey or black. *Iris* brown to reddish-brown or dark grey. *Bill* black, base of lower mandible silver grey. *Mouth* white, bluish-flesh, mauvish-flesh: tongue flesh white.

Geographical Variation. I can find no difference in colour to justify recognising *ashbyi* or *gilgandra,* though there is seasonal difference between pale worn birds and dark fresh-plumaged ones. Birds from the extreme southwest (*ashbyi* Mathews), are said to be larger, lighter in colour and with a bill "conspicuously longer", but ours from Newdegate (14) are in fact darker than most, the male has a bill of 31 mm, but this is no longer than others from Eyre (17), Granite Downs (68) or Nonning (79). The four Newdegate (14) birds do have slightly longer wings, that of the male (90 mm) being the longest measured, but the differences are not great enough to justify recognition of *ashbyi.* All our specimens are therefore regarded as *P. s. superciliosus* (Cowles, 1964).

Other Notes. BILL ABRASION. Birds from Aroona Valley (82) in the Flinders Range show considerable wear at the tip of the upper mandibles, probably as a result of probing in the rocky soil of this area. The abrasion is not evident in other babbler species or from other areas in Australia. In some cases the tip has been worn away 3-5 mm, giving the bill a blunt and rounded appearance instead of the usual fine point. NEST. The nest was dome-shaped with a side funnel entrance, constructed from sticks and lined with soft material. It was in a low stunted tree 4 ft from the ground and well concealed. A single bird was in the tree when the nest was discovered in the evening but next morning at about 07.45 hrs a party of several White-browed Babblers came in and around the trees. After they had left a single bird was flushed from the nest.

Parasites. See Appendix 1.

Pomatostomus halli (Frontispiece) Hall's Babbler

Specimens. QUEENSLAND. Tyrone (59) 15 May 63, 1 ø, 1 ø skel.

Specimens to other Museums. QM 1 ♂ (Holotype) Tyrone. NMV 1 fl (now a skin) Tyrone.

Field Notes. Found only at Tyrone (59). One or possibly two parties seen in mixed woodland and scrub consisting of Mulga, sandalwood, box and Gidgee, the area is noted as arid scrub (Williams, 1955). Since its discovery Hall's Babbler has been seen near Cunnamulla, south of the type locality (S. A. Parker & D. J. Nicholls *in litt*).*

Noisy when feeding in a party on the ground; also seen alighting in the low branches of trees where mutual preening took place. Very similar to the White-browed Babbler *P. superciliosus* in behaviour.

Food. Beetles.

Annual Cycle. The unsexed specimen had just completed full moult. Gonads were small in the male holotype.

* Specimens were mistnetted, banded, photographed and measured about 270 miles northwest of the type locality, at Opalton 137 km south of Winton, central western Queensland (Horton 1973). Horton also reports seeing this babbler still further to the northwest, on the Winton to Boulia road west of the Diamantina River. There are now specimens in the AM.

Impermanent Colours. *Legs* black. *Iris* brown to reddish-brown. *Bill* black with the ventral edge of the lower mandible grey. *Mouth* orange-grey or flesh.

SYLVIIDAE — Old World Warblers
B. P. Hall

Eremiornis carteri Spinifex-bird

Specimens. Western Australia. Chichester Range (3) 16 May 66, 1 fl. Mount Anderson (28) 8 – 10 May 68, 1 ♂, 1 fl (13.2 g).
Field Notes. In spinifex in Tampatha Creek, Chichester Range (3) and spinifex savanna at 100 ft at Mount Anderson (28). Only single birds seen, sometimes on ground, or in spinifex, or perched in a low bush. Carried the tail low in flight but cocked when hopping along the ground. The only call heard was a sharp alarm *tak*.
Annual Cycle. The male examined from Mount Anderson (28) was commencing moult in wings, tail and body. The gonads were very small.
Impermanent Colours. *Legs* greyish-pink. *Iris* brown. *Bill* grey or brown, paler below. *Mouth* pinkish-grey.

Cisticola exilis Golden-headed Fantail-warbler

Specimens. Western Australia. Derby (29) 29 May 68, 1 ♂ (8.3 g), 2 ♀ (6.8, 7.4 g), 2 ø (6.8, 8.0 g), 1 imm ♂ (7.2 g), 1 imm ø (7.2 g). Geikie Gorge (31) 2 June 68, 1 imm ♀ (6.5 g). Mount Bell (32) 19 June 68, 1 ♂ (7.8 g), 1 ø (7.4 g), 1 imm ♂ (8.1 g). Mount Caroline (33) 21 June 68, 1 ♀. Wyndham (39) 2 Aug 68, 1 ♂ (6.2 g). Northern Territory. Darwin (41) 5 – 14 Sept 68, 1 ♂ (7.1 g), 4 ♀ (6.2, 6.3, 6.6, 7.0 g). Oenpelli (43) 9 Oct 68, 1 ø (6.2 g). Mount Roper (45) 1 Nov 68, 2 ♀ (6.5, 7.2 g). Queensland. Glenore (49) 20 Apr 64, 2 imm ♀ (6.2, 6.9 g), 1 imm ø (6.3 g). Inkerman (50) 30 May – 5 June 64, 1 ♂ (7.8 g), 2 ♀ (6.1, 6.7 g). Victoria. Nelson (88) 30 Dec 62, 1 ♂, 1 ♀, 1 imm ø, 1 ø skel.
Distribution & Abundance. Small parties were fairly common in suitable habitat at all camps in the Kimberleys (31-35). At Mount Bell (32) 15-20 individuals were estimated to occupy 1½acres. In Queensland one was collected (but only the head retained) from among a party of *Malurus melanocephalus* at Glendower (55). Seen also at Grafton (94).
Habitat. Rank grassland and grassy patches in reed beds. In Queensland perhaps in less open areas; at Glendower (55) in pocket of grassland in open forest. At Grafton (94) in coastal heath near mouth of the Clarence River. At Darwin (41) collected only in the Noogoo swamp, where found once in close proximity to *C. juncidis,* no habitat difference being noted; but it was not appreciated at the time that two species were present and the fact that only one out of four collectors obtained both species argues some segregation, as was found by Givens & Hitchcock (1953). On the Roper River *exilis* was collected in a reed bed seven miles northeast of Mount Roper (45) and *juncidis* on an open grassy plain eight miles westnorthwest.

Behaviour. Although birds when called up would use the top of grass stems, bushes and occasionally trees for observation posts, after sustained flight they would always drop down into the base of the grass. In flight sometimes climbed steeply and jerkily to over 20 ft, at others flitted ahead of the pursuer just above the grass level, rarely flying more than 50 yds. The usual call was a soft *peep-peep,* but also a grating alarm note.
Food. Unidentified insects.
Annual Cycle. Of the young birds the one from Geikie Gorge (2 June) is the youngest; comparison with a specimen from the American Museum of Natural History examined by Lynes suggests it is not much more than six weeks old. One from Derby (29) 29 May is slightly older but mainly in juvenile plumage, while the other and those from Mount Bell (32) 19 June, Queensland (49) April, and Victoria (88) December were more advanced in moult to first year plumage. This shows the breeding season to be locally variable, which is supported by the adults. Gonad activity was shown only by the December male from Victoria, in worn breeding dress: May and June birds from both Western Australia and Queensland were completing moult into non-breeding dress: The August male from Wyndham (39) was in worn dress with moult starting (either non-breeding to breeding or, since gonads were minute, first year to adult): September to November Northern Territory birds were in worn non-breeding dress.
Impermanent Colours. *Legs* pale pinkish-brown. *Iris* light brown. *Bill* upper mandible in breeding dress black, otherwise brown: lower mandible paler. *Mouth* breeding male "yellow with black surrounding border", female yellow; non-breeding April to June, usually yellow; August to September, white, flesh-pink, yellow, black; immatures yellow.
Geographical Variation. No satisfactory discussion on *Cisticola* variation is possibl- without series taken throughout the year from each locality, but these birds fall reasone ably well into the subspecies recognised by Lynes (1930: 188-197), with some doubt only on the status of those from Western Australia.

C. e. exilis of the southeast is a dark, heavily-streaked race to which the Victoria birds belong. The more russet *C. e. diminuta* from Cape York is not represented in this series for the birds from Inkerman (50) on the Gulf of Carpentaria are clearly the pale *C. e. alexandrae* which Lynes believed to range from Glenore (49) across to the Kimberleys in semi-arid districts. Our Inkerman (50) birds are slightly paler on the hind neck, mantle and flanks than those in comparable plumage from the Kimberleys, and the difference shows also in the immatures from Glenore (49).

Lynes recognised that birds from the less arid parts of Arnhemland *C. e. lineocapilla* were more richly coloured than *C. e. alexandrae.* In series this is borne out by ours from Darwin (41) and Oenpelli (43), and might be more marked if fresh-plumaged specimens were available, but individually it is not easy to distinguish some of them from some Kimberley (29-33) specimens. The latter might best be regarded as intermediates between *lineocapilla* and *alexandrae.*
Parasites. See Appendix 1.

Cisticola juncidis — Streaked Grass-warbler

Specimens. NORTHERN TERRITORY. Darwin (41) 5 – 18 Sept 68, 2 ♂ (7.9, 8.2 g), 2 ♀ (7.9, 8.2 g), 1 fl (7.6 g). Mount Roper (45) 1 Nov 68, 1 ø (7.3 g).

Distribution & Abundance. At Darwin (41) found in parties in the Noogoo swamp and at Hope Inlet. Only a single bird seen at Mount Roper (45). This is a new locality for *C. juncidis,* which has not previously been found between Darwin (41) and Normanton.

Habitat. In the Noogoo swamp found in short reeds and long marsh grass: at Hope Inlet on dry mud flats near the coastal mangroves: at Mount Roper (45) in open grassland. (See Givens & Hitchcock, 1953, and notes on *C. exilis*).

Behaviour. Like *exilis* found sometimes perching, sometimes skulking in the grass, but the flight faster and usually longer, 50-100 yards, ending, as with *exilis,* with the bird pitching into the grass, from which it was difficult to flush a second time.

Food. Stomachs contained vegetable as well as insect remains, and included a wing 12 mm long.

Annual Cycle. Darwin (41) birds were in fairly worn non-breeding dress with inactive gonads. The Mount Roper (45) November bird very worn with moult to breeding dress just starting.

Impermanent Colours. *Legs* flesh-pink. *Iris* pale yellowish-brown. *Bill* upper mandible brown, greyer in November bird; lower mandible pinkish. *Mouth* September, pinkish or whitish; November pinkish with black palate.

Geographical Variation. Darwin (41) birds are topotypical of *C. j. leanyeri* Givens & Hitchcock, 1953, and the one from Mount Roper (45), allowing for the worn plumage, is similar.

Megalurus timoriensis Tawny Grassbird

Specimens. Northern Territory. Darwin (41) 14 – 16 Sept 68, 1 ♂ (18.5 g), 2 ♀ (15.6, 16.3 g), 1 imm ♀ (15.3 g). Oenpelli (43) 30 Sept 68, 1 fl.

Distribution & Abundance. All the Darwin (41) birds were collected in the Noogoo swamp 12 miles to the northeast, the habitat of both *Cisticola* species. In Queensland two immature birds were collected but not retained from Big Tableland (51) and Tully (54) in July, and a male (20.4 g) from Mount Dryander (57) in August.

Seen also at Grafton (94) near mouth of Clarence River and in parties of up to seven birds on three occasions at Thylungra (60).

Habitat. In long grass the edge of the swamp (41); in dead grass beneath thick *Pandanus* among sandstone hills (43); in grass in forest at 2,200 ft (51) or sea level (54, 57); on open plain (60) and coastal heathland (94).

Behaviour. Very furtive, once perched on small bush, occasionally coming to the top of grass stems, and dropping back quickly. When flushed flying low for 50 yards or so before dropping back into grass and creeping through. Scolding notes *tic-tic-tic tic,* uttered from cover.

Food. Tiny insects, once green Orthoptera.

Annual Cycle. Of the Darwin (41) adults the male is in fresh plumage, but the females are worn and one was starting moult. The young bird is much more rufous above and on the breast and flanks, and its skull was not fully pneumatised, nor were the skulls of the two young birds from Queensland, but no notes were kept on their plumage.

Impermanent Colours. *Legs* pale flesh. *Iris* light yellow-brown. *Bill* Northern Territory dark horn, paler below; Queensland, dark brown or blackish-brown, yellowish below. *Mouth* pinkish or white.
Geographical Variation. The Northern Territory specimens are referable to *M. t. alisteri*.

Acrocephalus stentoreus Reed-warbler

Specimens. Western Australia. Karratha (1) 30 May 66, 1 ♀ skel, 1 fl. Derby (29) 29 May 68, 4 ♂ (13.3, 15.1, 16.6, 21.6 g), 1 ♀ (18.7 g). Joint Hill (34) 4 July 68, 1 ♂. Lake Gladstone (36) 21 July 68, 2 ♂ (21.4, 21.7 g). Queensland. Inkerman (50) 28 – 29 May 64, 1 ♂ (20.0 g), 1 ø (8.9! g – probable error for 18.9 g).
Distribution & Abundance. Common where collected, but local. In South Australia heard near Furner (86) and two birds were seen at Bool Lagoon (87).
Habitat. Usually found in marshy areas supporting reeds and associated grasses. Not found at Noogoo Swamp, Darwin (41) where both species of *Cisticola* were co-existing with *Megalurus timoriensis*. In *Pandanus* at Joint Hill (34) along a flowing creek with reed beds, and at Lake Gladstone (36) in both fresh-water mangroves and reeds. Twice encountered in estuarine mangroves 8-10 feet high at Inkerman (50).
Behaviour. Generally keeps low in cover and utters short grating alarm notes. When undisturbed ventures out into the lower branches of nearby trees where it was sometimes seen to catch insects. In May the short sub-song was heard once at Karratha (1) though in the same month the Derby (29) population was silent. In full song at Lake Gladstone (36) in July.
Food. Insects.
Annual Cycle. No specimens showed any gonad activity. The birds from Derby (29) are in fairly fresh plumage while those from further east were in varying stages of moult.
Impermanent Colours. *Legs* blackish-grey to dull blue-grey, soles buff. *Iris* dark brown or brown. *Bill* upper mandible dark brown or black with pale cutting edge: lower mandible paler with a fleshy tint at base. *Mouth* orange-red to orange-yellow, once yellowish.
Geographical Variation. Mayr (1948) suggested that most populations from tropical northern Australia overlap, do not vary sufficiently to warrant subspecific separation and should be included under *A. s. cervinus* (=*A. s. sumbae* (see Mees, 1961a)). He did, however, tentatively retain *A. s. carterae* from Point Torment (30) suspecting that further specimens might confirm the larger bill size (24 mm) of the one male described. Our four adult males from Derby (29), bills 20-21 mm (20.5) do not support this; they have wings 70.5-76 mm (72.1) against 74.5-77 mm (76.0) for eastern birds (34-50) and are generally darker, possibly as a result of fresher plumage.

Cinclorhamphus cruralis Brown Songlark

Specimens. Western Australia. Chichester Range (3) 13 – 14 May 66, 1 ♂ skel, 1 fl. Mount Shenton (18) 2 – 3 Aug 65, 2 ♂ (70 g). Queensland. Moonlight Creek (47) 16 May 64, 1 imm ♂ (60.3 g). Inkerman (50) 26 May 64, 1 ♀ (33.2 g). Monkira (61)

2–7 June 63, 1 ♂, 1 ♀, 1 ♂ skel, 1 fl. SOUTH AUSTRALIA. Kingoonya (70) 20 June 65, 1 ♂. VICTORIA. Coleraine (89) 21 Jan 63, 1 ♂, 1 ♀.

Specimens to other Museums. NMV 1 ♂ Monkira (61).

Distribution & Abundance. Fairly common at Monkira (61) but not elsewhere. A single songlark, believed this species, was seen in the Nullarbor Plain north of Cocklebiddy (15) in July, and both species were seen at Karratha (1).

Habitat. Open country, grass, spinifex or Bluebush plains; once on a salt pan (47). On the Chichester Range (3) in same habitat as *Mirafra javanica* but preferring higher grasses.

Behaviour. Most birds were collected from the ground or perched on low bushes. Only the bird in the Nullarbor (15) was observed in display flight.

Food. Insects, including grasshoppers, and seeds.

Annual Cycle. The breeding season is clearly variable and not wholly synchronised with moult, for of the pairs collected in Victoria in January and southern Queensland (61) in June the females each had eggs forming in the ovary, while the January male was starting, and the June male finishing moult. The May female from Inkerman (50) had a slightly enlarged ovary and was finishing moult, while the immature from Moonlight Creek (47) (identified as such by its skull) is in fresh plumage: so was the June male from Kingoonya (70). The August males from Mount Shenton (18) had enlarged testes, both were in moult, one almost complete, the other only with fresh wings.

Impermanent Colours. *Legs* light pinkish-brown or greyish (89 ♂). *Iris* brown. *Bill* male variable (see below), female brown, paler below. *Mouth* adult male black, immature pinkish-grey; female fleshy.

Geographical Variation. Allowing for seasonal plumage differences the three females are alike, except that the one from Inkerman (50) is larger (wing 91, against 82 and 84 mm). The males all differ: the two worn birds from Victoria (89) and Western Australia (18) are alike above with grey, almost plain heads and rumps and a few pale edges to the mantle feathers; but the former has the underparts plain black and a black bill; and the latter has the throat white with some black spots, the breast grey, the abdomen black with some pale edges to the feathers, the under tail-coverts and flanks sandy: the bill dark brown, paler below. Of the three adults in fresh or mainly fresh plumage the one from southern Queensland (61) is richer brown on the head and mantle and is mainly black below with a white chin, some spots on the throat and narrow, pale edges to most feathers; the bill is grey-brown above, paler below. The one from Western Australia (18) is generally similar to the worn male from the same place in colour and pattern below, though most of the plumage is fresh – showing that this pattern is not seasonal; the bill is black. The Kingoonya (70) male, geographically intermediate between Mount Shenton (18) and Monkira (61), is also intermediate in pattern below, having a considerable amount of black on the lower throat, separated from the black of the abdomen by a band of grey-brown across the breast: the bill is dark grey, paler below.

The immature male from Moonlight Creek (47) is slightly paler above than other fresh-plumaged birds: below it has the throat white with some dark flecks at the base, the rest of the underparts greyish-brown, with a blackish patch in the centre of the lower breast and abdomen; the bill is dull olive-brown, pale below.

Individual variation in size of males is illustrated by the two from Mount Shenton (18) with wings of 103 and 114 mm.

A small series such as this is inadequate for a proper understanding of geographical, seasonal and individual variation but taken in conjunction with the few old specimens in the British Museum certainly suggests that, as Mathews believed, males from the southeast have more black below than others, and there is a progressive decrease northwards and westwards. *C. c. clelandi,* type locality Perth, is a name available for western birds but the systematics and movements of the eastern birds need full study before subspecific names can be applied.

Cinclorhamphus mathewsi Rufous Songlark

Specimens. WESTERN AUSTRALIA. Karratha (1) 29 May – 6 June 66, 1 ♂, 1 fl. Coolawanyah (4) 20 – 21 May 66, 1 ♂, 2 ♂ skel, 1 fl. Mount Anderson (28) 2 – 14 May 68, 5 ♂ skel (33.0, 37.5 g), 1 fl. Derby (29) 26 May 68, 1 ♂. Mount Caroline (33) 24 June 68, 1 ♀ (18.9 g). Joint Hill (34) 1 July 68, 1 ⌀ (35.1 g) (♂ on size). Mount Elizabeth (35) 7 July 68, 1 ♂ (32.7 g). QUEENSLAND. Moonlight Creek (47) 13 May 64, 1 ♂. Glenore (49) 16 – 20 Apr 64, 1 ♀ (23.5 g), 1 imm ⌀. Thylungra (60) 22 – 23 May 63, 1 ♂, 1 ♀, 1 imm ♀, 1 imm ⌀ (♀ on size). Monkira (61) 5 June 63, 1 fl. VICTORIA. Coleraine (89) 24 Jan 63, 1 ♀, 1 imm ⌀ (♂ on size).

Specimens to other Museums. NMV 1 ♂ Monkira (61).

Distribution & Abundance. Fairly common in the Kimberleys. Usually solitary but once in a party of six (89). In the southern half of the continent only seen where collected except for a single bird at the base of Mount Eveline in the Warburton Range (23).

Habitat. Often in gums along creeks: also in tea-trees in tropical savanna and in acacia savanna, always in vicinity of trees or bushes.

Behaviour. Most often seen perched, at various levels in the trees; once feeding in a tea-tree; also observed feeding on the ground, in spinifex and grass. Song uttered in flight or when perched, "a melodious high whistling interspersed with grating" (AH).

Food. Insects, including a caterpillar, and seeds, some believed to be spinifex.

Annual Cycle. Males at Karratha (1) and Coolawanyah (4) had slightly enlarged gonads and from their frequent song and territorial aggressiveness appeared to be in breeding condition (Harrison & Colston, 1969), otherwise all adults had inactive gonads. All were in moult except fresh-plumaged males from Karratha (1) and Mount Elizabeth (35) (respectively May and July) and a very worn female from Coleraine (89) in January. Of the four "immatures" (identified by their skulls) the April one from Glenore (49) is quite young, having a soft gape and some buff on the throat; one of the two from Thylungra (60) in May was very worn and starting moult, the other just completing moult, but both resemble the adult female from the same place except that they have no speckles on the breast (see below). The immature January male from Coleraine (89) has speckles on the upper breast and streaks on the lower breast, unlike any other specimen examined. If this is first year plumage, corresponding to first year plumage being assumed by the Thylungra (60) females, then breeding in 1963 must have been earlier in Victoria than further north.

Impermanent Colours. *Legs* pale fleshy-brown. *Iris* brown. *Bill* adult male black (1, 4, 47) or brown (elsewhere); female and immature male light brown, paler below. *Mouth* adult male black (1, 4) or "white grading to black" (47); other males and all females, yellowish.
Geographical Variation. No geographical variation in size or colour discernible in this series but seasonal variation in bill and mouth of males (see above); and individual variation in the amount of spotting on the breast, which is clear in two out of the four Kimberley males (all adult) and in the adult female from Thylungra (60), but faint or absent from all other birds except the spotted and streaked immature male from Coleraine (89). A good series is required for a full understanding of age and seasonal variations.

MALURIDAE (Part 1) — Australian Warblers except for Blue Wrens, Emu Wrens & Grass Wrens

B. P. Hall

From an examination of morphological and behavioural characters of the blue wrens and similar species (Harrison & Parker, 1965, Harrison, 1969a) it seemed to the authors that species of *Malurus, Stipiturus* and *Amytornis* are more closely related to each other than to the species of Australasian warblers with which they are usually grouped. Furthermore their characters suggest that these three genera probably belong with the babblers, Timaliidae, and are not a part of the Australasian warbler assemblage, although Sibley (1970) is of the opinion that the egg-white of *Malurus* shows sylviid affinities.

For this report we have retained the family name Maluridae for convenience and pending further study, but the two groups have been kept separate, the Australian warblers in Part 1, and the three wren genera in Part 2.

Gerygone olivacea White-throated Warbler

Specimens. Western Australia. Joint Hill (34) 27 June 68, 1 ♂. Mount Elizabeth (35) 14 July 68, 1 ♂ (7.2 g), 1 ♀ (6.9 g), 1 imm ♀ (6.2 g). Northern Territory. Mount Roper (45) 1 Nov 68, 1 ♀ (5.6 g). Queensland. Moonlight Creek (47) 2 – 14 May 64, 1 ø (7.2 g), 1 imm ♀ (6.0 g), 1 imm ø (6.3 g). Wernadinga (48) 20 May 64, 1 ♂, 1 imm ♂. Glenore (49) 21 Apr 64, 1 ♂ (7.1 g), 1 ♀ (6.6 g). Inkerman (50) 30 May – 2 June 64, 1 ♂ (7.2 g), 1 ♀ (6.6 g). Cape Gloucester (56) 6 – 7 Aug 64, 1 ♂ (8.5 g), 1 ♀ (6.8 g). New South Wales. Dunedoo (93) 19 Mar 63, 1 ♂, 1 fl. Grafton (94) 1 Apr 63, 1 ♀.
Specimens to other Museums. NMV 1 ♂ Dunedoo (93), 1 fl Grafton (94).
Distribution & Abundance. Nowhere common but seen also at Pine Creek (44), Borroloola (46) and Glendower (55).
Habitat. Woodland or light forest, often eucalypts; also in *Grevillea* at Mount Elizabeth (35), in tea-trees at Inkerman (50) and Glendower (55) and in woods of ironbark, Yellow Box and pine at Dunedoo (93). (Plate 5(b)).

Behaviour. Solitary or in twos or threes, usually in the canopy. Its voice "a sweet modulating song, descending in pitch, given whilst moving through scrub" (BB). The female from Glenore (49) was feeding a young cuckoo *Chrysococcyx minutillus*.

Food. Insects, including ants.

Annual Cycle. In Queensland young birds were seen being fed at Glendower (55) in March. The immatures collected elsewhere all seem to have been moulting from juvenile to first year plumage (still retaining the yellow throat) with the one from Mount Elizabeth (35) in July and one of the two from Moonlight Creek (47) in May apparently older, with blacker bills and fuller skull pneumatisation than the other two May birds.

Meise (1931: 326) has noted that moult in this species takes place throughout the year and this is borne out by our series, all of which except the April pair from Glenore (49) were in fresh or fairly fresh plumage with some new feathers coming in on the body, mostly on the throat. The Glenore (49) female is very worn, and the male was in full moult of wings and tail: the Dunedoo (93) March bird also had the wings in moult, nearly completed.

No gonad activity was shown.

Impermanent Colours. *Legs* black or dark grey, paler in young birds. *Iris* usually males dark red, females orange-red, young birds olive-brown, but recorded for all birds from Mount Elizabeth (35) as brown with white outer ring. *Bill* black, paler in young birds. *Mouth* adults black or dark grey: young birds golden yellow (47 & 48) or pale pinkish (35).

Geographical Variation. Meise (1931) recognises three races in Australia; (a) *G. o. olivacea* along the east coast; (b) *G. o. flavigaster* along the eastern side of the Gulf of Carpentaria, slightly smaller; (c) *G. o. rogersi* of northern Western Australia and Northern Territory in which the white patches on the inner webs at the base of the rectrices, characteristic of the eastern races, are grey and only just discernible. Our series demonstrates that *rogersi* extends eastwards to Moonlight Creek (47) without apparent intergradation with *flavigaster* which was found 100 miles further east at Wernadinga (48). In addition to the differences in the tail-pattern there seems a tendency for *rogersi* to have a finer bill than *flavigaster* or *olivacea,* though there is much individual variation: also the immature *rogersi* are brighter below than the single immature *flavigaster*. Adults of northern races are less olive on the back than *olivacea,* while the size difference between *olivacea* and *flavigaster* is confirmed by the weights of the Cape Gloucester (56) birds in comparison with those from the Gulf (48-50).

Gerygone mouki Brown Warbler

Specimens. Queensland. Big Tableland (51) 25 June – 11 July 64, 8 ♂ (5.5, 5.5, 5.5, 5.6, 5.8, 5.9, 5.9, 5.9 g), 1 ♀ (5.9 g), 2 ø (5.0, 5.5 g), 1 imm ♂ (5.7 g), 1 imm ø (5.6 g). Walter Hill Range (53) 17 – 23 July 64, 8 ♂ (5.0, 5.0, 5.0, 5.0, 5.4, 5.6, 6.2 g), 4 ♀ (5.1, 5.2, 5.3, 5.6 g), 1 ø (4.5 g), 1 imm ♀ (4.9 g), 1 imm ø (5.0 g).

Distribution, Abundance & Habitat. Common where collected in the rainforest between 2,000 and 2,300 feet, often on the edges of clearings. At Big Tableland (51) the forest contained large stands of Johnstone Hardwood.

Behaviour. Feeds at all levels from the undergrowth to the canopy. Usually solitary or with two or three others, but once found in a party of ten: once in a mixed party including *Meliphaga notata, Tregallasia capito,* and *Zosterops lateralis:* once with *Machaerirhynchus flaviventer* and once with *Sericornis magnirostris.*

Voice, a thin twitter and a song described as a "rich ascending series *Having-a-good-time* repeated without pause" (SAP).

Food. Insects.

Annual Cycle. The immature birds were identified as such only on the lack of complete pneumatisation of the skull. Two adults from Big Tableland (51) were just completing moult on 1 July, while all birds were in fresh (51) or fairly fresh (52) plumage and showed no gonad activity.

Impermanent Colours. *Legs* slate-purple to black with yellowish soles. *Iris* red or red-brown. *Bill* black. *Mouth* black.

Geographical Variation. The Walter Hill Range (53) forms part of the Atherton Tableland which is drained to the north by the Barron River. This is the type locality of *G. m. mouki,* a race greyer above and less suffused with colour on the underparts than the either *G. m. amalia* of southern Queensland or *G. m. richmondi* of New South Wales, the brownest race. The Walter Hill Range (53) series conforms with the description of *mouki* and may be regarded as typical. Big Tableland (51) is an isolated block of high country about 300 miles to the north: it is unexpected therefore to find the birds from there in series showing an approach to *amalia* (of which we have no specimens for comparison) being slightly browner above and more tinged with buff below. A review of all available specimens would be desirable.

G. mouki is the rainforest representative of the woodland *G. fusca* and mangrove *G. levigaster.* Since in this group ecology is more important than morphology in differentiating between the species we hesitate to follow Meise in regarding the Australian races as conspecific with the New Zealand *G. igata,* a widespread bird of forest and scrub, until direct comparisons have been made between them in the field.

Gerygone palpebrosa — Fairy Warbler

Specimens. QUEENSLAND. Inkerman (50) 29 May – 11 June 64, 6 ♂ (7.2, 7.2, 7.3, 7.5, 8.0 g), 4 ♀ (6.6, 6.9, 7.0, 7.3 g), 1 imm ♂ (7.0 g), 1 imm ♀ (6.5 g). Ayton (52) 20 – 29 June 64, 6 ♂ (7.1, 7.4, 7.6, 7.8, 7.9, 8.6 g), 3 ♀ (7.2, 7.5, 8.2 g), 2 imm ♂ (7.2, 8.0 g), 1 imm ♀ (6.8 g). Tully (54) 11 – 25 July 64, 4 ♂ (7.0, 7.1, 7.1, 7.4 g), 2 ♀ (6.3, 6.5 g), 2 imm ♀ (6.3, 6.5 g). Cape Gloucester (56) 6 – 7 Aug 64, 4 ♂ (8.0, 8.0, 8.2, 8.3 g), 2 ♀ (6.9, 7.6 g), 2 imm ♂ (7.1, 7.8 g). Mount Dryander (57) 1 – 13 Aug 64, 9 ♂ (7.2, 7.2, 7.5, 7.8, 8.0, 8.0, 8.1, 8.1, "12.5" g), 5 ♀ (6.6, 6.8, 7.0, 7.1, 7.2 g), 1 ø (7.9 g), 2 imm ♂ (7.7, 8.3 g).

Distribution, Abundance & Habitat. Common where found. At Inkerman (50) in mangroves at the mouth of the Staaten River and Salt Arm Creek. At Ayton (52) in rainforest along the Bloomfield River and in mangroves at Weary Bay. At Tully (54) in both dense forest and clearings. At Cape Gloucester (56) in mangroves and tea-tree forest. At Mount Dryander (57) in low, dry rainforest.

Behaviour. Usually feeding among blossoms and foliage in the canopy, but also found in small bushes and vines. Often solitary but also in small family parties or mixed bird parties including variously *Malurus amabalis, Machaerirhynchus flaviventer, Zosterops lutea, Myiagra ruficollis, Ptiloris victoriae, Monarcha trivirgata, Colluricuncla megarhyncha, Tregellasia capito, Pachycephala simplex, Rhipidura fuliginosa* and *Sericornis magnirostris.* Its song a rising and falling warble, also a soft twitter or "braying squeak", apparently either not frequently uttered or else not very penetrating as few collectors made notes on it.

Food. Insects, all unidentifiable but the mash in the stomach was sometimes, black, grey, brown, green or yellow, suggesting a varied diet.

Annual Cycle. The breeding season was apparently the same in all localities, for pneumatisation of the skulls of immature birds and the completion of moult of adults was progressively more advanced through June, July and August. All birds are mainly or entirely in fresh plumage except a female from Inkerman (50) which was in moult of body, tail and wings. Some males from Ayton (52) and Tully (54), and all from Cape Gloucester (56) and Mount Dryander (57) had enlarged testes but the females showed no significant gonad activity.

Impermanent Colours. *Legs* slate grey to black, soles pale. *Iris* bright red or orange-red; immature birds usually duller, sometimes with paler outer ring. *Bill* blackish-brown or black with pale tip: immatures with paler lower mandible. *Mouth* blackish, whiter in young birds.

Geographical Variation. Our series falls into three distinct groups differing in the colour of the upper breast and centre of the throat of the males; (a) from Inkerman (50) in which the pattern is dark brown contrasting sharply with the white moustachial streak and the yellow of the abdomen: (b) from Ayton (52) in which the pattern is consistently and strikingly paler brown but still sharply defined: (c) from Tully (54), Cape Gloucester (56) and Mount Dryander (57) in which the pattern is very pale olive-yellow, not well-defined and not very different from the pale yellow of the females.

The dark-throated birds from the Gulf of Carpentaria (50) match *G. p. personata* from Cape York and the pale-throated southern birds are clearly *G. p. flavida,* type locality Herbert River at about 18°29′S between Tully (54) and Cape Gloucester (56). It is more difficult to name the intermediates from Ayton (52), for the only name available on the east coast is *G. p. johnstoni,* type locality Johnstone River 17°32′S, under 40 miles north of Tully (54). Meise (1931: 328) uses this name for the populations from the Johnstone River north to Cairns of which we have no representatives, Ayton (52) lying about 100 miles further north. Unless there is an abrupt change in the 40 miles between Tully (54) and the Johnstone River it does not seem from our series that *johnstoni* can be used for the Ayton (52) birds, but the question can only be resolved by reassembling the relevant types and available material.

Gerygone magnirostris (excluding *tenebrosa*) Large-billed Warbler

Specimens. Northern Territory. Darwin (41) 22 Sept 68, 1 ♂. Mount Roper (45) 26–28 Oct 68, 3 ♂ (7.5, 7.6, 7.8 g), 1 ♀ (6.9 g), 1 juv ♂ (7.7 g). Queensland. Ayton (52)

20 – 27 June 64, 4 ♂ (7.0, 7.4, 7.5, 7.8 g), 2 ♀ (6.2, 7.1 g). Tully (54) 11 – 14 July 64, 1 ♂ (6.8 g), 2 ♀ (6.3, 6.7 g).

Distribution & Abundance. Only found where collected, except that one was shot but not retained at Cape Gloucester (56). Some of the Ayton (52) birds were collected from islands in the mouth of the Bloomfield River. Fairly common locally.

Habitat. Usually mangroves, but in northern Queensland (52, 54) also in dense rain-forest, twice in Bull Oaks. On the Roper River (45) in a riverine forest strip with mangroves beneath, flooded at high tide, the habitat more open than the mangroves in which it was found at Darwin (41).

Behaviour. Usually found singly, high in the canopy. The song was described as a descending scale of whistling notes, sometimes interspersed with chatter.

Food. Small insects.

Annual Cycle. The October juvenile from Roper River (45) indicates spring breeding, but birds from all localities were either just completing or had completed moult, which suggests that the breeding season in Queensland may be earlier than in Northern Territory.

Impermanent Colours. *Legs* blue-grey. *Iris* adults orange-brown to red; juvenile brown. *Bill* adults black or dark brown (twice); juvenile black with yellow gape. *Mouth* adults usually black, sometimes paler; juvenile yellow.

Geographical Variation. Birds from Ayton (52) in northern Queensland differ from those of Northern Territory *G. m. magnirostris* in being more olive above with the buff on the underparts richer and more extensive, almost forming a breast-band: they are also smaller (wing ♂ 54–55, ♀ 51–52, against ♂ 56–58, ♀ 52–54). They come from 40 miles north of the type locality of *cairnsensis* and answer to Mathews' description. The three Tully (54) birds from 100 miles south of Cairns are slightly paler and less buffy below and smaller again (wing ♂ 53, ♀ 50). They may prove to be separable.

Notes. We do not follow the CSIRO Index (1969) in regarding *tenebrosa* as conspecific with *G. magnirostris* for adult birds differ in their greater size, the fineness of the bill, the paler and plainer underparts, black legs and the white eye. Furthermore the young birds of *tenebrosa* have a marked yellow wash while those of *magnirostris* (as exemplified by the juvenile collected) have hardly a trace of yellow. Also *tenebrosa* is found in lower mangroves and may have a rather different call.

Gerygone tenebrosa — Dusky Warbler

Specimens. Western Australia. Karratha (1) 29 – 30 May 66, 2 ♂, 1 ♀, 1 imm ♂, 1 imm ø, 1 ♂ skel, 1 fl. Peg's Beach (2) 1 – 3 June 66, 3 ♀, 2 ♂ skel, 1 ø skel. Carnarvon (5) 14 June 66, 1 ♂. Point Torment (30) 30 May – 12 June 68, 1 ♂ (7.7 g), 1 ø (6.1 g), 1 imm ♂ (7.1 g).

Distribution & Abundance. Common where collected.

Habitat. At Karratha (1) in low mangroves on the tidal flats at the mouth of the Maitland River and on the edge of the shell beach at Peg's Beach (2). In fairly dense 12 ft high mangroves at Carnarvon (5). At Point Torment (30) where both this species and *G. levigaster* were collected in the same mangrove swamp, the latter was found on the edge and *tenebrosa* in the main swamp formed of thick but fairly low trees.

Behaviour. Found usually singly, feeding in the foliage. The song was described as a soft trill. One bird coming to a decoy flew into nearby mangroves and "hung upside-down with wings quivering" (WHB).
Food. Small insects.
Annual Cycle. The immature June birds from Point Torment (30) and Carnarvon (5) are in comparable plumage, while those from Karratha (1) in May appear somewhat younger, having more yellow below, so the breeding season appears to be the same everywhere. Adult birds, as would be expected, were mostly completing moult.
Impermanent Colours. *Legs* black. *Iris* adults white; young birds yellowish white. *Bill* adults black; young birds brownish. *Mouth* grey or black; young birds yellow and black.
Other Notes. Since *G. magnirostris* has been separated and retains the name Large-billed Warbler, *G. tenebrosa* can appropriately conserve the name Dusky Warbler.

Gerygone chloronota — Green-backed Warbler

Specimens. Western Australia. Joint Hill (34) 1 July 68, 1 ♀ (5.7 g). Northern Territory. Darwin (41) 12 – 20 Sept 68, 2 ♂ (6.6, 6.8 g), 1 imm ♂ (6.7 g). Mount Roper (45) 1 Nov 68, 1 ♂ (6.9 g).
Distribution, Abundance & Habitat. This species has not previously been found in Australia west of Parry's Creek near Wyndham (39) nor east of Yirrkala (136° 53′E) and only in the coastal plain under 500 ft. The single birds found at Joint Hill (34) and Mount Roper (45) therefore extend the range considerably. At Darwin (41) it was fairly common in the forest on the edge of mangroves in Hope Inlet, and was found also in the mangroves themselves; the Mount Roper (45) bird was in similar forest. In contrast the Joint Hill (34) female was in eucalypt woodland of a sandstone gorge in Manning Creek, 14½ miles south of Joint Hill peak, and at 1,800 feet. The Manning Creek runs into the Hann, which forms part of the upper Fitzroy river system, so this bird appears to represent a population in a very different habitat from others in Australia and isolated from them by the watershed of the Kimberleys. The habitat change is less surprising since in New Guinea the species is found in "open forest and secondary growth formations from the lowlands up to 1000 m" (Mayr, 1941: 123).
Behaviour. Very restless. The Kimberley (34) bird was picking at the underside of leaves in a low bush: a pair at Darwin (41) were also in low bushes, two others were in the topmost branches of forest trees, while the Roper (45) bird was about 20 feet up. A few high-pitched notes were heard from the young bird, but the others were silent.
Food. Insects.
Annual Cycle. All adults collected were in various stages of moult and apparently post-breeding. The July female (34) was in mixed plumage in wings and tail (but no feathers actually moulting) and had some new feathers coming in on the throat; the September adults from Darwin (41) were in full moult and the November male (45) was just completing moult. The immature bird did not have the skull fully pneumatised and is duller on the back than the adults, with less contrast between the green of the mantle and the grey of the head.

Impermanent Colours. *Legs* dark grey. *Iris* red or red-brown. *Bill* adults black; immature dark horn. *Mouth* black.

Geographical Variation. The Joint Hill (34) female is yellower on the mantle and paler on the head than the others. Below it has more yellow on the under tail-coverts and thighs, more buff wash in the centre of the breast and less grey at the sides of the breast than the males. It has been compared also with the series in the American Museum of Natural History and was found to differ from all others in the same characters above and in the amount of yellow on the under tail-coverts, but some specimens from Melville Island have as much or more buff on the underparts.

In view of the difference in habitat and in colour it seems likely that the Joint Hill (34) bird represents a new subspecies, but more specimens are needed and more needs to be known of its status in inland districts before it is named.

Gerygone levigaster — Buff-breasted Warbler

Specimens. WESTERN AUSTRALIA. Point Torment (30) 12 June 68, 1 ♂ (5.8 g), 1 ♀ (6.0 g). Wyndham (39) 31 July – 6 Aug 68, 4 ♂ (5.5, 5.7, 5.8 g), 2 ♀ (5.2, 6.3 g), 1 imm ♂ (5.2 g), 1 imm ♀ (5.8 g), 1 imm ø (5.1 g), 4 fl (5.1, 5.7, 6.2, 6.2 g). QUEENSLAND. Moonlight Creek (47) 4 – 12 May 64, 6 ♂ (6.2, 6.3, 6.4, 6.6, 6.9, 7.1 g), 1 ♀ (6.2 g), 1 ø (5.9 g), 7 imm ♂ (5.5, 5.6, 6.2, 6.2, 6.2, 6.4, 6.6 g), 1 imm ♀ (6.3 g), 8 imm ø (5.4, 5.5, 5.6, 5.7, 5.9, 6.0, 6.4, 6.5 g). Inkerman (50) 9 – 11 June 64, 4 ♂ (6.6, 6.2, 6.3, 6.8 g), 1 ♀ (6.2 g), 1 ø (6.0 g), 1 imm ø (6.6 g).

Distribution, Abundance & Habitat. The 5th Expedition found it fairly common in thin mangroves in Western Australia (30, 39) but apparently absent from the much taller and more extensive mangroves at Darwin (41) and on the Roper River (42). In Queensland very common in mangroves at Moonlight Creek (47) both in stunted bushes on the off-shore flats, in dense stands of tall trees (always outside the canopy), and in a single line of tall trees along the edge of creeks: once just outside mangroves in a mixed bird party in savanna woodland. Less common at Inkerman (50), in tall dense mangrove forests and bushes along the estuary.

Behaviour. Active at all levels in the trees; alone or in small groups, sometimes with *Zosterops lutea, Rhipidura fuliginosa* and *Myzomela erythocephala.* Inquisitive, coming readily to squeaking. Voice in Western Australia "very fine whistling notes, almost too high to hear," "slow descending warble", and in Queensland "beautiful song full of cadences"; these descriptions all applied to collected males; the small proportion of females collected suggests that they are more silent.

Food. Small insects including beetles and some white grubs. The stomach noted as muscular, with the lining easily detached.

Annual Cycle. Adults from all four localities were in various stages of moult, more fully advanced in those from Point Torment (30) and Inkerman (50). The young birds from Wyndham (39) in August have the underparts variously suffused with yellow and are in comparable plumage to some from Moonlight Creek (47) in May. Another from Moonlight Creek (47) on 8 May has only a trace of yellow below and is comparable to the June Inkerman (50) immature bird.

The Wyndham (39) males had the testes somewhat enlarged but otherwise no gonad activity was shown.

Impermanent Colours. *Legs* black, greyer in young birds. *Iris* adults varying between orange-brown and scarlet; young birds browner, twice noted as having a buff outer ring. *Bill* adults black with greyer base to lower mandible; young birds brownish but becoming black while still in immature plumage. *Mouth* adults black, young birds greyish, tinged with yellow or pink.

Geographical Variation. On locality the Western Australia birds represent *G. l. broomei* and the Queensland birds *G. l. mastersi,* while we have no specimens of *G. l. levigaster* from Northern Territory. Meise (1931: 364) separates the latter from the other two races on small differences in the tail-pattern, while he found *levigaster* and *broomei* browner, less olive than *mastersi.* Our series shows no consistent difference in these characters but the Queensland birds are slightly larger than those from Western Australia, the difference more apparent in weights than in measurements.

Gerygone fusca — Western Warbler

Specimens. WESTERN AUSTRALIA. Perth (7) 13 Feb 66, 1 ♀, 1 ♀ skel. Wonerup (8) 12 Mar 66, 1 fl. Metricup (9) 6 – 11 Mar 66, 2 ♂, 1 ♀, 1 imm ♂, 1 imm ♀ skel, 2 imm fl. Scott River (11) 18 Feb – 3 Mar 66, 2 ♂, 1 imm ♂, 1 imm ♂ skel, 1 fl. Newdegate (14) 12 Apr 66, 1 ♂, 1 imm ø. Van der Linden Lakes (24) 6 Sept 65, 1 ♂. Giles (25) 21 Aug 65, 1 ♂. QUEENSLAND. Tyrone (59) 12 May 63, 1 ♂, 1 ♀. NEW SOUTH WALES. Dunedoo (93) 23 Mar 63, 1 imm ♂.

Distribution & Abundance. Common locally in the southwest, but at other localities only seen singly.

Habitat. In the southwest in all types of woodland and forest. In Desert Oaks at the Van der Linden Lakes (24) and in mulga and gums by a creek near Giles (25). Mulga with some sandalwood and Gidgee trees at Tyrone (59), and mixed woodland at Dunedoo (93).

Behaviour. Feeds actively in the canopy. Found singly or in small parties, sometimes in company with thornbills and silvereyes. Its song was heard frequently at all times of the day in February at Perth (7) and Scott River (11) and described as "prolonged, rambling, relatively low-pitched and melodious" (CJOH). Serventy noted that the song of the Giles (25) bird differed slightly from those at Perth (7).

Food. Small insects.

Annual Cycle. The March young birds from the southwest and from New South Wales are in comparable plumage with a yellow wash on the underparts (in this plumage notably more ochre and less lemon than young birds of *levigaster*). All adults from the southwest (7-11) were in full moult except those from Newdegate (14) which had completed moult. The two August, September birds from the centre (24, 25) were fairly worn, and those from southern Queensland (59) in May fairly fresh. No gonad activity.

Impermanent Colours. *Legs* dark grey or black. *Iris* adults red or red-brown, young birds red (once), creamy-white (once), buff (once). *Bill* black with a grey base, paler in young. *Mouth* blackish.

Geographical Variation. Birds from the southwest (7-11) are greyer on the breast and throat and slightly darker above than those from the centre (24-25) and the east (59), and can be referred to *G. f. fusca*. Two other races are recognised, western birds *G. f. mungi* being paler and slightly smaller than *G. f. exsul* in the east; this division is supported by our three males, the Western Australian specimens having wings of 56 against 59 mm for the Queensland bird: they appear slightly paler above with less white round the eye, but this may be due to wear and the make-up of the skins.

The Newdegate (14) male is closest to *G. f. fusca* above but has a buff wash on the breast and matches *mungi* in size, wing 55 mm. It is perhaps best regarded as *G. f. fusca* x *mungi*.

Smicrornis brevirostris Weebill

Specimens. WESTERN AUSTRALIA. Coolawanyah (4) 18 May 66, 1 fl. Mount Barker (12) 22 – 24 Mar 66, 1 ♀, 1 ø, 2 ♂ skel. Newdegate (14) 12 – 14 Apr 66, 1 ♀, 1 ø. Cocklebiddy (15) 8 – 10 July 65, 2 ♂, 2 fl. Eucla (17) 1 – 5 July 65, 3 ♂, 3 ♀. Mount Shenton (18) 3 Aug 65, 1 ♂, 1 ♀. Mungilli Claypan (20) 13 Sept 65, 1 ♀, 1 ♂ skel. Lake Gruszka (21) 10 Sept 65, 1 ♀. Van der Linden Lakes (24) 6 Sept 65, 2 ♂, 1 imm ♂. Mount Anderson (28) 8 – 13 May 68, 1 ♀, 1 ø (5.2 g), 1 ♂ skel (5.1 g), 1 ♀ skel (5.2 g), 3 fl. Derby (29) 22 May 68, 1 fl (4.2 g). Joint Hill (34) 27 June 68, 1 ♀. QUEENSLAND. Moonlight Creek (47) 2 – 13 May 64, 2 ♂ (5.4, 5.6 g). Glenore (49) 11 – 23 May 64, 3 ♂ (5.0, 5.1, 5.1 g), 3 ♀ (5.0, 5.5, 6.5 g). Glendower (55) 12 Mar – 2 Apr 64, 5 ♂ (5.4, 5.4, 5.4, 5.5, 5.8 g), 3 ♀ (5.4, 5.5, 6.1 g). Banoona (58) 3 May 63, 1 fl. Monkira (61) 6 June 63, 1 ♂. SOUTH AUSTRALIA. Kingoonya (70) 19 June 65, 1 ♀, 1 ♀, 1 fl. Minippa (73) 1 Nov 64, 1 ♂. Nonning (79) 1 – 3 Apr 65, 1 ♂, 1 ♀ (5.5 g). Aroona Valley (82) 8 – 17 Apr 65, 4 ♂ (5.3, 5.8, 5.9, 6.0 g), 3 ♀ (5.1, 6.3, 6.9 g). Angepena (84) 20 Nov 63, 2 ♂. VICTORIA. Coleraine (89) 21 Jan 63, 1 fl. Little Desert (91) 9 Jan 63, 1 fl. NEW SOUTH WALES. Albury (92) 11 Mar 63, 1 ♂. Dunedoo (93) 23 Mar 63, 1 ♀, 1 ♂ skel, 1 fl. Tenterfield (95) 15 Apr 63, 1 ♂.

Specimens to other Museums. NMV 1 fl Tyrone (59) 12 May 63, 1 ♀ Monkira (61). AM 1 ♂ Banoona (58), 1 ♀ Grafton (94) 5 Apr 63.

Distribution & Abundance. In Western Australia present but local at most main camps except Karratha (1), Bush Bay (6) and Metricup (9) on the west coast, Warburton (23) and Giles (25) in the centre, and Wyndham (39) in the northeast. Not seen in Northern Territory, nor on the east coast of Queensland but fairly common near the Gulf of Carpentaria (47-49) and inland at Glendower (55). Absent from the central districts of South Australia (64-68), but common in the mallee at Kingoonya (70) and Nonning (79) and along the creeks at Aroona Valley (82). Not recorded at Furner (86) but present at all other camps in the southeast (87-95).

Habitat. Usually in eucalypts but also in tea-trees (49): mainly in the canopy but found once in the Kimberleys (28) keeping low in spinifex.

Behaviour. Usually in small parties of up to ten birds, active in searching the foliage, hovering and flitting from twig to twig. Sometimes in company with thornbills or pardalotes, inquisitive, unafraid and very vocal, calls included a liquid musical call and a throaty *tchik-tchik-tchik* like that of the Brown Thornbill (DG).

Food. Mainly small insects, including ants.

Annual Cycle. Definite evidence of August/September breeding in central Western Australia is provided by the young bird from the Van der Linden Lakes (25) and a female with large yolked oocytes in the ovary from the Mungilli Claypan (20). All birds from this central area (20–25) were in conspicuously worn plumage. In northern Queensland females from Glenore (49) in April had eggs forming, and the males here and at Moonlight Creek (47) had enlarged testes, but these birds were mainly in fresh plumage, completing moult on the body. Birds from Glendower (55) in April the same year and from Monkira (61) in June the previous year were in similar plumage to the northern birds but showed no gonad activity.

In the south of the country the Mount Barker (12) birds were in full moult of wings (once) and tails in March, and Newdegate (14) birds two weeks later were completing moult. The two males from Cocklebiddy (15) in July had enlarged gonads in contrast to birds from Eucla (17) collected the week before, in which the gonads were minute. In the Eyre Peninsula the Minippa (73) November male was in early stages of moult and the Nonning (79) April birds completing moult. Similarly in the Flinders Range the Angepena (84) November birds were starting moult in the wings and those from the Aroona Valley (82) in April were completing, or had completed, moult. The three specimens examined from New South Wales (92–94) were also in moult.

In summary it seems that the moult is either very protracted or the season varies locally or annually, and it is not necessarily synchronised with the breeding season.

Impermanent Colours. *Legs* varying shades of brown and grey, on the whole paler and pinker in the Yellow Weebills (see below), this difference visible also in the dried skins. *Iris* pale yellow or whitish, greyer in the young bird than in adults collected at the same time. *Bill* similar to the legs in variability, the lower mandible paler than the upper. *Mouth* greyish, perhaps darker in some birds from the southwest and pinker in northern birds.

Geographical Variation. Keast (1958g) recognises three main groups of Weebills, the Yellow, Pallid and Brown, the two latter intergrading. Within these groups he recognises respectively three, two and two races.

All our birds from the southwest of Western Australia across the continent to New South Wales clearly belong to the Brown Group. They show remarkably little variation throughout this range though a tendency in the western birds to be darker above and yellower below allows two races to be recognised, (a) *S. b. brevirostris* extending westwards to the Flinders Range (82–95) and Banoona (58), (b) *S. b. stirlingi* in Western Australia (12–18), while birds from Kingoonya (70) and the Eyre Peninsula (73, 79) are intermediate.

The worn birds from central Western Australia (20–25) fall within the range of *S. b. mathewsi* of the Pallid Group but in this plumage are not conspicuously different from worn specimens of *S. b. brevirostris*. The fluid specimen from Banoona (58) cannot be identified subspecifically but when collected was noted as a Brown not a Yellow Weebill.

The remaining specimens belong to the Yellow Group and show little variation except that Glendower (55) birds are darker above, especially on the heads, and answer the description of Keast's new race *S. b. cairnsi*, although he believed this to be confined to the area of high rainfall in northeastern Queensland and so it would not be expected at an inland locality. The others from northern Western Australia (28-34), northern Queensland (47, 49) and Monkira (61) in southern Queensland are *S. b. flavescens*. The Monkira (61) bird is of especial interest for this locality lies well within the range of the Pallid Group, as shown on Keast's map, and suggests that in this area the two groups may overlap without intergradation. The situation requires further study with particular attention to the breeding seasons, since our specimens of the Pallid Group were breeding in August/September and of the Yellow Group in April. It may eventually be desirable to recognise two species within the Weebills.

Acanthiza lineata — Striated Thornbill

Specimens. VICTORIA. Coleraine (89) 19 – 21 Jan 63, 2 ♂, 1 ♀, 1 ♀ skel, 1 fl. NEW SOUTH WALES. Tenterfield (95) 10 Apr 63, 2 ♂, 2 ♀, 2 ∅.
Specimens to other Museums. NMV 1 ♂ Tenterfield (95).
Distribution & Abundance. At Coleraine (89) in a party of about ten and at Tenterfield (95) in a large flock of over twenty. Observed also at Furner (86) and Albury (92).
Habitat. Coleraine (89), tall Red Gum and stringy-bark woodland; Tenterfield (95), open woodland and in open forest with tall bushes and secondary growth on a mountainous ridge, 2,800 ft.
Behaviour. All birds encountered were actively searching the outer foliage in tall trees, some in company with the Buff-rumped Thornbill *A. reguloides*.
Food. Insects.
Annual Cycle. All gonads were minute. The New South Wales birds were completing moult.
Impermanent Colours. *Legs* grey or brown. *Iris* greyish-white or light brownish-yellow. *Bill* grey or brown. *Mouth* grey. A greyish white iris and greyish bill and legs were noted only in all the Coleraine (89) birds and in one female from Tenterfield (95). The variation displayed might, therefore, have some geographical or a seasonal basis.
Geographical Variation. The birds from New South Wales are greener above and yellower below than those from Victoria, thus fitting with the accepted races *A. l. lineata* and *A. l. chandleri* respectively.

Acanthiza nana — Little Thornbill

Specimens. QUEENSLAND. Banoona (58) 29 Apr – 6 May 63, 2 ♂, 2 ♀. Tyrone (59) 15 May 63, 1 ∅.
Specimens to other Museums. NMV 1 ♂ Banoona (58).
Field Notes. Fairly common in small parties in mixed woodland with some pines and low scrub at Banoona (58). Found at Tyrone (59) in Mulga woodland. Also recorded from New South Wales at Dunedoo (93) and Grafton (94). Noted as being more arboreal than *A. pusilla*, and feeding 10 to 20 feet up in the trees.
Food. Small insects.

Annual Cycle. All birds had very small gonads and were in fresh plumage, one just completing moult 6 May.
Impermanent Colours. *Legs* black with greyish soles. *Iris* brown. *Bill* black. *Mouth* black.
Geographical Variation. These birds clearly belong to the pale inland race *A. n. modesta.*

Acanthiza pusilla (including *katherina*) Brown Thornbill

Specimens. WESTERN AUSTRALIA. Bush Bay (6) 12 – 18 June 66, 2 ♂. Perth (7) 12 Feb 66, 1 imm skel. Metricup (9) 6 – 15 Mar 66, 5 ♂, 1 fl. Margaret River (10) 14 Mar 66, 1 imm ♀. Scott River (11) 18 Feb – 1 Mar 66, 3 ♂, 3 ♂ skel, 2 ♀ skel, 5 ø skel, 1 fl. Mount Barker (12) 24 Mar 66, 1 ø. Green Range (13) 31 Mar – 8 Apr 66, 5 ♂, 1 ø. Newdegate (14) 12 – 14 Apr 66, 4 ♂, 2 ♀, 3 fl. Cocklebiddy (15) 8 – 10 July 65, 4 ♂, 2 ø, 1 fl. Eucla (17) 1 – 5 July 65, 4 ♂, 4 ♀, 1 ø, 1 ♀ skel, 1 fl. Mount Shenton (18) 3 Aug 65, 1 ♀. Lake Gruszka (21) 10 Sept 65, 1 ♂. Hunt Oil Camp (22) 9 Sept 65, 1 ♂. Warburton (23) 14 Aug 65, 2 ♂, 1 ♀. Van der Linden Lakes (24) 6 Sept 65, 5 ♂, 1 ♀. Giles (25) 21 Aug – 1 Sept 65, 2 ♂, 4 ♀, 2 ø, 1 ♂ skel. Walter James Range (26) 26 Aug 65, 1 ♂. QUEENSLAND. Mount Dryander (57) 1 – 3 Aug 64, 2 ♂ (7.9, 81. g). Banoona (58) 28 Apr – 6 May 63, 2 ♂, 2 ♀, 1 fl. SOUTH AUSTRALIA. Granite Downs (68) 4 – 11 June 65, 3 ♂. Nullarbor HS (69) 4 – 6 Oct 63, 2 ♂. Kingoonya (70) 10 – 19 June 65, 2 ♂, 1 ♀, 1 fl. Sceale Bay (75) 13 – 14 Oct 63, 1 ♀, 1 juv ø. Nonning (79) 30 Mar – 2 Apr 65, 2 ♂ (7.8, 8.6 g), 3 ♀ (6.8, 7.2, 8.1 g). Aroona Valley (82) 8 – 19 Apr 65, 7 ♂ (5.3!, 7.0, 7.0, 7.6, 7.7, 7.8, 8.7 g), 2 ♀ (7.0 g), 1 imm ♂ (8.0 g). Angepena (84) 21 Nov 63, 1 ♂. Furner (86) 7 – 9 Jan 63, 1 ♂, 1 fl. VICTORIA. Victoria Range (90) 30 Jan 63, 1 fl. NEW SOUTH WALES. Tenterfield (95) 12 Apr 63, 1 ♀.
Specimens to other Museums. NMV 1 ♂ Banoona (58). SAM 1 ♂ Nullarbor HS (69).
Distribution & Abundance. Found at most main camps in Western Australia south of the Tropic, but while common in the southwest, and at Eucla (15) and Cocklebiddy (17), it was curiously local in the central camps: for instance at Warburton (23) it was found only in the Gibson desert to the west of the mission, it was very common at the Van der Linden Lakes (24) to the north but scarce at Mount Shenton (18) and not seen at Carnegie (19). In South Australia it was noted as common at Nonning (79), Granite Downs (68) and in the Flinders Range (82), and "common but local" at Kingoonya (70) where it was only collected at the Amelira Rockhole 44 miles to the southwest. It was not seen at all in the northeastern part of the State or at Monkira (61) in south-eastern Queensland, but a few were seen in the other three camps of southwestern Queensland (59-61). It was common at most camps in the southeast (86-95) though only a few were seen at Nelson (88) and Coleraine (89). It was often found singly or in small family parties, but also joined mixed parties of small birds, including particularly other thornbills and Weebills.
Habitat. Extremely varied. In Western Australia from samphire scrub along the coast at Bush Bay (6), mulga in the central camps (18-26), mallee and scrub in the southeast

(14-17) to woodland, forest and thick bush in the extreme southwest (9-13). In South Australia associated particularly with native pines in Aroona Valley (83) and with mallee at Kingoonya (70), but found also in more open country; at Furner (86) in open woodland with dense scrub. The birds from northeastern Queensland (57) were in dry rainforest on the edge of a rocky creek, and in southern Queensland, New South Wales and Victoria mostly in woodland.

When not actually in forest or woodland the Brown Thornbill seems rarely far from some sizeable trees or good cover. Casuarinas, in particular the Desert Oak (*C. decaisneana*), eucalypts or native pines are mentioned frequently in the field notes, and these associations may account for the local distribution in some areas.

Behaviour. Usually encountered searching for food in trees and bushes at all levels in the foliage but seen feeding on the ground on at least two occasions. Its habit of cocking its tail, said to be characteristic, was observed frequently but by no means invariably, and Goodwin noted that tail-cocking to a fairly high angle seemed to occur chiefly when the bird was in a state of excitement (social or sexual?). He describes various calls, a *tchit*! *tchit*! of alarm, a soft, plaintive *sooee*! *sooee*! and a short song phrase.

Food. Chiefly insects, including beetles and ants, but also some seeds and buds.

Annual Cycle. In southwestern Western Australia (7-14) the majority of specimens collected were in moult and the immature birds from Perth (7) and Margaret River (10) on 12 February and 14 March indicate that birds were post-breeding. Further north both males from Bush Bay (6) in June had very enlarged testes, possibly indicating an earlier breeding season. Among the birds from the centre of the State there was gonad activity in August and September, many from Warburton (23) and the Van der Linden Lakes (24) having enlarged gonads, though at Giles (25), the intervening camp further west, the gonads were appreciably smaller. This central area had been having and did have some local rain before and during the visit of the expedition, and the local variation in the state of the gonads suggests opportunist breeding. In the southeast two males from Cocklebiddy (15) had slightly enlarged gonads.

The two males from northeastern Queensland (57) had very large testes in early August: those from the south (58) were in fresh plumage at the end of April, one just completing moult.

In South Australia the juvenile from Sceale Bay (75) in the Eyre Peninsula was probably hatched in September: at Nonning (79) at the end of March birds were in full moult, and they were finishing moult in the Aroona Valley (82) in April. The Furner (86) specimen in January was also in full moult. Elsewhere no specimens show moult or gonad activity.

Impermanent Colours. *Legs* black or dark grey. *Iris* generally red-brown – a narrow white outer ring, or part of a ring, was recorded on a number of specimens of both sexes from various localities and by different collectors, though the same collectors do not mention this feature for other specimens: it is difficult to know whether it was overlooked, or if there is individual variation in this character. *Bill* black or blackish. *Mouth* grey or black.

No impermanent colours were recorded for the one juvenile collected, but the immature bird from Perth (7) had a greyish eye, some pink towards the base of the bill and a yellow mouth, and that from Margaret River (10) a brown eye, a light base to the bill and a grey mouth.

Geographical Variation. Two groups are recognised. One, which occurs in the east, southeast and on the islands of the Bass Strait and Tasmania, has the upper parts olive green. Only four of our specimens belong to this group, the two from Mount Dryander (57) which have clear yellow on the flanks and abdomen and well-marked rufous foreheads, the Tenterfield (95) female, in which the flanks are more buffy and the chestnut of the forehead merges into the crown, and the moulting Furner (86) bird, which appears rather duller, but is not really suitable for comparison. The four birds fall into three recognised races, respectively *A. p. katherini* of northern Queensland, *A. p. pusilla* of New South Wales and *A. p. macularia* of Victoria.

(The Queensland *katherina* is regarded as a distinct species in the CSIRO Index 1969 but in view of the variation in habitat and colour elsewhere its differences do not seem great enough to warrant this.)

Of our series in the second group with greyer backs, the southern Queensland (58) birds stand out on the rich chestnut of their upper tail-coverts, light markings of the throat and breast and white flanks and abdomens: they are clearly referable to *A. p. albiventris.*

The other population in this group which is clearly racially identifiable is *A. p. leeuwensis* of the woodland and forest of the extreme southwest (9-13) in which birds have some olive-brown on the back, the throat and breast are heavily marked and the flanks and under tail-coverts are deep buff.

Our birds from Newdegate (14), Cocklebiddy (15), Eucla (17) and southern South Australia east to Lake Angepena (84) and some in the British Museum from Kalgoorlie and Southern Cross are greyer above than *leeuwensis* and the flanks and under tail-coverts are paler buff. They also differ quite sharply from those of Bush Bay (6) and the mulga areas of central Western and South Australia (18-26, 68) which are again paler on the back, less heavily streaked on the throat and breast, and paler on the abdomen and flanks, though not pure white as in *albiventris*. The Kingoonya (70) specimens are intermediate between those of the mulga areas and those of the south.

Mayr & Serventy (1938) separated southwestern birds immediately surrounding the range of *leeuwensis* as *A. p. apicalis* but lumped those from other areas of Western and South Australia as *whitlocki,* following Mack (1936), though recognising considerable variation amongst them. Our new material suggests that *apicalis* can be used for all birds from the mallee and other richer vegetation south of the mulga across to Lake Angepena (14, 15, 17, 69, 75, 79, 82, 84), and that *whitlocki* should be confined to the pale mulga birds.

Acanthiza robustirostris — Robust Thornbill

Specimens. WESTERN AUSTRALIA. Mount Shenton (18) 30 July – 2 Aug 65, 1 ♂, 1 ♀, 1 ø. Mungilli Claypan (20) 12 Sept 65, 1 ♂. Lake Gruszka (21) 10 Sept 65, 1 imm ♂. Hunt Oil Camp (22) 9 Sept 65, 1 ♂, 1 ♀, 1 ø. Warburton (23) 9 – 13 Aug 65, 5 ♂, 4 ♀. Giles (25) 24 Aug – 2 Sept 65, 3 ♂, 3 ♀. SOUTH AUSTRALIA. Granite Downs (68) 4 – 10 June 65, 3 ♂, 1 ♀, 1 ø.

Specimens to other Museums. WAM 2 fl Warburton (23).
Distribution & Abundance. Common at Carnegie (19) as well as at all the camps where it was collected except Mount Shenton (18), where it was rare.
Habitat. Always in mulga, sometimes with spinifex or saltbush, and on rocky soil or sand.
Behaviour. The Robust Thornbill occurs so frequently alongside the Chestnut-rumped Thornbill and they are so alike in appearance that particular attention was paid to comparisons between the two, but it was difficult to specify many ecological or behavioural distinctions. On the whole the Robust Thornbill seemed less gregarious, only six out of the thirty birds collected being found in parties (sometimes with Chestnut-rumped Thornbills) while the rest were either solitary or in pairs. Both species feed on the ground as well as in bushes and trees but the Robust Thornbill was not observed taking food on the wing. Its *see-see* call is lower pitched than that of the Chestnut-rumped Thornbill (Goodwin, 1967c: 244) and more like that of the Brown Thornbill, as is its harsh *tschritt:* it also has a sweet trisyllabic song phrase, heard uttered on one occasion by a male displaying to a female; the song was accompanied by a puffed out throat, raised feathers of the forehead and side to side movements of the head.
Food. Out of twenty stomachs examined, three contained caterpillars and two ants, while the remainder had nothing but unidentified insect fragments – in contrast to the Chestnut-rumped Thornbill (see below).
Annual Cycle. One bird at Mungilli Claypan (20) on 12 September was seen carrying food in its bill, apparently to feed young. Some birds from all the camps in Western Australia had the gonads enlarged, but those from Granite Downs (68) showed no gonad activity. One bird from Giles (25) was replacing some tail-feathers but no others were in moult and little can be learnt from the condition of the plumage. The immature male was diagnosed as such because the skull was not fully pneumatised, but it gives no other sign of immaturity.
Impermanent Colours. *Legs* greyish black; soles, white. *Iris* reddish-brown, sometimes noted as having a paler or white outer ring. *Bill* black. *Mouth* dark fleshy grey or black.
Geographical Variation. None, but some individual variation in the amount of grey in the crown and mantle.

Acanthiza uropygialis — Chestnut-rumped Thornbill

Specimens. Western Australia. Newdegate (14) 14 Apr 66, 1 ♂ skel. Mount Shenton (18) 30 – 31 July 65, 1 ♂, 3 ♀. Carnegie (19) 16 – 17 Sept 65, 3 ♂. Lake Gruszka (21) 10 Sept 65, 1 ♀. Hunt Oil Camp (22) 9 Sept 65, 5 ♂, 1 ♀, 1 ø. 80 m SW Warburton (23) 6 Aug 65, 1 ø. Warburton (23) 9 – 13 Aug 65, 2 ♂, 1 ♀, 1 ø, 5 fl. Van der Linden Lakes (24) 6 Sept 65, 1 ♂. Giles (25) 23 – 30 Aug 65, 5 ♂, 6 ♀. Walter James Range (26) 26 Aug 65, 1 ♂, 1 ♀. Queensland. Tyrone (59) 12 – 15 May 63, 1 ♂. Thylungra (60) 24 – 26 May 63, 1 ♂, 1 fl. South Australia. Etadunna (66) 18 – 22 May 65, 2 ♂, 1 ♀. Granite Downs (68) 4 – 11 June 65, 13 ♂, 3 ♀, 4 ø. Kingoonya (70) 16 – 23 June 65,

6 ♂, 1 ♀, 1 ø, 1 ♀ skel. Nonning (79) 29 – 31 Mar 65, 3 ♂ (6.3, 6.4 g). Aroona Valley (82) 8 – 15 Apr 65, 2 ♂ (6.07, 6.4, 6.4 g), 2 ♀ (6.09, 6.2 g). Angepena (84) 3 Dec 65, 1 imm ♂.

Specimens to other Museums. NMV 1 ♂ Tyrone (59). WAM 2 fl Giles (25).

Distribution & Abundance. The 3rd Expedition found the Chestnut-rumped Thornbill about the commonest bird in all the camps of the west centre (18-26, 68, 70, 79). Eastwards also common in the Flinders Range (82), but not so abundent as the Brown Thornbill; only two were seen at Oodnadatta (67), it was uncommon at Etadunna (66) and absent from Clifton Hills (65) and Innamincka (64). Also absent from Eucla (17) and Cocklebiddy (15), which is surprising since it was found in mallee and scrub at Newdegate (14).

Habitat. Chiefly mulga country, but there found equally in woodland, saltbush, spinifex, on rocky hillsides or sand-dunes and in creeks. Also in mallee at Newdegate (14), in native pines in the Aroona Valley (82) and on the edge of woodland and saltbush at Etadunna (66).

Behaviour. Feeds mostly at low levels in the trees or in scrub, but sometimes also at high levels or on the ground. Goodwin (1967c: 243) observed a party gathering at dusk, possibly to roost communally, in small Bluebushes: when flushed the birds flew to the bases of other bushes, not to the near-by trees. Usually found in parties, often with Brown or Yellow-rumped Thornbills, Whitefaces or Redthroats, and three birds were once seen half-heartedly mobbing a Grey Butcher-bird. Most food is taken from the branches of foliage but it sometimes makes aerial sallies after flying insects.

Various calls were noted: a soft *see-see* when seeking food, which becomes louder and sharper in flight, *tchit-tchit-tchit;* a high-pitched alarm note and a short song phrase.

Food. Mostly insects including (once) beetles; sometimes seeds and buds as well.

Annual Cycle. In Western Australia the Newdegate (14) April bird had the tail in moult. Some of the birds from Gibson Desert camps (19-22, 24) and the male from the Walter James Range (26) had enlarged gonads but there was no activity evident in any from Mount Shenton (18), Warburton (23) and Giles (25). One of the males from the Hunt Oil Camp (22) with enlarged gonads was in worn plumage with some moult on the throat, and the male from the Van der Linden Lakes (24), also with enlarged gonads, was replacing some primaries and rectrices.

The Queensland birds were in fresh plumage with no gonad activity.

In South Australia a summer breeding season is indicated by the December immature from Angepena (84). Consistent with this is evidence of moult being completed and no gonad activity at Nonning (79) and Aroona Valley (82) in March and April. Some June birds from Kingoonya (70) had slightly enlarged gonads, while none from Granite Downs (68) show gonad activity, and amongst them are four with either the wings or tail in moult.

Impermanent Colours. *Legs* black with grey soles. *Iris* white tinged lemon or creamy white. *Bill* black. *Mouth* dark grey to black.

Geographical Variation. In this series considerable individual variation in the amount of rufous on the crown, the speckling on the forehead and rufous on the rump almost obscures geographical variation. The worn plumage of some specimens, and soil-staining in those from "the Red Centre" add to the difficulties of comparison. The Queensland birds have, however, more dark tips to the feathers of the throat and breast than do others, and on this character alone are identified as the allegedly darker race, *A. u. uropygialis,* which Mayr and Serventy (1938) recognise in the wetter parts of the range in both the west and the east, and to which the Newdegate (14) bird (now a skeleton) should also belong. The remainder are referred to *A. u. augusta* except those from Etadunna (16) which seem intermediate between the two races.

The immature from Angepena (84) differs from adults in having broad chestnut edges to the secondaries. It also has a considerable amount of rufous on the crown. It seemed that this latter character, which is apparent in several other individuals, might be associated with age, but this cannot be so since a female in which it is particularly marked has a convoluted oviduct, showing that she has laid eggs.

Acanthiza iredalei — Samphire Thornbill

Specimens. WESTERN AUSTRALIA. Carnarvon (5) 15 June 66, 1 ø skel. Bush Bay (6) 13 – 18 June 66, 5 ♂, 2 ♀ skel, 2 fl. Cocklebiddy (15) 9 July 65, 1 ♂. Eucla (17) 2 July 65, 2 ♂, 5 ♀. SOUTH AUSTRALIA. Kingoonya (70) 25 June 65, 1 ♂, 1 ♀.

Distribution, Abundance & Habitat. In the Carnarvon area (5, 6) found in small parties in samphire flats or the neighbouring mangrove bushes along the edge of the beach. At Eucla (17) fairly common in the Bluebush and saltbush of the Nullarbor Plain 16 to 20 miles from the coast, but in similar country north of Cocklebiddy (15) only one pair was seen. Only one small party was seen at Kingoonya (70) in an open, stony area of Bluebush (Big Tank paddock, North Well) about 15 miles to the northwest of the town.

Behaviour. Seen feeding both on the ground and in low bushes, sometimes going right into the centre of the bush. The members of a party followed each other in loose order but covered the ground quickly, sometimes chasing and sparring. There was one fight in which two met about 20 ft up and fell interlocked, almost to the ground. Their general behaviour seemed to Goodwin very like that of *A. chrysorrhoa* and so were their calls, a "tinkling twitter", and a more musical *seu-seu-seu.*

Food. Unidentified insect fragments.

Annual Cycle. No birds were in moult. The pair from Kingoonya (70) had the gonads enlarged, and there was evidence of gonad activity in one male from Bush Bay (6), and a few birds from the Nullarbor (15, 17).

Impermanent Colours. *Legs* black. *Iris* pale yellow or whitish. *Bill* black. *Mouth* black.

Geographical Variation. This series shows a cline from west to east of decreasing pigmentation; birds from Bush Bay (6) have the lower back greener, the rump buffier and the underparts yellower than those from the Nullarbor (15, 17), while the Kingoonya (70) birds are the palest. Condon (1951a: 46) has shown that these pale birds may be separated as *A. i. morgani:* it is more difficult to name the two distinct populations in Western Australia without series in strictly comparable plumage or any specimens from Lake Way, the type locality of *A. i. iredalei.* It seems possible that these from an arid inland habitat would be closer to the Nullarbor series than to the coastal Bush Bay (6) birds, in which case *A. i. uranie* is available for the latter. If, on the other hand, the coastal birds resemble typical *A. i. iredalei,* then the Nullarbor birds are best regarded as intermediates between *iredalei* and *morgani,* closer to the latter.

Acanthiza reguloides — Buff-rumped Thornbill

Specimens. SOUTH AUSTRALIA. Furner (86) 7 Jan 63, 1 imm ♀, 3 fl. VICTORIA. Coleraine (89) 19 Jan 63, 1 ♂. Little Desert (91) 26 Jan 63, 1 fl. NEW SOUTH WALES. Albury (92) 11 Mar 63, 1 ♂ skel, 2 fl. Dunedoo (93) 19 Mar 63, 1 ♀. Grafton (94) 5 Mar 63, 1 ♂. Tenterfield (95) 10 Apr 63, 1 ♂, 1 ♀.
Specimens to other Museums. NMV 1 fl Furner (86), 1 ♀ Albury (92), 1 ♂ Tenterfield (95). AM 1 ♂ Dunedoo (93), 1 ♀ Tenterfield (95).
Field Notes. On one occasion a single bird was collected, but otherwise always encountered in small parties or flocks of at least 20 birds, sometimes with other thornbill species (*A. pusilla* and *A. lineata*). A large party at Little Desert (91) was found in low scrub with *Banksia* association, but elsewhere always in open woodland, with gums, casuarinas, stringy-bark or ironbark. Feeds at any level in the trees and occasionally in the grass as well. The "jingling, fairy-like" call seemed not unlike that of the Yellow-rumped Thornbill.
Food. Once a tipulid, otherwise unidentified insects.
Annual Cycle. Some specimens from Coleraine (89), Dunedoo (93) and Tenterfield (95) were completing moult; all gonads were very small; the immature bird from Furner (86) had some yellow in the gape and the skull was not fully pneumatised. It seems that breeding had been completed in all areas.
Impermanent Colours. *Legs* dark grey to black. *Iris* whitish sometimes with yellow tinge. *Bill* dark grey, paler at base of lower mandible. *Mouth* adult, grey; immature, yellow.
Geographical Variation. The three specimens from northern New South Wales (94, 95) differ from the southern three in having the forehead greyer, the mantle greener and the lower back, upper tail-coverts and under parts yellower. On locality they can be referred to *A. r. nesa* (see Mayr & Serventy, 1938: 253), while the Dunedoo (93) female is *A. r. reguloides*. The single adult from Coleraine (89) and the immature from Furner (86) resemble the latter in having the mantle and lower back grey-green, but are duller above and below. The immature bird has some chestnut edges to the wing-coverts. Mack (1936: 99) places birds from this area as *A. r. australis* while Mayr & Serventy include them with *A. r. reguloides*.

Acanthiza inornata — Western Thornbill

Specimens. WESTERN AUSTRALIA. Perth (7) 17 Nov 62, 1 imm ♀. Metricup (9) 7–15 Mar 66, 4 ♂, 1 ♀, 2 ♀ skel, 1 ø skel, 2 fl. Margaret River (10) 14 Mar 66, 2 ♂, 1 ♀. Green Range (13) 31 Mar 66, 1 ♂, 1 ♀.
Field Notes. Usually in small parties, sometimes with *Rhipidura fuliginosa* or *Acanthiza pusilla*. In sclerophyll forest on the Darling Range at Perth (7); secondary growth in

forest of Jarrah, Marri and casuarina at Metricup (9) and Margaret River (10) and in mallee and mixed woodland at Green Range (13). Noted as feeding both at low levels in the trees and in the canopy.

Food. Small insects.

Annual Cycle. The immature November bird from Perth (7) places the breeding season there in early spring. In the extreme southwest birds were in full moult (9-10) or finishing moult (13) in March.

Impermanent Colours. *Legs* dark grey to black, soles buff. *Iris* pale straw or yellow-cream. *Bill* dark grey, paler at base of lower mandible. *Mouth* greyish-black, sometimes with pinkish tinge. In the immature bird the legs and bill were dark horn, the eye pale grey, the mouth pale yellow.

Geographical Variation. The immature Perth (7) bird is browner on the back and rump than the others: on locality it should be *A. i. inornata* while the latter are *A. i. mastersi* but in these plumages no true comparisons are possible.

Acanthiza chrysorrhoa — Yellow-rumped Thornbill

Specimens. Western Australia. Perth (7) 15 Feb 66, 1 ♂ skel. Metricup (9) 16 – 17 Mar 63, 2 ♂, 1 ♀. Scott River (11) 18 – 26 Feb 66, 2 ø, 2 ♂ skel, 3 fl. Cocklebiddy (15) 8 – 9 July 65, 2 ♂, 1 ♀, 1 ø. Eucla (17) 1 – 5 July 65, 3 ♂, 1 ♀. Warburton (23) 12 Aug 65, 1 ♂, 1 ♀. Giles (25) 23 – 31 Aug 65, 3 ♂, 1 ♀. Queensland. Glendower (55) 17 – 27 Mar 64, 1 ♂ (8.8 g), 1 ø (7.3 g). Banoona (58) 30 Apr – 6 May 63, 2 ♂, 1 fl. South Australia. Granite Downs (68) 7 – 10 June 65, 1 ♂, 2 ♀, 1 ø. Nullarbor HS (69) 4 – 7 Oct 63, 2 ♂. Kingoonya (70) 17 – 24 June 65, 3 ♂. Sceale Bay (75) 14 Oct 63, 1 ♀. Nonning (79) 1 Apr 65, 1 ♂ (8.0 g), 1 ♀ (8.5 g), 1 ø (8.8 g). Aroona Valley (82) 8 – 19 Apr 65, 3 ♂ (8.4, 9.1, 9.5 g), 2 ♀ (8.8 g), 1 ø (8.2 g). Angepena HS (84) 3 Dec 63, 1 ♂, 1 ♀. Furner (86) 7 Jan 63, 1 fl. New South Wales. Albury (92) 11 Mar 63, 1 ♀. Grafton (94) 3 Apr 63, 1 ♂.

Specimens to other Museums. NMV 1 ♂ Thylungra (60) 22 May 63, 1 fl Furner (86). AM 1 ♂ Banoona (58).

Distribution & Abundance. In Western Australia present at all camps from Metricup (9) to Newdegate (14); fairly common in the area and at Cocklebiddy (15) and Eucla (17). A few were seen at Mount Shenton (18); more common again at Warburton (23) and Giles (25) but none were observed at any of the camps in the Gibson Desert (19-22, 24), nor did the 4th Expedition find it in their northern camps (1-4).

Not found at Monkira (61) in southwestern Queensland nor in the northeast of South Australia (62-67). Common but local at Granite Downs (68); only found twice at Kingoonya (70) but very common at Nonning (79) and Aroona Valley (82).

In the southeast recorded but not collected from Nelson (88), Victoria Range (90), Little Desert (91) and Dunedoo (93).

Habitat. Extraordinarily varied: in open woodland or forest including Jarrah and Marri (11), mallee (17), mulga (68), native pine (82), ironbark (55), as well as in fairly open grassland with scattered trees (60) and Bluebush and saltbush (15).

Behaviour. This species feeds more on the ground than do most thornbills though in trees as well, sometimes high up. It was usually met in small parties or in mixed flocks which might include other thornbills, *Petroica* sp., *Smicrornis* sp., *Zosterops, Pardalotus* sp.

Three different calls were noted; a melodious trilling song in descending scale, a tinkling twitter (probably contact calls, and sounding very like those of *Aphelocephala leucopsis*), and a *tchrik, tchrik* of alarm.

Food. Usually insects, including ants, but seeds were found in one stomach, and buds and vegetable material in another.

Annual Cycle. There was little gonad activity in the specimens collected, only two males from Kingoonya (70) in June having the testes significantly enlarged, while others from Cocklebiddy (15), Eucla (17), Warburton (23) and Giles (25) had them slightly enlarged in July and August.

March – April birds from the southwest (9, 11), northern Queensland (55) and central South Australia (79, 82, 84) were in various stages of moult.

Impermanent Colours. *Legs* black. *Iris* variable (see below). *Bill* black. *Mouth* dark grey to black.

The eye colour was recorded variously as "grey", "cream", "dark orange", "lemon yellow", "medium brown" or "brown"; or else as a mixture of colours, "grey-brown with white outer ring", "brown flecked yellow", "white flecked brown", "light brown, yellow outer ring", etc. This variability does not seem correlated to age, sex, season or locality.

Geographical Variation. Southwestern birds (9, 11) are dark above and have a considerable amount of white mixed with the pale buff of the underparts and under tail-coverts, the characters of *A. c. multi*. Our series supports the view of Mayr & Serventy (1938: 267) that three arid country races are recognisable, for the birds from Cocklebiddy (15) are paler than *A. c. multi* but resemble that race in the amount of white below, thus showing the characters of *A. c. alexanderi;* birds from the centre, Warburton (23) and Giles (25), are again paler and have a pale yellow wash on the abdomen and under tail-coverts (*A. c. ferdinandi*), while those from the east-central districts of South Australia (68, 70, 79, 82, 84) are deep buff below (*A. c. addenda*). It is more difficult to name the more southern birds from Eucla (17), Nullarbor HS (69), Kingoonya (70) and Sceale Bay (75) which seem intermediate between *A. c. ferdinandi* and *A. c. addenda* in the amount of buff below.

All specimens from the eastern states are greyer on the head and back and have more yellow on the rump. Those from Glendower (55) are pale lemon below with no tinge of buff, and they have whitish ear-coverts. On description they show an approach to *A. c. normantoni,* though they come from some distance southeast of the type locality and are not apparently so yellow on the back. The three from southern Queensland (55, 58) have darker ear-coverts, and the underparts are slightly darker lemon-yellow with a tinge of buff; they should on locality represent *A. c. chrysorrhea*. The single female from Albury (92), representing *A. c. sandlandi,* is less yellow below, resembling the indeterminate birds from Kingoonya (70), etc., in this character but it is slightly less olive above.

Parasties. See Appendix 1.

Sericornis frontalis (including *maculatus*) Southern Scrub-wren

Specimens. Western Australia. Bush Bay (6) 12 – 18 June 66, 2 ♂, 1 ♀, 1 ♂ skel, 1 ♀ skel, 2 fl. Wonerup (8) 12 Mar 66, 1 ♂, 1 ♀. Scott River (11) 28 Feb – 1 Mar 66, 1 ♂, 1 ♀ skel, 1 fl. Mount Barker (12) 21 – 28 Mar 66, 1 ♂, 1 ♂ skel, 1 fl. Green Range (13) 31 Mar – 4 Apr 66, 1 ♀, 1 ø (appears ♂). Newdegate (14) 18 – 19 Apr 66, 1 ♂, 1 ♀. Cocklebiddy (15) 4 Dec 62, 1 ♀: 8 July 65, 3 ♂. Eucla (17) 1 July 65, 1 ♂, 1 ♀. Queensland. Mount Dryander (57) 3 – 10 Aug 64, 1 ♂ (13.0 g), 2 ♀ (11.1, 12.0 g). South Australia. Sceale Bay (75) 14 Oct 63, 1 ♂, 2 ♀. Victoria. Nelson (88) 30 Dec 62, 1 imm ø. New South Wales. Tenterfield (95) 12 – 15 Apr 62, 1 ♂, 1 ♀.

Specimens to other Museums. SAM 1 ♀ Nullarbor HS (69) 25 Oct 64.

Distribution & Abundance. Fairly common in its specialised habitat where collected, but only seen elsewhere at Metricup (9).

Habitat. Typically in mallee and scrub on cliffs and dunes along the coast (Plate 2(f)) (8, 11, 15, 17, 69, 75, 88), extending at Scott River (11) and Metricup (9) into thickets and tea-tree scrub in open forest. At Mount Barker (12) around swamps and shrubby undergrowth in mixed woodland; in mallee on heathland at Green Range (13) and Newdegate (14).

At Tenterfield (95) collected in the undergrowth beside a creek in rainforest and also in dry scrub on the wooded slopes of the Gibraltar Range 18 miles to the west. At Mount Dryander (57) only seen beside two creek-beds in low, dry rainforest with a boulder-strewn floor.

Behaviour. Found singly, in pairs or sometimes small parties, skulking in the undergrowth or feeding in low bushes. In north Queensland found to be much shyer than *S. citreogularis*. Very vocal with a variety of calls: Goodwin describes a repeated *tssee-tsse-tsse* with variants, some like the "reeling" of *Malurus* sp. but much louder: also a chattering and a clearer musical phrase.

Food. Usually insects including beetles and caterpillars: once seeds as well, and at Mount Dryander (57) the stomachs of two of the three birds collected contained tiny snail shells.

Annual Cycle. Evidence of late winter or spring breeding in certain areas is given by the young December bird from Nelson (88) and by gonad activity in birds from Bush Bay (6) (where males were seen singing and chasing) in June and from Mount Dryander (57) in August. The former were in fairly worn and the latter in fresh plumage.

In contrast, all birds from the extreme southwest (8, 11, 12, 13) were in full moult March to early April, while at Newdegate (14) in mid-April moult had been completed.

Birds collected in October from Sceale Bay (75) and Nullarbor HS (69) and in December from Cocklebiddy (15) were in worn plumage. The pair from Tenterfield (95) in April were in fresh plumage.

Impermanent Colours. *Legs* brown or greyish brown, paler in birds from Queensland and New South Wales. *Iris* white tinged with pale yellow or grey. *Bill* greyish brown, the lower mandible paler with a whitish area at the base. *Mouth* pinkish, apparently whiter in birds from Newdegate (14), Cocklebiddy (15) and Eucla (17).

Geographical Variation. The Southern Scrub-wrens are divided broadly into two groups often considered species, the *frontalis* group of the eastern states with no streaking on the breast and at most only a little on the throat, and the *maculatus* group of the west, with heavily streaked throat and breast. The two groups apparently intergrade in the Mount Lofty region and for this reason they are tentatively regarded here as conspecific and given the common name of Southern Scrub-wren. Our specimens from northeast to southwest round the coast show the following characters: birds from Mount Dryander (57) have pale yellow underparts with no spotting on the throat, dark lores, a long superciliary stripe and conspicuous white tips and a broad black bar on the tail. They thus answer the description of *herbertoni* (see Mayr 1937a: 7) though they come from some 300 miles south of the type locality.

The Tenterfield (95) male has pale yellow underparts, a little streaking on the throat and only a trace of a black bar on the outer tail-feathers: the female from 18 miles west has less streaking on the throat and slightly more trace of a tail bar. They are best regarded as intermediate between *S. f. frontalis* (with a streaked throat and no tail-bar) and *S. f. tweedi* of northeastern New South Wales and southern Queensland (with little or no streaking on the throat and a well-marked tail bar).

The young bird from Nelson (88) is moulting from juvenile to adult plumage. It has some new, pale yellow feathers below, some with light brown tips and a few streaks on the throat. The superciliary stripe is indistinct, (see Mayr & Wolk, 1953: 68, for description of juvenile plumage). On locality it is assigned to *S. f. longirostris*.

The worn Sceale Bay (75) birds have a considerable amount of streaking on the throat and breast, and the flanks and underparts are washed with brown giving a dirty appearance; they come from the range of *S. f. mellori* (see Condon, 1951a: 48). They are browner above than the December female from Cocklebiddy (15), the only Western Australian bird in comparable plumage.

Coastal birds from the head of the Bight (15, 17, 69) are similar to those from Newdegate (15), having no yellow below, heavy streaking on the throat and breast, white abdomens and greyish flanks; they extend the range of *S. f. condoni* as outlined by Mayr & Wolk (1953: 68). Other birds from the southwest (8-13) have yellow underparts and are *S. f. maculatus*.

The three from Bush Bay (6) (*S. f. balstoni*) have only light streaking on the breast and throat and a pure white abdomen.

Sericornis citreogularis **(=*lathami*)** Yellow-throated Scrub-wren

Specimens. QUEENSLAND. Big Tableland (51) 1 – 2 July 64, 3 ♂ (16.8, 17.3, 18.1 g). Walter Hill Range (53) 21 July 64, 1 ♂ (16.8 g), 5 ♀ (15.9, 16.1, 16.2, 16.8, 17.1, 27.0! g).
Field Notes. Fairly common between 2,100-2,200 ft in the two areas in which it was collected, but not seen elsewhere. Seen alone or in pairs in rainforest, once in company with *Psophodes olivaceus*. Usually on the ground or around fallen trees and rotten stumps. One came hopping to squeaking, another was seen scratching the ground. Only once encountered outside forest, and then in tangled undergrowth of an artificial clearing in the Walter Hill Range (53). Most birds uttered a scolding chatter. The other

terrestrial forest *Sericornis, S. keri,* was collected, but not recognised, in the same general area of the Walter Hill Range (53) forest on the same day, but no detailed comparative notes were made of their habitats.

Food. Unidentified insect and small seeds. The stomachs were noted as very muscular: all contained some grit or small stones.

Annual Cycle. No gonad activity was shown. One bird on 21 July was replacing some rectrices in an otherwise worn tail but no moult was found in other birds, which were in indeterminate plumage. Three birds were recorded as having skulls not fully pneumatised but they show no sign otherwise of being immature.

Impermanent Colours. *Legs* whitish buff with some pink tinges. *Iris* brown (once, "dark, dull pink") possibly redder in males and yellower in females. *Bill* black. *Mouth* black with pinkish throat.

Geographical Variation. This series supports the view that northern birds are separable as *S. c. cairnsi* for they are slightly darker on the back and more richly coloured below than some *S. c. citreogularis* in the British Museum (admittedly old skins) from the Richmond and Tweed Rivers. It might be expected that there would be a cline of increasing pigmentation northwards but this does not seem to be the case for the Walter Hill Range (53) birds are darker and less olive above than those from Big Tableland (51), 150 miles further north. With only one male from the former and no females from the latter locality it is impossible to judge if there is any difference between the two populations on the underparts, for the sexes differ, males having the yellow of the throat brighter and extending further on to the breast and abdomen, while females have the patches on the flanks and breast browner and more extensive, sometimes forming an irregular band across the breast.

Other Notes. This bird must be known as *S. citreogularis* Gould 1838, and not *S. lathami* (Stephens 1817) for the latter name was based on Watling Drawing No. 180 which is indeterminable (Mathews, 1930: 611, and Hindwood, 1965: 86).

Sericornis magnirostris — Large-billed Scrub-wren

Specimens. QUEENSLAND. Big Tableland (51) 1 – 2 July 64, 4 ♂ (10.5, 10.8, 11.0, 11.4 g), 6 ♀ (10.0, 10.1, 10.3, 10.8, 11.0, 11.3 g). Ayton (52) 22 – 24 June 64, 1 ♂ (11.0 g), 1 imm ♂ (9.7 g). Walter Hill Range (53) 21 – 23 July 64, 1 ♂ (11.6 g), 4 ♀ (8.4, 8.9, 9.1, 9.5 g). Tully (54) 11 – 25 July 64, 5 ♂ (10.1, 10.4, 10.4, 11.4, 11.6 g), 3 ♀ (9.0, 9.1, 9.5 g). Mount Dryander (57) 1 – 16 Aug 64, 2 ♂ (10.3, 10.3 g). NEW SOUTH WALES. Tenterfield (95) 12 Apr 63, 1 ♂.

Distribution, Abundance & Habitat. Fairly common in rainforest in northern Queensland, either in lowlands by the Bloomfield River, Ayton (52) and Tully River (54) or between 2,100-2,300 ft on Big Tableland (51) and Walter Hill Range (53). In the same low dry rainforest as *S. frontalis* by the Gregory River, Mount Dryander (57), and in a blue gum in a forest clearing by a creek at 3,300 ft at Tenterfield (95). Not seen elsewhere.

Behaviour. Found at all levels in the trees and undergrowth but, in contrast to *S. keri* (see below), not on the ground: sometimes spiralling or zigzagging up vines or trunks of trees searching for food, or in parties of up to six birds chasing each other through the foliage. Once in mixed party including *Tregellasia capito, Rhipidura rufifrons,* and once with *Gerygone palpebrosa.* The usual call was a harsh chatter but a "high wheedling twitter" and a "merry chatter, incessant, of tiny musical notes" were also heard.

Food. Usually insects but also small snails, the largest shell found being 3 mm in diameter. The stomach noted as muscular.

Annual Cycle. All birds were in fresh or fairly fresh plumage; three males from Tully (54) and the two from Dryander (57) had enlarged testes but otherwise none showed significant gonad activity. The immature male from Ayton (52) in June was in first year plumage and the greater part of the skull was not pneumatised (Galbraith & Parker, 1969: 219).

Impermanent Colours. *Legs* pale brown with a pink or purple tinge. *Iris* adults usually orange-brown, once noted as having the inner ring bright red, outer ring whitish: immature olive-grey. *Bill* black. *Mouth* greyish-pink.

Geographical Variation. There is considerable individual variation in the amount of rufous on the head and olive-brown wash on the breast but the three birds from Mount Dryander (57) and Tenterfield (95) (*S. m. magnirostris*) are distinctly paler than those from further north (*S. m. viridior*). Among the latter Galbraith & Parker (1969: 223) have shown that in the Big Tableland (51) population the females are closer in size to the males than elsewhere, and this is illustrated by their weights as well as by most measurements.

Sericornis keri — Bellenden Ker (Atherton) Scrub-wren

(See Galbraith & Parker, 1969)

Specimens. QUEENSLAND. Walter Hill Range (53) 21 – 23 July 64, 1 ♀ (12.4 g), 1 ø (12.1 g).

Field Notes. Only found in the rainforest (Plate 9(a)) on the Walter Hill Range (53) above Cochable Creek at 2,100 and 2,200 ft. One bird was feeding low in ferns and the other came to squeaking, hopping across ground. They were not then recognised as distinct from *S. magnirostris* so that no particular note was made of their behaviour or environment.

Food. The stomachs contained "elytra of small black beetles" and "brownish insect paste".

Annual Cycle. The female had a flat, structureless ovary and the skull almost fully pneumatised though soft. The unsexed specimen had the skull fully pneumatised. Both were in fresh plumage but with some wear on the tips of the rectrices.

Impermanent Colours. *Legs* buff or pinkish-white. *Iris* deep red-brown. *Bill* black. *Mouth* pinkish-white.

Other Notes. Galbraith & Parker (1969) have shown that *S. keri,* long placed in the synonymy of *S. magnirostris,* is a good species, being heavier with longer legs and relatively shorter wings, consistent with its probably more terrestrial habits. It is also darker on the back and face.

Oreoscopus gutturalis Fern-wren

Specimen. Queensland. Big Tableland (51) 1 July 64, 1 ♀ (18.6 g).
Field Notes. One of two seen in a gully of dense rainforest at 2,200 ft, "hopping along ground and peering over log at collector". It came to squeaking.
Food. The stomach (at 11.20 a.m.) was empty.
Annual Cycle. The ovary was enlarged with oocytes forming, and the plumage worn.
Impermanent Colours. *Legs* light brown. *Iris* very dark brown. *Bill* black. *Mouth* black with greyish-pink palate.

Pyrrholaemus brunneus Redthroat

Specimens. Western Australia. Bush Bay (6) 13 June 66, 1 ♀ skel, 2 fl. Newdegate (14) 14 Apr 65, 1 ♀ skel, 1 fl. Eucla (17) 1 – 5 July 65, 2 ♂, 3 ♀, 1 ♂ skel; 7 Dec 62, 1 ø skel. Mount Shenton (18) 3 Aug 65, 1 ♂. Mungilli Claypan (20) 13 Sept 65, 1 ♀. Hunt Oil Camp (22) 9 Sept 65, 3 ♂, 1 ♀, 1 fl. Warburton (23) 9 Aug 65, 1 ♂. Giles (25) 23 Aug – 2 Sept 65, 5 ♂, 3 ♀, 1 ♀ skel, 1 fl. Walter James Range (26) 26 Aug 65, 1 ♂. South Australia. Granite Downs (68) 9 – 11 June 65, 2 ♂, 1 ♀. Nullarbor HS (69) 6 Oct 63, 1 ♀. Aroona Valley (82) 15 – 17 Apr 65, 2 ♀ (11.9, 12.01 g).
Distribution & Abundance. Very common at Bush Bay (6), common on the cliffs at Eucla (17), at the Hunt Oil Camp (22) and Giles (25), and fairly common at Granite Downs (68). Otherwise scarce and, except for one bird at Cocklebiddy (15), not seen except where collected.
Habitat. Low bushes and scrub in mallee (14, 17), mulga (18, 20, 22, 23, 25, 26, 28) or on rocky hillsides and the peaks of the Flinders Range (82).
Behaviour. Usually found foraging in low bushes or on the ground, sometimes also in the lower branches of trees, the quick movements rather similar to thornbills. On the ground moves in long hops. Birds at Giles (25) were obviously paired, elsewhere they were found singly, in pairs, or in small parties (17, 22). Most males were very vocal, singing continuously except when in the immediate vicinity of the female. The long warbling song sometimes included some mimicry (once of the alarm call of *Rhipidura fuliginosa*) and some harsh churring notes. The alarm call was a churring, scolding chatter, and other calls recorded by Goodwin were a loud *pee-pee-pee, tr, tr* and a soft *tchik-ick*.
Food. Mainly insects including beetles and ants, and sometimes small seeds swallowed whole. Some from the stomach of a female from the Aroona Valley (82) were identified as possibly *Calandrinia* sp. The stomach of the Warburton (23) bird contained only seeds and some tiny black stones.
Annual Cycle. At Giles (25) and the Walter James Range (26) birds were breeding at the beginning of September, all having enlarged gonads, one female with eggs formed. They were in worn plumage. In the Gibson Desert (17, 20, 22) the males but not the females had enlarged gonads. A male and female from Eucla (17) had enlarged gonads and the plumage was mixed. All from Granite Downs (68) and the Aroona Valley (82) had small gonads and were in fresh or fairly fresh plumage, one of the latter completing moult in the tail.

Impermanent Colours. *Legs* dark brown, grey or black. *Iris* red-brown frequently paler with an outer ring (see below). *Bill* dark brown, grey or black, paler below. *Mouth* grey sometimes with pink tinge. Of the 15 males and 9 females of which the iris was examined no paler outer ring was noted in seven males and three females: in three males and one female the outer ring was a paler brown than the inner iris, and in the remainder the outer ring was whitish. The latter birds came from various localities and included breeding and non-breeding birds, so this variation seems not to be either seasonal or geographical.

Geographical Variation. No geographical variation is currently recognised but our birds differ appreciably in the colour of the flanks and under tail-coverts, and to a lesser extent in the darkness of the upper parts and the richness of the throat of the males. The latter character especially seems chiefly seasonal, the fresh-plumaged Granite Downs (68) males having the darkest throats, but flanks of birds from Eucla (17) are brown, from Granite Downs (68) are pinkish-buff, while those from Nullarbor HS (69) and Aroona Valley (82) are intermediate, and this cannot be due to wear. On the other hand the birds in worn plumage from the central camps (18-26) have pale pinkish-buff flanks and before fading may have been similar to those from Granite Downs (68) as would be expected from the ecology. Various names have been given by Mathews to different populations, some of which he later discarded, and without material in fresh plumage from the various type localities no proper revision can be made. However it seems logical to expect the variation to follow the demarcation between different vegetation zones; this would admit an eastern race *P. b. brunneus* extending westwards possibly to include the Aroona Valley (82) birds; a paler race with pinker flanks in the central mulga areas which would include all our birds from the central camps (18-26) and Granite Downs (68), for which the correct name is probably *P. b. pallescens,* type-locality Bore Well, East Murchison (synonum *P. b. centra*): and a darker race with browner flanks to the south and west of the mulga-eucalypt line (Serventy & Whittell, 1967: 54) for which *P. b. kalgoorlie* (synonym *P. b. killigani*) is available and which would include our birds from Eucla (17) and possibly Nullarbor HS (69).

Parasites. See Appendix 1.

Hylacola pyrrhopygia — Heath-wren

Specimens. New South Wales. Tenterfield (95) 12 Apr 66, 1 imm ø.

Field Notes. Found at 3,300 feet in the hills 14 miles ESE of Tenterfield (95) on the edge of a swamp in open woodland of stringy-bark and casuarina, with some rocks. This bird, the only one seen, flew out from a bush and alighted on a small rock.

Food. Insects.

Annual Cycle. This bird was completing moult of wings, tail and body; the skull was noted as not fully pneumatised and the gonads were not found.

Impermanent Colours. *Legs* light brown. *Iris* very light brown. *Bill* dark brown paler at tip.

Geographical Variation. The only other specimen in the British Museum is the type of *pyrrhopygia* – a very old, faded, ex-mounted specimen, useless for comparison. The

Tenterfield (95) bird might on locality belong to either the inland race, *H. p. magna*, or the typical race.

Hylacola cauta Mallee Heath-wren

Specimens. WESTERN AUSTRALIA. Newdegate (14) 14–16 Apr 66, 1 ♂, 1 ♀, 1 ♂ skel, 1 fl.
Field Notes. All specimens came from near Lynwood Homestead, 12 m east of Newdegate in lignum scrub amongst mallee. One bird collected was solitary, one was in company with Brown Thornbills, the other two were paired. All were either on or near the ground. One was heard scolding. Seen but not collected at Little Desert (91) in scrub mallee.
Food. Insects, including small beetles.
Annual Cycle. The two preserved as skins were in very fresh plumage with small gonads.
Impermanent Colours. *Legs* grey or black. *Iris* light brown with yellowish-white (2 ♂) or white (1 ♀) outer ring. *Bill* dark grey or black. *Mouth* flesh.
Geographical Variation. With no comparative material these birds must be assigned to the Western Australia race *H. c. whitlocki*.

Calamanthus fuliginosus Field-wren

Specimens. WESTERN AUSTRALIA. Bush Bay (6) 12 – 18 June 66, 2 ♂, 1 ♀. Southern Cross, 220 m E of Perth (7) 2 Dec 62, 1 ø. Newdegate (14) 18 Apr 66, 1 ♀ flat skin and skel, 1 fl. Eucla (17) 2 July 65, 1 ♂. VICTORIA. Nelson (88) 4 Jan 63, 1 ♂, 1 ♀. Coleraine (89) 8 July 65, 1 ♀.
Distribution & Abundance. Only those collected were encountered at any locality except Bush Bay (6) where several were seen.
Habitat. Low bushes on the edge of the beach at Bush Bay (6); in scrub on the edge of the main Perth-Adelaide road, 60 miles east of Southern Cross; mallee scrub, Newdegate (14); arid Bluebush on the Nullarbor Plain 21 miles north of Eucla (17); coastal scrub on the Glenelg estuary, Nelson (88); open heath on the Victoria Ranges, near Coleraine (89).
Behaviour. Usually found skulking in low branches of bushes or on the ground, but singing males were prominently perched, once on the top of a bush seven feet up and on three other occasions on lower branches. The song was described as a short, repetitive, musical phrase. Birds at Southern Cross, Newdegate (14) and Nelson (88) were paired.
Food. The stomach of the Eucla (17) bird contained insect debris, a small snail and grit.
Annual Cycle. In Western Australia the April Newdegate (14) female was in fresh plumage and had a small ovary; all birds taken in June and July were in fairly fresh plumage with enlarged gonads, especially large in the Eucla (17) male. The December bird was in worn plumage. The January birds from Victoria were in moult.
Impermanent Colours. *Legs* pale fleshy brown. *Iris* brown in specimens from Victoria, elsewhere golden brown (6) golden-buff and yellowish-cream (14), medium-brown with yellow-brown outer rim (17). *Bill* dark brown, paler below. *Mouth* pinkish-grey in Western Australia, black or grey in Victoria.

Geographical Variation. Two groups are recognised in the Field-wren, the olive-backed *fuliginosus* group from the southeast and southwest, and the rufous-backed *campestris* group from the arid areas, with intergrading populations in South and Western Australia. The Newdegate (14) and Victoria birds are olive-backed, the latter more heavily-streaked above and showing marked sexual dimorphism (plainly visible in the field), the male having the supercilium and the throat white and the abdomen pale yellow, these areas being buffy in the female. On locality they can be referred respectively to *C. f. montanellus* (S.W. Australia) and *C. f. albiloris* (Victoria).

The birds from Bush Bay (6) and Eucla (17) belong to the *campestris* group, and in spite of the distance separating the two localities they are very similar, the Eucla (17) bird being only slightly more rufous on the head and ear-coverts. If races are to be recognised within this group the Bush Bay (6) birds match fairly well with topotypical *C. f. rubiginosus* from Point Cloates, while the Eucla (17) bird might possibly prove to be *C. f. suttoni*, a form with a rich rufous head and ear-coverts from the northern Eyre Peninsula (Condon, 1951a: 51).

The Southern Cross bird is intermediate, more rufous on the head and rump than *montanellus* and more olive on the back and yellower below than *rubiginosus*. Several names have been given to intermediates in Western Australia, of which the oldest is *C. f. carteri*.

Parasites. See Appendix 1.

Aphelocephala leucopsis — Whiteface

Specimens. Western Australia. Cocklebiddy (15) 8 July 65, 1 ♂. Eucla (17) 2 – 5 July 65, 1 ♂, 2 ♀. Mount Shenton (18) 30 July – 3 Aug 65, 4 ♂, 2 ♀, 1 ♂ skel, 1 ♀ skel, 1 fl. Carnegie (19) 16 – 17 Sept 65, 4 ♂, 1 ♀. Mungilli Claypan (20) 12 Sept 65, 2 ♂, 1 ♀, 1 ø, 1 juv ♂, 1 juv ø. Lake Gruszka (21) 10 Sept 65, 1 ♂, 3 ♀. Lake Throssel, 160 m SW of Warburton (23) 5 Aug 65, 1 ♂. Warburton (23) 7 – 10 Aug 65, 6 ♂, 1 ♀, 1 ø, 1 ♂ skel, 1 ♀ skel, 2 fl. Giles (25) 20 – 31 Aug 65, 3 ♂, 5 ♀. South Australia. Cordillo Downs (63) 6 May 65, 1 ø. Etadunna (66) 20 May 65, 1 ♀. Oodnadatta (67) 31 May 65, 1 ♂, 1 ♀, 1 ø. Granite Downs (68) 5 – 10 June 65, 6 ♂, 2 ♀, 1 ø. Nullarbor HS (69) 29 Oct 64, 1 ♂. Kingoonya (70) 16 – 25 June 65, 9 ♂, 6 ♀, 1 ø. Nonning (79) 29 Mar – 1 Apr 65, 2 ♂, 1 ♀, 2 ø. Parachilna (81) 12 Apr 65, 1 ø. Aroona Valley (82) 10 – 13 Apr 65, 2 ♂. Wirrealpa (83) 15 Apr 65, 1 ♀, 1 ø. Angepena (84) 21 Nov 63, 1 ♂. Victoria. Coleraine (89) 19 Jan 63, 1 ♂, 1 ♀. New South Wales. Albury (92) 11 Mar 63, 1 ♀.

Distribution & Abundance. Although an especial watch was kept for it, the Whiteface was not seen along the Warburton Track between Lake Throssel (160 m SW of Warburton) and Warburton (23) itself; nor in the single days collecting at the Hunt Oil Camp (22) and the Van der Linden Lakes (24). Its apparent absence suggests a genuine distributional gap between the chestnut-flanked western *A. l. castaneiventris*, which was collected at Lake Throssel and was common at Mungilli Claypan (20), Lake Gruszka (21) and Mount Shenton (18), and the brown-flanked central form which was common at Warburton (23) and Giles (25), and was present, though less common, at Cocklebiddy (15) and Eucla (17).

In South Australia the 3rd Expedition found Whitefaces very common in the western and central camps (68, 70, 79, 82) but only a few were seen at either Oodnadatta (67) or Etadunna (66) and none at Clifton Hills (65) or Innamincka (64). A few were found at Cordillo Downs (63) in the extreme northeast of the state but the first expedition did not collect it at any of their camps in southern Queensland, although Thylungra (60) and Monkira (61) are less than 200 miles from Cordillo Downs (63).

Habitat. Dry, open woodland, especially Mulga; scrub in dry creeks and on the coast; bushes and trees on Bluebush, saltbush and spinifex plains; native pines in the Flinders Range (82). Found on both stony and sandy soil. There was no reason readily apparent for its absence from the country west and north of Warburton (23) but at Clifton Hills (65) and Innamincka (64) there were few suitable areas of small trees and bushes between the open gibber and the taller trees of the creeks.

Behaviour. Most of the Whitefaces were collected from small parties. They seem to feed exclusively on the ground, turning over bits of dry dung and other debris by pushing the bill underneath and pushing up, in the same manner as do turnstones, starlings and others. The gait is a hop or "half-walk (perhaps a hop with one foot in front of the other)" (DG). When disturbed the birds fly immediately to a bush or tree but there were several instances of the presumed mate returning when a bird had been shot.

Some birds at Kingoonya (70) in June appeared to be showing beginnings of breeding behaviour, with much chasing of twos and threes through trees, both pursued and pursuer often stopping a few inches apart to indulge in emphatic bill-wiping, at the same time erecting the head feathers to display the white face prominently to the front (DG).

A characteristic tinkling twitter on the move seemed to be a contact call, but there was also a song phrase and a soft *tk-tk-tk-tk*. The voice of the chestnut-flanked western birds, *A. l. castaneiventris,* seemed to Goodwin (1967c: 244) to be slightly sweeter and less strident than those of the brown-flanked central and eastern forms, but the substance of the calls was identical.

Food. Stomachs examined contained grit, insects (including a caterpillar once and beetles), various seeds (often swallowed whole) and leaves of a small succulent plant. At Warburton (23) three different sorts of seeds were found in the stomachs of both a common Whiteface and a Banded Whiteface collected at the same time.

Annual Cycle. In Western Australia all birds at Mount Shenton (15) and Lake Throssel had the gonads enlarged, the females with eggs in the ovaries, while those taken six weeks later from Lake Gruszka (21), Mungilli Claypan (20) and Carnegie (19) included two juveniles, and adults had small gonads and were mostly completing moult.

In contrast the brown-flanked birds from Warburton (23) and Giles (25) taken in the intervening six weeks give no clear evidence of breeding, the plumage being worn and the gonads showing little activity, except that three females from Giles (25) had somewhat enlarged ovaries.

The four brown-flanked birds from Eucla (17) and Cocklebiddy (15) were again different, being in mainly fresh plumage, though showing some gonad activity.

In South Australia the evidence is confusing and mostly indefinite. In 1965 birds from Nonning (79) and the Flinders Range area (81, 82, 83) in April were apparently post-breeding with small gonads and some completing moult, yet the Angepena (84) male taken in November 1963 was starting to moult its primaries. June birds from Kingoonya (70), where the pre-breeding behaviour was noted, had slightly enlarged gonads.

The three birds from Victoria and New South Wales were in moult with small gonads.

Impermanent Colours. *Legs* greyish-black. *Iris* white with either a yellow, greyish or brownish tinge: juveniles greyish white. *Bill* black. *Mouth* dark grey or black: juveniles yellow.

Geographical Variation. The chestnut flanks of the populations from central Western Australia (18-21 and L. Throssel), *A. l. castaneiventris,* distinguish them immediately from central and eastern birds in appearance and it has already been shown that they differ slightly in voice, there may be a definite distributional gap between them, and in 1965 they had different breeding seasons. If observations in other years confirm our own, it is evident that there can be no interbreeding between chestnut and brown-flanked birds and it may be more realistic to regard *castaneiventris* as a species.

Among the central and eastern populations it is difficult to distinguish geographical variation because of (a) the effects of soil-staining on all birds taken on red sand, (b) the difference between the greyer fresh plumage and the browner worn plumage, and (c) the individual variation in the amount of colour on the flanks. Nevertheless there is clearly a tendency for eastern birds, *A. l. leucopsis,* to have less brown on the flanks than those of the centre, *A. l. whitei,* with which we would include the populations of eastern Western Australia (15, 17, 23, 25), western South Australia (67-70), certainly those from Nonning (79) in the Gawler Ranges, and possibly the single female from Etadunna (66), though a series from here is required for positive identification. The Cordillo Downs (63) bird agrees best with *A. l. leucopsis*. Except in respect of the Gawler Range birds, this accords with the distribution mapped by Keast (1957a: 41), but we do not find the differences as clear-cut as he suggests. For this reason it seems impractical to distinguish a further subspecies, *A. l. pallida,* intermediate between *leucopsis* and *whitei,* as has been done by Condon (1962a: 124).

Aphelocephala nigricincta — Banded Whiteface

Specimens. WESTERN AUSTRALIA. Warburton (23) 9 – 12 Aug 65, 5 ♂, 2 ♀, 1 ø. Van der Linden Lakes (24) 6 Sept 65, 1 ♀. Giles (25) 25 Aug – 2 Sept, 4 ♂, 3 ♀.

Distribution & Abundance. At Warburton (23) and near the camp at Giles (23) the Banded Whiteface was not as common as *A. leucopsis,* though at the Van der Linden Lakes (24) and the west end of the Rawlinson Range (where three of the Giles (25) specimens were collected) it was the only whiteface seen. Elsewhere it was only encountered once, by the roadside at William Creek, about 100 miles south of Oodnadatta (67). All birds collected were in small parties or in pairs.

Habitat. Typically on more open ground than *A. leucopsis,* either sand, gravel or clay, with Bluebush or saltbush, spinifex, Prickleweed, and scattered mulga, but the two species were once found together where there were thicker clumps of Mulga in anopen area.

Behaviour. Usually found seeking food on the ground, though in the comparatively short time it was studied it was not observed turning over debris in the manner of the common Whiteface. It may hop quite slowly, or very rapidly, with frequent short pauses, and it seemed very unafraid of man or guns. Goodwin (1967c: 245) saw a display flight in which the bird flies up and then descends with slowly flapping wings uttering a short song-phrase which incorporated its usual contact call – a tinkling twitter, rather softer and sweeter than that of the *leucopsis* – with a piping call.

Food. Mainly seeds swallowed whole, on one occasion including the same three species as were found in the stomach of *A. leucopsis;* on two occasions insect remains were found as well, including those of ants, which were not among the insects identified in *A. leucopsis.*

Impermanent Colours. *Legs* dark grey. *Iris* white. *Bill* black or dark grey, paler at the base. *Mouth* usually grey but recorded as "dull, pale flesh" once in a male with minute testes, and once in a female with an enlarged ovary.

Annual Cycle. The gonads of nine of the sixteen birds were much enlarged, which, with the display flights seen at Warburton (23) make it likely that some birds at least were on the point of breeding in August and September. Yet all birds were in fresh plumage or completing moult (a male with enlarged testes completing body moult and a female with an enlarged ovary completing tail moult). This was in contrast to, the condition of the common Whiteface taken in the same period (see above).

Chthonicola sagittata Speckled Warbler

Specimens. New South Wales. Dunedoo (93) 18 – 23 Mar 63, 1 ♂, 1 ø, 2 juv ø 1 ♂ skel. Tenterfield (95) 12 Apr 63, 2 ♂.

Specimens to other Museums. NMV 1 ♂ Dunedoo (93), AM 1 ♂ Dunedoo (93).

Distribution & Abundance. Several seen also at Victoria Range (90), Albury (92) and Banoona (58).

Habitat. Scrub in woodland, usually associated with stony ridges or gullies: at Dunedoo (93) between 1,600 and 1,700 ft (Plate 5(b)); at Tenterfield (95) between 2,800 and 3,300 ft.

Behaviour. A party found feeding on the ground flew immediately to surrounding trees when disturbed, from which they uttered a subdued chatter. In another party (all collected) two adults were feeding two juveniles in low scrub.

Food. Usually insects, once seeds. The adult bird was offering a small cricket to the juvenile.

Annual Cycle. The Dunedoo (93) juveniles were out of the nest but not fully grown. All adults were either completing moult or in very fresh plumage with minute gonads.

Impermanent Colours. *Legs* adults light blue-grey, to greenish-brown, juveniles light blue with fleshy tint. *Iris* brown. *Bill* dark grey or brown. *Mouth* adults fleshy-grey, juveniles bright yellow.

Geographical Variation. The Tenterfield (95) birds are paler on the mantle and have broader pale centres to the feathers of the crown than those from Dunedoo (93), but as two males from southern Queensland in the British Museum show individually the same degree of variation there is no reason not to refer all specimens to *C. s. sagittata.* The juveniles are similar to the adults except that the streaks above and below are not so sharply defined.

MALURIDAE (part 2) — Blue Wrens, Emu-wrens & Grass-wrens
C. J. O. Harrison

Malurus cyaneus — Superb Blue Wren

Specimens. QUEENSLAND. Banoona (58) 29 Apr 63, 1 ♂, 2 ♀, 1 imm ♂, 1 fl. SOUTH AUSTRALIA. Elliston (76) 15 Oct 63, 2 ♂, 2 ♀. Furner (86) 7 – 9 Jan 63, 2 ♂, 1 ♀, 1 ø. NEW SOUTH WALES. Dunedoo (93) 25 Mar 63, 1 ø. Grafton (94) 1 – 5 Apr 63, 1 ♀, 1 ø, 2 imm ♂. Tenterfield (95) 15 Apr 63, 1 ♂.
Specimens to other Museums. NMV 1 ♂ Grafton (94). AM 1 ♂ Banoona (58), 1 ♂ Grafton (94).
Distribution & Abundance. Found at all the camps within its range; in parties of 2-10 birds.
Habitat. Associated with bushes and low cover, frequently on the edge of woodland or swamp, also in bushes and low tea-tree in forest, in reeds in dry reed-bed, and in mixed low scrub with tall grass.
Behaviour. Usually remained in low bushes, or fed on the ground; once noted flying up to the top of a small tree, about 20 ft up.
Food. Insects.
Annual Cycle. Males from Eyre Peninsula (76) in mid-October, and from Furner (86) in January were in full breeding plumage, the last with large testes. A female from Furner (86) at the same period was breeding, with an enlarged oocyte in the ovary and an undamaged egg in the oviduct. A male from Grafton (94) in early April was half-moulted into eclipse plumage, with minute testes; and one from Tenterfield (95) in mid-April and another from Banoona (58) in early May are wholly in eclipse, while females from these regions show undeveloped ovaries.
Impermanent Colours. *Legs* adult male dark brown to black; female light brown, reddish brown, or flesh; immature male brown to dark brown. *Iris* dark brown. *Bill* adult male black; female reddish brown; immature male black, or dark brown with black towards the tip on upper mandible. *Mouth* adult male blackish; female flesh, pink, white or dirty white.
Geographical Variation. There is a clinal trend towards deeper blue dorsal plumage and larger average size from north to south. The mainland birds were subdivided into a southern form, *M. c. australis* and a paler northern form, *M. c. cyanochlamys;* but since new specimens from the Eyre Peninsula (76) and Furner (86) in southeast South Australia cannot be obviously separated from some specimens from Grafton (94) and Sydney in New South Wales and Banoona (58) in southern Queensland, *australis* would appear to be invalid.

Malurus melanotus — Black-backed Blue Wren

Specimens. QUEENSLAND. Tyrone (59) 10 May 63, 2 ♂.
Field Notes. Tyrone (59) was the only collecting locality within the range of this species. Specimens were taken from parties of two and four in low sandalwood bushes and Gidgee in Mulga woodland.
Food. Insects.

Annual Cycle. One male was moulting into almost complete breeding plumage, with gonads only a little enlarged. The other was in full eclipse plumage with minute gonads.
Impermanent Colours. *Legs* black with sandy-coloured soles. *Iris* dark brown. *Bill* black. *Mouth* grey.

Malurus callainus Turquoise Wren

Specimens. WESTERN AUSTRALIA. Hunt Oil Camp (22) 9 Sept 65, 1 ♂, 1 ♀. Warburton (23) 13 Aug 65, 1 ♂, 2 ♀. Giles (25) 24 Aug – 2 Sept 65, 4 ♂, 3 ♀, 1 imm ♂, 1 ♂ skel. SOUTH AUSTRALIA. Granite Downs (68) 4 – 10 June 65, 5 ♂, 4 ♀. Kingoonya (70) 16 – 18 June 65, 3 ♂, 1 imm ♂. Nonning (79) 30 Mar 65, 1 ♂. Angepena (84) 1 Dec 63, 2 ♂, 1 ♀.
Distribution & Abundance. Within its range widespread in areas of suitable habitat, being noted at all camps where this was present. Usually in pairs or threes, more rarely in parties of up to six individuals. More than one male apparent in parties.
Habitat. The basic necessary element in the habitat appears to be low shrubby growth. The species occurred in mulga where shrubby growth alive or dead, occurred. Also in sparse mulga on stony ground or in spinifex, in dead trees around a claypan, in low scrub, and once in saltbush.
Behaviour. Seen feeding in and under low bushes and around the roots of trees. The species seemed less shy than other *Malurus* species, more ready to show itself, and responded inquisitively to squeaking noises. The males had a loud reeling trill, similar to that of *lamberti* but lower-pitched and less sibilant. Trills were also heard from a young male in eclipse plumage, and females were heard to utter a shivering, sibilant trill, softer but similar to that of the male.
Food. Small insects; including ants.
Annual Cycle. At Nonning (79) in late March an adult was in full eclipse plumage. At Granite Downs (68) in early July three females showed no moult but small ovaries, and of five adult and one immature males, two showed very scattered blue feathers and the rest were eclipse, but none showed any indication of active moult and all the testes were small. At Kingoonya (70) in mid-June, of four males collected, one immature and one adult were in full eclipse and the other two in full breeding plumage, but all had tiny testes. At Warburton (23) in mid-August a male, one-third moulted into breeding plumage, had testes asymmetrically enlarged, and two females showed partial enlargement of the ovary. At the Hunt Oil Camp (22) in September and at Giles (25) in late August to early September, birds were in breeding plumage with partially to completely enlarged gonads. One male was in full eclipse plumage. At Wertaloona (85) in December two males were completing a moult into breeding plumage.
Impermanent Colours. *Legs* males dark brownish-grey to black; females dark brown to grey; immature males dark grey. *Iris* adults dark brown to blackish-brown; immature male dark grey-brown. *Bill* male black; female light reddish-brown; immature male upper mandible dark dusky-brown with red-brown cutting edge, or black; lower mandible red-brown with dark tip, or dark brown. *Mouth* grey, whitish or fleshy-white; on one male blackish-grey with fleshy-white tongue and pink-tinged back of palate.

Geographical Variation. All specimens were from within the range of the typical subspecies. In males in breeding plumage there is a distinct clinal increase in the depth of colour of the breast and belly from a paler blue on the eastern side of the range near Lake Frome (85), to deep blue near Warburton (23) in the west. There is a similar but less striking increase in the depth of colour on the throat.

Malurus splendens — Splendid Blue Wren

Specimens. WESTERN AUSTRALIA. Perth (7) 20 Nov 62, 14 Feb 66, 1 ♂ skel, 1 ø skel. Wonerup (8) 12 Mar 66, 3 ♂, 1 imm ♂. Metricup (9) 6 – 17 Mar 66, 2 ♂, 1 ♀. Scott River (11) 18 Feb – 1 Mar 66, 1 ♀, 1 imm ♂, 1 ♂ skel, 1 imm ♂ skel, 2 ø skel, 1 ♂ fl, 2 ø fl. Mount Barker (12) 24 – 26 Mar 66, 1 ♂, 1 ♀, 1 imm ♂.
Distribution & Abundance. Only found in moister areas bordering the southwest forest zone. In small parties of five to seven birds.
Habitat. In low bushy sandhill scrub near a lake, in bushes along a stream in Tuart and peppermint forest, in tea-tree scrub by streams, and in tea-tree scrub in a dry swamp in a paddock.
Behaviour. Feeds low in bushes, occasionally on open ground among bushes. Odd birds, usually males, would at times go up to 20-30 ft in trees, but otherwise a shy and skulking species.
Food. Insects, including small beetles.
Annual Cycle. Males in late February and March were completing moult into eclipse plumage, and there was some evidence of moult on females. All gonads were small.
Impermanent Colours. *Legs* males blackish-brown or black; females and immature birds dark brown to palish grey, with sandy-coloured soles. *Iris* dark brown. *Bill* males black; females and immature birds, light brown or flesh-brown on the upper mandible, lighter brown or flesh-coloured on lower mandible. *Mouth* males black or flesh-coloured; females and immatures flesh-coloured.
Geographical Variation. A doubtful paler inland form *aridus* was recognised by Mack (1934). The present material all relates to the nominate form, but shows individual variation in intensity of colour and some evidence of a clinal trend towards lighter dorsal colour in inland localities.

Malurus leucopterus (including *leuconotus*) — Blue-and-white Wren

Specimens. WESTERN AUSTRALIA. Chichester Range (3) 12 – 14 May 66, 2 ♂, 2 imm ø, 2 ♂ skel, 1 ♂ skel, 3 fl. Coolawanyah (4) 19 – 21 May 66, 1 ø, 1 ♀ skel, 1 fl. Carnarvon (5) 18 June 66, 1 ø. Eucla (17) 2 July 65, 1 ♂. Carnegie (19) 16 – 17 Sept 65, 3 ♂. Mungilli Claypan (20) 13 Sept 65, 1 ♂. Warburton (23) 5 – 12 Aug 65, 2 ♂, 1 ♀, 1 imm ♀, 1 ø. QUEENSLAND. Glendower (55) 22 Mar 64, 3 ♂ (7.9, 8.3, 8.3 g). Thylungra (60) 28 May 63, 1 ♂, 2 imm ø. Monkira (61) 7 June 63, 1 ♀. SOUTH AUSTRALIA. Innamincka (64) 29 Apr 65, 4 ♂. Clifton Hills (65) 10 – 12 May 65, 7 ♂ (7.7, 8.4, 9.2 g), 2 ♀ (8.4 g), 1 ø. Etadunna (66) 17 – 18 May 65, 8 ♂, 3 ♀, 1 ø. Oodnadatta (67) 1 June 65, 1 ♂. Nullarbor HS (69) 27 Oct 65, 1 ♂. Kingoonya (70) 24 June 65, 3 ♂, 1 ♀. Nonning (79) 31 Mar 65, 1 ♂ (7.9 g). Parachilna (81) 12 Apr 65, 1 ♂ (8.8 g), 1 ♀ (8.1 g). Aroona Valley (82)

9 Apr 65, 1 ø (6.5 g). Wirrealpa (83) 15 Apr 65, 1 ♂ (7.9 g). Angepena (84) 18 – 25 Nov 63, 2 ♂, 1 ♀.

Distribution & Abundance. This species was found at all the main camps within its range and the specimens from Glendower (55), northern Queensland, represent a northeasterly extension of the known range. It occurred in parties of from two to twenty birds, usually with only one male in breeding plumage to a party, but on one occasion (70) one completely blue and one partly blue male recorded in a party of eight.

Habitat. Typically a bird of arid areas with low and sometimes sparse cover. Found on spinifex plains, often associated with low scattered bushes; in samphire marshes; in saltbush and Bluebush areas with or without scattered bushes or Mulga; in sparse saltbush or low scrub on sand dunes; and in sandy areas in low scrubby growth associated with saline lakes or dry creekbeds. At Glendower (55) in open grassland near savanna woodland. In open areas prefers the neighbourhood of bushes or larger cover.

Behaviour. The birds spent most of the time on or near the ground when feeding, but would mount up into bushes or low trees to feed or to look around; and when pursued they tended to take to these. They would also sit in bushes sunning and preening in the early mornings. Although apparently tame they were not easy to approach once alarmed. The call-notes noted were a soft short *tsit-tsit-tsit* contact call; a more abrupt *trit-trit-trit* possibly in alarm; a high-pitched *tsree-tsree* like a similar call of *M. lamberti* but higher in pitch; and a trilling song like a speeded-up and more musical series of these calls *trawee-trawee-trawee-trawee-trawee-trrrp*.

Food. Stomachs contained insect remains; small beetles and caterpillars. "Insect eggs" were noted at Carnegie (19) and Warburton, "?seeds" at Warburton (23) and "seeds" at Wirrealpa (83). It seems possible that these might all have been pupae or parts of insects.

Annual Cycle. The annual cycle in this, as in other *Malurus* species, is difficult to summarise. Breeding may be related to rains and not correlated with moult, and males in eclipse plumage, in parties with males in breeding plumage, may show apparently mature testes, and moult from one brown plumage to another.

In Western Australia, at Chichester Range (3) in mid-May after unseasonable rains (Harrison & Colston, 1969) a male was in full breeding plumage, another showed no breeding plumage but large testes, and two recently-fledged young were collected and others seen. There were no positive indications of breeding or moult in specimens collected in July – September, except for a male half-moulted into breeding plumage in September (13); others were in breeding plumage with enlarged testes (19, 20).

In Queensland in March a young bird was collected (60) and one male at Glendower (55) was beginning to moult from breeding plumage. In South Australia a male at Wirrealpa (83) was half-moulted into eclipse plumage in April, and one at Innamincka (64) was just beginning to moult into eclipse plumage. Three others, in eclipse, also showed signs of moult from one eclipse plumage to another. At Clifton Hills (65) in May all birds were in eclipse but one male still showed a few blue feathers of a previous breeding plumage and one was in full breeding plumage.

Impermanent Colours. *Legs* males greyish-brown to black; females pinkish-brown or light brown to pale grey; immature males intermediate between female and male;

juveniles blue-grey or light blue-grey. *Iris* brown to very dark brown. *Bill* male black; females pink to pale brown; immature and possibly eclipse males like those of females but becoming more light brown, developing black at tips and nostrils, the dark colour finally spreading along the culmen and then to whole bill. Some birds identified as adult males with fully ossified skulls are in eclipse plumage with brown bills and there might be either a reversion to this colour by such males during eclipse periods, or such birds may not yet have reached the hormonal threshold necessary for bill-colour change. *Mouth* males white, grey or dark grey, sometimes flesh-tinted; females flesh or pink with greyish or whitish tint; immature males intermediate; juveniles yellow with bright yellow at gape, changing to pale yellow with white at the gape.

Geographical Variation. Little variation. Only one mainland form now recognised. Breeding males show a tendency towards a darker colour at the western extreme of the range; supporting the contention that the almost black isolate of the west coast islands, *M. leucopterus leucopterus*, is a subspecies; the mainland bird therefore should be known as *M. leucopterus leuconotus*.

Malurus lamberti — Variegated Wren

Specimens. Western Australia. Peg's Beach (2) 3 June 66, 1 fl. Chichester Range (3) 17 – 23 May 66, 3 ♂, 1 ♀, 1 ♂ fl. Bush Bay (6) 13 June 66, 1 ♀, 2 ♀ skel. Mungilli Claypan (20) 13 Sept 65, 2 ♂. Lake Gruszka (21) 10 Sept 65, 1 ♂. Warburton (23) 14 Aug 65, 1 ♂, 1 ♀. Van der Linden Lakes (24) 6 Sept 68, 1 ♂. Giles (25) 21 – 30 Aug 65, 1 ♀, 1 imm ♂. St George's Range (27) 8 June 68, 1 ♂ fl. Mount Anderson (28) 3 – 14 May 68, 5 ♂ (8.5, 8.5 g), 4 ♀ (6.6, 7.4, 8.0 g), 3 imm ♂ (8.1, 8.5, 8.5 g), 1 ♂ skel, 2 ♂ fl. Derby (29) 25 Apr 26 – 27 May 68, 3 ♂ (7.4, 7.9, 8.2 g), 1 ♂ fl. Point Torment (30) 12 June 66, 1 ♀ (6.4 g). Mount Bell (32) 20 June 68, 1 fl (7.4 g). Northern Territory. Mount Roper (45) 27 Oct, 1 Nov 68, 2 ♂ (6.6, 8.0 g), 2 ♀ (6.7, 7.8 g). Borroloola (46) 7 – 15 Nov 68, 1 ♂. Queensland. Moonlight Creek (47) 4 – 12 May 64, 5 ♂ (7.3, 7.4, 7.9, 8.0 g). Glenore (49) 17 – 23 Apr 64, 6 ♂ (6.9, 7.3, 7.4, 7.9, 8.1, 8.4 g), 2 ♀ (7.1, 7.3 g), 2 imm ø (7.1, 7.2 g). Glendower (55) 16 – 30 Mar 64, 6 ♂ (7.6, 7.7, 7.9, 8.0, 8.1, 8.7 g), 1 ♀ (8.1 g), 4 imm ♂ (7.0, 7.7, 8.2, 8.4 g). South Australia. Innamincka (64) 30 Apr – 1 May 65, 1 ♂, 1 imm ♂. Clifton Hills (65) 11 May 65, 1 ♂, 1 ♀. Etadunna (66) 15 – 21 May 65, 3 ♂, 4 ♀. 1 ø. Oodnadatta (67) 31 May – 1 June 65, 3 ♂, 1 ♀. Nonning (79) 29 Mar 65, 1 ♀ (8.1 g), 1 ø (8.1 g). Parachilna (81) 12 Apr 65, 1 ♀ (7.5 g). Aroona Valley (82) 10 Apr 65, 1 ♂ (9.0 g), 1 ♀ (7.6 g). Angepena (84) 24 – 29 Nov 63, 4 ♂. Victoria. Little Desert (91) 26 Jan 63, 1 ♂. New South Wales. Grafton (94) 1 Apr 63, 1 ♂.

Distribution & Abundance. Occurred normally in small parties of from three to five individuals, usually with only one apparent male. More rarely occurred in pairs, or larger parties of up to ten individuals.

It was widely distributed and appears to have occurred at all camps where suitable habitat was present. The relative abundance is not clearly indicated. Specimens were collected on single occasions at many camps, but with greater frequency at camps in the western Kimberleys and northern Queensland.

Habitat. A wide variety of habitats from gibber desert (66) to mangroves (3, 30, 47) in which the apparent common factor was the presence of bushy cover at intervals, usually in association with watercourses, the overall distribution being within relatively dry areas.

Behaviour. Generally seen feeding low down in or around the bases of bushes, or searching among dead stems. Once recorded among the leaves of a tall open tree (49). The males were recorded darting from the base of one bush to another, constantly moving on as the rest of the party followed and caught up. The birds were inquisitive, investigating sounds made by the observer and staying to scold if one of a party was shot. The male in breeding dress was recorded as usually being more cautious than birds in dull plumage, but one was seen perched conspicuously on the top of a shrub (23) singing. *M. lamberti* was seldom far from cover and fed less on the ground than did *M. leucopterus*. In dry sandstone country it was mainly in shrubs, and although it was seen occasionally on the rocks, it appeared on these less frequently than did *M. dulcis*.

At Glendower (55) and Borroloola (46) small parties of this species and of *M. melanocephalus* were apparently associating together.

The most frequent call noted was a whirring, reeling or trilling call, sometimes described as harsh; usually given from cover, once recorded during flight. It was audible for a remarkably long distance.

The song of the male was found to be more complex than that of *M. caillanus* but similar in tone, and calls as softer and more sibilant than those of the latter: a soft high-pitched *tsi-tsi-tsi,* and *tsrreee* rising in pitch, were similar to those of *M. leucopterus* but a scolding *trree* alarm note was louder and deeper in tone: a scolding male called *trr-trr, trr-tri-trit, trree-trr-trr:* a female used a loud plaintive *tseep* and thinner thornbill-like *see:* a solitary bird called with a loud plaintive *tew-tew-tew* (DG).

Food. Insects, including very tiny ones, small beetles, small greenish beetles, grasshoppers, ants, wasps, and dark insects with legs at least 5 mm long. Once (81) "small white seeds" – see *M. leucopterus*.

Annual Cycle. This is difficult to determine in *Malurus* from randomly collected specimens. Our large series suggests that in the social breeding groups subdominant birds may be at different stages of moult and gonad development from the dominant pair. The following notes are confined to positive evidence of breeding and of active moult.

WESTERN AUSTRALIA. In May, a party were seen at Tambrey (3) with recently fledged young and a male near by, at Tampatha Well (3), was just completing moult into breeding plumage; at Derby (29) males had just completed moult into breeding plumage; and at Mount Anderson (28) four males were in non-breeding plumage, one was half moulted into non-breeding plumage, and three were just completing moult into breeding plumage. In September (20, 21, 24), of four males two were in breeding plumage and two moulting into breeding plumage.

QUEENSLAND. In March, of six adult males at Glendower (55) one was just completing moult into breeding plumage and five others were in various stages of moult into non-breeding plumage. The female was also undergoing a complete moult. In April, three males at Glenore (49) were in, or just completing moult into, breeding plumage;

two males were beginning to moult out of breeding plumage; and young birds were collected. In May, of five males at Moonlight Creek (47) three were in breeding plumage and two were just beginning to moult into non-breeding plumage.

SOUTH AUSTRALIA. In May one male from both Innamincka (64) and Etadunna (66) were just beginning to moult out of non-breeding plumage.

VICTORIA. In January a male at Little Desert (91) was in breeding plumage with enlarged testes.

NEW SOUTH WALES. An adult male was just completing moult into breeding plumage.

Impermanent Colours. *Legs* males brown with reddish and purplish tints; females brown with greyish tint. *Iris* brown or dark brown. *Bill* males black in breeding plumage tending to be browner in non-breeding plumage; females, reddish or yellowish brown. *Mouth* whitish with flesh or greyish-pink tint.

Immature birds are like the female in the impermanent colours, but the male approaching maturity shows variable and increasing amounts of black on the bill.

Geographical Variation. *M. l. lamberti* from east of the Great Dividing Range is a distinct population, recognisable by the deep blue, not violet-blue, on back and nape of the breeding male, and the darker, warmer brown of females and immature birds.

The population to the west of this, from the Great Dividing Range and some of its eastern ridges in southern Queensland to the west coast, and from southern Australia to the Gulf of Carpentaria, appears to be a single variable entity. No consistent local size differences are apparent. Condon (1951a) suggested a racial separation based on ventral colouring, but this does not hold good for the array of specimens examined here. The most striking difference between individuals is in the blue colour of the forehead, ear-coverts and crown. These vary from light blue to dark blue, and associated with this change there may be a loss of the slightly azure tint on the ear-coverts and an intensification of the violet tint on nape and back. The extreme examples are obviously different, but the variation is not only between one locality and another but also between individuals. It was possible to find a range of specimens showing a complete gradation from one extreme to the other. These did not form distinct localised clines, but overall there is a general tendency for birds to show lighter foreheads and more azure ear-coverts in the central desert regions and towards the northern limits of distribution, and for deeper coloured individuals to occur on the peripheral areas to east and west. This is not, however, sufficiently well-defined to permit nomenclatorial expression. Mack (1934) reduced the recognised subspecies within this range from nine to three. Of the latter Mathews' *bernieri* from Bernier Island, Western Australia, is not larger nor more blue and violet in colour than some examples from further east, and his *mastersi* was based on a Northern Territory specimen with the lighter blue forehead. I do not consider these to be valid races and the specimens within this extensive range can best be treated as a single, slightly variable population under the earliest name *M. l. assimilis* North, 1901.

Parasites. See Appendix 1.

Malurus elegans — Red-winged Wren

Specimens. Western Australia. Scott River (11) 24 – 28 Feb 66, 1 ♂ skel, 1 ♀ fl. Mount Barker (12) 21 – 23 Mar 66, 1 ♂, 1 ø. Green Range (13) 6 Apr 66, 1 ♀ skel.
Field Notes. A skulking species restricted to the extreme southwest corner. Found in infrequent family parties. At Scott River (11) in thick low bushes among shrubby secondary growth in a clearing in a zone of Jarrah – Karri – Marri forest, about 100 yards from the river. At Mount Barker (12) in small swamp overgrown with tea-trees, reeds and tussock grass and surrounded by trees, among cleared paddocks. At Warriup, near Green Range (13), in thick rushes and tea-trees bordering a lake in the coastal dunes.
Food. Insects.
Annual Cycle. Gonads were very small. Birds at Scott River (11) were in moult; one male (now a skeleton) in heavy moult showed about four-fifths breeding plumage, but the direction of plumage change was not recorded. The unsexed Mount Barker (12) specimen shows some wing-moult. The male from Mount Barker (12) in eclipse is black on the lores where females show chestnut-red.
Impermanent Colours. *Legs* dark brown to dark grey or black. *Iris* reddish-brown to dark brown. *Bill* brown in one female, otherwise black in both sexes. *Mouth* pink to flesh-coloured.

Malurus pulcherrimus — Blue-breasted Wren

Specimens. Western Australia. Newdegate (14) 12 – 16 Apr 66, 4 ♂, 2 ♀, 1 ø, 2 ♂ skel, 1 ♂ fl, 1 ♀ fl.
Field Notes. In family parties of six to ten individuals. In small areas of low shrubby growth bordering salt lakes and pools in mallee scrub.

Like other blue wrens tends to keep low in cover. In one case several of a party performed the rodent-run distraction display when other adults in the party were shot.
Food. Insects.
Annual Cycle. Most specimens show evidence of moult on wings, tail and body. Males were just commencing moult into breeding plumage from eclipse. They have black lores. Those with no other breeding plumage show a fine ring of tiny blue feathers surrounding the eye, and on one specimen this has spread to a broader ring with scattered blue feathers appearing on ear-coverts, crown and mantle. Females show chestnut-red on lores and on a narrow ring round the eye. The unsexed bird has the chestnut-red on lores and eye-rim much reduced and a few tiny bluish-white feathers around the eye, and may be a young male moulting from juvenile to first male plumage.
Impermanent Colours. *Legs* dark grey to black, blackish-brown (one male). *Iris* brown to dark brown. *Bill* adult male black, others dark red-brown or blackish brown becoming black towards the tip, paler towards the base of the lower mandible; female upper mandible reddish-brown to dark brown, blackish at tip, lower mandible reddish-brown, paler at base. *Mouth* pale grey to grey near bill, light flesh colour towards throat.

Malurus amabilis Lovely Wren

Specimens. QUEENSLAND. Ayton (52) 24 – 30 June 64, 6 ♂ (7.9, 7.9, 8.1, 8.6, 9.0, 9.8 g), 2 ♀ (8.1, 8.9 g). Tully (54) 11 – 20 July 64, 3 ♂ (8.7, 8.8, 9.3 g), 1 ♀ (8.1 g), 1 ø (8.2 g).
Distribution & Abundance. Usually recorded in pairs, but once three together, and only found where collected.
Habitat. On the edge of rainforest or in open places such as the top of a ridge; and in suitable cover, such as low bushes, felled trees, or logs.
Behaviour. Usually in or near low cover, but once 20 feet up and once three were seen in foliage at the top of a tall tree in company with *Gerygone palpebrosa*. The birds were inquisitive, coming to investigate squeaking noises made by the collectors.
Food. Insects, including caterpillars.
Annual Cycle. Adult males and females collected were in fresh, unworn plumage suggesting the beginning of a breeding season, the former having full breeding plumage. Females showed partly enlarged, granular ovaries; three fully adult males from Ayton (52) had half to two-thirds grown testes, those of the two from Tully (54) were rather smaller. Males in eclipse plumage were distinguishable from females by chestnut edges to tertials and some scapulars, and whiter undersides. Of two younger males from Ayton (52) with particularly pneumatised skulls one was in full eclipse, the other was beginning to moult into breeding plumage. A more adult bird, with vestigial areas of incomplete pneumatisation, from Tully (54), was about one-third moulted into adult male breeding plumage.
Impermanent Colours. *Legs* usually light purple or dull purplish-brown, but sometimes dark brown, ochraceous, or yellowish-buff, the soles usually buff to light yellow. *Iris* blackish brown. *Bill* black. *Mouth* adult males usually black, subadult males grey; females grey, light grey, greyish-white or purplish-white, with whiter or pinkish throat.
Geographical Variation. Mack (1934: 114) separated birds in the southern part of the range, from Cardwell to about Cooktown, as a subspecies *M. a. clarus,* stating that they were a lighter blue. An examination of the collected material, together with that already in the British Museum and specimens borrowed from the Queensland Museum, indicates that although there is a detectable difference in the depth of blue colouring between the most southerly and most northerly examples, birds from near the centre of the range, such as those from Ayton (52), show some intermediacy and the overall difference appears to be clinal in character and too slight to justify nomenclatorial recognition, even if this could be justified for end-points of a cline.

Malurus dulcis White-flanked (=Lavender-flanked) Wren

Specimens. WESTERN AUSTRALIA. Mount Bell (32) 22 June 68, 3 ♂ (8.6, 8.8, 8.8 g), 1 ♀ (7.9 g). Mount Caroline (33) 25 June 68, 3 ♂ (8.6, 8.9, 9.3 g), 1 ♀ (8.1 g), 1 imm ♂ (8.2 g), 3 fl (8.2, 8.2, 8.2 g). Joint Hill (34) 4 July 68, 1 ♂. Mount Remarkable (37) 26 July 68, 2 ♂ (8.5, 8.6 g). Lissadell Hill (38) 29 July 68, 1 ♂ (8.4 g). NORTHERN TERRITORY. Oenpelli (43) 30 Sept – 17 Oct 68, 5 ♂ (7.5, 7.6, 8.2, 8.2, 8.3 g), 3 ♀ (7.7, 7.9, 8.7 g), 1 imm ♂ (7.8 g), 1 ø (7.9 g).

Distribution, Abundance & Habitat. Encountered in small parties of three to eight individuals where sandstone ridges, and in one case granite, were broken by gorges, cliffs or slopes; in association with which a growth of small shrubs and trees, and often spinifex, occurred among broken rocks or in gullies or fissures, usually in the lower and more sheltered places.

Behaviour. Birds moved low through the vegetation, once ten feet up in a tree, sometimes pausing at intervals to congregate before moving on. They frequently settled on and moved over the rocks, at time on almost vertical surfaces. The full-plumaged males tended to be more vocal than the dull-coloured birds. High-pitched squeaking notes were heard, and short sharp notes which increased in frequency during excitement. On one occasion a very "fine" chatter, a repetitive *tic-tic-tic-tic* and an extremely high-pitched, barely-audible single note were recorded. The song of the male was variously described as a "trill" or "reeling". The birds were often secretive, but showed curiosity and would approach in response to squeaking notes by the observer. There were from one to three males showing breeding plumage with each group.

Food. Insects.

Annual Cycle. In June at Mount Bell (32) two out of three males were moulting into breeding plumage, and at Mount Caroline (33) one out of three. At the former gonads were small in all birds, at the latter partially but not fully enlarged. In July males at other camps (34, 37, 38) had more enlarged gonads: a male in breeding plumage at Mount Remarkable (37) showed incompletely grown tail feathers. A male at Lissadell Hill (31) had full breeding plumage on the body, but only one or two new feathers on the wings. In October at Oenpelli (43) males in full breeding plumage had very large testes, but females showed ovaries which, although a little enlarged, had very small oocytes.

Impermanent Colours. *Bill* adult males black; females and immature males reddish-brown or orange-brown (once "pale"). *Iris* brown or dark brown. *Legs* pinkish-grey to brown. *Mouth* adult males white to blackish, once noted as light on lower part, grey on upper; females and immatures white or flesh-coloured.

Geographical Variation. Males are very similar dorsally to *M. lamberti* but differ ventrally in having white, not pale buff, on the posterior flanks. The lavender wash on the flanks described as characteristic of the species by Mathews is nowhere obvious, although faintly indicated on two of the Oenpelli (43) specimens. One male from that locality has a distinct pale buff colour on the posterior flanks although this might be retention of a female character rather than evidence of affinity. If a common name is required it would be preferable to call this species "white-flanked" rather than "lavender-flanked". The males are similar from both Kimberleys and Northern Territory except that the latter have more of a bluish wash on the wing feathers making the wings appear a little darker.

The plumage of females and immatures confirms the two races – *M. d. dulcis* from Northern Territory, and *M. d. rogersi* from the Kimberleys region of Western Australia. These birds and males in non-breeding plumage are blue-grey above, immature birds being drabber and more grey. *M. d. rogersi* specimens appear consistently a little paler in overall colouring. Below, females and immatures have a pale buffish wash while males are almost white, and specimens of *M. d. dulcis* birds show a slightly greyish tint on belly and flanks as well. The most distinctive character is white feathering on lores and around the eye-rim on *M. d. dulcis* and chestnut-red on these areas on *M. d. rogersi*. This is present on young males as well as females. It seems possible that it might be absent from males of *rogersi* in non-breeding plumage but specimens in critical plumage are not available.

Malurus melanocephalus — Red-backed Wren

Specimens. WESTERN AUSTRALIA. Mount Anderson (28) 29 Apr – 8 May 68, 2 ♂, 1 ♀, 2 imm, 1 ø, 2 ♂ skel (7.2 g). Derby (29) 25 May 68, 1 ♂ fl (8.6 g). Geikie Gorge (31) 2 June 68, 1 ♂ (7.4 g). Mount Bell (32) 21 June 68, 1 ♂, 1 ♀. Joint Hill (34) 28 June 68, 2 ♂ (6.7, 7.0 g), 1 ♀ (6.5 g). Mount Elizabeth (35) 13 July 68, 1 ♂ fl (7.6 g), 2 fl (7.1, 7.2 g). Wyndham (39) 2 Aug 68, 1 ♀ (7.0 g). QUEENSLAND. Glenore (49) 13 Apr 64, 2 ♂ (5.7, 7.2 g), 1 ♀ (6.3 g). Inkerman (50) 1 June 64, 1 imm ♂ (6.5 g).

Distribution, Abundance & Habitat. Sparsely but generally distributed in Western Australia and north Queensland camps in suitable country. In small parties of two to seven, each with only one apparent male. Usually in very tall grasses of tropical savanna woodland, frequently near a stream or creek where birds were in grasses or shrubs. At Mount Anderson (28) birds were, however, found in an area of spinifex and low bushes. In northern Queensland often in tea-trees near water; once (47) in undergrowth overgrown with wild passionfruit under tea-trees; once (56) in tall grass bordering mangroves.

Behaviour. Generally shy and secretive, moving and feeding low in cover where possible. Call described as a low rattling note, and the song as a metallic descending trill.

Food. Insects, including small beetles.

Annual Cycle. At Glenore (49) in mid-April a pair were in breeding plumage with large gonads. A male was taken on the same day with enlarged gonads but only a few feathers of breeding plumage. Males from Western Australia showed a range of moults from eclipse plumage, with a few feathers of breeding plumage, in late April at Mount Anderson (28), through intermediate stages in May, to complete breeding plumage in June, although gonads were still small in late June. Females from these areas showed evidence of moult in May and more in June, and the ovaries were not enlarged by late June.

Impermanent Colours. *Legs* males ochraceous to reddish-brown; females ochraceous or light brown; immatures, pinkish or flesh-coloured. *Iris* dark brown. *Bill* males black; females light brown, tending to almost white at the base in some instances, the lower mandible whitish or pale pink; immature males, browner than those of females with a pale lower mandible. *Mouth* males dark grey with pinkish throat; females and immatures pale pink to whitish.

Geographical Variation. Two well-marked races are recognisable, the mainly red-backed *cruentatus,* and the orange- or tangerine-backed nominate race which occurs north of Proserpine and Torrens Creek in Queensland. An orange-backed bird in the British Museum collected by Elsey is labelled Gilbert River. Since Elsey travelled right up the river and over the range (Macdonald & Colston, 1966) this may have come from the upper reaches, but it suggests that the nominate form may reach the upper edge of the Gulf drainage system. Outside the range of *melanocephalus* all birds have red backs, though from west to east there is an increase in the amount of orange tint in the red. This caused Mack (1934) to recognise an intermediate form *M. m. pyrrhonotus* between Cardwell and Cairns, but since the variation is clinal this action hardly seems warranted. Our specimens are all referred to *M. m. cruentatus*.

Malurus coronatus — Purple-crowned Wren

Specimens. WESTERN AUSTRALIA. Joint Hill (34) 2 July 68, 1 ♂ (11.8 g), 1 ♀ (11.2 g). NORTHERN TERRITORY. Borroloola (46) 8 – 17 Nov 68, 3 ♂ (9.7, 10.0, 11.0 g), 3 ♀ (9.7, 9.9, 10.0 g), 1 imm ♂ (10.9 g), 1 ♀ skel, 3 fl (9.3, 9.4, 10.5 g).

Distribution & Abundance. Only recorded at these two camps. Very local. Usually in parties of three to five birds, once seven or eight.

Habitat. In *Pandanus* grove by river at Joint Hill (34); at Borroloola (46) in six foot high grass by river, and in thick vegetation of dead shrubs, *Pandanus* and grass on the edges of lagoons.

Behaviour. Rather shy, moving through cover two or three feet off the ground; twice seen about 15 ft up in a tree. They tend to move rapidly. One party of five had two birds in male breeding plumage. The voice was described as a monotonous ticking or bubbling trill, and occasionally a harsh rattle; seemingly louder than that of other *Malurus* species. On 11 November at Borroloola (46) three hen-plumaged birds were seen in tall dead grass by a creek, a quarter to a half mile from the nearest obvious water, showing interest in a half-constructed nest, a domed structure about three and a half feet up in a low prickly bush. The nest was mainly of grass, with some fine twigs in the coarser outer structure, and fine grass as an inner lining, and it had the entrance facing east. Since it was full of droppings it was probably an old nest or one used for roosting. A similar nest was found about 30 yards away (DJF).

Food. Insects; including small beetles.

Annual Cycle. The male from Joint Hill (34) shows partial moult into breeding plumage on forehead and crown and some scattered body moult; while the female shows moult of some body feathers and rectrices. Three males from Borroloola (46) have small testes and show final stages of moult into breeding plumage. An immature male has the ear-coverts dark grey rather than black, and a dark blue-grey forehead, while the central rectrices have bases in sheath. The females show evidence of head moult and ovaries not very enlarged.

Impermanent Colours. *Legs* light brown. *Iris* dark brown. *Bill* males black, females very dark chocolate brown, almost black, with the lower mandible paler in one instance. *Mouth* males pink to pale grey; females pink.

Geographical Variation. The species is distributed in a narrow zone from the Gulf to the Kimberleys. Mack (1934) recognised two races, arbitrarily dividing the range down the centre. Specimens from Borroloola (46) and Joint Hill (34) do not show the alleged differences in head colour, and the only specimen showing the blue on the head which Mack attributed to the female is an immature male. A blue-grey wash on the back, most distinct on the upper mantle, is absent from Kimberley males but present on males from Borroloola (46) and is the principal character for Mack's *M. c. caeruleus*. One male from Borroloola (46) appears to lack this grey, but the mantle feathers appear to be old feathers of the non-breeding, eclipse plumage. The grey is also absent from the immature male and may therefore be a character of the breeding plumage only.

Stipiturus malachurus Southern Emu-wren

Specimens. WESTERN AUSTRALIA. Metricup (9) 7 – 17 Mar 66, 2 ♀, 1 ♀ skel. Scott River (11) 18 Feb – 1 Mar 66, 1 ♂, 1 imm ♂, 1 ♀ skel. Mount Barker (12) 23 Mar 66, 1 ♂, 1 ♀. Green Range (13) 6 Apr 66, 1 ♂ fl. Newdegate (14) 18 Apr 66, 1 ♂ fl. VICTORIA. Victoria Range (90) 30 Jan 63, 1 ♂, 1 ♀. NEW SOUTH WALES. Tenterfield (95) 12 Apr 63, 10 ♂, 4 ♀.

Specimens to other Museums. NMV 2 ♂, 1 ♀ Tenterfield (95). AM 1 ♂ Tenterfield (95).

Distribution & Abundance. A very difficult species to detect, rarely showing itself. The fact that it was noted at all main camps within its range, including Nelson (86), Furner (88) and Coleraine (89), and Grafton (94) indicates that it is widespread, and might be abundant.

Habitat. This is a bird of very low cover, typical of rough and slightly swampy grass in which it hides. It occurred on open heathland in Victoria; in heathland swamp in New South Wales; and in Western Australia was usually associated with swampy growth in heathland clearings, by lakes or in paddocks; at Metricup (9) in small patches of scrub among woodland (see Harrison, 1969b) and at Newdegate (14) in low heath scrub and mallee.

Behaviour. Skulked in low grass, usually keeping to the ground, very difficult to move if any cover was available, but if flushed flying very low for a short distance. Call a thin, shrill squeak.

Food. Small insects.

Annual Cycle. Gonads were small in all birds, and birds from all localities showed some evidence of moult on body, wings and tail.

Impermanent Colours. *Legs* males yellowish-buff to reddish brown, soles yellow; females sandy yellow or light orange buff, soles yellow. *Iris* brown, reddish brown, or hazel brown. *Bill* males brown with pink base to lower mandible, blackish brown, or medium grey to black; females dark brown with light grey base to lower mandible, or grey-brown with buff base to lower mandible. *Mouth* male yellow, pinkish-buff, light grey or blackish; female yellow.

Geographical Variation. Most alleged subspecies are very poorly differentiated. On our material eastern birds *S. m. malachurus* can only be distinguished from *S. m. westernensis* of the southwest on the colour of the forehead and ear-coverts of males. Specimens in the British Museum from South Australia do not appear to differ from *S. m. malachurus*.

Amytornis textilis Western Grass-wren

Specimens. WESTERN AUSTRALIA. Giles (25) 20 Aug – 2 Sept 65, 11 ♂ (22.0, 23.0 g), 2 ♀ (19.0, 21.0, 21.0 g), 1 ♀ skel (21.0 g), 1 ♀ fl. Walter James Range (26) 26 Aug 65, 1 fl.

Distribution & Abundance. Only encountered on the slopes of the Rawlinson Range (25) and Walter James Range (26) in Western Australia. Usually seen in pairs or parties of up to five birds; once eight to ten.

Habitat. On rocky hillsides with broken terrain and boulders; usually with scattered spinifex or shrubs.

Behaviour. (See Goodwin, 1967c). Similar to *Malurus* in appearance and general behaviour; hops rapidly with tail cocked, flies low and reluctantly; inquisitive but timid. Song sometimes delivered from a rock, a reeling trill, more "shivering" or sibilant than that of *Malurus* species, sometimes interspersed with calls. Callnotes, a high-pitched ventriloquial single note *seet* and a louder, sharper *tchip* or *tchip-ip*. When a shot bird uttered distress calls a group of birds performed a distraction display around the collector, running with head and tail low and wings drooped and scraping the ground.

Food. Seeds and small insects, including Mulga seeds and ants.

Annual Cycle. Gonads small but apparently enlarging a little. No sign of moult.

Impermanent Colours. *Legs* males dark brown, to dark grey; females brown. *Iris* brown or dark brown. *Bill* males dark grey or dark brown to black; females brown. *Mouth* males yellow to grey or grey-mauve with yellow throat or tongue, females very dull mauve.

Geographical Variation. The population of the central ranges, to which these belong, was named *A. t. purnelli* by Mathews. Keast (1958e) separated Everard and Musgrave range birds as *everardi,* with larger bills (♂ 10.5 – 10.8, ♀ 10.2 – 10.5 mm) against *purnelli* (♂ 9.6 – 10.4 mm). As our *purnelli* have bills ♂ 10.2 – 11.5, ♀ 10.2 – 10.9 mm it does not seem that two races can be recognised.

Parasites. See Appendix 1.

Amytornis striatus Striated Grass-wren

Specimens. WESTERN AUSTRALIA. Chichester Range (3) 22 May 66, 2 ♂ (+1 ♂ flat skin from skel), 1 ♂ skel, 1 fl. Carnegie (19) 14 Sept 65, 1 imm ♀.

Distribution & Abundance. The only birds seen were a group of four (3) and two (19).

Habitat. On low undulating sandstone hills of the Chichester Range (3) with exposed rocky slabs and shallow gullies, covered with spinifex, and with scattered bushes and a few white gums. At Carnegie (19) on open spinifex plain between sandhills, with scattered small trees and bushes.

Behaviour. Inquisitive but skulking, the party moving very fast through low cover of spinifex and loose boulders. The Carnegie (19) bird was seen hopping with tail erect. Calls, a harsh *tchritt, tchritt* like that of a Brown Thornbill but lower in pitch, with *Malurus*-like tone; also a soft *see* and a whispered soft song were probably this species.

Food. A large quantity of beetles and other insects (3); seeds, with sand grains in the gizzard (19).
Annual Cycle. Males at Chichester Range (3) showed slightly enlarged gonads and some moult of primaries (two birds) and outer rectrices (one bird). The immature female at Carnegie (19) had a slightly enlarged ovary.
Impermanent Colours. *Legs* males dark grey; immature female slate grey. *Iris* brown. *Bill* males dark grey; immature female dark grey upper mandible and blue-grey lower mandible, both with paler cutting edges. *Mouth* males grey with fleshy throat; immature female black with white markings, and yellowish tongue.
Geographical Variation. The Chichester Range (3) birds appear to be typical *A. s. whitei* and the Carnegie (19) bird to be *A. s. oweni* (see Keast, 1958e). It is smaller than other specimens of this race (wing 53, cf 56-61 mm) but this is not surprising as it is an immature female.

Amytornis woodwardi White-throated Grass-wren

Specimens. NORTHERN TERRITORY. Oenpelli (43) 9 – 17 Oct 68, 2 ♂ (34.1, 37.3 g), 7 ♀ (30.1, 30.1, 32.8, 33.5, 34.3, 34.4, 34.8 g), 1? imm ♂ (32.0 g).
Distribution, Abundance & Habitat. Seen in parties of up to seven birds by the East Alligator River near Oenpelli (43), the only camp within the range of the species. They were found in eroded sandstone hills with terraces, faults, and rocky gorges, which had a vegetation of spinifex, grasses, low shrubs, and some trees. They appeared to be local although more extensive areas of apparently suitable habitat were present.
Behaviour. The birds moved with tails cocked. Generally very secretive, keeping to low cover in rocks and spinifex, but inquisitive and likely to investigate squeaking calls or shots. Flight was usually short and low. A deep *tic-tic-tic* alarm call was separable from those of *Malurus* species though the vocabulary as a whole was similar. High pitched peeps were used as contact calls, while the song-phrase was a little twittering song uttered from rocks or branches 3 ft up.
Food. Insects, including ants, seeds and vegetable matter, including some one-inch long grass fragments. Grit was present in the gizzard.
Annual Cycle. Gonads generally small; ovaries may have been slightly enlarged in two birds and testes in one male. The skulls of four birds were incompletely pneumatised, least so in the one identified as immature. Slight moult on tail and body of some birds.
Impermanent Colours. *Legs* male brownish-grey or grey; females pale brown to blackish. *Iris* brown. *Bill* males dark bluish-grey or very dark grey; females upper mandible blackish-grey to dull black, lower mandible blue-grey at the base, blackish at the tip. *Mouth* males purplish or blue; females blackish; ? immature male pale yellow.

Amytornis housei (see Freeman 1970) Black Grass-wren

Specimens. WESTERN AUSTRALIA. Joint Hill (34) 2 – 4 July 68, 3 ♂ (29.0, 29.3, 30.3 g), 4 ♀ (24.0, 27.4, 27.9 g), 1 juv ♀, 1 ♂ fl (30.9 g).
Specimens to other Museums. WAM 1 ♂ (30.5 g), 1 ♀ (27.5 g) Joint Hill (34).

Distribution & Abundance. Limited to the type locality. Collected in two areas nine and a half miles apart; in parties of up to seven birds.
Habitat. A large area of sandstone with large broken boulders and gullies between, with spinifex, near a river gorge. A high sandstone outcrop rising from savanna woodland about a half by a quarter mile and 200-250 feet above the surrounding country, with many gullies, crevices and ridges, and studded with small trees and grass.
Behaviour. The birds moved in small, active and fairly dispersed parties; moving by both hopping and running, with the tail usually held fairly low but frequently cocked momentarily. They tended to keep low in cover or rocks and vegetation but at intervals individuals, particularly males, might mount small rocks. When moving among rocks they appeared to prefer to descend and ascend on foot rather than fly from the top of one rock to another; and they used the wings only for infrequent low flights across gullies or downhill. They showed some initial curiosity and little obvious fear. The calls noted were a soft, high-pitched whistle, a purring note, a harsh rattling note, and a ticking reel.
Food. Of six stomachs examined all contained insect remains and seeds of *Sorghum* (cf. *S. stipoideum*): other identified seeds were *Setaria* (cf. *S. apiculata*) and *Panicum* (cf. *P. seminudum*), each in four stomachs; small numbers of Cyperaceae seeds in two, and a single seed of Euphorbiaceae (*Antidesma* sp.).
Annual Cycle. A juvenile with partly-grown flight feathers and still food-begging was collected from a group containing two females and a male. In general gonads were small; slightly enlarged in one male and three females. Plumages appeared worn, and one female was moulting the central rectrices.
Impermanent Colours. *Legs* males greyish brown to almost black; females dark brown, dark slate grey, blue-grey; juvenile dark grey. *Iris* brown to dark brown, noted as russet for one female. *Bill* males black, sometimes with dark grey base to the lower mandible; females blackish-grey paler at the base of the lower mandible; juvenile blackish-grey with light grey lower mandible. *Mouth* males yellowish to yellowish-grey, dark grey with creamy tongue, or blue-grey; females yellowish with tongue yellow, or dark grey with creamy tongue; juvenile, mouth and tongue yellow. In view of the juvenile colour the yellow tint of the mouths of some adults probably indicates a degree of immaturity.

EPHTHIANURIDAE* — Australian Chats
C. J. O. Harrison

Ephthianura albifrons White-fronted Chat

Specimens. Western Australia. Scott River (11) 20 Feb – 1 Mar 66, 2 ♂, 1 imm ♂, 1 ♂ skel. Mount Barker (12) 25 – 28 Mar 66, 2 ♂, 2 ♂ skel, 1 ♀ skel, 1 fl. Eucla (17) 2 July 65, 1 ♂, 1 ♀. Victoria. Little Desert (91) 26 Jan 63, 2 fl.
Distribution & Abundance. Also seen at Green Range (13) and Newdegate (14) in Western Australia; Kingoonya (70) and Bool Lagoon (87) in South Australia; and

*See **Emu 70,** 1970: 205 for spelling.

Nelson (88) in Victoria. Seen in pairs or parties and small flocks of up to 40 birds. Also seen in mixed flocks with *E. tricolor* and *E. aurifrons*.
Habitat. Coastal lagoon bordered by reeds, heath and sparse dune scrub; grass paddocks; saltbush and Bluebush plain; and areas of low sparse scrub in semi-arid country.
Behaviour. Feeds on the ground, walking, or running in short bursts. Perches freely on stems, bushes or trees when disturbed. Call noted low and slightly nasal.
Food. Insects, and some seeds.
Annual Cycle. An adult male had almost completed moult by January (91). Immatures were moulting into adult plumage and adults were in fresh plumage by February – March (11-12). Gonads were enlarged in a male in July (69) but not in a female.
Impermanent Colours. *Legs* adults black, immatures blackish-grey. *Iris* adult males white, females pinkish; immatures light brown through pinkish buff and pinkish orange to pink. *Bill* adults black; immatures black with grey base. *Mouth* adults white; immatures, black to yellowish-brown.

Ephthianura tricolor — Crimson Chat

Specimens. WESTERN AUSTRALIA. Chichester Range (3) 13 Apr 66, 2 ♂ skel, 1 imm skel. Mount Shenton (18) 2 – 3 Aug 65, 4 ♂, 1 ♀, 1 ♀ skel, 1 fl. QUEENSLAND. Moonligh, Creek (47) 9 – 13 May 64, 1 ♀ (10.0 g), 1 ♀ skel (10.3 g), 1 ø skel (9.8 g), 3 fl (9.8, 9.8t 10.0 g). Thylungra (60) 25 Apr 63, 1 imm ø. SOUTH AUSTRALIA. Innamincka (64) 29 Apr 65, 2 ♂, 3 ♀, 1 ø. Etadunna (66) 20 May 65, 1 ♂, 2 ♀. Nullarbor HS (69) 7 Dec 62, 1 ♂ skel, 1 ♀ skel; 5 – 10 Oct 63, 1 ♂, 1 imm ♂. Angepena (84) 18 Nov 63, 1 ♂.
Specimens to other Museums. NMV 1 ♀ Nullarbor HS (69).
Distribution & Abundance. Also recorded at Coolawanyah (4), Warburton (23), Van Der Linden Lakes (24), Mount Anderson (28), Oodnadatta (67), and between Normanton and Glenore (49) on 22 May 64 although none were seen in the same area 6 – 27 April. Usually in loose flocks of three to thirty birds, and occasional pairs were usually near others.
Habitat. Mostly open areas of short herbage or bare ground with bushes and trees nearby. Noted on grassy or bare patches in spinifex; open sandy areas and areas with flowers after rain in sparse mulga; creekbeds with shrubs and trees in arid country; grassy saltpans near savanna woodland; grassy plains, sandhills with sparse vegetation, saltbush and Bluebush.
Behaviour. (See also Goodwin, 1967c). Sociable, not only flocking with its own kind, but also with *E. aurifrons* and *E. albifrons*. It feeds mainly on the ground, walking like a wagtail (*Motacilla* sp.) and also in shrubs. The displaying male has been seen swooping towards the female with extended wings slightly raised and slowly waved, and with the red crown feather fully erected. The voice is a tinkling twitter when the flock takes flight. Calls noted were a throaty *tch-tch-tch-tch,* a Whiteface-like ringing *cheng-cheng-cheng* (usually prefaced by a rather nasal *tchu-tchu*), a *tchu* note and varient, especially a disyllabic call frequently uttered. One alarm call is a thin, high-pitched *seee-ee.*
Food. Insects, including ants, small beetles and caterpillars, seeds recorded at Chichester Range (3) and possibly at Mount Shenton (18).

Annual Cycle. Although the male has an eclipse plumage all males in moult at different times of year were moulting into the red breeding plumage. At Chichester Range (3), Thylungra (60), and Innamincka (64) in April males were moulting into red plumage, with gonads a little enlarged at the first locality, minute elsewhere. Females at the first locality appeared to have recently moulted. In May at Moonlight Creek (47) gonads were small or a little enlarged, and at Etadunna (66) a male (possibly immature from plumage) and female had almost completed moult but gonads were minute. At Mount Shenton (18) in August birds were in breeding plumage with gonads enlarged but one male was still in latter stages of moult with smaller gonads. At Nullarbor HS (69) in December birds were in, or coming into breeding plumage except for one male only just beginning body moult.

Impermanent Colours. *Legs* medium grey to black. *Iris* pale yellow to white, usually yellowish-white; some shades of pale brown to yellow may indicate immaturity. *Bill* dark grey to blackish grey, base of lower mandible often light grey to whitish; dark brownish bills with fleshy tint to base and lower mandible may indicate immaturity. *Mouth* dull yellow.

Geographical Variation. The plumage appears very subject to wear and fading and I would agree with Keast (1958f) in recognising no subspecific variation.

Other Notes. Some females show a pink colour on the tips of some of the breast feathering. The male from Etadunna (66) appears to have moulted into breeding plumage but shows only pink tips to breast feathers like those on some females and a light reddish cap resulting from limited red pigment on otherwise brown feathers. This may constitute a subadult plumage.

Ephthianura aurifrons Orange Chat

Specimens. SOUTH AUSTRALIA. Innamincka (64) 29 Apr – 1 May 65, 1 ♂, 5 ♀. Clifton Hills (65) 11 – 12 May 65, 3 ♂ (10.5, 11.1 g), 1 ♀ (10.2 g). Etadunna (66) 14 – 20 May 65, 4 ♂, 1 ♀. Granite Downs (68) 7 June 65, 1 ♂, 2 ♀. Nullarbor HS (69) 4 – 5 Oct 63, 1 ♂, 3 imm ♂, 1 imm ♀. Kingoonya (70) 22 June 65, 1 ♂. Parachilna (81) 12 Apr 65, 5 ♀ (9.9, 9.9, 10.2, 10.5, 11.4 g). Angepena (84) 18 Nov 63, 1 ♂, 1 imm ♂.

Distribution & Abundance. Also at Aroona Valley (82).

Habitat. Open sandy areas and sandhills, with sparse low bushes; dry part of a lake bed with some herbage; open Bluebush and saltbush flats.

Behaviour. Usually seen walking and feeding on the ground. Sociable; seen in mixed flocks with *E. tricolor* and *E. albifrons,* and with or near parties of *Taenopygia guttata.* Usually in arid country, but when water is present will drink, feed along its edge, and wade. Tends to be wary and flies low for short distances when disturbed. Will perch freely on the tops of bushes or low trees, and tends to perch after flight when alarmed. Call, a single, rather long-drawn chirp often uttered during flight.

Food. Insects, including beetles and Diptera; vegetable remains once (66).

Annual Cycle. It is uncertain if the male has an eclipse plumage (Keast, 1958f). The four males of the present series not in breeding plumage are moulting into it and could be immature birds attaining adult plumage.

Birds taken in April and May had small gonads. Birds from Parachilna (81) in early April and Clifton Hills (65) in early May showed some moult. At Granite Downs (68) in early June birds had slightly enlarged gonads and a single male from Kingoonya (70) in mid-June had enlarged testes. October and November birds from Nullarbor HS (69) and Angepena (84) are in fresh plumage, and an immature male was commencing moult into adult plumage.

Impermanent Colours. *Legs* grey to blackish-grey. *Iris* orange-brown to reddish-brown; twice described as red with yellowish or whitish rim. *Bill* adult males black or brownish-black; females and immatures black-tipped but pale grey or flesh-tinted at gape and base of lower mandible. *Mouth* yellow, tinted grey or flesh in younger birds.

Ashbyia lovensis — Gibber-bird

Specimens. Queensland. Monkira (61) 6 – 8 June 63, 1 ♂, 1 ♀. South Australia. Clifton Hills (65) 10 May 65, 1 ♂, 1 ♀.

Distribution, Abundance & Habitat. Not recorded elsewhere. In pairs or singly in open country near gibber, bare or sparsely covered with low vegetation (61), or surrounded by saltbush (65).

Behaviour. They showed little fear, even when one of a pair was shot. They were seen walking about in search of food on the ground, or standing upright on a small elevation, with the yellow breast showing.

Food. Insects, including large green caterpillars; seeds, and vegetable matter.

Annual Cycle. Gonads were enlarged in the Monkira (61) male and Clifton Hills (65) female, but small in others. Both females show a little body moult.

Impermanent Colours. *Legs* pearl grey to dark grey. *Iris* cream, yellowish-white, or pale yellow. *Bill* black. *Mouth* blackish-grey, violet, or black.

MUSCICAPIDAE — Old World Flycatchers

I. C. J. Galbraith

Microeca leucophaea — Jacky Winter

Specimens. Western Australia. Eucla (17) 1 July 65, 1 ♂. Mount Anderson (28) 1 – 19 May 68, 1 ♂ (12.5 g), 1 imm ø. Derby (29) 21 – 25 May 68, 1 ♂ (11.5 g), 1 ♀ (11.1 g), 1 imm ♀ (10.7 g), 1 skel. Point Torment (30) 12 June 68, 1 fl. Mount Bell (32) 14 June 68, 1 ♂, 1 fl (12.5 g). Joint Hill (34) 27 June – 2 July 68, 1 ♀, 1 ø (12.4 g). Northern Territory. Borroloola (46) 8 – 12 Nov 68, 1 ♂ skel (11.8 g), 1 ♀ skel (11.9 g), 2 fl (11.9, 12.1 g). Queensland. Glenore (49) 24 Apr 64, 1 ø (12.4 g). Banoona (58) 30 Apr 63, 1 ♀, 1 ø. Tyrone (59) 10 May 63, 1 ♂, 2 ♀. Thylungra (60) 22 May 63, 1 ♀ skel. Monkira (61) 2 – 4 June 63, 2 ♂, 1 ♀. South Australia. Kingoonya (70) 19 June 65, 1 ♂, 1 ♀. Nonning (79) 29 Mar 65, 1 ♂, 1 ♀. Furner (86) 7 – 8 Jan 63, 1 juv ♂, 1 ♀ skel. Victoria. Little Desert (91) 26 Jan 63, 1 juv, 1 fl. New South Wales. Dunedoo (93) 18 – 25 Mar 63, 1 ♂, 1 imm ♂, 1 fl. Grafton (94) 1 – 5 Apr 63, 1 ♂, 1 ♀, 1 ø, 1 fl.

Specimens to other Museums. AM 1 ♂ Dunedoo (93). NMV 1 ♂ Monkira (61), 1 ♀ Grafton (94).

Distribution & Abundance. Also recorded from Mount Caroline (33), Mount Remarkable (37), Pine Creek (44), Glendower (55), Victoria Range (90), Albury (92) and Tenterfield (95). Only one seen at both Eucla (17) and Mount Shenton (18), few at Mount Anderson (28), and Kingoonya (70), recorded as common locally in woodland at Borroloola (46) and Nonning (79), very common at Derby (29).
Habitat. Savanna woodland and mallee, often in denser vegetation especially near water, patches of woodland in shrubland, and edges of denser woodland. Once in mangrove swamp (30).
Behaviour. Perches high or low, often in exposed positions. Often inactive during the heat of the day, or sallying out after insects and returning to the same or a different perch. Occasionally on the ground. Usually solitary but sometimes in parties of up to six. Usually silent, occasionally a short, sharp alarm call, once apparent song *pretty-pretty-pretty* (BPH). A juvenile was apparently begging to *Artamus cyanopterus*.
Food. Small winged insects, beetles and grasshoppers.
Annual Cycle. Three males had enlarged testes on 2 and 24 May and 4 June (47, 29, 61). Young were in juvenile plumage on 7 and 26 January (86, 91), completing the moult into first year plumage on 18 March (93), and in the latter plumage on 1 and 25 May (28, 29). The younger juvenile (26 January) had an unpneumatised skull, and the first-year bird on 1 May a partly pneumatised one, but the moulting and the other first-year birds had fully pneumatised skulls. Most adults were in moult between 19 March and 24 April (93, 79, 94, 49).
Impermanent Colours. *Legs* black or dark grey, brown or brownish at Borroloola (46) and once each at Derby (29) and Nonning (79). *Iris* brown, usually dark but recorded as from mid-brown to black. *Bill* black occasionally greyish or brownish, lower mandible often paler; unusually pale (light brown with very light brown lower mandible) only in one first year young (28). *Mouth* flesh (greyish to pinkish) once dark grey (34), yellow or yellowish at Borroloola and Tyrone (46, 59) and occasionally elsewhere; light yellow to orange yellow in juveniles and first-year birds.
Geographical Variation. The adult skins agree with the subspecies recognised by Keast (1958i):
M. l. pallida northern Australia (28, 29, 32, 34, 49). Smaller and somewhat paler dorsally than nominate *leucophaea,* but tails have less white than some of the latter.
M. l. leucophaea southeastern Australia (58, 59, 93, 94). The specimens from southeastern Queensland have the most white in the tail of any examined, with the third rectrix mainly white and a large white tip on the fourth.
M. l. barcoo interior. Monkira (61), well to the north of the subspecies range indicated by Keast. Similar in size to *leucophaea* but markedly paler, slightly paler than *pallida.*
M. l. assimilis southern mallee areas (17, 70, 79). Markedly darker and browner below, browner and slightly darker above, than *leucophaea.*
Other Notes. The younger juvenile 26 January (91) can only just have fledged. It has a very small bill and its wings and tail are only partly grown. Apart from its remiges, which are like the adult, and its upper wing-coverts, which are browner, the dark parts of its plumage are blackish fuscous, and all the contour feathers of its upperparts have large white terminal spots. The older juvenile 7 January (86) is of virtually adult

size and proportions. The blackish fuscous has foxed to a warm brown, especially upon the crown, and many additional contour feathers of more adult coloration have grown, though they seem to belong to the same feather generation since intermediate stages are present. The oldest feathers below are all-white, above all olivaceous or with a vague fuscous area at the tip. In both juveniles the upper wing-coverts have white tips extending slightly up the shafts, and those rectrices whose tips are white have them smudged with fuscous.

The first-year birds closely resemble adults, but retain juvenile primary-coverts, outer secondary-coverts, and rectrices.

Parasites. See Appendix 1.

*Microeca brunneicauda** — Brown-tailed Flycatcher

Specimens. WESTERN AUSTRALIA. Point Torment (30) 30 May 68, 2 imm ♂ (12.7, 13.5 g).

Field Notes. Found only in a mangrove swamp subject to regular flooding, mostly very dense. Moving about unobtrusively at the edge of mangroves close to the ground, thought to be feeding on swarms of mosquitoes (BB). Solitary or in small parties. Not encountered elsewhere.

Annual Cycle. Both specimens were in first-year plumage, one completing the moult from juvenile plumage.

Impermanent Colours. *Legs* black. *Iris* dark brown. *Bill* dark brown above, paler below (pale brown, or flesh-coloured with a brown tip). *Mouth* with a yellowish wash.

Other Notes. The primary-coverts and outer secondary-coverts have white spots as in the first-year plumages of other species. The few juvenile feathers in the rump and upper tail-coverts of one specimen are dull reddish-brown with white tips.

According to Hill (1910) the species lives entirely on small crabs and other forms of marine life. Unfortunately the gut contents were not examined.

Since there is no good evidence that this form occurs east of Napier Broome Bay (see Storr, 1967: 70), it can be accepted as the geographical representative of *M. flavigaster*.

Microeca flavigaster — Lemon-breasted Flycatcher

Specimens. NORTHERN TERRITORY. Darwin (41) 10 Sept 68, 1 fl (12.9 g). Oenpelli (43) 5 Oct 68, 1 fl. Pine Creek (44) 27 Aug – 2 Sept 68, 1 ♂ (12.7 g), 2 ♀ (11.7, 11.8 g). QUEENSLAND. Inkerman (50) 27 – 30 May 64, 2 ♂ (11.9, 12.5 g), 3 ♀ (10.6, 11.5, 11.7 g), 1 ø (12.2 g), 1 imm ø (14.6 g). Ayton (52) 25 June 64, 1 ♀ (13.1 g). Tully (54) 11 – 18 July 64, 4 ♂ (13.5, 13.9, 14.2, 16.0 g), 3 ♀ (13.9, 14.2, 14.4 g), 1 imm ♂ (15.9 g). Cape Gloucester (56) 5 Aug 64, 1 ♂ (14.2 g).

Distribution & Abundance. Not recorded elsewhere, not encountered on 'islands' at Moonlight Creek (47). Plentiful in appropriate habitats at Darwin (41) and Inkerman (50), few seen at Oenpelli (43) and Pine Creek (44).

* Parker (1973) shows that this form should be regarded as a race of *M. flavigaster* under the name *M. f. tormenti* Mathews, 1916 (=*brunneicauda* of authors, not Campbell, 1902).

Habitat. At Darwin (41) in monsoon forest, and to a much less extent in the adjacent mangroves. At Oenpelli (43) in trees around a lagoon, and flowering trees on the blacksoil plain. At Pine Creek (44) along a dry creek with pools of standing water, in savanna woodland. At Inkerman (50) mainly in and near the densest vegetation, partly over a swamp, on the 'island' near Salt Arm Creek. On the east coast (52, 54, 56) in open forest, woodland and clearings, near rainforest or layered woodland.

Behaviour. At all heights though most often in the crowns of trees, sometimes on bare branches and several times over water. Often inactive, but sometimes investigating flowers and foliage or sallying out after insects. Solitary or in pairs, occasionally small parties. Once two different birds, at an interval of 25 minutes, were with *Rhipidura fuliginosa, Myiagra rubecula* and *Melithreptus albogularis* (50).

Voice seldom recorded, but then as a pretty little song like that of *M. leucophaea,* a varied whistling call (IM), a short and simple musical song of about seven notes (SAP), and phrases lasting about four seconds reminiscent of a Whitethroat *Sylvia communis* (BB & DJF).

Food. Insects of various sizes, including ants and other Hymenoptera, beetles and caterpillars.

Annual Cycle. Males had enlarged testes on 27 May (two individuals one with greatly enlarged testes) and on 5 and 28 August. Young birds were in first-year plumage on 27 May (in body-moult, apparently into the adult plumage) and 11 July. Single adults were in body-moult, on 27 and 30 May, 25 June, 14 July and 2 September.

Impermanent Colours. *Legs* dark grey, brownish or olivaceous to black, feet often blacker than shanks, soles yellowish, buffy or dirty white. *Iris* dark brown. *Bill* sepia to black above, paler below especially basally (brown, buff or whitish). *Mouth* dull yellow to orange-yellow, occasionally mottled with grey.

Geographical Variation. Although, as Rand (1940) and Deignan (1964) warn, the subspecific characters of this species are liable to be much affected by wear, foxing and soiling, our specimens are all in reasonably fresh and apparently comparable condition – in marked contrast to the few skins already in the British Museum. They fall into colour groups which do not entirely agree with the classifications adopted by Rand and Keast (1958i) and in the absence of suitable comparative material it seems best to list them by locality.

Pine Creek (44). Small, dorsally dark, no yellow on throat and sides of head, breast and flanks greyish – agrees with nominate *flavigaster.*

Inkerman (50). Intermediate in wing-length (though apparently low in weight), pale dorsally, with much yellow pigment. The underparts are rich yellow; the upperparts of some specimens a golden olive; eye-ring, lores and superciliary yellow and the throat and sides of the head washed with yellow; and the breast and flanks pale olivaceous. These are the characters of *terraereginae* said to be best-developed at Cape York (Rand), and indeed one specimen from the Olive River does approach this series. Thomson (1935) describes his specimens from the Coleman and Edward Rivers, 80-110 miles north of Inkerman, as pale yellow below with a brownish tinge, but does not make it clear how they differ in this from his east coast specimens from the Stewart River.

Ayton (52). The single specimen matches the Inkerman (50) series dorsally, and the Pine Creek (44) series in the amount of yellow in the underparts except that its throat is washed with yellow. Its throat-patch is more clearly demarcated from the darker olivaceous breast and flanks than in any other specimen.

Tully (54) and Cape Gloucester (56). Largest, agreeing in colour best with the Pine Creek (44) series (though some specimens have a little yellow on the throat and sides of the head), but with the throat-patch less clearly demarcated and continuing well down on to the breast as an irregular whitish wash. Presumably referable to *laetissima*.

Other Notes. The first-year young differ from adults in the same characters of the upper wing-coverts as those of *M. leucophaea,* while the vague tips to their outer rectrices are buffy, rather than whitish as in adults.

Petroica multicolor — Scarlet Robin

Specimens. WESTERN AUSTRALIA. Perth (7) 27 Apr 66, 1 ♂ skel. Metricup (9) 11 Mar 66, 1 ø. Scott River (11) 19 – 20 Feb 66, 1 ♀, 2 ♂ skel, 1 ø skel. VICTORIA. Coleraine (89) 22 Jan 63, 2 fl. NEW SOUTH WALES. Albury (92) 11 Mar 63, 1 imm ♂. Tenterfield (95) 10 – 12 Apr 63, 1 ♂, 1 ♀, 1 imm ♂.

Specimens to other Museums. NMV 1 fl Coleraine (89).

Field Notes. Found in various habitats with perches, from a fence at the edge of an open paddock to one along a track through a forest. Also recorded at Mount Barker (12), Green Range (13), Furner (86), Nelson (88) and Victoria Range (90).

Food. Insects, including beetles and small insects.

Annual Cycle. An individual in the dull plumage sexed as male on 11 March (92) was moulting head, body and secondaries. Its new feathers are not those of the adult male, so that it had presumably just lost the juvenile plumage.

Two adults were in moult on 19 February, and two females on 20 February and 10 April.

Impermanent Colours. *Legs* black. *Iris* reddish to very dark. *Bill* black. *Mouth* yellow to orange.

Geographical Variation. These belong to the two Australian subspecies, which are indistinguishable in the adult male plumage but distinct in the dull plumage (Keast, 1958i): *P. m. campbelli* in the southwest (9, 11), with dark throat and blackish upperparts; and *P. m. boodang* in the southeast (89, 92, 95) with pale throat and brownish upperparts.

Other Notes. As Disney (1969: 18-19) remarks, first-year males of *Petroica* do not seem to be separable from first-year and adult females. Although the shape of the rectrices and the colour of the wing-bars vary considerably in this dull plumage (which is red beneath only in this species), the variation seems to be individual. Females for all species of this genus are therefore listed as adults. Disney (1969: 38-39) gives diagnostic characters of the dull plumage.

Petroica goodenovii — Red-capped Robin

Specimens. WESTERN AUSTRALIA. Chichester Range (3) 12 – 23 May 66, 1 ♀, 1 imm ♂, 1 ♀ skel. Newdegate (14) 14 – 15 Apr 66, 1 ø, 1 ♂ skel. Eucla (17) 2 July 65, 1 ♂. Mount

Shenton (18) 30 July 65, 3 ♂, 2 ♀. Carnegie (19) 16 Sept 65, 1 imm ♂. Giles (25) 24 Aug – 1 Sept 65, 1 ♂, 1 ♀, 2 imm ♂, 1 ♂ skel. QUEENSLAND. Tyrone (59) 10 May 63, 1 ♂, 1 ♀. Thylungra (60) 23 – 25 May 63, 1 ♀, 1 imm ♂, 3 fl. SOUTH AUSTRALIA. Etadunna (66) 17 – 22 May 65, 1 ♂, 1 ø, 1 imm ♂. Oodnadatta (67) 31 May 65, 1 ♀. Granite Downs (68) 7 June 65, 1 ♂, 1 ♀, 1 ø. Nullarbor HS (69) 4 Oct 63, 1 ♂. Kingoonya (70) 16 June 65, 1 ♂, 1 ♀. Nonning (79) 29 – 30 Mar 65, 1 ♂ (8.5 g), 1 ø (9.4 g), 1 imm ♂ (8.2 g). Parachilna (81) 12 Apr 65, 1 ♂ (9.4 g). Aroona Valley (82) 8 – 12 Apr 65, 2 ♂ (8.7, 9.9 g), 3 imm ♂ (9.2, 9.3, (?), 9.8 g). Angepena (84) 3 Dec 63, 1 imm ♂.

Specimens to other Museums. AM 1 ♂ Tyrone (59). NMV 1 ø Tyrone (59), 1 fl Thylungra (60).

Distribution & Abundance. Also recorded from Cocklebiddy (15), the Gibson Desert (20, 22-24), Banoona (58) and Clifton Hills (65). Only one seen at Clifton Hills (65), two at Carnegie (19), a few at Cocklebiddy and Eucla (15, 17). Recorded as scarce at Van der Linden Lakes (24), elsewhere fairly common (17, 25, 66, 67, 68), common (79), or very common (18, 23, 70, 82).

Habitat. Mulga especially and also mallee, Bluebush, saltbush and other scrub associations and occasionally savanna woodland.

Behaviour. Keeps low in bushes and trees or feeds on the ground, flitting about with cocked tail. Usually solitary, or an adult male with one or more dull-plumaged birds. Twice with parties of thornbills, once with a male Whiteface. Voice seldom recorded: a trill, a melodious subdued song suspected to come from a young male, and a ticking alarm note. A male gave the ticking note followed by a trill *tk tk-tree* while erecting its red forehead feathers and quivering its tail (DG). Flying young gave a long sibilant *seee* very like the call of young European Robins *Erithacus rubecula*.

Food. Insects including beetles, caterpillars and large winged forms.

Annual Cycle. Males had more or less enlarged testes on 30 July (18) – three birds – 27 August (in first-year plumage) and 1 September (25). Pairs or what were probably family parties were noted on 30 March, 10 and 25 May, 7 June, 2 and 30 July and 31 August (79, 59, 60, 68, 17, 18, 25), and a pair with three flying young on 20 September near Willuna, southwestern Australia.

Young were in juvenile plumage at Nullarbor HS (69) on 4 October (now in South Australian Museum) and at Angepena (84) on 3 December, and moulting into the first-year plumage on 8 and 12 April (82). Males in first-year plumage were taken in March (79), April (82), May (3, 60, 66), August (25) and September (19). One of the moulting young had its skull fully and the other not fully pneumatised, while among those in juvenile plumage the proportion was two to four.

Adult males were in moult on 29 March and 12 April (79, 81, 82), and retained shreds of sheath on 2 July (17).

Impermanent Colours. *Legs* black or dark grey. *Iris* brown, usually dark or very dark, once grey. *Bill* black or occasionally dark brown, base sometimes yellowish, gape of young birds often yellow. *Mouth* yellow (bright lemon-yellow to dark chrome-yellow), once orange. *Orbital skin* of flying young yellow.

Geographical Variation. As Keast (1958i) concludes, the only variation seems to be in the intensity of coloration of the dull plumages, which he correlates with humidity.

The palest of our birds are from Chichester Range (3) and Thylungra (60) and the darkest from Granite Downs (68), Kingoonya (70) and Nonning (79).

Other Notes. Disney (1969: 18-19) gives diagnostic characters of the dull plumage. The juvenile (84) differs in having more open and slightly softer plumage, the dorsal feathers with whitish shaft-streaks and blackish fringes at the sides of the tips, those of the breast with fuscous fringes giving a mottled rather than evenly grey appearance, with the lower flanks more or less white.

Two males from Giles (25) and Granite Downs (68), show a mixture of adult and first-year coloration which is not due to moult, and appear to be in a retarded adult and a progressive immature plumage respectively, as discussed by Mayr (1934) for island individuals and populations of *P. multicolor*. The adult has considerable yellow among its ventral red pigmentation. Similar though less marked tendencies are shown by a few of the other specimens.

Petroica rosea — Rose Robin

Specimens. New South Wales. Tenterfield (95) 15 Apr 63, 1 ♂.

The single specimen, in adult male plumage and with small testes, was alone in ironbark and scrub. Its legs were dark brown with pale orange soles, iris brown, bill black, and mouth orange. Its stomach contained insects. Keast (1958i) recognises no geographical variation in this species, while Disney (1969) gives diagnostic characters of the dull plumage.

Petroica cucullata — Hooded Robin

Specimens. Western Australia. Chichester Range (3) 20 May 66, 1 fl. Bush Bay (6) 18 June 66, 1 fl. Green Range (13) 1 Apr 66, 1 ♀ skel. Mount Shenton (18) 30 July – 2 Aug 65, 2 ♂, 1 ♀, 1 ♂ skel, 1 ♀ skel. Carnegie (19) 16 Sept 65, 1 ♂. Lake Gruszka (21) 10 Sept 65, 1 ♀, 1 imm ♂. Warburton (23) 13 – 14 Aug 65, 1 ♀, 1 ♀ skel. Giles (25), 1 ♂ (21 g), 1 ♀, 1 fl. Northern Territory. Pine Creek (44) 29 Aug 68, 1 ♂ (19.8 g). Mount Roper (45) 1 Nov 68, 1 ♂ (17.3 g). Queensland. Banoona (58) 6 May 63, 1 ♂. Tyrone (59) 12 May 63, 1 fl. South Australia. Granite Downs (68) 5 – 11 June 65, 4 ♂ (24.7 g), 2 ♀ (23.5 g). Kingoonya (70) 17 – 22 June 65, 1 ♂, 1 ø (♀ by inference), 1 imm ♂. Nonning (79) 29 Mar 65, 1 ♂. Furner (86) 8 Jan 63, 1 imm ♂. Victoria. Coleraine (89) 22 Jan 63, 1 fl. New South Wales. Dunedoo (93) 20 – 25 Jan 63, 1 ♂, 1 fl.

Specimens to other Museums. NMV 1 fl Tyrone (59).

Distribution & Abundance. Also recorded between Willuna and Sandstone, Western Australia 20 September and at Mungilli Claypan (20), Hunt Oil Camp (22), Glendower (55), Thylungra (60), Monkira (61), Victoria Range (90), Little Desert (91). Only one specimen encountered at Bush Bay (6), two at Mount Roper (45) and three at Pine Creek (44). Recorded as fairly common at Carnegie (19), Granite Downs (68) and Kingoonya (70), common at Nonning (79), very common at Mount Shenton (18), Warburton (23) and Giles (25).

Habitat. Mulga and other shrub associations and open woodland.

Behaviour. Often perched on exposed sites such as dead trees and posts, flying to ground or low branches to feed, and returning. Single or in pairs. Once with a party of Whitefaces *Aphelocephala leucopsis,* once with a mixed party of this species and Red-capped Robin *Petroica goodenovii,* White-browed Babbler *Pomatostomus superciliosus,* Yellow-rumped Thornbill *Acanthiza chrysorrhoa* and Redthroat *Pyrrholaemus brunneus.* Voice only twice recorded – a cheerful-sounding short phrase from an adult male and a long-drawn twittering trill from a presumed female.

A female made a distraction display at Goodwin, presumably to protect a fledgling or nest which he did not discover. She chattered harshly as he approached a prickly bush, flew towards him and suddenly fell to the ground. There she 'hobbled' away at a right angle from her previous line of flight, with back and rump feathers erected and wings drooping. She hopped on dead branches on the ground and faced him with all body plumage fully erected and wings drooping, then flew to a nearby tree and did not display again.

Food. Insect remains, occasionally identified as ants, beetles, a grasshopper one inch long and larvae; once seeds.

Annual Cycle. Adult males had enlarged testes on 30 July (18) in August (18, 44, 25), on 16 September (19) and 1 November (45). Birds were apparently paired on 20 March (93), 30 and 31 July (18), 13 August (23) and 1 November (45). The female mentioned above performed its distraction display on 20 September (between Willuna and Sandstone, Western Australia). Adult males accompanied young ones on 8 January (86 – adult male by inference) and 17 June (70). The first of these young was in moult from juvenile into first-year plumage, the second in first-year plumage, and one on 10 September (21) apparently in moult (perhaps halted as discussed for *Colluricincla harmonica*) from first-year to adult male plumage. The state of the first's skull was not recorded, while the second's was not fully pneumatised.

Adults were in various stages of moult on 25 and 29 March, 1 April, 6 May, 22 June and 30 July (93, 79, 13, 58, 70, 18). The adult males with enlarged testes on 30 August, 16 September and 1 November showed a few pinfeathers. The skins from Granite Downs (68) in June are more or less greasy (as is one from Warburton (23) in August) and two of the greasiest were noted on internal examination as being very fat.

Impermanent Colours. *Legs* black or occasionally dark grey. *Iris* shades of brown from brownish black to chestnut. *Bill* black occasionally greyish or brownish. *Mouth* pale (dirty white, pink, flesh, cream or yellowish), apparently black near tomia; young birds had yellow gapes.

Geographical Variation. As Keast (1958i) points out, southern birds are larger than northern, without any indication of a discontinuity which would justify the recognition of races. Our males from the Northern Territory (44, 45) have no more white on their tails than those from elsewhere.

Peneoenanthe pulverulenta — Mangrove Robin

Specimens. WESTERN AUSTRALIA. Point Torment (30) 30 May – 12 June 68, 2 ♂ (21.9, 23.7 g), 1 ♀ (17.5 g), 1 imm ♀ (18.5 g), 1 fl (17.8 g). Wyndham (39) 31 July – 9 Aug 68, 4 ♂ (20.8, 21.3, 21.7 g), 2 imm ♂ (18.1, 23.1 g), 1 imm ♀ (17.8 g), 1 imm ♀?

(17.9 g), 2 imm ø (18.1 g), 3 fl (16.7, 16.8, 17.0 g). QUEENSLAND. Inkerman (50) 8 – 9 June 64, 2 ♂ (21.5, 22.8 g), 3 ♀ (17.4, 18.2, 18.5 g), 2 ø (17.2, 22.1 g), 1 imm ♂ (22.7 g). Ayton (52) 27 June 64, 4 ♂ (21.0, 21.9, 22.9, 24.7 g), 5 ♀ (17.9, 18.1, 18.8, 19.4, 20.6 g), 3 imm ♂ (16.9, 22.0, 22.9 g), 1 imm ø (20.2 g).

Distribution & Abundance. Not encountered elsewhere, despite intensive collecting in mangroves at Darwin (41), Mount Roper (45), Moonlight Creek (47) and Cape Gloucester (56). Locally common.

Habitat. Mangroves, usually in thickets near the edge (Plate 8(a)). Near Inkerman (50) encountered only near the edge of tall mangrove forest, in one of the mangrove localities visited.

Behaviour. Usually perched low, from the mangrove roots up to about 10 ft, occasionally coming down to the mud to feed – several specimens had mud on faces, bills and feet. Confiding and inquisitive. Solitary or in small parties, up to 15 being seen together at Wyndham (39). Twice two adult males were in company. Tails are flicked open to show the basal white patches, and wings are also flicked occasionally. Voice recorded only once in Queensland, a male singing a very pretty and varied song (IM); in Western Australia, as a long whistle (HB) a plaintive whistle (AH) and a peep followed by a whistle (DJF) with or without snatches of melody, as well as a harsh repeated *chack* or burring.

Food. Small crabs, and insects, including beetles, green ants and wasps.

Annual Cycle. Some adult males had greatly enlarged testes on 27 June and 4 and 6 August. The first of these was in company with an adult female, as was one with small testes on 9 June.

A young bird was moulting from the juvenile into the first-year plumage on 27 June when three were in the latter plumage, as were six between 31 July and 8 August. Others were apparently completing the moult from first-year into adult plumage on 30 May and 8 June. All these young birds had more or less incompletely pneumatised skulls, except for one which was damaged and one in first-year plumage with a pneumatised but thin skull; whereas all the recorded adult skulls were completely pneumatised.

Adults were in moult on 30 May, 8 and 9 June (five specimens) and 27 June (four specimens). Most of those from Queensland (June) had fresher wings than those from Western Australia (late May, early August), while two from Wyndham (39) in August show signs of arrested moult, as discussed for *Colluricincla harmonica*.

Impermanent Colours. *Legs* black, dark grey or sepia; juvenile warm brownish grey, feet blackish. *Iris* brown, usually dark, reddish to blackish or occasionally greyish or olivaceous. *Bill* black; juvenile dull black, base blackish horn above. *Motuh* yellowish from fleshy to bright orange-yellow, tips and tomia black, which sometimes extends over most of the mouth.

Geographical Variation. As Keast (1958i) indicates, dorsal coloration is slightly paler in the west and darker in eastern Queensland; but in this character the slight breaks between our series are between Point Torment (30) and Wyndham (39), and between Inkerman (50) and Ayton (52), with none apparent between Wyndham (39) and Inkerman (50). However, Queensland birds are separable on the decidedly greater amount of black on the face. Without Northern Territory material for comparison, it is impossible to suggest a better division into what are at best poorly-marked subspecies, *P. p. leucura* (Queensland), *P. p. cinereiceps* (Western Australia).

Other Notes. The young bird moulting out of the juvenile plumage (52) had a deformed bill with crossed mandibles.

The juvenile body feathers are brown without pale spots, grading into silvery-grey bases on back and breast and long whitish bases on the flanks. The upper wing-coverts and secondaries have buff terminal spots and the rectrices are somewhat narrower and more pointed than those of adults, and these feathers are retained into the first year when the body-plumage is indistinguishable from the adult.

Parasites. See Appendix 1.

Heteromyias cinereifrons — Grey-headed Robin

Specimens. QUEENSLAND. Walter Hill Range (53) 23 July 64, 1 ♂ (40.5 g).

Field Notes. Several seen at this locality, but nowhere else. The single specimen was alone on the floor of tropical rainforest at 2,100 feet, where it ran along the ground, jumped onto a log, and came to the collector's squeaking. Other single birds were seen fleetingly in the substage, moving quietly about.

Food. The stomach contained insect remains, and the bird was recorded as very fat.

Impermanent Colours. *Legs* whitish-buff. *Iris* blackish-brown. *Bill* blackish-brown, with a white tip. *Mouth* mottled white and bright yellow.

Poecilodryas superciliosa — White-browed Robin

Specimens. WESTERN AUSTRALIA. Geikie Gorge (31) 5 June 68, 1 ♀ (17.8 g). NORTHERN TERRITORY. Mount Roper (45) 27 Oct 68, 1 ♂ (21.3 g). Borroloola (46) 14 – 15 Nov 68, 1 ♂ (21.8 g), 1 imm ♀, 1 ♀ skel (17.5 g), 1 fl (20.1 g).

Distribution & Abundance. Not encountered elsewhere, and apparently uncommon at these localities.

Habitat. Riverside vegetation, and also mangroves at Mount Roper (45).

Behaviour. Solitary or in pairs which at Borroloola (46) kept very much to themselves (BB). Usually low in thickets. Very secretive at Mount Roper (45), somewhat less so at Borroloola (46). Flick tails up and down, or freeze on alarm, when disruptive pattern makes them difficult to see in dappled thickets. Short, quick, direct flight, usually low. Voice a loud whistle with ventriloquial effect, usually three short clear notes, the middle one once recorded as lower (WHB), or three short clear whistles followed immediately by a longer drawn out descending note (DJF).

Food. Insects, including grasshoppers.

Annual Cycle. A male had somewhat enlarged testes on 27 October, and another was in company with a female with developing oocytes on 15 November. A young female was completing the moult from juvenile into first-year plumage on 14 November.

Impermanent Colours. *Legs* dark grey or brown to black; young greyish. *Iris* dark brown. *Bill* black or sepia. *Mouth* black, tongue and floor of mouth yellow or pale orange; young all dull yellow.

Geographical Variation. These all belong to the western, rufous-flanked, form

P. s. cervininventris, within which Keast (1958i) finds no appreciable geographical variation.

Other Notes. The few juvenile feathers of the young bird are soft and rufous. Its first-year plumage is very like that of adults in colour, but with a rufous tinge to the edges of the secondaries, the white tips of the outer ones smaller and vaguer, and the white tips of the central rectrices reduced to buffy streaks, while the wings and tail are softer.

Eopsaltria australis (including *griseogularis*) Yellow Robin

Specimens. Western Australia. Wonerup (8) 12 Mar 66, 1 juv ø, 1 fl. Metricup (9) 6 Mar 66, 1 ♂ skel, 1 fl. Margaret River (10) 14 Mar 66, 1 ♂, 1 ♂ skel, 1 fl. Scott River (11) 19 – 24 Feb 66, 1 imm ♂, 1 ♂ skel, 2 ♀ skel. Queensland. Tully (54) 14 – 18 July 64, 4 ♂ (16.1, 19.9, 23.2, 23.4 g), 1 ♀ (18.4 g), 1 ♀? (17.5 g), 2 imm ♀ (16.1, 17.0 g). Mount Dryander (57) 3 Aug 64, 1 ♂ (19.6 g). South Australia. Port Lincoln (77) 16 Oct 63, 1 ♂, 1 ♀. Nonning (79) 30 Mar 65, 1 imm ♂ (21.7 g). Furner (86) 8 Jan 63, 1 imm ø. Victoria. Victoria Range (90) 31 Jan 63, 1 imm ♂, 1 imm ø. New South Wales. Dunedoo (93) 18 – 24 Mar 63, 3 ♂, 1 ♀, 1 fl. Grafton (94) 5 Apr 63, 1 ♀. Tenterfield (95) 15 Apr 63, 1 ♀ skel.

Specimens to other Museums. NMV 1 ø skel Dunedoo (93).

Distribution & Abundance. Also recorded at Mount Barker (12), Newdegate (14), Streaky Bay (74) and Nelson (88): fairly common at Nonning (79).

Habitat. Various types of woodland and forest, including pine plantations, and second growth in clearings. Not in rainforest at Ayton (52) or Tully (54), but one in dry rainforest at Mount Dryander (57).

Behaviour. Mainly recorded as perched on twigs and branches (once lengthwise). Inquisitive, coming to collectors' squeaks. Mostly solitary, or in pairs or small parties. Seen with Golden Whistler *Pachycephala pectoralis*, Grey Fantail *Rhipidura fuliginosa* and *Acanthiza*. Voice seldom recorded, a mewing chatter, a harsh triple rasp and a subdued piping.

Food. Insects, including black beetles and ants with golden-brown heads.

Annual Cycle. Four of five adult males between 14 July and 3 August had enlarged testes (54, 57). A young bird, in company with adults and with unpneumatised skull, was beginning to moult from juvenile into first-year plumage on 12 March (8). Three others (one with adults and one with a fully pneumatised skull) were completing this moult on 8 and 31 January (86, 90). Two specimens were in what seems to be first-year plumage on 14 July (54), and three apparently in moult from this into adult plumage on 24 February and 30 March (11, 79). Of the four skulls noted in this group, only one on 14 July was incompletely pneumatised.

All six adults between 14 March and 5 April (10, 93, 94) were in moult.

Impermanent Colours. *Legs* dark grey to black – except in north Queensland (54, 57) where they were various hues of brown with the feet and front of shanks tending to be blacker – soles buffy; juvenile paler grey. *Iris* various hues of dark brown – females possibly redder in the *griseogularis* group. *Bill* black; juvenile dark horn, first year intermediate. *Mouth* very varied: pale, with dull yellowish, pink or purple hues, giving way to blackish areas of varying extent towards the tips of the bill.

Geographical Variation. The main division is between the western *griseogularis* group, with white throats and grey gorgets, and the eastern *australis* group, in which the yellow of the underparts extends high on the throat, and the gorget is reduced to an olive wash. Within each group, variation is mainly in the colour of rump and upper tail-coverts, from olive to yellow. The specimens fall into subspecies as recognised by Keast (1958i):

E. a. griseogularis southwestern Australia (8, 10, 11), throat white, gorget grey, rump olive-yellow.

E. a. rosinae Eyre Peninsula (77, 79) rump olive.

E. a. viridior southeastern South Australia and Victoria (86, 90), yellow high on throat and rather pale with considerable olive wash, rump olive.

E. a. chrysorrhoa northern New South Wales (93, 94), yellow richer and less olive, below, rump yellow.

E. a. magnirostris northern Queensland (54, 57), bill longer, yellow deeper. Our specimens show no consistent separation from *chrysorrhoa* in wing-length.

Other Notes. We follow Condon (1951a) and Keast (1958i) in regarding *australis* and *griseogularis* as conspecific.

The specimens moulting from the juvenile plumage show that the first-year is very like the adult plumage. The only difference seems to be in the rectrices, which tend to be narrower and more pointed. Those specimens mentioned as in or moulting from the first-year plumage, without any juvenile feathers, are tentatively separated on this characteristic.

Eopsaltria georgiana — White-breasted Robin

Specimens. WESTERN AUSTRALIA. Metricup (9) 6 – 8 Mar 66, 1 imm ♀, 1 imm ø. Margaret River (10) 14 Mar 66, 2 ♂, 1 ♂ skel, 1 fl. Scott River (11) 23 Feb – 1 Mar 66, 1 imm ♂, 1 ♂ skel, 1 fl.

Distribution & Abundance. Not encountered elsewhere.

Habitat. Shrubbery of varied kinds, sometimes sparse, including growth along a stream through a pine planatation, second growth in forest, and low scattered patches on coastal dunes.

Behaviour. Recorded as scolding, boldly in forest second growth, furtively from the cover of isolated scrub. Mainly solitary.

Food. Not recorded.

Annual Cycle. Young birds were completing the moult from juvenile to first-year plumage on 23 February and 6 and 8 March. The skull of one was recorded as incompletely pneumatised. Both adults on 14 March were in moult.

Impermanent Colours. *Legs* black, soles yellow (one first-year dark grey, soles sandy). *Iris* dark brown (first year paler ?). *Bill* black. *Mouth* yellow.

Other Notes. The first-year birds, with traces of brown juvenile plumage, are otherwise distinguished from presumptive adults only by a tendency towards white spots at the tips of the upper wing-coverts and white tips to the secondaries.

Tregellasia capito Pale-yellow Robin

Specimens. QUEENSLAND. Ayton (52) 20 – 25 June 64, 5 ♂ (13.9, 14.5, 14.8, 15.1, 15.6 g), 3 ♀ (12.7, 13.0, 14.5 g), 2 imm ♂ (13.2, 15.6 g), 2 imm ♀ (12.4, 13.2 g). Walter Hill Range (53) 21 – 23 July 64, 7 ♂ (13.3, 13.4, 13.4, 13.6, 13.7, 14.7, 14.7 g), 1 ♀ (11.4 g). Tully (54) 11 – 13 July 64, 3 ♂ (14.2, 14.2, 14.5 g), 1 ♀ (12.5 g).
Distribution & Abundance. Encountered only in the main block of north Queensland rainforest (51-54), where it was very common.
Habitat. Tropical rainforest.
Behaviour. In substage shrubs, on fallen trees, and running about on tree trunks like a tree-creeper *Climacteris*. Solitary or in pairs. Came to squeaking collectors. Voice seldom recorded: a small grating or harshly scolding note, monotonously repeated, and a harsh scolding buzzing from a netted bird when handled.
Food. Insects, including small black beetles and small black ants; one stomach full of what seemed to be tiny red-brown seeds.
Annual Cycle. All eight adult males between 23 and 25 June and 11 and 13 July (52, 54) had enlarged testes (one only slightly) but only one of seven between 21 and 23 July (53) had them appreciably enlarged. One adult female on 24 June had developing oocytes, and one on 25 June a large yolk. Five adults with small gonads had skulls recorded as from 'not pneumatised' to 'fully pneumatised but soft'. Young birds between 20 and 25 June and on 13 July (52, 54) were in first-year plumage: three of the five had fully pneumatised skulls. No specimen was in moult.
Impermanent Colours. *Legs* dull yellow to ochraceous, feet yellower and fronts of shank browner, soles dull yellow. *Iris* dark brown, from sepia to olivaceous (once yellow-brown); first year sometimes paler and greyer. *Bill* black (occasionally blackish-brown), lower mandible sometimes horny or brown towards gonys. *Mouth* orange to yellow, tips and edges greyish or blackish .
Geographical Variation. These specimens belong to the northern race *T. c. nana* which has the lores and eye-ring washed with rufous. The series from Walter Hill Range (53) differs from the lowland ones (52, 54) in having the olive of flanks and breast decidedly darker. The type of *barroni* Mathews is from the Barron River, probably on the Atherton Tableland, and may agree in this respect though Mathews' description does not indicate this.
Other Notes. A young juvenile of this species, and one of *T. leucops* from New Guinea moulting into the next plumage, both in the British Museum, confirm that our young specimens are in the first-year plumage. This is very like the adult, except for variable cinnamon tips to the secondary coverts and edges to the inner secondaries. The rectrices have slightly less angled points, the primaries perhaps slightly broader tips, and the flight-quills are marginally softer. Judging from the two British Museum specimens, the juvenile plumage of this species is much less uniform than that of *T. leucops,* the dorsal feathers having whitish shafts, pale-buff shaft-streaks, and dark brown edges.

RHIPIDURIDAE — Fantails

I. C. J. Galbraith

Rhipidura fuliginosa Grey Fantail

Specimens. WESTERN AUSTRALIA. Karratha (1) 29 May 66, 1 ♀. Peg's Beach (2) 3 June 66, 1 ♂. Bush Bay (6) 14 June 66, 1 ♂. Margaret River (10) 14 Mar 66, 1 imm ♀. Scott River (11) 18 Feb 66, 1 ♂ skel, 1 fl. Green Range (13) 4 Apr 66, 1 ♂. Derby (29) 26 – 27 May 68, 2 ♂ (7.4, 7.6 g). Point Torment (30) 30 May 68, 1 ♂ (7.3 g), 3 fl (6.7, 7.1 g). Wyndham (39) 31 July – 6 Aug 68, 2 ♂ (6.4 g), 1 ♀, 2 fl (6.6 g). QUEENSLAND. Moonlight Creek (47) 8 – 14 May 64, 1 ♂ (8.0 g), 1 ♀ skel (7.4 g), 1 ø skel (8.0 g). Glenore (49) 23 Apr 64, 1 ♂ (8.0 g). Inkerman (50) 3 June 64, 1 ♀ skel (7.4 g). Banoona (58) 28 Apr – 6 May 63, 2 ♀. Tyrone (59) 15 May 63, 1 ♂. SOUTH AUSTRALIA. Innamincka (64) 3 May 65, 1 ♂. Aroona Valley (82) 8 – 15 April 65, 1 ♀ (7.8 g), 1 imm ø (8.0 g). Furner (86) 7 Jan 63, 1 fl. NEW SOUTH WALES. Albury (92) 12 Mar 63, 1 ♂. Dunedoo (93) 23 Mar 63, 1 ♂, 1 imm ♀. Grafton (94) 30 Mar – 1 Apr 63, 2 ♂ skel, 1 ø skel. Tenterfield (95) 11 Apr 63, 1 ø skel.

Distribution & Abundance. Also recorded from Perth (7), Mount Barker (12) and Newdegate (14), Walter Hill Range (54), Clifton Hills (65), Nelson (88) and Victoria Range (90). Noted as absent from Darwin (41) and Oenpelli (43) and as uncommon at Innamincka (64) and Aroona Valley (82), and only one specimen encountered at Clifton Hills (65).

Habitat. Tree associations from forest to savanna woodland. Collected only in and near mangroves and riverside vegetation towards the north of Western Australia (1-6, 29-39), and perhaps the most ubiquitous bird in mangroves on the Queensland gulf coast (47, 50). However, it was there equally common in woodland, while on the north-east coast of Queensland (51-54, 56, 57) it occupied rainforest, open forest, woodland and mangroves.

Behaviour. Mainly low in the substage, flitting out after insects. Usually confiding and inquisitive, coming very close in response to collectors' squeaks, but occasionally wary. Solitary or in pairs.

Food. Insects – mainly unidentified but once beetles.

Annual Cycle. Males had enlarged testes on 15 May, 31 July and 6 August (59, 39). A young bird was completing the moult from juvenile plumage on 14 March (10), another was in first-year plumage on 15 April (82), and a third was moulting into adult plumage on 23 March (93). Only the first and third of these had incompletely pneumatised skulls, like two adults on 26 and 29 May (29, 1). An adult was completing moult on 23 March (93).

Impermanent Colours. *Legs* black, occasionally dark grey or brownish. *Iris* dark brown or brown (once grey 82). *Bill* black, base often more or less whitish or pinkish below, and occasionally above. *Mouth* yellow to orange (once white) in Western Australia (1-39); white, pale grey or pinkish (once yellow, twice yellowish) elsewhere.

Geographical Variation. All these specimens and almost all those already in the British Museum were collected during the winter. Southern populations of this species are now known to winter far to the north (Ford, 1971; Storr, 1973). In northern Queensland most of the individuals in and near mangroves, rainforest and other closed habitats were probably resident *phasiana* and *frerei*, those in more open habitats wintering *alisteri*. The skins fall into three groups:

R. f. preissi, southwestern Australia (10, 13). Small with broad grey gorget.

R. f. phasiana (=*subphasiana*), northwestern Australia (1-6, 29-39). Small and pale with narrow grey gorget. There is considerable variation in the intensity of ventral ochraceous, and our series are inadequate to show to what extent this may be geographical.

R. f. alisteri, eastern Australia (47-93). Larger in the north and darker, gorget broad and black. The specimen from Innamincka (64) and one from Lithgow, New South Wales have black but narrow gorgets.

Other Notes. The juvenile plumage differs from the adult in being looser in texture, and in having brownish upper wing-coverts and breast and brown tips to the dorsal feathers. The first-year plumage seems to differ from the adult only in retaining the juvenile upper secondary-coverts.

Rhipidura rufifrons — Rufous Fantail

Specimens. WESTERN AUSTRALIA. Wyndham (39) 8 – 9 Aug 68, 2 ♂ (8.9, 9.4 g), 1 fl (8.9 g). NORTHERN TERRITORY. Mount Roper (45) 28 Oct 68, 1 ♂ (9.7 g). Borroloola (46) 15 Nov 68, 1 ♀ (9.5 g). QUEENSLAND. Moonlight Creek (47) 8 – 15 May 64, 1 ♂ (10.0 g), 1 imm ø (9.9 g). Ayton (52) 19 – 30 June 64, 2 ♂ (10.2, 11.2 g), 2 ♀ (9.3, 9.8 g), 2 ø (9.2, 10.0 g), 1 imm ♂ (9.0 g). Tully (54) 13 – 15 July 64, 3 ♂ (10.0, 10.4, 10.4 g), 1 imm ♂ (10.0 g), 1 imm ♀ (9.0 g). Cape Gloucester (56) 6 – 7 Aug 64, 1 ♂ (9.9 g), 1 ø (8.7 g). Mount Dryander (57) 1 – 12 Aug 64, 1 ♂ (10.7 g), 2 ♀ (8.6, 91. g), 2 ø (9.1, 9.9 g). NEW SOUTH WALES. Grafton (94) 30 Mar – 1 Apr 63, 1 imm ♀, 1 fl.

Distribution & Abundance. Also recorded at Geikie Gorge (31), Albury (92) and Tenterfield (95). A few single birds at Mount Roper (45), and moderately common at Geikie Gorge (31) and Grafton (94); very common in true rainforest in northern Queensland (51-54). Not encountered at Inkerman (50) and apparently unknown from north of Normanton on the Gulf of Carpentaria.

Habitat. Mangroves at Wyndham (39), Mount Roper (45) and Moonlight Creek (47). Riverine forest with dense green understory at Borroloola (46). Also recorded from monsoon forest in the northern Territory by Deignan (1964) and Storr (1967). In north-eastern Queensland in (and near) rainforest (51-54, 57) and in mangroves (52, 56). At Grafton (94) in open forest and tea-tree woodland.

Behaviour. Mainly in dense low understory. Several recorded as feeding on fallen logs or the ground, one as feeding along the edge of the incoming tide. Very active, constantly fanning tail and flying out after insects, inquisitive and confiding. Solitary, occasionally in pairs, one flock of five (94). Once in a flock of *Rhipidura*, *Lalage* and *Malurus* (52), once with a group of *Monarcha leucotis* (57). Voice seldom recorded: repeated single sharp notes or high-pitched chirps.

Food. A small dragonfly one inch long was the only identifiable insect among usually black remains in the stomachs.

Annual Cycle. A male on 8 August (39) had enlarged testes. A female on 1 April (85) was in what appears to be the juvenile plumage, but had a fully pneumatised skull.

Five birds were in the first-year plumage on 30 June and 1 August (52, 54, 57) and had fully pneumatised skulls, though two were recorded as thin or very soft. Two were moulting from the first-year plumage on 15 May and 27 June (47, 52) of which one had the skull apparently little pneumatised and the other fully pneumatised but very soft. Three in moult on 8 May, 9 August and 28 October (47, 39, 45) could have just lost the first-year plumage, and only the last had a fully pneumatised skull, while one in full adult plumage on 15 July (54) had the skull fully pneumatised but soft.

Impermanent Colours. *Legs* very variable in hue and tone – brownish tending towards purple, pink, buff or grey, and from pale to almost black; soles pale. *Iris* dark brown, from blackish to deep chestnut. *Bill* black or blackish brown, lower mandible sometimes paler, gonys often and base of upper mandible occasionally whitish. *Mouth* whitish or pale grey, usually washed with yellow, pink or purple.

Geographical Variation. These specimens fall into the two Australian subspecies recognised by Keast (1958i), which are rather distinct and assigned to different subspecies-groups by Mayr & Moynihan (1946):

R. r. dryas, northwestern Australia to the Gulf of Carpentaria (39, 45-47).

R. r. rufifrons, eastern Australia (52, 54, 57, 94). Larger but with a shorter tail, the rectrices more square-cut. The dorsal rufous extends slightly higher on the back, and much lower on the tail (over half the length on both webs of the central rectrices, rather than less than a third on the inner webs of the outer rectrices, where it is concealed by the coverts). The white tail-tips much shorter, greyer and less well defined. On the average the flanks are more rufous and the sides of the breast greyer.

Keast does not mention one of the most trenchant differences, the extent of rufous on the tail. The difference in scaliness of the breast which he describes is not apparent in these specimens. The two from Moonlight Creek (47) and other Gulf birds in the American Museum of Natural History do not show appreciable intermediacy with nominate *ruficollis,* which is not to be expected if the two forms are separated by a gap from Normanton to the Watson River. However, Keast records the single specimen from the latter locality (not seen by ICJG at the AMNH in 1966) as belonging to *dryas* though with tendencies towards *rufifrons.*

Other Notes. A specimen in what must be the juvenile plumage (94) differs from adults in its slightly softer plumage, reduced black on the breast, reduced and paler rufous on the forehead, rufous edges to the upper wing-coverts, much deeper and more extensive rufous on the underparts, and pale rufous tips to the rectrices. Though a few specimens in the British Museum, of various subspecies of *R. rufifrons,* approach it in most of these respects, none shows the rufous tail tips.

The first-year plumage differs from the adult only in retaining the rufous-edged upper wing-coverts.

Parasites. Seven of the skins (including 4 of 5 from Mount Dryander (57)) had mallophagan eggs (not preserved) at the bases of the throat-feathers.

Rhipidura rufiventris — Northern Fantail

Specimens. Western Australia. Geikie Gorge (31) 3 – 5 June 68, 2 ♂ (13.1, 13.6 g),

2 ♀ (13.3 g), 1 imm ♂ (13.4 g). Mount Bell (32) 14 – 22 June 68, 1 ♂ (12.4 g), 1 ♀ (12.4 g), 2 fl (12.5, 13.0 g). Mount Caroline (33) 24 June 68, 1 ♂ (13.2 g), 1 ♀ (12.5 g). NORTHERN TERRITORY. Darwin (41) 13 Sept 68, 1 ♂ (12.2 g). Oenpelli (43) 9 – 14 Oct 68, 2 ♂ (12.1 g), 1 ♀. Pine Creek (44) 2 Sept 68, 1 imm ♀ (13.7 g), 1 fl (12.7 g). Mount Roper (45 26 Oct – 1 Nov 68, 1 ♂ (12.2 g), 1 ♀ (13.3 g), 1 juv (12.3 g), 1 ♂ skel (12.7 g), 3 fl (11.6, 12.8 g). QUEENSLAND. Moonlight Creek (47) 12 May 64, 1 ♂ fl (12.0 g), 1 ♀? fl (10.7 g). Inkerman (50) 11 June 64, 2 fl (11.8, 12.3 g).

Distribution & Abundance. Only four birds noted at Oenpelli (43). Recorded as fairly numerous at Darwin (41) and Pine Creek (44), plentiful at Mount Roper (45).

Habitat. Mainly in rather dense and tall vegetation near water – woodland in gorges and along rivers and creeks, and the more open areas of estuarine mangroves. Also in savanna woodland some way from water.

Behaviour. Usually among the lower branches – either darting about on feeding flights, during which it seldom returns to the same perch; or more or less inactive, usually in deep shade, especially during the heat of the day. Single or (frequently) in pairs, but not associating closely with other species. One attacked by a *Halcyon macleayii* darted down, twisting and turning, and then returned to its original perch.

Voice: a repeated *chip-chip-chip;* a quiet *chack* or *cluck* thought to be an alarm note; and rather loud, disjointedly descending song phrases, described as a single note between two double ones, and as like the amplified song of a *Gerygone* (BB).

Food. Insects, including small flies and ants.

Annual Cycle. A nest was found about 13 September (41), in which one egg was laid at around 08.15 hrs on each of two successive mornings, but it was empty on the third day. Most males and females taken in June (31, 32) were apparently paired but had undeveloped gonads except for one female which had the ovary slightly enlarged. Males had much enlarged testes on 31 September (41) and 12, 14 and 26 October (45), the birds on the last date being apparently paired.

The juvenile on 28 October (45) had growing wings and tail and a completely unpneumatised skull. Birds in what is probably the first-year plumage had a partly pneumatised skull on 4 June (31) and a fully pneumatised skull and somewhat enlarged testes on 2 September (44). Others, listed as adult but in what may be a second-year stage of plumage, on 4 and 5 June (31) and 14 October (43) had fully pneumatised skulls, and the last was one of those with enlarged testes. No specimen was in moult.

Impermanent Colours. *Legs* dark grey or dark brown to black, once bluish; juvenile light grey. *Iris* dark to very dark brown, apparently olivaceous. *Bill* dark brown to black, lower mandible apparently paler and sometimes with a whitish base; juvenile brown, paler basally. *Mouth* white or pale greyish, purplish, pinkish, fleshy or yellowish; juvenile yellow.

Geographical Variation. All Australian populations belong to *R. r. isura,* within which the only detectable variation is an irregular one in bill width (Keast, 1958i). This cannot be studied on our small samples, but the bills seem to be significantly wider at Mount Bell (32) than at Geikie Gorge (31).

Other Notes. The juvenile has growing wings and tail, a small narrow bill and soft plumage. This is coloured like adults ventrally, except that there is less grey on the breast and this is mixed with buff rather than streaked with white. Dorsally it is warmer grey than adults, with rich brown scalloping increasing from the nape to the upper tail-coverts. All the upper wing-coverts are broadly tipped with rich buff, and the narrow edges of the remiges are buff rather than white. The rectrices are softer than those of the adult and slightly more pointed. First-year birds seem to retain only the primary-coverts and outer secondary-coverts of the juvenile plumage. Three specimens have juvenile primary-coverts and fully adult secondary-coverts, and may be in a second-year phase of plumage.

Parasites. See Appendix 1.

Rhipidura leucophrys — Willie Wagtail

Specimens. WESTERN AUSTRALIA. Perth (7) 10 Feb 66, 1 ø skel. Metricup (9) 17 Mar 66, 1 ♀. Scott River (11) 19 – 22 Feb, 1 ♂ skel, 1 fl. Mount Shenton (18) 3 Aug 65, 1 ♀, 1 ♂ skel. Carnegie (19) 17 Sept 65, 1 ♀. Hunt Oil Camp (22) 9 Sept 65, 1 ♂. Derby (29) 25 May 68, 1 imm ♂. Lissadell Hill (38) 28 – 29 July 68, 1 ♀, 1 fl (15.3 g). NORTHERN TERRITORY. Oenpelli (43) 6 Oct 68, 1 ♂ (19.4 g). Borroloola (46) 8 Nov 68, 2 fl (15.8, 16.6 g). QUEENSLAND. Moonlight Creek (47) 14 May 64, 1? ♂ skel (16.5 g), 1 ø skel (16.5 g). Glendower (55) 31 Mar 64, 1 fl (18.5 g). Thylungra (60) 23 May 63, 1 imm ♀. SOUTH AUSTRALIA. Innamincka (64) 29 Apr 65, 1 imm ♂. Etadunna (66) 18 May 65, 1 ♂. Nonning (79) 29 Mar – 3 Apr 65, 1 ♂ (19.3 g), 1? ♀ (18.3 g). Aroona Valley (82) 17 Apr 65, 1 ♂, 1 imm ♂ (19.7 g). Angepena (84) 29 Nov 63, 1 juv ♀. NEW SOUTH WALES. Grafton (94) 1 – 5 Apr 63, 2 ♀, 1 ♂ skel, 1 fl.

Specimens to other Museums. NMV 1 ♀ Thylungra (60).

Distribution & Abundance. Recorded at every locality for which a list was compiled, except Mungilli Claypan (20), Geikie Gorge (31), Lake Gladstone (36), Victoria River (40), Darwin (41), Banoona (58), the Nullarbor Plain (north 15; 17), Coleraine (89), Little Desert (91) and Tenterfield (95). Only two recorded at Carnegie (19), a few at Cocklebiddy (15) and near the coast at Eucla (17), and several at Mount Shenton (18). Recorded as scarce at Van der Linden Lakes (24), elsewhere fairly common (23, 25, 28, 37, 44, 46, 68, 70, 79), common (45, 55, 66, 67), or very common (64, 65, 82).

Habitat. Savanna and any other habitat with open but not bare ground and some perches.

Behaviour. The species characteristics of feeding on the ground as well as on the wing, and of waving the tail laterally, were noted, though comments are few because of its familiarity. Solitary, in pairs or in loose parties. Noted as following man, cattle and Emus closely, and as joining bird parties and trying to rob smaller birds such as *Acanthiza* of food (DG).

Food. Insects, including small black beetles, probably ants, and medium-sized insects (head-capsules 5 mm across, legs 12 mm long).

Annual Cycle. Adult males on 9 September (22) and 6 October (43) had enlarged testes, and an adult female with convoluted oviduct on 17 September (19) had a somewhat enlarged ovary. A young female on 29 November (84) was in full juvenile plumage. The state of its skull was not recorded. Birds in first-year plumage were taken in April (64, 82) and May (29, 60), and the skulls of the two last were recorded as fully pneumatised. Adults were in moult in February (11), March (9, 79) and April (94).

Impermanent Colours. *Legs* black or dark grey. *Iris* dark brown, once grey. *Bill* wholly black. *Mouth* pale and fleshy, more or less grey, pink or yellowish. Colours were not recorded for the juvenile.

Geographical Variation. Keast (1958i) recognises northern and southern subspecies *R. r. picata* and *leucophrys,* based on a size-cline. Our few measurements suggest that the difference even between the northern and southern extremes may be insufficiently constant for recognition.

Other Notes. The juvenile plumage differs from the adult in having the blue-black replaced by dark brown (also glossy), with pale brown tips concentrated behind the superciliary stripe, on the scapulars and upper wing-coverts, and on the inner secondaries. The white superciliary stripe, under tail-coverts and white tips to the throat feathers are tinged with brown. The rectrices may be narrower than those of adults, but are otherwise similar. First-year birds seem to retain only the primary-coverts and outer secondary-coverts of the juvenile plumage.

Parasites. Juv ♀ (84) mallophagan eggs (not preserved) on throat.

MONARCHIDAE — Monarch Flycatchers
I. C. J. Galbraith

Myiagra rubecula — Leaden Flycatcher

Specimens. Western Australia. Geikie Gorge (31) 2 – 5 June 68, 4 ♂ (10.2, 11.2 g), 3 ♀ (2 sexed by plumage; 10.4, 10.6 g), 1 fl (♂ by plumage, 10.1 g, dehydrated). Mount Caroline (33) 24 – 25 June 68, 1 ♂ (11.7 g), 1 ♀ (10.0 g). Northern Territory. Victoria River (40) 11 Aug 68, 1 ♂ (11.7 g). Darwin (41) 7 Sept 68, 1 fl (♂ by plumage, 12.7 g). Pine Creek (44) 27 Aug 68, 1 ♀ (11.0 g). Mount Roper (45) 1 Nov 68, 1 ♀ skel (10.0 g). Queensland. Inkerman (50) 26 May – 5 June 64, 3 ♂ skel (10.5, 11.6, 12.7 g), 1 ♀ skel (11.2 g), 2 fl (♂ by plumage, 12.3, 12.5 g). Tully (54) 18 July 64, 1 ♀ (13.3 g), 1 imm ø (11.6 g). Mount Dryander (57) 3 Aug 64, 2 fl (1 ♂ by plumage, 11.8 g; other 11.5 g). New South Wales. Albury (92) 12 Mar 63, 1 imm ♂.

Distribution & Abundance. Also recorded from Oenpelli (43), Borroloola (46), Moonlight Creek (47) and Dunedoo (93). Only one pair at the lagoon at Borroloola (46), two pairs at Victoria River (40), and a few individuals at Darwin (41) and Oenpelli (43); but the species was fairly numerous at Pine Creek (44) and locally in mangroves at Mount Roper (45).

Habitat. Savanna woodland, denser trees beside rivers and creeks (and swamps), and mangroves at their mouths – all specimens taken from the Kimberleys to the Gulf (31-50) were near water.

Behaviour. Usually in trees and shrubs, noted as feeding there more often than as flying out after insects and returning to the perch. Inquisitive and sometimes confiding. Solitary or occasionally in pairs – spaced as though territorial – one with *Rhipidura* *fuliginosa* and *Zosterops lutea*. A bird in the plumage of an adult female reacted to two *Seisura inquieta* by hunching with tail down, wings quivered and bill gaping, when they came close and apparently fed it. When they were further off it called and caught insects on the wing (WHB). An adult male attacked a green snake in the top of the tree, calling excitedly (WHB). Another, returning from flights after insects, would crouch low parallel to the branch and flutter its wings (WHB). Voice seldom recorded. The bird attended by *Seisura* responded by imitating the scissor-grinding and *cheeky-cheeky-cheeky* calls of that species, while the one attacking a snake gave a high-pitched whistle. Mewing and short rasping notes, and a soft grating as of pebbles rubbed together, were also noted.

Food. Insects, including grub 5 mm long.

Annual Cycle. Two males on 4 June (31) had somewhat enlarged testes, though the species breeds in spring and summer (Keast, 1958i, Storr, 1967). A young male was in first-year plumage on 12 March (92), and an unsexed bird in what seems to be corresponding plumage, though considerably worn, on 18 July (54). These are the only specimens recorded as having incompletely pneumatised skulls. Adults were in various stages of moult in early June (31).

Impermanent Colours. *Legs* black or dark grey (bluish or once brownish). *Iris* blackish brown. *Bill* blue-grey variably overlaid with black (centred on the tips and along the culmen), base occasionally whitish or brownish. *Mouth* grey to black around tomia, grading to whitish, cream or fleshy pink within; first-year male orange.

Geographical Variation. Western birds (31-45) have the small bill of *M. r. concinna* (see Keast, 1958i). While, within the eastern large-billed populations, those males which breed in the south have less intensely black lores (*M. r. rubecula*) than those breeding in northern Queensland (*M. r. yorki*), the difference is slight and both forms are found in the north during the winter. Our only relevant specimens, males in fluid (50, 57), seem to have the paler lores of *rubecula*.

Other Notes. The young male (92) presumably in first-year plumage, differs from adult females in its buffy (rather than whitish and grey) edges to remiges and upper wing-coverts, though with conspicuous white tips; deeper rufous on the underparts, extending on to the upper belly; and slightly more pointed rectrices. The other young bird (54) is worn so that its wings and tail are less distinctive, while the deep rufous is sharply cut off at the breast.

The female skeleton (BM(NH) no.S/1965.5.11) from Inkerman (50) has an unusually wide bill, and may be a misidentified *M. ruficollis*.

Myiagra ruficollis — Broad-billed Flycatcher

Specimens. Western Australia. Wyndham (39) 5 Aug 68, 1 imm ♂ (11.6 g). Queensland. Moonlight Creek (47) 6 May 64, 1 imm ♀? (12.3 g). Inkerman (50) 29 May – 11 June 64, 2 ♂ (11.5, 11.9 g), 1 ♀ (11.7 g), 1 imm ♀? (10.6 g), 1 imm ø (10.0 g).

Distribution & Abundance. Not recorded elsewhere. These specimens were almost the only ones encountered.

Habitat. Mangroves at the mouths of rivers and creeks.

Behaviour. Rather inactive, keeping under dense cover. Inquisitive, coming very close to squeaking observers, with tail flicking. Solitary, apart from an adult with one young. Voice: only rasping notes noted.
Food. Insects (only one record).
Annual Cycle. An adult male was accompanied by a young bird in first-year plumage on 11 June. Other young were in first-year plumage on 6 May and 8 June, and all three had partly pneumatised skulls. Another was beginning to moult from this plumage on 5 August, and its skull was fully pneumatised. No adult was in moult.
Impermanent Colours. *Legs* black; blue-grey in some first-year birds. *Iris* dark brown tending to olivaceous in some first-year birds. *Bill* black above, lower mandible blue-grey; in first-year birds mostly black or blackish-brown above, base of lower mandible pale buff. *Mouth* pale grey, more or less bluish, grading to pink within; bright yellow or orange to yellowish flesh in first year.
Geographical Variation. Keast (1958i) recognises none, and follows previous authors in assigning Australian and New Guinea populations to *M. r. mimikae.*
Other Notes. Since both may be found in mangroves, it is not clear how this species may be distinguished in the field from female and immature *M. rubecula.* In the hand, specimens from the Northern Territory westwards are easily separable on the smaller and narrower bill of *M. rubecula concinna.* On the Gulf coast of Queensland, however, there may be some overlap in bill form (especially since first-year birds of both species have decidedly narrower bills than adults), though even the broadest bills of *M. rubecula* tend to taper more evenly from the base. Judging from our limited material, *M. ruficollis* differs from female *M. rubecula* in the following characters of plumage. Both sexes have blue-grey not blackish auriculars, and more white below the eye. Adult males have glossy bluish crowns, blacker wings and tails without pale edges (though with small vague whitish tips to the outer rectrices), and more rufous throats and breasts. Adult females have pale lores, and more off-white under the tail: the outermost rectrices are more or less uniformly whitish, not grey with narrow whitish edges and shafts. First-year young of the two species differ as do the adult females.

Adult males of *M. ruficollis* seem to differ from their females in the characters which distinguish them from female *M. rubecula* (the difference in colour of the rectrices being most marked), and in lacking the pale lores. Young birds, in what is presumably the first-year plumage, differ from adult females in the same characters of wings and tail as do those of *M. rubecula.*

Myiagra alecto — Shining Flycatcher

Specimens. Western Australia. Point Torment (30) 12 June 68, 1 ♂ (20.1 g), 1 ♀ (18.0 g). Northern Territory. Darwin (41) 10 – 16 Sept 68, 2 ♂ (18.3, 22.3 g), 1 fl. Mount Roper (45) 26 Oct – 1 Nov 68, 1 ♂ (22.0 g), 3 ♀ (21.1, 21.9, 21.9 g), 6 fl (♂ plumage 21.5, 22.1 g; ♀ plumage 19.0, 20.1, 23.4 g). Queensland. Ayton (52) 27 June 64, 1 ♂ skel (19.6 g), 1 ♀ skel (22.0 g), 1 fl (♀ plumage, 20.4 g). Tully (54) 15 July 64, 1 ø (♀ plumage, 19.0 g).
Distribution & Abundance. Also recorded at Pine Creek (44), three birds. Not

encountered elsewhere in northern Queensland (47-51, 53, 55-57). Fairly common at Darwin (41) and Mount Roper (45).

Habitat. Estuarine or coastal mangroves, or adjoining forest, and rainforest beside a creek.

Behaviour. Usually on the ground or near it, in thickets or among open mangrove roots. In pairs or apparently solitary. Females respond to squeaking by appearing suddenly, and remaining close though wary until scared off by a sudden movement, whereas males usually remain out of sight (BB). Twice recorded as mobbing the collector (AH), one bird being in male plumage. Recorded as feeding on the ground, and in flight low over fresh water. A pair recorded as keeping close together as they moved about with bobbing tails.

Song of a male recorded as a rising swelling scale, stopping abruptly and unexpectedly, repeated several times – unlike any other in the mangroves (BB). Also a double-noted whistle from a male (DJF). Females in company with males twice recorded as giving a plaintive *peep-peep-peep,* once after the male had been shot (DJF). Otherwise grating calls from both sexes, described as harsh or soft, like two pebbles rubbed together, or the calls of *Rhipidura leucophrys* or *Chlamydera.*

Food. Unidentified insects, once "very small", and isopod crustaceans identified from a part 4.5 mm across.

Annual Cycle. Most birds seemed to be paired (mid-June to late October). Males had enlarged testes on 14 and 16 September and (very large) 28 October, and a female a somewhat enlarged ovary on the latter date.

A specimen in fluid, not in adult male plumage, was recorded as having been in moult on 1 November (45), but has not been re-examined to determine whether or not it was adult.

Impermanent Colours. *Legs* dark grey, more or less bluish or purplish (not west of Mount Roper (30, 41) ?), sometimes black in adult males; soles yellow. *Iris* dark brown (sepia or a warmer colour). *Bill* blue-grey or grey-blue, usually with black tips. *Mouth* brilliant vermilion to orange, blackish round tomia.

Geographical Variation. Our specimens, with those in the British Museum, confirm Keast's (1958i) conclusion, that populations in King Sound (Point Torment (30)) and northern Queensland are distinct, with those from the Northern Territory intermediate. However, they suggest some changes in the diagnostic characters.

M. a. tormenti King Sound (30). Bill long and narrow. Our only female has the crown glossy blue-black like those of other races (not grey-brown as described by Keast) and white under-tail coverts.

M. a. nitida Northern Territory (41, 45). Bill narrow but somewhat shorter, female under-tail coverts white. One female (45) has the crown more or less grey, as described by Keast for *tormenti.*

M. a. wardelli northern Queensland (54). Bill decidedly shorter, and broader basally. Male belly no more glossy than in *nitida.* Female under-tail coverts more or less rufous.

Other Notes. In collections the gloss of the black plumage seems to change colour, tending to become purplish rather than greenish.

Parasites. See Appendix 1.

Seisura inquieta Restless Flycatcher

Specimens. WESTERN AUSTRALIA. Mount Barker (12) 22 – 28 Mar 66, 1 ♂ skel, 1 ø skel, 2 fl. Mount Anderson (28) 3 – 7 May 68, 1 imm ♂, 1 ø skel. Derby (29) 27 May 68, 1 imm ♀ (12.8 g). Geikie Gorge (31) 5 June 68, 1 ♀ (14.1 g), 1 ♀? (13.9 g). Mount Caroline (33) 26 June 68, 1 imm ♀ (12.2 g). Lake Gladstone (36) 21 July 68, 1 ♂. NORTHERN TERRITORY. Oenpelli (43) 1 – 7 Oct 68, 3 fl (14.9, 15.9, 16.9 g). Mount Roper (45) 28 Oct – 2 Nov 68, 2 fl (14.5, 15.6 g). Borroloola (46) 16 Nov 68, 1 fl (14.7 g). QUEENSLAND. Moonlight Creek (47) 14 May 64, 1 ø skel (13.2 g), 1 fl (14.3 g). Inkerman (50) 6 June 64, 1 ♂ skel (11.8 g). SOUTH AUSTRALIA. Nonning (79) 30 Mar 65, 1 ♂ (23.5 g). NEW SOUTH WALES. Albury (92) 10 – 11 Mar 63, 1 ♂, 1 imm ø, 1 ♀ skel.

Specimens to other Museums. NMV 1 fl Coleraine (89) 22 Jan 63.

Distribution & Abundance. Recorded also at Scott River (11) Green Range (13) and Newdegate (14), throughout the Kimberleys (30, 32, 34, 35), at Pine Creek (44), Innamincka (64) and Kingoonya (70), and widely in the southeast (86, 90-95). Few at Mount Anderson (28), Oenpelli (43), Borroloola (46) and Kingoonya (70), fairly common at Pine Creek (44) and Nonning (79), and common at Geikie Gorge (31) and Mount Roper (45).

Habitat. In or near trees or shrubs, usually near water or at the edge of woodland.

Behaviour. At all levels from the ground to the treetops, moving on from one bush or tree to another. Clinging to vertical stems and reeds, one foot above the other like a Reed Warbler *Acrocephalus* (BB), hovering in front of foliage with wings wide spread (ICJG) and hovering Kestrel-like for 15 to 20 seconds above short grass, dropping to the ground and returning to a perch to eat (PRC). Solitary or in pairs. Grating calls ('scissor grinding' like metal on stone), buzzing calls, a harsh *chacking* call while mobbing the collector, and a very strident single whistle.

Food. Seen to catch what seemed to be a grasshopper by dropping into short grass. Gut contents seldom recorded, usually indeterminable insect paste but including spiders, flies and probably grasshoppers.

Annual Cycle. The bird (now in fluid) collected at Mount Roper (45) on 28 October was making frequent visits to a nest on a horizontal branch at about 30 ft; this was cup-shaped with a good deal of spiders' web incorporated in the outer rim. A young bird with incompletely pneumatised skull was apparently moulting into the first-year plumage on 3 May (28) while others in this plumage on 11 March, 27 May and 26 June (92, 29, 33) had fully pneumatised skulls. An adult on 30 March (79) and birds of unknown age on 22 January (89) and 23 March (12 – two birds squabbling together) were in moult. Adults were in fresh plumage on 11 March, 5 June and 21 July (92, 31, 45), the last with slightly enlarged gonads.

Impermanent Colours. *Legs* black, occasionally brownish or greyish; soles whitish. *Iris* dark brown, once hazel in a bird (now in fluid, NMV) of unknown age (those in first-year plumage did not differ from adults). *Bill* usually black above with bases of tomia tending towards blue-grey, blue-grey below with black tip; first-year birds (and others ?) tending towards all-black. *Mouth* whitish, pale grey or fleshy, tending to blue-grey at tip.

Geographical Variation. Keast's (1958i) measurements suggest that the populations of the southwest overlap too much in size with those of the southeast to be separated racially as *westralensis,* so that two subspecies should be recognised:

S. i. inquieta southern Australia (12, 79, 89, 92).

S. i. nana northern Australia from the Kimberleys to the Gulf of Carpentaria (28, 50): much smaller in all dimensions, and (despite Keast) apparently with more glossy backs.

Specimens collected in northern Queensland in 1970 (see Appendix 3) confirm that *nana* occurs near Normanton, and bring the northern known limit of nominate *inquieta* to Julia Creek at 20°40'S. The gap between them is thereby reduced to 200 miles. Though it seems likely that there is some barrier between them, Keast (1958i, 1961) is not justified in stating positively that they are isolated. He does not comment on the fact that his measurements of a single specimen from Alexandria on Barkly Tableland associate it with *inquieta* rather than *nana.*

Other Notes. The first-year plumage differs from the adult mainly in its wings and tail. The former are brownish with no blue gloss, with buffy or white tips to the coverts and inner secondaries and narrow white edges to the outer secondaries. The rectrices are somewhat browner and less glossy, and decidedly more square-cut than those of the adult. The upper parts are less glossy, the difference being especially marked on the upper mantle. All phases show a variable wash of buff on the breast, which on the average is best developed in the first-year plumage and least so in the adult male.

It is doubtful whether this species should be kept generically distinct from *Myiagra alecto,* unless both are to be regarded as monotypic genera.

Machaerirhynchus flaviventer — Boat-billed Flycatcher

Specimens. Queensland. Big Tableland (51) 1 July 64, 2 ♀ (9.4, 9.6 g). Ayton (52) 21 – 30 July 64, 2 ♀ (8.6, 9.7 g), 1 imm ♂ (10.1 g), 1 imm ♀ (9.8 g). Walter Hill Range (53) 21 – 30 July 64, 2 ♂ (9.7, 10.5 g). Tully (54) 14 – 20 July 64, 1 ♂ (10.3 g), 1 imm ♂ (9.8 g).

Distribution & Abundance. Encountered only at these localities of the main rainforest block of north Queensland, where it was fairly common.

Habitat. Rainforest, occasionally moving into the edge of adjacent open forest.

Behaviour. In the canopy and substage, once recorded as making flycatcher-like sallies (LRP). Not noted as associating much with its own species, but often in mixed flocks including species of *Malurus, Gerygone, Rhipidura,* monarch flycatchers, *Colluricincla, Myzomela, Meliphaga* and *Ptiloris.* One chased by *Rhipidura rufifrons.* Voice seldom recorded, and only as *tizzz-tizzz.*

Food. Insects, including a small fly and probably ants.

Annual Cycle. A male and a female in what is probably the first-year plumage on 21 and 30 June (52) were the only specimens to have incompletely pneumatised skulls, unlike another male in this plumage on 14 July (54).

Impermanent Colours. *Legs* grey varying in tone, usually bluish but sometimes more purplish. *Iris* apparently variable, from warm brown to olivaceous or grey; adult males perhaps more uniformly blackish brown. *Bill* black with whitish tips. *Mouth* black, throat yellowish or pinkish.

Geographical Variation. All four populations presumably belong to the southern form *M. f. secundus*, as the males from the Tully area (53, 54) certainly do. The first-year male from Ayton (52) is less blackish above than the one from Tully (54); but no comparative material is available to show whether or not the females from this area approach those of the Cape York Peninsula *M. f. flaviventer* in having paler upperparts (see Keast, 1958i).

Other Notes. The young males, presumably in first-year plumage, differ from adult females in having much deeper yellow on the underparts, concentrated on the lower throat which is white in females, and in having duller and less clearly defined green edges to the secondaries. The young female differs from adults only in the latter character.

Arses kaupi — Australian Pied Flycatcher

Specimens. QUEENSLAND. Ayton (52) 22 – 25 June 64, 6 ♂ (2 sexed by plumage; 13.2, 13.6, 13.7, 13.8, 14.8, 15.5 g), 1 ♀ (113.5 g). Tully (54) 14 – 18 July 64, 4 ♀ (1 sexed by plumage; 11.9, 11.9, 12.9, 13.4 g), 1 imm ♀ (13.1 g).

Distribution & Abundance. Encountered only in the main rainforest block of northern Queensland (51-54), where it was moderately common.

Habitat. Rainforest and nearby open forest.

Behaviour. Nearly always on tree trunks and branches, often clinging to vertical surfaces or the undersides of branches, or fluttering or creeping up – repeatedly compared to *Climacteris*. Solitary, once with *Monarcha trivirgata, Rhipidura rufifrons* and *Meliphaga notata*. A netted bird made a harsh angry grating when handled, and wounded birds gave a soft *chrr-chrr* and a repeated *twouldn't-it*.

Food. Insects, including two small beetles.

Annual Cycle. A young bird was in first-year plumage on 18 July (54), but had a fully pneumatised skull, while one in adult male plumage on 22 June (52) had not completed pneumatisation.

Impermanent Colours. *Legs* various shades of dark bluish and purplish grey; soles buffy or dull yellow. *Iris* dark brown, usually blackish but sometimes reddish. *Bill* pale grey-blue, tomia paler or yellowish, tip or distal culmen occasionally black; first-year creamy white tinged with blue, tip black. *Mouth* grey-blue, usually darker and greyer than bill; first-year greyish white. *Bare skin round eye* bright blue tending towards violet.

Other Notes. The first-year plumage differs from that of the adult female in having the wings, tail and breast brown; the rump merely paler and clearer than the brownish grey mantle into which it grades (though contrasting with the brownish grey upper tail-coverts) and the pale area not extended onto the upper flanks; the throat dusky without clear white or black; and the crown with scarcely any gloss. The female differs from the male in having the white areas (except of the belly) considerably reduced, with black fringing on the nape and lower throat, and the black areas less glossy.

Parasites. See Appendix 1.

Monarcha melanopsis Black-faced Flycatcher

Specimens. New South Wales. Grafton (94) 1 Apr 63, 1 imm ♂.

Field Notes. Common at all rainforest localities in northern Queensland (51–54, 57).

The specimen, beginning to moult from first-year into adult plumage and with its skull not fully pneumatised, was feeding in a thick clump of trees in tea-tree woodland. Its stomach contained insects.

Impermanent Colours. *Legs* blue-grey. *Iris* dark brown. *Bill* black, base and gape yellow.

Other Notes. In the CSIRO Index (1969) this bird is regarded as conspecific with *M. frater* of the New Guinea mountains, though the available evidence suggests that they both breed in the Cape York Peninsula (Keast, 1958i). Even if this is not true, they are quite distinct (see Rand & Gilliard, 1967), and may not even be especially closely related.

Monarcha trivirgata Spectacled Flycatcher

Specimens. Queensland. Ayton (52) 20 – 30 June 64, 5 ♂ (12.2, 12.7, 13.0, 13.5 g), 3 ♀ (12.0, 12.4, 12.6 g). Tully (54) 13 – 20 July 64, 1 ♂ (13.1 g), 2 ♀ (13.3, 13.4 g), 1 ø (12.5 g). Cape Gloucester (56) 6 Aug 64, 1 ♀ (12.8 g), 1 ø (12.7 g), 1 imm ♂ (13.0 g). Mount Dryander (57) 1 – 11 Aug 64, 5 ♂ (7.3 (!), 11.3, 12.5, 12.9, 14.1 g), 2 ♀ (11.8, 13.2 g), 1 ø (12.6 g), 1 ♀ skel (12.2 g).

Distribution & Abundance. Encountered at all localities on the east coast of north Queensland (51-54, 56-57), and very common at Ayton (52) and Mount Dryander (57).

Habitat. Rainforest and mangroves, occasionally nearby open forest. The mangroves in which it was found at Cape Gloucester (56) must have been at least three miles from the nearest rainforest, which was presumably of the dry type in which the species was abundant at Mount Dryander (57).

Behaviour. At all heights, often low, and commonly caught in nets at Ayton (52); once investigating dead leaves on the ground. Solitary or in pairs or small groups, or in mixed flocks with *Rhipidura rufifrons* and other flycatchers, *Colluricincla megarhyncha, Myzomela obscura* and *Ptiloris victoriae*. Only harsh rasping and scolding notes recorded, especially a triple *chrrk-chrrk-chrrk,* but also single notes, chattering and growling.

Food. Insects, including ants and medium-sized insects (orthoptera ?); also a small snail was found in each of two stomachs.

Annual Cycle. A young bird in first-year plumage on 30 June (52) had one juvenile feather, and an incompletely pneumatised skull. Another in this plumage on 6 August (56) was in company with a female in adult plumage, and both had pneumatised but soft skulls. Three birds in adult plumage between 20 June and 20 July had incompletely pneumatised skulls, while over a third of the pneumatised skulls were noted as soft, thin or transparent.

Impermanent Colours. *Legs* dark grey to blackish, often more or less blue, or purplish (only in males ?). *Iris* dark brown. *Bill* blue-grey or grey-blue tending to be paler on tomia, darker basally and below, with black tip (though this may be pale); one first-year bird, pinkish-blue basally, black distally. *Mouth* blue-grey (often matching

bill), occasionally mixed with yellow, tongue fleshy; first-year, yellowish white.
Geographical Variation. These specimens belong to the race *gouldi,* which is also found in winter in the Cape York Peninsula, where the white-flanked *albiventris* is resident (Keast, 1958i). Birds from south of the Burdekin Gap (56, 57) are on the whole paler rufous below.
Other Notes. The first-year plumage differs from the adult in having the facial pattern grey instead of black, with more or less whitish tips to the feathers of throat and forehead and a whitish crescent in front of and below the eye; in being paler rufous, especially at the side of the throat but with a deep zone on the upper breast; and in having softer flight quills, the remiges and upper wing-coverts brownish and the rectrices more pointed. Adult females generally have paler rufous and less intensely black facial patterns than adult males.
Parasites. See Appendix 1.

Monarcha leucotis — White-eared Flycatcher

Specimens. QUEENSLAND. Big Tableland (51) 1 July 64, 1 ♀ (11.8 g). Ayton (52) 22 – 26 June 64, 1 ♂ (11.6 g), 2 imm ♂ (12.1, 12.8 g). Tully (54) 15 July 64, 1 ♂ (11.3 g), 1 ♀ (10.8 g), 1 imm ⌀ (10.4 g). Cape Gloucester (56) 5 – 7 Aug 64, 2 ♂ (11.2, 11.9 g), 3 ♀ (10.8, 11.4, 12.2 g). Mount Dryander (57) 3 – 12 Aug 64, 4 ♂ (1 sexed by plumage; 11.4, 11.5, 11.7, 11.7 g), 4 ♀ (1 sexed by plumage; 10.9, 11.0, 11.2, 11.8 g), 1 imm ♂ (11.3 g).
Distribution & Abundance. Encountered only where collected. Although regarded as a rather rare species, it was common at Mount Dryander (57).
Habitat. Rainforest, and occasionally nearby open forest, and mangroves; one in thicket woodland some distance from either (56). Common only in second-growth and at the edge of dry rainforest, along creeks on the lower slopes of Mount Dryander (57).
Behaviour. Usually in the crowns of low trees, or hovering outside the foliage to feed. Solitary or occasionally in small parties, recorded as associating with *Rhipidura rufifrons, R. fuliginosa, Monarcha trivirgata* and *Zosterops lateralis.* Buzzing calls like those of *M. trivirgata,* including a thinner version of the triple *chrrk-chrrk-chrrk.* Songs recorded as a fluting whistle (both alone and following the scolding buzz), and a plaintive triple phrase – a rising and falling note separated by a short high one (SAP).
Food. Stomachs contained insect remains (very various in colour), including black beetles 3 mm long, legs (of orthoptera ?) c.25 mm long, and large green insects.
Annual Cycle. An adult male, adult female and first-year young were taken together on 3 August. First-year birds and an adult male on 22 and 26 June had incompletely pneumatised skulls. Young birds in first-year plumage except for some flight feathers, but with no sign of moult, on 15 July and 3 August had fully pneumatised skulls, though that of the later specimen was still soft.
Impermanent Colours. *Legs* black or blackish grey, sometimes tinged bluish or purplish, soles pale greyish. *Iris* blackish brown, occasionally warm brown or olive-grey. *Bill* black sometimes tinged with brown, base grey more or less tinged with blue or

lavender, more extensive on lower mandible – sometimes bill mainly grey with black tip; first-year, base buffy white, gape sometimes pale dull yellow. *Mouth* grey like base of bill, sometimes mottled with whitish; first-year whitish to pale warm yellow.

Other Notes. The first-year plumage is more like the adult than those of most pied monarchs. It differs in lacking any black and in having the greys paler and browner, with a distinct brown wash on breast and flanks; the facial pattern reduced, tinged with grey and less clearly defined and with the front of the crown hoary; and the flight feathers softer with slightly more pointed rectrices. Adult females differ from males in having the mantle, and sometimes the crown, more or less grey rather than glossy black, and the white spots in front of the eye extending on to the lores (which may be hoary rather than pure white). However the female from Big Tableland (51) agrees in both respects with males.

This species is geographically represented by the closely similar Moluccan *M. pileatus,* and by several island forms commonly recognised as generically distinct – *Neolalage banksiana* of the New Hebrides and (as its first-year plumage shows) *Heteranax mundus* of the Tenimber group; and probably *Mayrornis* of Polynesia and *Chasiempis* of Hawaii.

PACHYCEPHALIDAE — Whistlers & Shrike-thrushes

I. C. J. Galbraith

Pachycephala pectoralis — Golden Whistler

Specimens. WESTERN AUSTRALIA. Wonerup (8) 12 Mar 66, 1 fl. Metricup (9) 7 – 9 Mar 66, 2 ♀ skel, 1 fl. Scott River (11) 18 Feb – 1 Mar 66, 1 imm ♂, 1 imm ♀, 1 imm ø, 3 ♂ skel, 1 ♀ skel, 2 fl. Green Range (13) 4 – 6 Apr 66, 1 imm ♂, 1 ♀ skel. Newdegate (14) 15 Apr 66, 1 ♂. Eucla (17) 5 July 65, 1 ♀. QUEENSLAND. Big Tableland (51) 1 – 2 July 64, 6 ♂ (1 sexed by plumage) (23.4, 23.7, 23.8, 24.1, 24.5, 25.0 g), 2 ♀ (22.5, 23.5 g). Ayton (52) 22 – 27 June 64, 2 imm ø (23.5, 25.4 g). Tully (54) 11 – 18 July 64, 5 ♂ (21.2, 23.0, 24.8, 24.8, 25.1 g), 2 ♀ (21.1, 23.0 g), 3 imm ♂ (24.0, 24.4, 25.9 g), 1 imm ♀ (23.7 g), 2 imm ø (22.4, 23.5 g). Mount Dryander (57) 3 – 12 Aug 64, 1 imm ♀ (22.3 g), 3 imm ø (21.6, 22.0, 22.6 g). NEW SOUTH WALES. Dunedoo (93) 18 – 24 Mar 63, 1 imm ♂, 1 imm ø. Grafton (94) 1 – 3 Apr 63, 1 ♀, 1 imm ♂, 1 imm ♀, 1 fl. Tenterfield (95) 10 – 12 Apr 63, 1 ♂, 1 ø (♀ or imm ♂).

Specimens to other Museums. AM 1 ♂ Grafton (94). NMV 1 imm ♀ Grafton (94), 1 imm ♂ Tenterfield (95).

Distribution & Abundance. In Western Australia encountered only in the southwest (8-14), where noted as very common at Metricup (9), and at Eucla (17). Here a male was seen and a female collected in the coastal mallee scrub, thus extending the known range over 400 miles eastwards from Esperance. Serventy & Whittell (1967) incorporated this record. The species may be expected throughout the coastal mallee strip, which extends into South Australia as far east as the gap near Fowlers Bay (McEvey & Middleton, 1968: 209).

Found in all forested localities on the east coast of northern Queensland – a few immature birds at Ayton (52) and Mount Dryander (57), common at Big Tableland

(51), Walter Hill Range (53) and Tully (54). Big Tableland (51) is the northernmost locality known for the species in Australia.

Encountered at all localities in northeastern New South Wales (93-95).

Habitat. Found in open forest, and especially trees and thickets near water, in extreme southwestern Western Australia (8-11), and further east in or near mallee (13, 14 & 17). At Eucla (17) this was in the narrow coastal belt of low mallee scrub. In ironbark, tea-tree and other woodland and open forest of northern New South Wales (93-95). In northern Queensland in dry rainforest (57) and in true rainforest and the nearby open forest (51-54), with one first-year bird in forest surrounded by mangroves on a tiny island in the mouth of the Bloomfield River (52).

Behaviour. Usually seen feeding or perched, in vegetation at any height under a closed canopy. In rainforest (51, 53, 54) males in adult plumage were often conspicuous when perched and singing high under the canopy, while dull-plumaged birds usually skulked at lower levels. Often solitary, but sometimes apparently in loose family parties, or associating with other insectivorous birds (*Pachycephala rufiventris, P. simplex, Colluricincla megarhyncha, Tregellasia capito, Acanthiza* species, "flycatchers and warblers").

Food. Mainly soft-bodied medium-sized insects, including stick-insects, grasshoppers, earwigs and caterpillars, but including small beetles and spiders. Often some seeds. A male was seen beating a large stick insect on a fallen log (17), eating the parts which broke off and finally swallowing the body.

Annual Cycle. The only specimens recorded as having somewhat enlarged gonads were two adult males on 1 – 2 July (57). A male was seen feeding a fully-fledged young bird on the wing on 9 March (9).

In all months of collection, some first-year birds had not completed their moult but retained the juvenile inner secondaries: one of two in February (11), one of two in March (93), two of three in April (13 & 94), two of two in June (52), three of five in July (54) and two of four in August (57). Except in late July and August, these had some juvenile greater upper wing-coverts; and up to April juvenile feathers in the nape, sides of throat, auriculars and/or upper tail-coverts also. The state of the skull was not recorded in February, while it was incompletely pneumatised in other first-year birds except for two which had not completed their moult (July, August) and two which had (July).

A second-year female still had a few juvenile secondaries and feathers in the nape on 1 March (11), and this and another in July (54) had the lower mandible paler. The only second-year male was taken in July (54). The only second-year or adult bird with an incompletely-pneumatised skull was an adult male in July (54), in which the large number of small, jagged and irregularly-distributed 'windows' and very fragile inner layers showed an abnormal condition. No specimen was recorded as having enlarged gonads.

Adult males were recorded as moulting remiges, rectrices and body-feathers in February (11), while one in July (51) was replacing a single outer rectrix.

Impermanent Colours. *Legs* dark grey, brownish, purplish or greenish to black, soles pale greyish or buff to orange. *Iris* pale to deep red-brown; usually darker in first-year young (once olive). *Bill* black in adult male; dark grey or brownish to black in adult female and second-year male; brownish with base of lower mandible white tinged with pink or brown, gape yellowish, in first-year young. *Mouth* white tinged with pink, mauve or grey; tinged or streaked with yellow in first-year young; a variable black area towards the tip of the bill in adult males only. *Lower eyelid* (few recorded) greyish or grey-green, rim black.

Geographical Variation. The skins examined agree with the subspecies as recognised by Mayr (1954):

P. p. occidentalis, southwestern Western Australia extending to Eucla (11, 13, 14, 17). Adult females differ from those of the South Australian *fuliginosa* in their paler underparts and greyer throats and upperparts.

P. p. pectoralis. The grey tail and whitish under tail-coverts of the unsexed first-year bird from Dunedoo (93) show it to have been a female of this race. It remains to be determined where between Dunedoo (93) and Grafton (94) the female characters change, whether the increase in black on the adult male tail coincides with this, and whether the changes are sharp or gradual.

P. p. ashbyi. An adult male from Tenterfield (95) with little olive at the base of the tail, and an adult and a first-year female from Grafton (94) with olive tails and yellow under-tail coverts, agree with two adult females in the British Museum from the Richmond River in showing that this race extends into northern New South Wales (Galbraith, 1956: 200). The greyish tails of a few *ashbyi* and *queenslandica* are clearly due to wear and fading. The four specimens from Mount Dryander (57) are in first-year plumage, but agree with those of *ashbyi* (94) rather than *queenslandica* (52, 54) in having on the average less brownish underparts, like the adult females. It remains to be determined whether the Burdekin Gap (Keast, 1961: 406, 435) serves as the barrier between the two forms.

P. p. queenslandica. The females and young males from north of the Burdekin (51, 52, 54) have the browner underparts of this race.

Other Notes. Differences between the dull-coloured plumages have been briefly noted (Galbraith, 1967), the first-year and male second-year plumages corresponding respectively to the "I Phase" and "II Phase" of Mayr (1932a & b). Though first-year young retain most of the juvenile flight feathers, they moult the innermost three secondaries; and though they moult most of the wing-coverts, they retain the primary-coverts.

The suggestion that this species remains to be discovered in the Cape York Peninsula (Galbraith, 1967) fails to take into account its geographical interaction with *P. simplex,* which presumably indicates competition between them. With *P. melanura* removed, the ranges of the two species interdigitate most precisely (see Peters, 1967). The only overlap is in the rich rainforest block of northern Queensland, between Cooktown and Cardwell. Here *P. simplex* seems to be commoner in the lowlands and *P. pectoralis* at higher altitudes.

Pachycephala melanura Mangrove Golden Whistler

Specimens. Western Australia. Karratha (1) 30 May 66, 1 imm ♀. Peg's Beach (2)

3 June 66, 1 ♂, 1 imm ♀, 1 imm ♀?. Point Torment (30) 30 May – 12 June 68, 2 ♂ (19.2, 20.7 g), 1 ♀ (18.7 g), 1 imm fl. Wyndham (39) 31 July – 9 Aug 68, 3 ♂ (20.3, 20.3, 22.2 g), 1 ♀ (24.2 g), 1 imm ♂? (22.8 g), 4 imm ♀ (20.3, 21.1, 21.2, 21.9 g). NORTHERN TERRITORY. Mount Roper (45) 26 – 28 Oct 68, 2 imm ♂ (21.5, 22.9 g), 1 imm ♀ (23.2 g). QUEENSLAND. Moonlight Creek (47) 6 – 15 May 64, 2 ♂ (24.3, 24.7 g), 1 ♀ (26.0 g), 1 imm ♂ (21.4 g). Inkerman (50) 28 May – 11 June 64, 1 ♀ (22.6 g), 2 imm ♂ (20.5, 21.8 g), 1 imm ♀ (21.6 g), 1 imm ø (22.5 g). Cape Gloucester (56) 7 Aug 64, 1 imm ♂ (21.5 g).

Distribution & Abundance. A single bird in dull plumage was seen in the mangroves on Oyster Creek, Carnarvon (5) on 16 June 1966 (PRC) though the species was not otherwise encountered during several days search in the extensive mangroves of Carnarvon (5) and Bush Bay (6). This record predictably extends the known range by over 170 miles south from Exmouth Gulf, and the species may yet be found in the small mangrove area at the mouth of the Wooramel River another 75 miles southward. *P. pectoralis* is now known from Tamala near the head of Denham Sound (Serventy & Whittell, 1967), but the two species are probably separated by a gap of at least 50 miles (Galbraith, 1967).

Apart from this sighting, the species was collected wherever it was found, and was nowhere common. Three of the localities reduce notable gaps in known range: Wyndham (39) between Napier Broome Bay and the Daly River; Moonlight Creek (47) between the McArthur River and Normanton (the mangroves at Borroloola (46) were not worked); and Inkerman (50) between Normanton and Cape York. The specimen from Cape Gloucester (56) is the first to be recorded from the east coast south of Cape Grenville on the Cape York Peninsula, though old specimens from Springcliff (Galbraith, 1967) and Percy Island (QM no.011257) extend the range a further 150 miles to the south.

Habitat. Invariably in mangroves forming a closed canopy, though these varied in height from low bushes to tall trees, in situation from offshore shell-beds to the banks of tidal rivers and creeks, in extent from fringing strips to extensive swamps, and in structure from spaced-out bushes with interlacing crowns and clear well-lit understory to dense dark thickets.

Behaviour. Usually called up by the collector's squeaks and whistles, when it would move inquisitively about within a few yards of him, or found quietly perched a few feet from the ground. Once very actively hunting insects. Usually silent or making soft whistles and calls, though a male moulting into the adult plumage uttered a musical whistle with sharp notes (August). Usually solitary, except for the family party noted below, but two young birds were associated with separate feeding parties of *Zosterops lutea* and one of them also with *Myzomela erythrocephala* and *Rhipidura fuliginosa* (June).

Food. Usually insects, including grasshoppers and dragonflies; once crabs and other marine invertebrates.

Annual Cycle. An adult female had an ovary 15 mm long with a developing oocyte on 31 July (39) and an adult male testes 9 mm long on 8 August (39), while the gonads of most other adults were slightly enlarged. An adult male and female with small gonads and a young bird moulting into the first-year plumage were taken together on 30 May (30).

Of eleven young birds in first-year plumage (6 May to 26 Oct), all but two (30 May, 3 June) still had traces of the juvenile plumage, mainly among the inner secondaries and the secondary-coverts. Five had the skull incompletely and two completely pneumatised. Seven were in moult from the first-year to the second-year plumage (11 June to 28 Oct), and all but two of these had the skull fully pneumatised. One male was in the last stages of moult from the second-year into the adult plumage (4 Aug).

All eleven birds in full adult plumage had fully pneumatised skulls. One male (9 August) showed no sign of moult, while the others (8 May to 8 August) had at least a few pin-feathers or sheathed bristles on the chin or throat.

Impermanent Colours. *Legs* pale grey, often bluish, to black, without discernible trends of geography, age or sex; horn-coloured in one first-year bird. *Iris* red to reddish, purplish or yellowish brown, sometimes dark: no discernible trends. *Bill* black; in younger birds especially, the base of the lower mandible and more or less of the rest of the bill is paler – either brownish black, or dull greyish or yellowish. *Mouth* pale, greyish, pinkish or yellowish; or mainly black in half the adult males and one of four adult females.

Geographical Variation. Variation in this species is under study, using specimens from several institutions, and will be discussed in a separate publication. Some of Mayr's (1954) conclusions will have to be modified, while an attempt to summarise the variation (Galbraith, 1956) is misleading. The following notes relate only to Harold Hall skins.

P. m. bynoei. The adult male from Peg's Beach (2) belongs to this race. The undescribed female is very like that of *melanura,* and the three first-year females from Karratha (1) and Peg's Beach (2) agree closely with an adult except in their juvenile wings and tails.

P. m. melanura. The specimens from Point Torment (30) belong to this race.

P. m. hilli. The only character which reliably separates this race from *'violetae'* is the olive at the base of the male tail. The three adult males from Wyndham (39) show this character to a much slighter degree, and they and the adult female are otherwise inseparable from *'violetae'*.

P. m. 'violetae'. Populations from New Guinea and Cape York (*spinicauda*) may not be distinct from those of the Northern Territory (*violetae*), in which case all should be known as *spinicauda* (Storr, 1967). The specimens from Mount Roper (45), Moonlight Creek (47) and Inkerman (50) belong to this race.

P. m. subsp. Although the young male from Cape Gloucester (56) cannot be separated from comparable specimens of *'violetae'*, the damaged adult male in the Queensland Museum from the Percy Islands suggests that the southern populations may be distinct.

Other Notes. The moult into first-year plumage involves the same feathers as in *P. pectoralis.*

The species seems more prone than *P. pectoralis* to show 'progressive' and 'retarded' characters (Mayr, 1933), as an individual variant affecting parts of the male plumage. Three males which were partly in the second-year plumage have this mottled with black on the crown, and one of these also has black in the upper tail-coverts, and another yellow on the hind-neck. One of these three (39) on 4 August was moulting into the adult plumage, but the other two (45), on 26 & 28 October were moulting into the 'progressive' second-year plumage, thus proving that this is not the third-year plumage recorded by Wilson (in press). Three males in full adult plumage (2, 30, 47) have white barbs towards the tips of otherwise black feathers of the nape, and one of these also on the forehead.

Pachycephala rufiventris — Rufous Whistler

Specimens. WESTERN AUSTRALIA. Peg's Beach (2) 3 June 66, 1 fl. Coolawanyah (4) 15 May 66, 1 imm ø. Bush Bay (6) 17 June 66, 1 fl. Perth (7) 17 Nov 62, 16 Feb 66, 1 imm ♂, 1 skel. Newdegate (14) 12 Apr 66, 1 skel. Eucla (17) 1 July 65, 1 imm ♂. Mount Shenton (18) 31 July – 3 Aug 65, 2 ♂, 1 imm ♂. Carnegie (19) 16 Sept 65, 2 ♀ (24 g). Mungilli Claypan (20) 13 Sept 65, 1 ♀. Lake Gruszka (21) 11 Sept 65, 1 ♂. Warburton (23) 10 – 13 Aug 65, 2 ♂, 1 ♀, 1 imm ♂, 1 imm ø. Giles (25) 21 – 28 Aug 65, 3 ♂ (1 sexed by plumage), (25, 25 g), 1 ♀, 1fl. Walter James Range (26) 27 Aug 65, 1 ♂ skel. Mount Anderson (28) 2 – 14 May 68, 1 ♂, 2 ♀ (22.3 g), 1 ♂ skel, 1 ♀ skel. Derby (29) 21 – 25 May 68, 1 ♂ (21.4 g), 1 imm ♂ (21.4 g), 1 skel, 2 fl. Geikie Gorge (31) 4 – 5 June 68, 1 imm ♂ (13.9 g), 1 ♀ skel (21.5 g). Mount Caroline (33) 21 June 68, 1 ♂. Joint Hill (34) 27 June – 1 July, 2 ♂ (20.1 g), 1 ♀ (21.5 g). Mount Elizabeth (35) 6 July 68, 1 ♂. NORTHERN TERRITORY. Pine Creek (44) 28 – 29 Aug 68, 1 ♂ (22.5 g), 1 imm ♀ (20.6 g). Mount Roper (45) 28 Oct – 1 Nov 68, 1 ♀ (18.9 g), 3 imm ♀ (c.14, 19.2, 20.5 g). QUEENSLAND. Moonlight Creek (47) 4 May 64, 1 imm ♀ (18.6 g). Glendower (55) 16 Mar – 1 Apr 64, 1 ♂, 2 imm ø (20.3, 21.6 g). Banoona (58) 28 Apr – 6 May 63, 2 ♂, 1 ♀, 1 fl. Thylungra (60) 25 – 26 May 63, 1 ♂, 1 imm ø. SOUTH AUSTRALIA. Innamincka (64) 27 – 28 Apr 65, 1 ♀, 1 imm ♀. Etadunna (66) 18 – 20 May 65, 2 imm ♂, 1 imm ø. Granite Downs (68) 7 June 65, 1 ♂. Kingoonya (70) 16 – 20 June 65, 1 ♂, 1 ♀, 1 imm ♀, 1 imm ø. Nonning (79) 31 Mar – 1 Apr 65, 1 ♂, 1 imm ♂. Aroona Valley (82) 10 – 19 Apr 65, 2 ♂, 1 ♀. Angepena (84) 1 – 3 Dec, 2 ♂. Furner (86) 9 Jan 63, 1 ♂. VICTORIA. Coleraine (89) 19 Jan 63, 1 ♂. Little Desert (91) 26 Jan 63, 1 fl. NEW SOUTH WALES. Albury (92) 10 – 11 Mar 63, 1 ♂, 1 fl. Dunedoo (93) 18 – 23 Mar 63, 3 ♂, 1 ♀, 1 imm ♀, 2 fl. Grafton (94) 30 Mar – 5 Apr 63, 1 ♀, 1 imm ø, 1 ♂ skel, 1 ♀ skel, 2 fl.

Specimens to other Museums. NMV 1 ♂, 1 ♀ Banoona (58). SAM 1 ♂, 1 ♀ Angepena (84).

Distribution & Abundance. Recorded at most other main camps (9, 22, 27, 30, 32, 36, 43, 46). The specimen from Eucla (17) was the only one seen, and only a few were seen at Metricup (9), Oenpelli (43), Mount Roper (45 – in mangroves and hillside woodland), Etadunna (66) and Granite Downs (68); fairly common at Carnegie (19) and Nonning (79) and locally at Borroloola (46); common at Warburton (23), Giles (25), Geikie Gorge (31), Pine Creek (44) and Kingoonya (70).

Habitat. Occupying all habitats with trees or shrubs, except dense forest (Plate 5(b)).
Behaviour. Usually seen feeding among foliage or perched, occasionally noted as inquisitive and seldom as shy – Goodwin suspected that it was more skulking in open country. Usually in pairs, when the rich song of the male was often noted, or solitary; other intraspecific associations are noted under 'annual cycle'. Occasionally in mixed flocks with other insectivorous birds, especially *Aphelocephala, Rhipidura* and *Acanthiza,* but also *Colluricincla, Petroica, Lalage, Climacteris* and *Dicaeum.*
Food. Insects, especially beetles and also grasshoppers and caterpillars; occasionally seeds, and twice 'grass'.
Annual Cycle. Adult females in company with adult males were taken on 1 July, 12 August and 13 September (34, 23, 20). None of their gonads was enlarged, though two of the males were noted as singing. Adult males were noted as responding by song to dull-coloured individuals on 14 May and about 24 July (28, 18), the second of these giving an unusual two-syllabled whistle and frequently 'pivoting' several times like a displaying Goldfinch *Carduelis carduelis* (DG). Adult males with enlarged testes, most of which were noted as singing, were taken on 3, 27 and 29 August and 9 and 19 January (18, 26, 44, 86, 89); and another male, moulting from first-year into second-year plumage on 17 November (7), also had enlarged testes and was singing. The only female with an enlarged ovary was taken on 16 September (19), with a developing yolk 21 mm across. Birds in first-year plumage and in company with adult males were taken on 18 March (55, 93) and 21 May (29, a young male): one of these records (55) includes three other dull-plumaged individuals in the party. Other records of conspecific associations could be of either pairs or family parties.

Young birds completing the moult from juvenile into first-year plumage were taken on 1 April, 26 May and 20 June (94, 60, 70), three in fresh first-year plumage on 18 March, 21 May and 5 June (93, 29, 31), and others in worn first-year plumage from 31 March to 1 November. Older young beginning the moult into second-year plumage were taken on 15 May and 17 November (4, 7) and one completing it on 16 March (55), while males in worn second-year plumage were taken from 27 April to 1 July. The state of the skull was recorded for only one of the young moulting into first-year plumage, in which pneumatisation was incomplete, as it was in two of the three in fresh first-year plumage (21 May, 5 June) and one in worn (18 March) (55). Pneumatisation was recorded as complete for the other birds in first-year plumage, and most of the older ones.

Adults (which may include females in, or moulting from, second-year plumage) were taken in worn plumage between 7 June and 19 January; in worn plumage with the throat in moult on 1 and 28 April; in various stages of moult between 10 March and 1 July; five males in fairly fresh plumage with the throat in moult from 1 April to 28 June, and another on 29 August (44) which was singing and had enlarged testes; and fairly fresh plumage on 28 April and 6 May. No clear annual cycle emerges from these data, and the situation is no doubt confused by individual, geographical and annual variation. It appears that moult may begin on the throat, instead of ending there as is normal.
Impermanent Colours. *Legs* dark grey to black, occasionally bluish or brown; soles

whitish (seldom noted). *Iris* usually red-brown to crimson in adult males (once very light yellow, almost white (28)), dark brown in other birds. *Bill* almost always wholly black in adult male and adult female; in second-year male usually so; in first-year birds horn-coloured, brownish or greyish to black (once pink); lower mandible and bases tending to be paler and gape to be yellower in young birds. *Mouth* very variable; whitish to blackish with young birds tending to be paler, with or without bright or dull yellow (once orange).

Geographical Variation. Mayr (1954) summarises the variation of this species in Australia, which in the absence of barriers seems to be wholly clinal, and stresses that only birds in the same state of wear can be compared. No useful purpose would be served by attempting to place our specimens, many of which are much worn, in arbitrarily-delimited subspecies (cf. Storr, 1967: 52). They agree in general with the pattern of variation described by Mayr, with the palest birds in central and interior northeastern Western Australia (18-25, 31-35). Nearer the coast the two adult males from Mount Anderson (28) and Derby (29) are strikingly darker and there is probably a similar difference between inland Pine Creek (44) and coastal Mount Roper (45) in the Northern Territory. The adult male from Glendower (55), geographically nearest to the type locality of the palest race *pallida* at Normanton, is one of the darkest.

Two adult males, taken at Mount Shenton (18) on 3 August, differ in the relative widths of white throat-patch and black gorget, and in the depth of ventral coloration, to a degree that would be taxonomically important between populations of *P. pectoralis*.

Other Notes. Freeman noted that adult males in strong sunlight twice appeared to be rich yellow beneath, like *P. pectoralis* or *P. melanura*. On the second occasion both he and Hiller saw this colour, though aware that it was an illusion.

The sequence of plumages appears to be as in *P. pectoralis* (Galbraith, 1967). However, there is much less difference in the coloration, texture and shape of the feathers so that the identification of the various plumages is less certain. A male and an unsexed bird are intermediate between plumages of adult males and of second-year birds or adult females, with reduced ventral streaking, indications of black gorgets, and blackening on the ear-coverts and round the eyes. According to Mayr (1954) this mode of variation is shown by females, as well as by second-year males as in other dimorphic *Pachycephala*. In general the second-year and adult female plumage differs from the first-year in having the rectrices less pointed, the remiges and rectrices blacker, the upperparts pure grey without an olivaceous wash, the underparts deeper and more rufous less ochraceous, and the streaking of throat and breast narrower, less drop-shaped and less extended over the flanks and belly. However, there is a good deal of overlap. None of the young has more than traces of the rufous dorsal plumage, and none shows clearly the white ventral plumage, characteristic of juveniles.

The pale olivaceous wash on the grey upperparts, characteristic of the first-year plumage, appears under the microscope to be due to carotenoid rather than phaeomelanic pigment. This is certainly true of the olive edges to the rectrices of a first-year specimen of *P. r. meeki* from Rossel Island (Louisiade Archipelago), which also has rufous edges to the remiges. Both these characters serve to emphasise the close relationship of *P. rufiventris* to the *P. pectoralis* superspecies.

Pachycephala inornata Gilbert Whistler

Specimens. South Australia. Kingoonya (70) 19 June 65, 1 ♂, 1 ♀. Nonning (79) 5 Apr 65, 1 ♀.

Distribution & Abundance. These were the only specimens recorded. One at Kingoonya (70) was near the Amelira rock-hole, the other a dozen miles to the north-east.

Habitat. The edge of a dense mulga strip, a mallee and oak association, and open eucalypts.

Behaviour. Seen feeding on or near the ground, but flying frequently into a tree or shrub for a few moments (DG). Once in a large loose-knit party with *Colluricincla harmonica* (which attacked and chased it short distances several times), *P. rufiventris, Seisura inquieta, Rhipidura leucophrys, Ephthianura albifrons* and *E. aurifrons,* and *Artamus.*

Food. Insects, including ants and orthoptera.

Annual Cycle. None of the specimens had enlarged gonads. The one taken in April (79) was completing moult, while those in June (70) were in more or less worn plumage.

Impermanent Colours. *Legs* male black; females grey. *Iris* male red-brown; females dark brown. *Bill* male black; females olive grey and dark brown with base paler. *Mouth* pale and fleshy, edges of gape in one female dull yellow.

Geographical Variation. These specimens must belong to the eastern nominate race. The localities of British Museum specimens purporting to be from Western Australia are suspect.

Other Notes. It seems not to have been pointed out that females of this species are very like those of *P. pectoralis* from the same areas. In the hand they can best be distinguished by their more decidedly rufous wing-bends, under wing-coverts and axillaries; white not yellow circlet of feathers round the vent; pale lores; uniformly coloured throat; belly and flanks grey like the breast with a central whitish or pale rufous stripe, rather than uniform rufous; browner upperparts; and longer and more rounded rectrices. In the field their partly terrestrial habits may be the best distinguishing feature.

Pachycephala lanioides White-breasted Whistler

Specimens. Western Australia. Karratha (1) 29 – 30 May 66, 2 imm ♀. Peg's Beach (2) 3 June 66, 1 ♂, 1 imm ♀, 1 fl. Carnarvon (5) 14 – 16 June 66, 1 ♂, 1 imm ♀, 1 skel. Point Torment (30) 30 May – 12 June 68, 3 ♂ (42.0, 43.8 g), 1 ♀ (42.8 g), 1 imm ♀ (44.5 g), 2 fl (46.5, 46.5 g). Wyndham (39) 31 July – 8 Aug 68, 4 ♂ (39.0, 39.5, 40.7 g), 2 ♀, 4 imm ♂ (33.1, 35.3, 35.7, 37.4 g), 1 skel (36.0 g), 2 fl (36.7, 37.0 g). Northern Territory. Mount Roper (45) 28 Oct – 1 Nov 68, 4 ♂ (38.7, 40.8, 43.4, 44.3 g), 1 ♀ (38.7 g), 1 skel (39.5 g). Queensland. Moonlight Creek (47) 4 – 15 May 64, 2 ♂ (35.9, 36.7 g), 3 ♀ (29.9, 31.4, 33.7 g), 2 juv (20.4, 20.6 g).

Distribution & Abundance. Recorded only where collected, apparently commoner in the Kimberleys and Northern Territory (30, 39, 45) than in the midwest or Queensland (1, 2, 5, 47). All but the last locality are within the known range of the species (Mees, 1964), but the specimens from Moonlight Creek (47) are the only ones definitely

known from east of the Roper River. Not encountered in the mangroves near Darwin (41) and Inkerman (50).
Habitat. Always in mangroves, variable in form as for *P. melanura*.
Behaviour. Usually moving about on low branches or actually on the mud (feet and bill often muddy). Skulking but inquisitive, easily called up by collectors' whistles and squeaks. At Point Torment (30) dull-plumaged birds were much more common than adult males, and two or three of them would always appear before a single male, whereas at Wyndham (39) few such birds were seen (AH).

Adult males appeared to be holding territories, perching higher in the trees to sing (DJF). The territories at Wyndham (39) seemed to be no more than 50 yards long (BB), but those in the midwest and Queensland (1–5, 47) were probably several hundred yards long (CJOH, ICJG). The few interspecific associations noted seemed to result casually, from loose parties of other birds passing through territories held by this species.
Food. Crustacea, especially small crabs; less commonly insects, especially grasshoppers.
Annual Cycle. Two fully adult males on 4 and 6 May (47), four between 31 July and 8 August (39) and three on 28 October (45) had more or less enlarged testes; whereas three on 14 and 15 June (30, 2, 5), and one on 1 November (45) had not. Two of six females in adult plumage, on 4 May (47) and 6 August (39), had somewhat enlarged ovaries. Males were recorded as singing in May, June, August and November. A softly-trilling male moulting into adult plumage, and a female moulting into second-year plumage, were taken together on 30 May (30). An adult female was taken with two juveniles on 8 May (47).

The juveniles had growing wings, tails beginning to grow, and pin-feathers in their body-plumage. Their skulls were largely unpneumatised. A young bird was completing the moult into first-year plumage on 4 August (39), and others were in this plumage, more or less worn, on 29 and 30 May (1) and 8 August (39). Older young were moulting into second-year plumage on 30 May and 3 and 15 June (30, 2, 5). None of these young birds was recorded as having enlarged gonads or a completely pneumatised skull. Males which were moulting into the adult plumage on 30 May and 3 June (30, 2) had undeveloped testes, but the skull of the one recorded was fully pneumatised.

Most of the adults were moulting, or were in mixed worn and fresh plumage, with no correlation apparent between date and stage of moult. Only two females in worn plumage on 8 and 15 May (47) and one male in fairly fresh plumage on 12 June (30) show no sign of moult. Adults at Wyndham (39) between 31 July and 8 August seem to have been undergoing a pause in the moult: all were in mixed worn and fresh plumage, but only one has a few pin-feathers. No adult is recorded as having an incompletely pneumatised skull.
Impermanent Colours. *Legs* dark grey, sometimes bluish or brownish, to black; juvenile lilac tinged with grey or brown. *Iris* dark brown to chestnut or red; juvenile dull olivaceous. *Bill* black in adults, base of lower mandible greyish in younger birds; whole bill brown or horn with yellow gape in juveniles and some first-year birds. *Mouth* mainly black (though tongue apparently pale) in adults, especially males, pale areas greyish or suffused with red; bright orange-yellow in juveniles, fading in first-year birds.

Geographical Variation. Our material supports Mees (1964a) rather than Mayr (1954), in showing a major break between birds with brownish dull plumages southwest of the Eighty Mile Beach and birds with greyish ones with paler ochraceous underparts to the northeast. We follow his division into three subspecies, while suspecting that variation in the northeast (*lanioides* and *fretorum*) may prove to be irregular and local.

P. l. carnarvoni (1, 2, 5), Carnarvon to the Eighty Mile Beach, probably intergrading slightly with *lanioides*. Dull plumages brownish or tawny (with carotenoid) above, rich ochraceous below.

P. l. lanioides (30), known only from Broome to King Sound, but probably to be discovered on the northern coast of the Kimberley Division. Dull plumages grey or olivaceous above, much paler than *carnarvoni* below; but while one of our adult females from Point Torment (30) has the breast, the upper belly and flanks very grey, in the other they are decidedly buffy.

P. l. fretorum (39, 45, 47), Wyndham (39) in the northeastern Kimberleys to Karumba on the Queensland gulf coast. Our material suggests that the smaller size of *fretorum* will not prove consistent enough for subspecific separation, while the dull plumages show considerable variation in colour which may well be geographical. Two adult females from Wyndham (39) are intermediate between those from Point Torment (30), while the four first-year birds are still more variable. One adult female from Mount Roper (45) is closer to *carnarvoni* in its brownish and ochraceous coloration. Two adult females and a first-year bird from Moonlight Creek (47) are pale and sandy-coloured above and below, with shaft-streaking virtually confined to the upper breast.

Other Notes. The specimens from Moonlight Creek (47) make it probable that De Vis' lost types of *fretorum* did come from Kimberley, now Karumba, still further to the east (see Mayr, 1954, Mees, 1964a).

The juvenile plumage is white below with club-shaped fuscous shaft-streaks, and the feathers of the back are pale grey to whitish basally and pale rufous terminally, with similar streaks. The edges of the secondaries and most of the upper wing-coverts are rufous, but those of the major primary-coverts and alula are narrower and more ochraceous, while the very narrow outer edges of the primaries are almost white.

A first-year female from Karratha (1) has some breast feathers irregularly blotched with deep rufous. As in *P. rufiventris,* birds in the dull plumages tend to have a faint suffusion of carotenoid pigment dorsally, especially towards the upper tail-coverts, while an adult male from Peg's Beach (2) has one decidedly olive covert. Both these traits tend to link the species with the *pectoralis* superspecies.

As in *P. rufiventris,* the first and second-year plumages are not clearly differentiated.

Pachycephala simplex — Brown Whistler

Specimens. NORTHERN TERRITORY. Darwin (41) 11 – 21 Sept 68, 3 ♂ (18.0, 18.1 g), 3 ♀ (16.2, 17.1, 24.8 g), 2 imm ♂ (18.0, 18.4 g), 1 imm ♀? (16.0 g), 1 imm ø (16.8 g), 2 ♂ skel (18.1, 18.6 g), 3 ♀ skel (16.8, 17.9, 19.6 g), 4 fl (17.0, 18.4, 20.3, 21.0 g). QUEENSLAND. Ayton (52) 20 – 29 June 64, 5 ♂ (17.2, 17.5, 17.6, 18.1 g), 4 ♀ (16.2, 16.4, 17.8, 18.0 g), 2 imm ♂ (15.5, 18.3 g), 1 imm ø (16.9 g). Tully (54) 11 – 25 July 64, 2 ♂ (16.1, 17.2 g), 2 ♀ (15.5, 16.0 g), 1 imm ♂ (14.9 g), 1 imm ♀ (15.4 g).

Distribution & Abundance. Encountered only at Hope Inlet near Darwin (41) and in the main rainforest block of north Queensland, where it was very common. In the latter it was only collected in the lowlands (2, 4), but as a common lowland species may have been ignored during the short excursions to higher altitudes (51, 53).
Habitat. At Hope Inlet (41) in mangroves and nearby monsoon forest. At Ayton (52) in dense rainforest, one in a patch surrounded by mangroves on a tiny island in the mouth of the Bloomfield River. At Tully (54) in tall rainforest, low dense rainforest, and *Casuarina* and other open forest.
Behaviour. At all heights, from the undergrowth to the inside of the canopy. Alone or in pairs, believed to be holding loose territories. In Queensland twice associated with *P. pectoralis* (once also with *Colluricincla megarhyncha*), and once in a mixed flock of *C. megarhyncha, Monarcha trivirgata, Rhipidura rufifrons, Tregellasia capito, Gerygone palpebrosa* and *Ptiloris victoriae*.
Food. Insects, including grasshoppers, beetles, flies, caterpillars and other larvae; a spider and a small seed.
Annual Cycle. Between 20 June and 25 July (52, 54) the gonads of most birds in adult plumage were slightly developed, and between 10 and 19 September (41) they were mostly further developed – those of two males and one female considerably so.

Of two young birds in first-year plumage on 14 and 18 July (54), one retained a single juvenile inner secondary but had a completely pneumatised skull (the skull of the other was damaged but probably not completely pneumatised). Three between 10 and 20 September (41) were still in various stages of moult from juvenile into first-year plumage, while one had completed this moult: their skulls were incompletely pneumatised, in rough correlation with their stages of moult. Three between 21 and 27 June (52) were in various stages of moult into adult plumage: one skull was damaged, one bird largely in first-year plumage had the skull incompletely pneumatised, one largely in adult plumage had it completely pneumatised but soft. An adult at the same time and place also had a soft skull, and it was noted that the bony layers (especially the inner) were thin and the trabeculae incomplete.

Twelve adults between 20 June and 25 July (52, 54) were in moderately worn plumage. Five between 10 and 12 September (41) were in very worn plumage, and a sixth was in moult.
Impermanent Colours. *Legs* pale to very dark grey, sometimes bluish – apparently more purple in Queensland (52, 54); blackish brown in one adult (54), olive-brown in one first-year bird (41): soles yellowish buff. *Iris* various hues and tones of brown. *Bill* black, occasionally brownish or with greyish base to lower mandible; brownish with pale lower mandible in some young birds. *Mouth and tongue* pale, often pinkish or purplish, with or without variable amounts of grey or blackish spreading from tomia and palate (no regularities of geography or sex); yellowish to orange, sometimes including gape, in some young birds. Adult coloration of the bill and mouth is apparently attained during the first year in the Northern Territory (41) and during the second in Queensland (52, 54).
Geographical Variation. The skins agree with the two distinctive subspecies recognised by Mayr (1954):

P. s. simplex in the Northern Territory (41), without yellow carotenoid pigment, so that the upperparts are brownish and the underparts whitish.

P. s. peninsulae in northern Queensland (52, 54), with carotenoid pigment except on head, so that the upperparts are olive and the underparts pale yellow.

Other Notes. The juvenile plumage, like that of *P. melanura* and the Australian forms of *P. pectoralis,* is wholly rufous (except for the rectrices and remiges, which are those retained during the first year). The rufous is dark dorsally and on the upper breast, very pale on the under tail-coverts. The moult into the first-year plumage involves the same feathers as in *P. pectoralis.*

Schodde & Hitchcock's (1968) conclusion – that *simplex* should be separated specifically from *peninsulae* (=*inornata*) and the New Guinea forms of what they continue to call *P. griseiceps* – is based in part on what are apparently the very specimens needed to confirm the opposite conclusion. Their two females from Brown River, which they regard as *dubia,* clearly have the carotenoid pigmentation that *dubia* lacks. Brown River lies between Galley Reach and the upper Laloki River, the area in which Mayr & Rand (1937: 175) suggested that *dubia* would prove to intergrade with the carotenoidal *perneglecta,* as it was already known to do with *jobiensis* on the north coast. Thus *dubia* is unquestionably conspecific with the *griseiceps* group. It shares with *simplex* the lack of carotenoid which Schodde & Hitchcock emphasise, whose geographical distribution should suggest that this is a labile character. In its melanic pigmentation *dubia* differs much more from the *griseiceps* group than *simplex* does, while the differences between *simplex* and *peninsulae* are small compared to those between conspecific females of the *P. pectoralis* superspecies, even within Australia.

Parasites. See Appendix 1.

Grey Shrike-thrush

Colluricincla harmonica (including *brunnea* and *rufiventris*)

Specimens. WESTERN AUSTRALIA. Coolawanyah (4) 20 May 66, 1 ♂. Metricup (9) 6 Mar 66, 1 fl. Scott River (11) 23 – 24 Feb 66, 1 ♂ skel, 2 fl. Mount Barker (12) 21 Mar 66, 1 ♀ skel. Newdegate (14) 17 Apr 66, 1 ♂. Eucla (17) 1 – 5 July 65, 1 ♀, 1 imm ♀, 1 imm ø. Mount Shenton (18) 2 Aug 65, 1 imm ♀. Mungilli Claypan (20) 13 Sept 65, 1 ♂. Lake Gruszka (21) 11 Sept 65, 1 ♀, 1 imm ♀ (60 g). Warburton (23) 10 – 14 Aug 65, 2 imm ♀, 1 ♂ skel. Van der Linden Lakes (24) 6 Sept 65, 1 ♀. Giles (25) 20 – 27 Aug 65, 2 ♂, 1 ♀, 1 imm ♀. Walter James Range (26) 26 Aug 65, 1 ♂ (58 g), 1 ♀, 1 ♀ skel. Mount Anderson (28) 27 Apr – 17 May 68, 2 ♂, 2 fl. Mount Bell (32) 19 – 21 June 68, 3 ♂ (64.1, 66.2, 67.2 g), 1 imm ♂? (50.7 g). Joint Hill (34) 28 June – 4 July 68, 2 imm ♀ (57.3 g). Mount Elizabeth (35) 8 – 10 July 68, 2 ♂ (61.4, 65.0 g), 1 ♀ (64.5 g), 1 imm ♀ (69.6 g), 1 fl (62.3 g). Mount Remarkable (37) 26 July 68, 1 ♂. Lissadell Hill (38) 28 – 29 July 68, 1 ♂ (58.3 g), 1 ♀ (62.3 g), 1 imm ♀ (55.8 g), 1 fl (65.1 g). Wyndham (39) 6 Aug 68, 1 fl (65.2 g). NORTHERN TERRITORY. Oenpelli (43) 12 Oct 68, 1 ♀ (58.5 g). Pine Creek (44) 29 Aug – 2 Sept 68, 2 fl (61.2, 67.6 g). Mount Roper (45) 1 Nov 68, 1 imm ♀ (55.6 g). Borroloola (46) 8 – 14 Nov 68, 1 imm ♂ (51.6 g), 1 ♂ skel. QUEENSLAND. Tully (54) 18 July 64, 1 ♀ (58.0 g). Tyrone (59) 15 May 63, 1 ♀, 1 imm ♂. SOUTH AUSTRALIA. Innamincka (64) 3 May 65, 1 ♀ (48.5 g). Oodnadatta (67) 31 May 65, 2 ♀. Granite Downs (68) 10 June 65, 1 ø (♂ by plumage). Nullarbor HS (69) 4 – 8 Oct

63, 27 – 30 Oct 64, 3 ♂ (1+trunk skel), 1 ♀ (+trunk skel). Kingoonya (70) 20 – 23 June 65, 2 ♂, 1 imm ♂. Nonning (79) 29 Mar – 1 Apr 65, 3 ♂ (62.7 g), 2 ♀ (63.1, 65.5 g). Aroona Valley (82) 9 – 12 Apr 65, 1 ♂ (59.0 g), 2 imm ♀ (52.5, 59.2 g). Angepena (84) 28 Nov 63, 1 ♂ (+ trunk skel). VICTORIA. Victoria Range (90) 31 Jan 63, 1 fl. NEW SOUTH WALES. Dunedoo (93) 18 – 24 Mar 63, 1 ♂, 1 imm ♀, 1 skel, 1 fl. Grafton (94) 1 – 4 Apr 63, 2 fl. Tenterfield (95) 15 Apr 63, 1 ♀.

Specimens to other Museums. AM 1 ♀ Tyrone (59). NMV 1 skel Dunedoo (93). WAM 1 ♂ Lake Gruszka (21), holotype of *serventyi*.

Distribution & Abundance. Also at Chichester Range (3), Green Range (13), Cocklebiddy (15), Carnegie (19), St George's Range (27), Derby (29), Geikie Gorge (31), and Glendower (55). Recorded as very common at Giles (25), common at Eucla (17), Mount Remarkable (37) and Nonning (79), fairly common at Warburton (23), Innamincka (64) and Oodnadatta (67), while several were seen at Carnegie (19), a few at Cocklebiddy (19) and Granite Downs (68), and one at Mount Shenton (18).

The northern and southern populations seem to be connected (see Parker, 1970), as would be expected from the vegetation map (Williams, 1955).

Habitat. All types of habitat with trees or shrubs, from rainforest (54) to spinifex grassland with sparse scrub (4) and Desert Oaks on sandhills (24); and once on bare granite ledges (38). Twice in mangroves at the edge of savanna woodland (39, 45). Often near water.

Behaviour. Though most of the specimens were taken in trees and bushes, at all heights, the species was noted as feeding much on the ground under them, very like many European thrushes (*Turdus*) which it resembles much more in life than in the hand, and giving upward flirts of the tail like those of *T. merula* and *T. pilaris* though less emphatic (DG). Usually solitary, sometimes in pairs or occasionally small parties of up to four. A few records in mixed parties with various species, including *Aphelocephala, Microeca leucophaea, Climacteris affinis, Philemon citreogularis* and *Taeniopygia guttata*.

The melodious song and whistled calls were recorded frequently from adults of both sexes and immature birds, and resemblances to those of *Pachycephala rufiventris* were noted by several observers.

Food. Mainly insects, especially orthoptera (including grasshoppers, crickets, mantises and stick-insects), beetles and caterpillars, but also including ants, ichneumon flies and froghoppers. Vegetable matter, including seeds (once the principal contents), berries and once short pieces of grass stem, was recorded in a minority of gizzards. An immature female was seen searching for insects under bark (93) (RB).

Annual Cycle. Adult males were taken with testes slightly enlarged on 19 and 24 March (93), 20 May (4), 8 and 10 July (35) and 26 and 27 August (26, 25); and further enlarged on 11 and 13 September and 14 November (21, 20, 46). The one on 8 July was apparently paired with an adult female whose ovary was undeveloped. (Both were in a state of halted moult, see below.) A female gathering nesting material on 1 April (79) had a small ovary and was completing its moult. The adult male on 10 July was with two young birds which were in halted moult from first year into adult plumage; and another on 28 Nov (84) was with one completing its moult into first-year plumage.

Young birds were completing the moult into first-year plumage on 18 March (93), 9 and 10 April (82), 15 May (59), 28 July (38), 5 October (69 – examined by GSC) and 28 November (84); were in first-year plumage on 4 July (34) and 2 August (18); were in moult into adult plumage on 21 and 23 June (32, 70) and 1 July (17); were in body-moult with moult of the flight feathers halted on 10 August (23) and 11 September (21); and were at a halt in moult on 28 June (34), 5 and 10 July (17, 35), 14 and 23 August (23, 25) and 1 November (45). Most of them had the skull recorded as pneumatised, the only one shown as incompletely pneumatised being the female in moult into the adult plumage on 11 September which had large 'windows'. On the other hand, the skulls of two adults were recorded as incompletely pneumatised: a female with pneumatisation almost complete on 31 May (67), and a male recorded as showing no pneumatisation on 4 May (28) though in the skin the skull feels rigid.

Adults were in moult on 19, 29 and 30 March (93, 79), 1, 15 and 27 April (79, 95, 28), 15 and 31 May (59, 67) and 14 November (46 – reported by collector). Of these, the one on 27 April was in an early and the remainder in late stages of moult. Others on 4 May (28), 28 July (38) and 30 October (69) showed replacement in parts of the plumage, while other parts were in the halted stage now to be discussed.

In many of these specimens a group of secondaries is more or less clearly fresher than the outer and inner ones, though there is no sign of feather growth. The secondaries involved are the fourth-from-innermost and a variable number outwards, which seem to have been moulted more recently than the neighbouring secondaries and primaries. The fact that these are the remiges best shielded from wear and fading makes this difficult to determine in fully adult birds; but when the worn remiges belong to the first-year plumage, the difference in colour and sometimes in shape makes it difficult to deny that the central secondaries do belong to a later feather generation. The simplest, though unexpected, explanation for such a pattern of wear seems to be that a group of secondaries is often moulted and replaced some time before the rest of the wing. This state of halted moult is shown by two-thirds of the adults which were not actively moulting, from Oenpelli (43), the Walter James Range (26) and Nullarbor HS (69) westwards, in every month from May to October.

Impermanent Colours. *Legs* pale to usually dark grey (often bluish but occasionally brownish or olivaceous) or black. *Iris* brown, usually dark and/or reddish; but light at the Walter James Range (26) and in nearby individuals (21, 25), and occasionally red in the northwest (37, 39, 43-45). *Bill* black in adult males (rarely greyish); others usually grey (especially adult females), horn or brownish – but individuals including females moulting from juvenile plumage had the upper mandible (82, 93) or whole bill (12, 17, 59, 64) black. *Mouth* black or occasionally grey in adult males, grey or occasionally blackish in adult females, tinged with pink or yellow in some young birds – one adult female yellow (95).

Geographical Variation. Individual differences, the effects of wear, and the inadequacy of comparative material, make it unsatisfactory to divide this material more finely than as follows:

C. h. rufiventris Southern areas of Western Australia and western South Australia (14, 17, 67, 69-70, 79). Dark and grey with buff on the underparts restricted, without a superciliary stripe. The three adult males from Nullarbor HS (69) agree with females, and with one in the British Museum from Albany, in having shaft-streaks on throat and breast. The female from Oodnadatta (67) and first-year male from Kingoonya (70), regarded as *serventyi* by Macdonald (1968a), show mixed characters, while there are normal *rufiventris* from both localities.

C. h. serventyi Macdonald (1968a). Middle latitudes of Western Australia and north-western South Australia (4, 18, 20-21, 23-26, 68), and much of Northern Territory (Parker, 1970). Greys paler and browner and replaced by buff higher on the breast, females and first-year birds with a buff superciliary running over the loral spot. It is interesting that the narrow neck of stony desert around Lake Breaden seems to present no barrier.

C. h. brunnea. The northern parts of Western Australia and Northern Territory (28, 32, 34-35, 37-38, 43, 45-46). Larger, still paler and browner, little buff on adults especially males, streaking in males confined to throat and upper breast, superciliary deeper buff. The two males from Mount Anderson (28) show an approach to *serventyi* in smaller size and slightly deeper buff below.

C. h. harmonica (sensu lato). Queensland, eastern South Australia, and New South Wales (54, 59, 64, 82, 84, 93, 95) and Victoria. Contrast between brown mantle and grey head and rump, especially in males, increasing southwards; shaft-streaking of females and first-year birds very distinct and extending to upper belly; throat whitish: little or no buff in adult under tail-coverts and mid-belly; superciliary deep buff in first-year birds, whitish or obsolete in females. The females from Tyrone (59) and Innamincka (64), assigned to *anda* by Macdonald, are not pale. Those from Aroona Valley (82), which he mentions variously as subadult and adult, are completing the moult into first-year plumage, in which the buff edges to their dark throat feathers may not be unusual.

It is not clear whether *superciliosa* of the Cape York Peninsula is properly associated with *brunnea,* represents merely the end of a cline from *harmonica,* or is a distinct form in its own right. It is distinctive in the white superciliary of the male, while a first-year female from the Olive River, described by Macdonald, is like those of *harmonica* but with the ventral grey replaced by warm buff. The female from Tully (54) is little different from other northern *harmonica.*

Differences between the successive forms *rufiventris, serventyi* and *brunnea* are not profound. Traces of intergradation, whether primary or secondary, are apparent (67 & 70, 28), and there is probably a continuous series of populations northwards from the Eyre Peninsula, between the Gibson and Simpson Deserts, to the Northern Territory coast. The situation in northern Queensland is not clear, but there is nothing to suggest that the quite similar forms involved behave as distinct species.

In South Australia on the other hand, the markedly different *harmonica* and *rufiventris* seem to approach one another closely without reported intergradation (e.g., Terrill & Rix, 1950, Condon, 1951a, 1962a, 1968, Mayr in Peters', 1967, Macdonald, 1968a). Unfortunately the only locality near the presumed boundary which has been cited is Aroona Valley (82) by Macdonald. The greater part of the long potential boundary between the eastern and western forms is occupied by barriers of desert, treeless lakebeds and sea, leaving only two corridors each less than 50 miles wide – from the Simpson Desert and Lake Eyre to Lake Torrens, and from Lake Torrens to Spencer Gulf. Aroona Valley (82) is behind Lake Torrens, some 60 miles from the southern corridor, where it should be well shielded from gene-flow. The effects of the physical barriers would be reinforced by any ecological differences between the two forms, and by selection acting to restrict the spread of hybrid genotypes. Until the interactions of the two forms near the corridors have been critically studied, it is certainly premature to suggest the existence of an intrinsic isolating mechanism.

Other Notes. Reasons for following Mayr (in Peters', 1967) and most recent systematists, in regarding *harmonica, brunnea* and *rufiventris* as conspecific, are discussed above.

Discussion of the variation in this species by Macdonald (1968a) is to some extent invalidated by failure to distinguish the first-year plumage. The characters of the first-year flight feathers are as in *Pachycephala,* and the partial moult of the wing when this plumage is assumed seems to take the same course (see under *P. pectoralis*). The body plumage is much like that of the adult female but with tendencies (variably expressed in different parts of the range) to be more distinctly streaked and buffy below and to have deeper buff superciliaries. The only soft juvenile body plumage I have seen (on a young *harmonica* from Tea Tree Gully, South Australia) is whitish below with very broad brown shaft-streaks, and grey above.

The only reliably sexed males I have examined which closely resemble adult females are the whole series from Nullarbor HS (69), while none is moulting from the first-year plumage into one which can be identified as 'male' or 'female'. The adult male plumage may therefore represent either an additional plumage acquired in the third year as in *Pachycephala,* or (as seems more probable) the homologue of the adult female plumage.

The remiges of this species are much subject to fading, evidently by light, so that the exposed tips and edges are very pale, contrasting sharply with those parts of the feathers which underlie their neighbours.

Parasites. See Appendix 1.

Colluricincla woodwardi — Brown-breasted Shrike-thrush

Specimens. Western Australia. Mount Caroline (33) 25 July 68, 1 ♀ (57.0 g). Joint Hill (34) 3 – 4 July 68, 2 ♂ (52.0 g). Mount Elizabeth (35) 15 July 68, 1 ♂ (60.0 g). Northern Territory. Oenpelli (43) 17 Oct 68, 1 ♀ (47.1 g), 1 juv ♂ (55.0 g). Mount Roper (45) 1 Nov 68, 1 ♂ (49.7 g), 1 ♀ (49.7 g), 1 imm ♂ (52.5 g). Borroloola (46) 9 Nov 68, 1 ♀ (51.8 g).

Distribution & Abundance. Also recorded at Geikie Gorge (31) and Victoria River (40). Victoria River (40) and Mount Roper (45) considerably extend the range of the Arnhem Land populations beyond the limits given by Storr (1967). Few recorded at Oenpelli (43) and Mount Roper (45), but fairly numerous in its specialised habitat at Borroloola (46).

Habitat. Invariably in very rugged sandstone country, with trees, bushes and spinifex. Nearly always in river gorges near fresh water, but at Mount Roper (45) the small sandstone outcrop was a mile from the tidal reaches of the Roper River.

Behaviour. Unlike other *Colluricincla,* this species spends almost all its time on and among the rocks, and often flees on foot (DJF). Very furtive and shy, but with an exceptionally loud and beautiful song which rings among the cliffs – compared to those of the North African Blue Rock Thrush *Monticola solitarius* (BB) and the European Nightingale *Luscinia megarhynchos* (DJF). Reported singly and in groups of two or three.

Food. Insects, especially grasshoppers, and a skink.

Annual Cycle. A nest containing two or three young was found on 9 November (46), and the young had gone by the 12th. The nest was on a ledge 20 ft up a south-facing vertical sandstone face under a slight overhang, where it was reached by the sun only after 17.00 hrs. The rim of the fairly deep cup was almost flush with the rock surface. The nest was fairly well woven of grasses, some of which protruded at tangents to the rim (AH).

A juvenile beginning body-moult into the first-year plumage on 17 October (43) was about 30 yards from two birds in adult plumage. It is the only specimen whose skull was not fully pneumatised, having four small 'windows'. A male on 1 November (45) was in what may be first-year plumage, much worn.

Adults were in rather fresh plumage on 25 July and 3 July (33-34), while all those on 1 and 9 November (45-46) were very worn. A female on 15 July (35) was in somewhat worn plumage, beginning to moult on the nape.

Impermanent Colours. *Legs* dark grey or brown to black. *Iris* brown – dark, reddish or golden. *Bill* black in most adult males, grey with the upper mandible darker and sometimes brownish in females, brown in the juvenile with the upper mandible darker. *Mouth* pinkish purple to black with the males averaging darker, juvenile flesh pink with yellow gape. Bare patch behind the eye recorded once as grey.

Geographical Variation. Though our material includes only one adult female from Western Australia and one adult male (much worn) from the Northern Territory, there are indications that the populations are distinguishable, with *C. w. assimilis* of the eastern Kimberleys (33-35) having a larger and whiter throat patch, darker and greyer breast and richer rufous underparts than *C. w. woodwardi* of Northern Territory (43, 45-46). The female from Borroloola (46) *'didimus'* agrees well with those of *woodwardi,* but may have been deeper rufous ventrally before fading.

Other Notes. Its terrestrial habits, broad rounded rectrices, long upper tail-coverts, and more extensive bare patch behind the eye, distinguish this species from other *Colluricincla* and from *Pachycephala.*

The juvenile (and therefore first-year) flight-feathers differ less than those of other *Colluricincla* and *Pachycephala* from their adult homologues. The rufous outer edges of the secondaries and upper wing-coverts are dark, and so narrow that their replacement by drab edges in the adult specimens could be due to wear and fading, while the centres of the feathers are dark. The rectrices differ in being narrower with more parallel edges (whereas the wide inner webs of adults tend to narrow proximally), rather than more pointed. The juvenile body plumage is soft and paler than the adult, but otherwise similar in colour – except that only the upper throat is markedly paler than the breast, and is buff; while instead of being grey and buff with distinct dark shaft-streaks, the breast feathers are fuscous with vague buff edges. The male (45) provisionally identified as in first-year plumage is very worn. The remains of its rectrices are like those of worn adults, but the outer edges of its central secondaries have a deep rufous tinge, while its throat and breast are almost concolorous with its belly apart from their dark shaft-streaks. Its underparts, especially the thighs and under tail-coverts, are deeper rufous than those of a similarly worn adult male.

Parasites. See Appendix 1.

Colluricincla parvula — Little Shrike-thrush

Specimens. Northern Territory. Darwin (41) 7 – 18 Sept 68, 3 ♂ (40.8, 42.6, 42.8 g), 1 fl. Oenpelli (43) 2 Oct 68, 1 ♂. Mount Roper (45) 28 Oct – 1 Nov 68, 2 ♀ (38.8, 42.1 g), 1 imm ♀ (36.6 g), 1 ♂ skel, 1 ♀ skel.

Distribution & Abundance. Recorded also from Geikie Gorge (31). Fairly numerous in its habitats. Mount Roper (45) extends the known range over 200 miles south from Yirrkala (Storr, 1967).

Habitat. Mangroves, often overhung by forest, on coast (41) or along river (45), dense monsoon vegetation in gorges (43).

Behaviour. High (15-30 ft) in mangroves at Hope Inlet, Darwin (41) but low and often on the mud at Mount Roper (45) (BB). Apparently skulking and revealed mainly by its loud song, but responding to collectors' squeaks (as recorded by Deignan, 1964). Solitary or in groups of two, but with other individuals often close by.

Food. Insects, including grasshoppers and beetles, once with seeds about 2 mm across, once nothing but vegetable matter.

Annual Cycle. Males had somewhat enlarged testes on 7 and 18 September (41) and 1 November (45), the two last being apparently paired. A young bird was in first-year plumage, not yet complete, on 28 October (45). It was the only specimen with an incompletely pneumatised skull. No birds were in heavily worn plumage, nor in wing or tail moult, but some had a few pin-feathers in nape or throat on 7 and 18 September (41, the males with enlarged testes), 2 and 28 October (43, 45).

Impermanent Colours. *Legs* blue to blue-grey or purple-grey, once grey brown; soles yellowish. *Iris* brown, mostly dark, once reddish; first-year mid-brown. *Bill* grey-brown to black in males; greyish or blue-grey with whitish or flesh-pink mandible in females; first-year brown with pale blue-grey mandible. *Mouth* blue-black (once pinkish-purple) in males; flesh pink or bluish in females; first-year tinged pale blue-grey with cream tongue.

Geographical Variation. Deignan (1964) denies that the species varies geographically, though Mayr (in Peters', 1967) recognises *conigravi* from Admiralty Gulf in the Kimberleys. Our series of skins from Darwin (41) and Mount Roper (45) are all males and all females respectively.

Other Notes. Sexual dimorphism is rather conspicuous ventrally: the white throats and mid-bellies of the males contrast with their deep buff breasts and flanks; thereas the throat-patches of the females are smaller and washed with buff, and their concolorous flanks and bellies contrast only slightly with darker and greyer gorgets.

The first-year female is somewhat paler than adults, and has somewhat warmer outer edges to the remiges and decidedly more rufous upper wing-coverts though without clearly-marked rufous edges. Its rectrices are of the characteristic pointed shape with more flexible shafts. Patches of soft dull rufous plumage behind the eyes, and the fact that none of the secondaries or upper coverts are of the adult coloration, indicate that this bird had not completed its moult from the juvenile plumage.

This appears to be the geographical representative of *C. megarhyncha*.

Colluricincla megarhyncha Rufous Shrike-thrush

Specimens. QUEENSLAND. Inkerman (50) 28 May – 11 June 64, 5 ♂ (33.5, 34.4, 35.4, 35.8, 38.5g), 4 ♀ (33.2, 34.0, 35.4, 37.7 g), 2 imm ♀ (30.1, 32.5 g). Ayton (52) 20 – 24 June 64, 1 ♀ (35.3 g), 1 imm ø (31.8 g). Tully (54) 11 – 25 July 64, 2 ♂ (30.9, 33.4 g). Cape Gloucester (56) 8 Aug 64, 1 ♂ (36.4 g). Mount Dryander (57) 3 – 11 Aug 64, 2 ♂ (37.6, 39.7 g).

Distribution & Abundance. Encountered also on Big Tableland (53) and probably in the Walter Hill Range (53), and thus at all localities on the east coast of northern Queensland, as well as at Inkerman (50). Moderately common in appropriate habitats.

Habitat. At Inkerman (50) only in mangroves at the mouths of rivers and creeks. On the east coast in forest or dense tropical layered woodland.

Behaviour. Usually found low in trees or undergrowth, or on the ground; silent or calling harshly, once a short rich warbling song. Solitary or in twos. Two adult females and two *Meliphaga unicolor* were chasing *Zosterops lutea* with rasping whistles (50). One in a party of *Ptiloris victoriae* (51), another in a feeding party of *Meliphaga lewinii* and *Zosterops lateralis* (57).

Food. Insects, apparently usually small, but including a small mantis, a large locust, beetles and a grub 40 mm long; and a very small crab 6 mm across.

Annual Cycle. Males from 3 to 11 August had somewhat enlarged testes (56, 57). No sign of breeding was noted. A discarded specimen was recorded as an immature in moult on 1 June, and two were in somewhat worn first-year plumage, one with a few pin-feathers in the crown, on 9 and 11 June (50). Another was completing the moult into this plumage on 24 June (52). These all had incompletely pneumatised skulls, in contrast to the adults except for a female with thick but single-layered 'windows' on 29 May and a male with small 'windows' behind the eyes on 9 June (50).

Adults between 28 May and 20 June were in moult, with moult of the flight feathers mostly completed by 8 June. The first taken had impermanent colours which suggest youth.

Impermanent Colours. *Legs* dull tones of purple; first-year whitish to pale grey, tinged with pink or purple. *Iris* brown, usually rather pale and yellowish or reddish; first-year olive-grey to darker brown. *Bill* very variable, whitish tinged with purple, pink or brown to blackish-brown, upper mandible more or less darker than lower; all first-year upper mandibles in shades of brown, one with bright yellow gape. *Mouth* pale, tinged purple, or pink, sometimes yellow or brown; first-year dull to bright yellow. *Lower eyelid* pale grey or grey-brown.

Geographical Variation. The specimens fall into three distinct groups, which according to the ranges given by Mayr (in Peters', 1967) presumably take the following names:

C. m. normani Inkerman (50). Pale ventrally, with the throat not much paler than the belly, and greyish dorsally.

C. m. parvissima Ayton (52) and Tully (54). Much deeper rufous ventrally, with the throat-patch smaller and more contrasting, and browner dorsally with outer edges of the remiges decidedly rufous. Old specimens suggest that populations on islands in the Torres Strait and Dunk Island (*griseata*) are very similar, and that those of Cape York itself agree with the latter in being slightly paler.

C. m. gouldii Cape Gloucester (56) and Mount Dryander (57). Like *normani* but with a white throat-patch and more distinct grey shaft-streaks in the lower throat, and slightly more olivaceous above. Very distinct from *parvissima,* from which it is separated by the Burdekin Gap (Keast, 1961).

Other Notes. In the Inkerman (50) series the females average distinctly paler ventrally than the males, though with some overlap. Adults vary in the amount of rufous in the wing (especially of *parvissima*) and the shape of the tail, so that a few are not easily separated from first-year birds.

Colluricincla boweri — Bower Shrike-thrush

Specimens. QUEENSLAND. Big Tableland (51) 2 July 64, 1 ♂ (47.9 g). Walter Hill Range (53) 23 July 64, 1 ♀ (43.0 g). Tully (54) 25 July 64, 1 ♀ (42.4 g).

Distribution & Abundance. These were the only specimens encountered.

Habitat. In rainforest. Though this is considered to be a mountain species, the specimen from Tully (54) was in the lowlands.

Behaviour. All were in the dense canopy and alone, except that one was near *Meliphaga frenata* and other honeyeaters. One seemed to be noisily dropping fruit to the ground, though its stomach contained only insect remains. No voice was recorded.

Food. Insects, including beetle elytra, a large ant and a larva.

Annual Cycle. All three specimens were in fresh adult plumage, with fully pneumatised skulls and small gonads.

Impermanent Colours. *Legs* slate grey, female more purplish, soles ochraceous. *Iris* red-brown, male lighter. *Bill* blackish-brown to black, female greyish basally. *Mouth* pale grey, mixed with white in male, one female purplish. These apparent differences between the sexes may be merely individual.

Other Notes. The adult male and female plumages seem to differ, though no other sexed adult was available to supplement this inadequate sample. The male lacks the rufous superciliary stripe and the circlet round the eye (which is pale beneath), these feathers being dark grey. The edges of its remiges are warm olivaceous, not bright rufous. A female in what appears to be first-year plumage (BM(NH) 1901.3.20.105) agrees with the adult females, except that its flight feathers are softer and paler, and its rectrices more pointed.

FALCUNCULIDAE — Shrike-tits
I. C. J. Galbraith

Falcunculus frontatus Shrike-tit

Specimens. Western Australia. Scott River (11) 28 Feb 66, 1 ♂. Mount Barker (12) 21 Mar 66, 1 ♂. Victoria. Coleraine (89) 21 Jan 63, 1 imm ♀. Victoria Range (90) 30 – 31 Jan 63, 1 imm ♀ (sexed by plumage), 1 fl. New South Wales. Albury (92) 11 Mar 63, 1 ♂ (sexed by plumage), 1 ♂ skel. Grafton (94) 5 Apr 63, 1 ♂.

Field Notes. Apparently not encountered elsewhere, and nowhere common.

Seen at all heights in trees, in sclerophyll forest, woodland and lightly timbered country. Usually foraging in or under bark, once tearing loose bark from trunks. Singly or in parties of up to four.

Food. Insects.

Annual Cycle. No indications of breeding. One of the immature females taken in January (89) seems to be in moult from the first-year to the adult plumage. The other (90) shows juvenile feathers in nape, scapulars and throat; but apart from juvenile innermost secondaries (which had been replaced in the first) its wing feathers seem to be adult.

Adults from 28 February to 21 March were in moult, that on 5 April in fresh plumage.

Impermanent Colours. *Legs* grey or blue-grey, soles whitish. *Iris* brown or red. *Bill* black. *Mouth* black, mottled, light brown or flesh.

Geographical Variation. These specimens fall into the two southern races recognised by Mayr (1953a):

F. f. frontatus of eastern Australia (89, 90, 92, 94), with yellow bellies and little yellow in wings and tail.

F. f. leucogaster of southwest Australia (11, 12) with white bellies and much yellow in wings and tail.

Oreoica gutturalis Crested Bell-bird

Specimens. Western Australia. Hunt Oil Camp (22) 9 Sept 65, 1 ♂ (63 g). Warburton (23) 11 – 13 Aug 65, 2 ♀, 1 imm ♂, 1 ♀ skel. Giles (25) 23 – 31 Aug 65, 1 ♂ (64 g), 1 ♀ skel. Queensland. Glendower (55) 19 Mar 64, 1 ♂ (63.1 g). Tyrone (59) 12 May 63, 1 ♂. South Australia. Oodnadatta (67) 31 May 65, 1 ♀. Granite Downs (68) 4 – 11 June 65, 1 ♂, 1 ♀ skel. Kingoonya (70) 17 – 27 June 65, 2 ♀, 1 fl. Yardea (72) 11 Dec 62, 1 ♀ skel. Nonning (79) 31 Mar – 3 Apr 65, 1 ♂ (66.8 g), 1 imm ♂ (sexed by plumage, 66.9 g), 1 ♀ (61.0 g).

Specimens to other Museums. SAM 1 ♀ Lake Acraman (71) 3 Nov 64.

Distribution & Abundance. Also recorded at Newdegate (14), Cocklebiddy (15), Mount Shenton (18), Carnegie (19), Mungilli Claypan (20), Lake Gruszka (21), and Van der Linden Lakes (24). Only one each seen at Cocklebiddy (15) and Glendower (55), recorded as fairly common at Mount Shenton (18), common at Warburton (23), very common at Carnegie (19) and Giles (25).

Habitat. In mulga country, sometimes among other types of shrubs and trees (22, 23,

25, 68 & 70), in various types of open woodland (55, 59, 67, 70, 79), and in low scrub (71, 72, 79).

Behaviour. Solitary or in pairs. Hopping on the ground, perched in bushes or trees, or flying between them. Two birds singing antiphonally, one giving *tik-tik-tik* the other coming in with a bell-like *tonk tonk* at the end; also a prolonged harsh chattering (DG).

Food. Seeds, including those from *Aristida* and *Cassia* species. Insects, including beetles, hairy caterpillars and an ichneumon.

Annual Cycle. Males had enlarged testes on 19 March (55), 12 May (59, not greatly enlarged) and 31 August (25). The first two were completing moult, while the last was apparently paired. A female on 31 May (67) and a male on 4 June (68) were apparently paired, and the former was singing antiphonally.

A male and female on 11 and 12 August (23) had extensively unpneumatised skulls (and the female an unconvoluted oviduct). Their plumage is worn, that of the male apparently the first-year plumage whereas that of the female is more like that of adults. An unsexed bird on 31 March (79) was moulting from what seems to be the male second-year plumage.

Adults with small gonads were in moult on 1 and 3 April (79), 27 June (70) and 3 November (71).

Impermanent Colours. *Legs* dark grey to greyish. *Iris* male orange-yellow to orange-red (one second-year yellow, one first-year dull yellow), female yellow to yellow-brown. *Bill* male black, female and first-year dark grey to dark brown (one black). *Mouth* male black, female and first and second years dirty white to bright yellow.

Geographical Variation. Variation among these specimens seems to be mainly individual, and none seems to belong to the poorly-marked northern race *pallescens* recognised by Mayr (1953b). The specimens from Queensland (55, 59) have distinctly olivaceous inner secondaries as noted by Mayr, but are almost equalled in this respect by individuals from South Australia.

Other Notes. The juvenile plumage, of a specimen in the British Museum, is more uniform in colour and more rufous than that of adult females, with the throat and breast concolorous and only the mid-belly markedly darker. Its flight feathers are browner, and there is less difference in length between the short outer secondaries and the long primaries and central secondaries. The male recorded as in first-year plumage (23) also has a dark throat, but its worn flight-feathers are like those of adult females. The specimen recorded as a second-year male (79) resembles adults, except that its gorget is dark grey rather than black, and its white throat and forehead patches are smaller.

Parasites. See Appendix 1.

Sphenostoma cristatum Wedgebill

Specimens. QUEENSLAND. Thylungra (60) 25 May 63, 1 ♀. SOUTH AUSTRALIA. Parachilna (81) 12 Apr 65, 1 ♂ (44.5 g), 1 ♀ (36.2 g). Wirrealpa (83) 15 Apr 65, 1 ♀ (37.4 g), 1 skel (40.1 g). Angepena (84) 18 – 27 Nov 63, 1 imm ♀, 1 ♀ (trunk skels preserved).

Specimens to other Museums. NMV 1 imm ♂ Thylungra (60).

Distribution & Abundance. The song of the species was constantly heard, and a single pair seen, at Bush Bay (6), while the species was also seen at Mundiwindi in mid-central Western Australia (CJOH) and at Copley, near Leigh Creek, central South Australia.
Behaviour. Singly or in parties of up to six, moving about on the ground with long hops, or skulking in bushes. Once in a party with *Aphelocephala* and *Taenopygia guttata*. One recorded as singing sweet and strong notes, *kitty lin-tol* repeated (PRC).
Food. Small seeds, once fragments of large ones, once other vegetable matter.
Annual Cycle. A male with enlarged testes on 12 April (81). A party of five birds, thought to be a family, containing a young bird and a female with small ovary, on 25 May (60). A young bird, apparently moulting into first-year plumage, on 18 November (84).

All the skin specimens examined were in moult (heavy, except for a female on 15 April (83) which was completing the moult), but the young bird from Thylungra (60) was noted as in fresh plumage.
Impermanent Colours. *Legs* grey, one adult male black. *Iris* brown or grey. *Bill* light brown to black. *Mouth* light yellow.
Geographical Variation. No subspecies are currently recognised (Deignan in Peters', 1964, Condon, 1968) and the present material does not reveal the purely clinal variation mentioned by Keast (1961).
Other Notes. The immature (first-year ?) plumage is softer than the adult, paler below with the throat whiter and the breast not markedly darker than the belly, more cinnamon especially on the edges of the secondaries and upper wing-coverts, and with the feathers of the crest broader, more square-cut and with less distinct shaft-streaks.

Psophodes olivaceus Eastern Whipbird

Specimens. QUEENSLAND. Big Tableland (51) 1 – 2 July 64, 4 ♂ (63.2, 65.7, 69.0, 73.4 g), 1 ♀ (58.5 g). Walter Hill Range (53) 21 – 23 July 64, 2 ♀ (48.1, 54.6 g), 1 imm ♀ (50.0 g).
Distribution & Abundance. The calls of this species were heard on all sides at these two localities, and at Tully (54).
Habitat. Always on the floor of rainforest, among dense undergrowth (Plate 9(a)).
Behaviour. Solitary or in pairs or threes, hopping while feeding on the ground or on logs, or running from thicket to thicket. A chattering call, and the characteristic accelerating crescendo of ventriloquial pulsing whistles, ending in an explosive whipcrack. The whistles and whipcrack were recorded from a female specimen. The immature bird was in company with a *Sericornis citreogularis*.
Food. Stomachs contained insect mash and fragments, including beetle elytrae and tiny pupae, and opilionids.
Annual Cycle. Three of the males had enlarged testes, while the oocytes of the adult females were somewhat enlarged. The young bird (in first-year plumage ?) had an incompletely pneumatised skull. The plumage of all the specimens was fairly fresh, except for somewhat worn tips to their tails, while two adults had single rectrices in sheath.

Impermanent Colours. *Legs* dark brown or sepia. *Iris* dark brown (dull to reddish or purplish). *Bill* black, two of three adult females with greyish gape. *Mouth* blackish near tomia, dull yellow and flesh-pink internally; immature greenish-yellow.
Geographical Variation. These specimens belong to the race *lateralis* of the Cooktown-Cardwell block of rainforest, slightly distinguished by differences in proportions and coloration (Keast, 1968) from those further south.

SITTIDAE — Nuthatches
G. S. Cowles

Neositta chrysoptera (including *leucocephala, pileata, leucoptera* and *striata*) Sittella
Specimens. Western Australia. Metricup (9) 10 Mar 66, 2 ♂. Scott River (11) 26 Feb 66, 1 fl. Mount Shenton (18) 2 – 3 Aug 65, 2 ♂ (13.5 g), 1 ♀. Carnegie (19) 16 Sept 65, 2 ♂, 1 ♀, 1 ♀ skel. Van der Linden Lakes (24) 6 Sept 65, 1 ♀, 1 ♂ skel. Giles (25) 27 Aug 65, 2 ♂, 1 ♀. Walter James Range (26) 26 Aug 65, 1 ♂, 1 ♂ fl. Mount Anderson (28) 12 Mar – 28 Apr 68, 3 ♂ (12.2, 12.3, 12.6 g), 2 ♀ (11.8 g), 1 ø, 1 ♀ fl (12.4 g). Derby (29) 25 May 68, 2 fl (11.4, 11.5 g). Mount Caroline (33) 25 June 68, 2 ♂ (12.0, 12.1 g), 1 fl (12.2 g). Mount Elizabeth (35) 10 – 13 July 68, 1 ♀ (11.8 g), 1 imm ♂ (12.2 g), 1 fl (12.8 g). Northern Territory. Pine Creek (44) 28 – 30 Aug 68, 3 ♂ (10.5, 11.5, 11.8 g), 1 ♀ (12.0 g), 1 imm ♂ (10.8 g), 1 skel (10.1 g), 1 fl (10.5 g). Borroloola (46) 9 Nov 68, 1 fl (11.6 g). Queensland. Moonlight Creek (47) 7 – 13 May 64, 4 ♂ (10.8, 11.0, 11.0, 11.1 g), 6 ♀ (10.1, 11.0, 11.2, 11.6, 11.8, 12.3 g), 1 imm ♂ (10.7 g). Wernadinga (48) 21 May 64, 5 ♂ (11.1, 11.2, 12.1, 12.2, 12.5 g), 3 ♀ (11.9, 12.0, 12.4 g), 2 ø (10.8, 12.1 g), 1 imm ♀ (12.4 g). Glenore (49) 13 – 18 Apr 64, 3 ♂ (9.8, 9.8, 10.3 g), 5 ♀ (10.5, 10.8, 10.8, 12.0, 12.4 g). Glendower (55) 12 – 17 Mar 64, 2 ♂ (11.2 g), 1 ♀ (11.2 g), 1 ø (11.9 g). Banoona (58) 30 Apr 63, 1 ♂, 1 ø. South Australia. Granite Downs (68) 5 June 65, 1 ♂, 2 ♀. Kingoonya (70) 19 June 65, 1 ♂, 2 ♀, 1 ø, 1 ♂ skel. Nonning (79) 1 – 3 Apr 65, 3 ♂ (13.4, 14.2, 14.5 g), 2 ♀ (13.4, 14.1 g). New South Wales. Albury (92) 11 Mar 63, 2 ♂, 1 ♀, 1 ♂ skel. Grafton (94) 1 Apr 63, 1 ♂, 1 ♀. Tenterfield (95) 12 Apr 63, 1 ♂.
Specimens to other Museums. NMV 1 ♀ Albury. SAM 1 ♂, 1 ♀, Minippa (73), 1 Nov 64.
Distribution & Abundance. Noticeably absent from the coastal camps except in the southwest at Metricup (9) and Scott River (11) and at Derby (29) in the northwest.

Noted as fairly common at Mount Shenton (18), Carnegie (19), Pine Creek (44), Kingoonya (70) and Nonning (79).
Habitat. Usually savanna woodland and tropical savanna woodland. Otherwise forest (9, 11, 95), open Mulga (18, 19), Desert Oaks in red sand-hill country (24).
Behaviour. Noisy feeding parties were seen moving about from tree to tree in groups of 2-12, 20 seen once (95). Cling to tree trunks at all angles; upon alighting in a tree sometimes at once turn and descend the trunk head first. Voice noted in *leucoptera* as a shrill chattering, a high pitched peeping or a soft *per-weet* or *tick-tick*. In *striata* the call has been noted as *tweety-tweet* repeated frequently. In *pileata trit-trit* was thought to be

the alarm call, while a high piping *psee-psee* was given in flight. These descriptions were made by different collectors and do not necessarily indicate subspecific differences.

Food. Insects.

Annual Cycle. The only positive sign of breeding was at Carnegie (19) where an egg was found in the oviduct of a bird collected on 16 September. Testes were slightly enlarged at Metricup (9) in March (in an immature bird); Glenore (49) in April; Mount Anderson (28) and Moonlight Creek (47) in May; Mount Caroline (33) in June; Giles (25) in August. In early March at Glendower (55) full moult was in progress and at Albury (92) moult was finishing. Specimens in immature or juvenile plumage were at Metricup (9) in March; Moonlight Creek (47) and Wernadinga (48) in May; Mount Elizabeth (35) in July; and Pine Creek (44) in August.

Impermanent Colours. *Legs* yellow, duller in young birds. *Iris* generally some shade of yellow or orange, or yellow with red inner ring (24, 25), browner in young birds and some females, but some geographical (or seasonal?) variation is apparent; in particular descriptions indicate that adults from Northern Territory (44, 46) and northwest Queensland (47, 48) have paler, less orange eyes than those from the Kimberleys (28-35) and in young birds they are greyish rather than brown, although all these birds belong to *N. c. leucoptera:* also birds of *N. c. pileata* from southern South Australia (70, 73, 79) seem to have redder eyes than those from central Australia. *Bill* varying geographically: in southeast *N. c. chrysoptera* (92) and intergrades (58), mainly brown with pink round nostrils: in west and centre *N. c. pileata* (9-26, 68-79), mainly brown with yellow base: in north *N. c. leucoptera, striata, magnirostris* adults mainly yellow with black or dark brown tip, immatures greyish; those members of *leucoptera* in which the iris is paler (44, 46, 47, 48) may possibly also have the bill slightly paler yellow at the base and browner at the tip. *Mouth* flesh, whitish or yellow. *Orbital skin* yellow or orange-pink (58); immatures grey.

Geographical Variation. Greenway (in Peters', 1967) has been followed, all races being considered to belong to the one species, *chrysoptera*. Our birds fall into the recognised races as follows:

N. c. chrysoptera. The three specimens from Table Top Mountain, Albury (92).

N. c. magnirostris. The specimens found in the open forest habitat at Glendower (55) fit the description of *magnirostris* (Ingram, 1908a, b) previously only known from Inkerman Station, eastern Queensland (not our Inkerman (50) on the Gulf of Carpentaria). Other specimens in the British Museum which appear to be similar are five from Prairie, 20 miles to the south east of Glendower (55), one from Torrens Creek east of Prairie and one "N. Australia" presented by Sir T. Mitchell in 1847. It would seem from the present data that *magnirostris* extends in a narrow band inland from Inkerman Station (approximately 50 miles SW of Townsville and about 10 miles from the Burdekin River), westward to just east of Hughenden at Glendower (55). Ingram gives the eye colour as cream but the Glendower (55) specimens have pale orange eyes.

N. c. pileata. Specimens from the southwest (9, 11) and central Western (18-26) and South Australia (68-79) are referable to *pileata* and are reasonably uniform in colour though showing some variation in eye colour (see above).

N. c. leucoptera. Specimens from the Kimberleys (28, 29, 33, 35), Northern Territory

(44, 46), and northwest Queensland (47, 48) show geographical variation in the colour of the eye and possibly the bill (see above) but all are referable to *leucoptera*. Birds from Wernadinga (48) have lighter backs than those from other localities but this is probably due to wear and some bleaching effect. At Pine Creek (44) the underparts are darker but this is probably due to soiling of the plumage since many of the trees in the area were blackened by fire. There is no sign of *leucoptera* having any contact with *striata* which is present at Glenore (49) about 80 miles to the east separated by the "Carpentaria Gap".

N. c. chrysoptera x *N. c. leucocephala*. The specimens from Banoona (58), Grafton (94) and Tenterfield (95) have been described by Macdonald (1969b). Their crowns range from greyish-brown to grey and they are thought to represent a zone of intergradation between the brown-headed *chrysoptera* and the white-headed *leucocephala*.

Other Notes. Immature birds have the feathers of the crown tipped with white giving a streaked appearance and differ from adults in the colour of the legs, bill, eye and orbital ring.

Parasites. See Appendix 1.

CLIMACTERIDAE — Australian Tree-creepers
C. J. O. Harrison

Climacteris picumnus (including *melanota*) Brown Tree-creeper

Specimens. QUEENSLAND. Glenore (49) 14 – 21 Apr 64, 1 ♂ (27.5 g), 2 ♀ (23.8, 25.0 g). Inkerman (50) 2 June 64, 1 ♂ (24.5 g), 1 ♀ (23.8 g). Glendower (55) 19 – 20 Mar 64, 3 ♂ (26.0, 27.3, 30.5 g). Tyrone (59) 12 May 63, 1 ♂, 3 ♀. Thylungra (60) 22 – 26 May 63, 1 ♀, 1 ø. SOUTH AUSTRALIA. Innamincka (64) 27 – 28 Apr 65, 1 ♂, 2 ♀ (27.1, 28.4 g). Aroona Valley (82) 8 Apr 63, 1 ♀. Wirrealpa (83) 15 Apr 65, 1 ♂ (30 g). Angepena (84) 29 Nov 63, 1 ♂, 1 ♀. VICTORIA. Coleraine (89) 19 Jan 63, 1 ♀, 1 imm ♂. NEW SOUTH WALES. Dunedoo (93) 18 – 25 Mar 63, 1 ♂, 4 ♀, 1 ♀ skel, 1 fl. Tenterfield (95) 10 – 15 Apr 63, 1 ♂, 1 ♀, 1 ♀ skel.

Specimens to other Museums. NMV 1 ♂, 1 ♀ 1 fl Dunedoo (93). AM 1 ♀ Tyrone (69).

Distribution & Abundance. Generally distributed in small numbers in suitable habitats. In addition to above localities also seen at Albury (92). Occurred singly, or in pairs, or (? family) parties of up to five birds.

Habitat. Open woodland, sometimes mixed with denser woodland or low scrub. Also in trees along the courses of rivers and creeks, usually where some water was present.

Behaviour. Usually seen ascending the trunks of trees, most frequently fairly low down but occasionally high in a tree. Twice noted on ironbarks; and twice as spiralling up tree-trunks. Calls noted were a loud noisy *pink pink,* possibly in alarm; a triple whistle; and a high-pitched *cheep*.

Food. Insects, including ants and beetles; small white seeds recorded once (64) but as suggested elsewhere these might have been eggs or pupae.

Annual Cycle. Gonads were generally small but one female from Glenore (49) in April shows a slightly enlarged ovary. Birds from Aroona Valley (82), Wirrealpa (83) and Innamincka (64) in April were in body moult. An immature bird was taken at Coleraine (89) in January (89) and a male from Glendower (55) in March was moulting into adult plumage.
Impermanent Colours. *Legs* adults grey, dark grey, black; immature grey. *Iris* dark brown. *Bill* adults black, dark brown or dark grey to black, lower mandible sometimes whitish with dark tips, possibly indicating younger birds; immature slate-grey on upper mandible, whitish yellow on lower. *Mouth* adults dark grey to greyish-white, flesh or pink in the throat, partially immature birds showing some yellow; immatures yellow.
Geographical Variation. The typical nominate form occurs north to a line westwards from the Burdekin River, north Queensland. North of this a dark form, *melanota,* occurs, to which specimens from Glenore (49) and Inkerman (50) belong: those from Glendower (55) are of the nominate subspecies but show some intermediate characters (see also Appendix 4), so it is best to treat *melonota* as a subspecies.

Climacteris rufa Rufous Tree-creeper

Specimens. WESTERN AUSTRALIA. Scott River (11) 22 Feb 66, 1 skel. Mount Barker (12) 27 Mar 66, 1 ♂ skel, 1 fl. SOUTH AUSTRALIA. Nonning (79) 29 – 30 Mar 65, 1 ♂ (33.2 g), 3 ♀ (29.6, 31.5 g).
Distribution & Abundance. Only recorded where collected. Found singly or in pairs.
Habitat. Woodland along creek in heathland (11); large gums among saplings by small swamp in paddocks (12); large trees bordering open shrubby woodland, and mallee by a dam (79).
Behaviour. Ascending trunks of trees in search of insects. Once, calling with a single note at regular intervals. Call a shrill *chik.*
Food. Insects. Possibly seeds once (or pupae?).
Annual Cycle. Gonads were small. Birds in moult were taken at Mount Barker (12) and Nonning (79) in March.
Impermanent Colours. *Legs* blackish-brown, dark grey, grey. *Iris* dark brown. *Bill* dark brown, blackish brown upper mandible and fleshy-brown lower mandible. *Mouth* pinkish-grey or white with some yellow in the throat.
Geographical Variation. Eyre Peninsula birds have been separated as a subspecies, *orientalis,* on a basis of paler colour and larger size. Our specimens, and others in the British Museum, show no colour difference. The size ranges of the two forms as given by Keast (1957c) overlap and the only difference is in the means of the measured parts.

Climacteris leucophaea White-throated Tree-creeper

Specimens. QUEENSLAND. Ayton (52) 2 July 64, 1 ♂ (18.7 g), 1 ♀ (16.4 g). Tully (54) 23 July 64, 2 ♂ (17.3, 18.3 g). SOUTH AUSTRALIA. Furner (86) 7 – 8 Jan 63, 1 ♂, 1 imm ♂, 1 fl. VICTORIA. Coleraine (89) 19 Jan 63, 1 ♀. NEW SOUTH WALES. Albury (92) 11 Mar 63, 1 ♂ skel. Dunedoo (93) 18 – 24 Mar 63, 1 ♂, 1 ♀, 2 ♂ skel, 2 fl. Grafton (94) 1 – 5 Apr 63, 3 ♂, 3 ♀. Tenterfield (95) 10 – 12 Apr 63, 1 ♂, 1 ♀, 1 ♂ skel.

Specimens to other Museums. NMV 1 ♂ skel, Albury (92). AM 1 ♂, 1 ♀, Grafton (94).
Distribution & Abundance. Only seen where collected. Usually singly or in pairs, once in a small flock.
Habitat. Northern Queensland birds were in tropical rainforest on the edge of clearings. Southern birds in medium to open woodland; noted in Spotted Gum and ironbark, and in stringy-bark woodland.
Behaviour. Usually seen ascending tree-trunks searching for insects, once spirally; noted probing under bark and taking ants from trunk. Seen at heights from 3-50 ft; usually on larger limbs, once on a small sapling. Calls noted were a short, sharp *tweet* uttered repeatedly, a high-pitched *cherp,* and loud piercing notes.
Food. Insects, including beetles, small ants.
Annual Cycle. Gonads small in southern birds in January to April, and slightly enlarged in north Queensland birds in July. Some birds from Grafton (94) and Tenterfield (95) in April show slight body moult.
Impermanent Colours. *Legs* black, grey, dark brown. *Iris* dark brown. *Bill* black, with some grey or whitish area towards the base of the lower mandible on some, probably more immature, birds, yellowish on one; the lower mandible with a small white tip in *minor. Mouth* flesh-coloured, greyer in *minor;* some yellow colour on more immature birds.
Geographical Variation. The nominate form occurs from Victoria to mid-Queensland. A smaller, darker, form, *C. l. minor,* occurs on the eastern side of the Queensland peninsula, in the rainforest region. Specimens from Tully (54) and Ayton (52) are of this subspecies, the remainder of the nominate subspecies.

Climacteris erythrops — Red-browed Tree-creeper

Specimens. NEW SOUTH WALES. Tenterfield (95) 12 Apr 63, 2 ♂, 2 ♀.
Specimens to other Museums. NMV 1 ♀ Tenterfield (95).
Field Notes. Recorded only at the one locality. Birds seen singly or in pairs in tall woodland in mountain area with stringy-bark and other gums. Feeding on trunks of trees. Voice – a rippling call quite distinct from those of *C. picumnus* and *C. leucophaea,* and a high-pitched hissing.
Food. Insects.
Annual Cycle. Gonads were small. Specimens showed body moult.
Impermanent Colours. *Legs* male black; female grey. *Iris* dark brown. *Bill* male black; female blackish-grey or black. *Mouth* yellow.
Geographical Variation. Our specimens are of the nominate form.

Climacteris affinis — White-browed Tree-creeper

Specimens. Western Australia. Mount Shenton (18) 30 July 65, 2 ♂. Mungilli Claypan (20) 13 Sept 65, 1 ♂. Lake Gruszka (21) 10 Sept 65, 1 ♂, 3 ♀. Warburton (23) 9 – 13 Aug 65, 2 ♂, 1 ♀, 1 skel, 1 fl. Giles (25) 25 Aug 65, 1 ♂. QUEENSLAND. Tyrone (59) 10 – 15 May 63, 3 ♂, 1 ♀. SOUTH AUSTRALIA. Granite Downs (68) 7 – 10 June 65, 2 ♂, 2 ♀. Angepena (84) 2 Dec 63, 1 ♀.

Specimens to other Museums. NMV 1 ♂ Tyrone (59).
Distribution, Abundance & Habitat. Not noted elsewhere. Usually solitary, occasionally in twos or threes in open Mulga woodland, often sparse and mixed with shrubs, or in Mulga and gums along dry creeks.
Behaviour. Seen seeking food by ascending the trunks of living and dead trees, or searching in their hollows, or on the ground on fallen tree-trunks. Like others of the genus climbs without using the tail as support, and frequently perches across branches in typical passerine fashion. Rather silent but a soft *seep, seep* was heard from a male after a female had been shot.
Food. Insects, mostly ants, also beetles and grubs.
Annual Cycle. Gonads were mostly small but testes slightly enlarged in males at Lake Gruszka (21) and Mungilli Claypan (20) in September, and very enlarged in a male at Mount Shenton (18) in July. No evidence of moult, but the dorsal plumage of some birds from Western Australia showed marked fading and wear.
Impermanent Colours. *Legs* males black, females dark grey to black. *Iris* very dark brown, dark grey-brown. *Bill* black with greyish base in younger birds. *Mouth* greyish-flesh, to grey and black, in most, yellow in some (? immature) birds.
Geographical Variation. Birds of the nominate form from eastern Australia have grey upper tail-coverts. Specimens from Angepena (84) and Tyrone (59) are of this type. The remaining specimens are of *C. a. superciliosa* with brown upper tail-coverts. Two of the three specimens from Granite Downs (68) show a greyish tint on the brown upper tail-coverts.

Climacteris melanura Black-tailed Tree-creeper

Specimens. WESTERN AUSTRALIA. Coolawanyah (4) 15 – 19 May 66, 1 ♂, 2 ♀, 1 imm ♂, 1 imm ♀, 1 ♂ skel, 1 ♀ skel, 2 fl. Mount Anderson (28) 11 – 18 May 68, 4 ♂ (38.2, 41.0 g), 2 ♀ (29.3, 33.6 g), 2 skel, 1 ♂ fl, 2 ♀ fl. Derby (29) 21 May 68, 1? imm skel. Mount Bell (32) 19 Jan 68, 1 ♂ (34.6 g), 1 ♀ (34.4 g). Mount Elizabeth (35) 14 July 68, 1 ♀ skel (30.3 g). NORTHERN TERRITORY. Borroloola (46) 11 – 12 Nov 68, 2 ♂ (30.1, 32.2 g), 1 ♂ skel (31.5 g), 3 fl (27.7, 32.4, 34.4 g). QUEENSLAND. Moonlight Creek (47) 13 May 64, 1 ♂ (31.7 g), 1 ♀ (30.2 g). Wernadinga (48) 19 May 64, 1 ♀ (34.6 g).
Distribution & Abundance. Also recorded at Geikie Gorge (31) in Western Australia; and Oenpelli (43) and Pine Creek (44) in Northern Territory.
Habitat. In zone of tall gums on river flats (4, 35, 46), in trees along creeks with or without water in savanna woodland (28), and in more open savanna woodland bordering an escarpment (32).
Behaviour. Seen feeding mostly on live and dead tree-trunks and larger branches, spiralling up and probing in crevices. Once recorded spiralling down a tree-trunk. Flight from one tree to another low and undulating, pale wing-patches conspicuous. Feeding parties tend to scatter rather widely. Calls noted variously were a loud metallic *ting* or *ching,* quite different from the shriller *chik* of *C. rufa,* a quick *peep-peep* uttered during feeding and flight, and a high-pitched, sharp and penetrating whistle.
Food. Insects; including beetles, ants and large smooth larvae.

Annual Cycle. Gonads mostly small. Two females from Coolawanyah (4) and one from Moonlight Creek (47) in May had ovaries slightly enlarged; and males from Borroloola (46) in November had enlarged testes. Birds from Coolawanyah (4), Mount Anderson (28) and Derby (29) in May showed some moult.
Impermanent Colours. *Legs* adults dark grey to black, rarely brown, soles pale grey or whitish; immatures grey. *Iris* adults olive-brown to dark brown, immatures light brown. *Bill* adults black, sometimes greyer or browner basally or on lower mandible; immatures dark grey. *Mouth* adults vary from flesh through grey to black, occasionally noted as white in females; immatures yellow, with white gape.
Geographical Variation. The nominate subspecies occurs from the Kimberleys east to northern Queensland. According to Macdonald (1966) records of this species from northern Queensland east of the Carpentaria Gap are unsatisfactory. These specimens may constitute the first authentic record for northern Queensland. Specimens from Coolawanyah (4) are of the subspecies *wellsi* of the Hammersley region of Western Australia in which both sexes are rufous and not dull brown below. An immature male from the above locality is, however, blackish with paler streaks over most of the underside.

DICAEIDAE — Flower-peckers

G. S. Cowles

Dicaeum hirundinaceum Mistletoe Bird

Specimens. Western Australia. Chichester Range (3) 20 May 66, 1 fl. Eyre (16) 8 July 65, 1 ♂. Warburton (23) 9 – 13 Aug 65, 1 ♂, 1 ♀, 1 ♂ skel, 2 ♀ skel, 1 fl. Giles (25) 20 – 23 Aug 65, 1 ♂ (10.0 g), 1 fl. Mount Anderson (28) 27 Apr – 12 May 68, 1 ♂, 3 ♂ skel (8.6, 8.6, 8.7 g), 1 fl (8.6 g). Derby (29) 1 June 68, 1 fl. Geikie Gorge (31) 4 June 68, 1 fl (7.7 g). Mount Caroline (33) 24 June 68, 2 fl (8.9, 9.4 g). Joint Hill (34) 1 July 68, 1 ♂ (9.0 g). Mount Remarkable (37) 28 July 68, 1 fl (8.5 g). Northern Territory. Oenpelli (43) 29 Sept 68, 1 fl. Queensland. Glendower (55) 11 Mar 64, 1 ♀ (8.0 g). Tyrone (59) 15 May 63, 1 fl. Thylungra (60) 25 – 26 May 63, 3 fl. South Australia. Innamincka (64) 29 Apr 65, 1 imm ♀. Etadunna (66) 18 – 22 May 65, 1 imm ♀, 1 fl. Lake Acraman (71) 3 Nov 64, 2 ♂. Nonning (79) 29 Mar – 1 Apr 65, 1 ♂ (9.3 g), 1 imm ø (9.2 g). New South Wales. Tenterfield (95) 12 Apr 63, 1 ♂.
Specimens to other Museums. NMV 1 fl Thylungra (60).
Distribution & Abundance. Encountered widely but apparently absent from some central camps in South and Western Australia (65, 68, 18) and from Darwin (41), but at short camps where it was not recorded it may have been overlooked (17, 21, 22, 24, 40, 67).
Habitat. This highly nomadic bird is usually found wherever fruiting mistletoe (family Loranthaceae) is present and was recorded from a variety of habitats; sclerophyll forest, sclerophyll woodland, open savanna woodland, savanna grassland and semi-arid scrub, low mallee and mulga.

Behaviour. Single birds were often seen singing from high perches in tall trees. Small groups of from two to five birds were seen feeding on the ripe mistletoe berries within a radius of half a mile. The alarm or excitement call is a sharp *trrit-trrit* or a throaty *krik-krik,* given in flight, and the contact call is *see* or *seep* (DG).
Food. Stomachs contained berry remains and undigested seeds.
Annual Cycle. From March to May young birds from South Australia (64, 66, 79) were in post-juvenile moult. During the same months body moult was present in adults from New South Wales (95), and wing, tail and body moult was present in adults from the Kimberleys (28) and northern Queensland (55). The only birds with enlarged gonads were two males from Warburton (23) in August, others collected at the same time were in body moult.
Impermanent Colours. *Legs* dark grey, dark brown or black. *Iris* blackish-brown. *Bill* upper mandible, dark grey or black: lower mandible dark grey to brown. *Mouth* salmon pink, flesh-colour in the centre.
Geographical Variation. There appears to be little geographical variation in this species. Keast (1958a) notes that specimens taken from the northern fringe of the range have shorter tails on average but among our birds there is no apparent geographical variation in the length of the tail.

Pardalotus punctatus — Spotted Pardalote

Specimens. Western Australia. Mount Barker (12) 27 Mar 66, 1 ♂, 2 ♀ skel. New South Wales. Albury (92) 11 Mar 63, 1 imm ♂, 1 imm fl. Grafton (94) 5 Apr 63, 1 ♂. Tenterfield (95) 15 Apr 63, 1 ♀.
Specimens to other Museums. NMV 1 ♀ Dunedoo (93) 25 Mar 63.
Field Notes. Seen in small numbers. In eastern Australia restricted to the forest and woodland areas both east and west of the Great Dividing Range. In Western Australia found in the saplings of secondary growth in an area cleared for agriculture, with tall trees which had been left standing.

Frequently seen feeding high in the foliage, often with other pardalote species. Single birds were sometimes heard calling loudly from dense foliage. The call is unexpectedly loud for such a small bird.
Food. Insects.
Annual Cycle. The young birds from Albury (92) were in post-juvenile moult. The Tenterfield (95) female had the ovary slightly enlarged.
Impermanent Colours. *Legs* light brown to black. *Iris* dark brown, dark grey. *Bill* blackish-brown to black with pale buff or fleshy base. *Mouth* grey, grey-pink, pale pink or yellow.
Geographical Variation. Specimens from the two isolated populations, in Western Australia (12) and in New South Wales (92, 93, 94, 95), show no appreciable differences and support the view of Salomonson (1961a) that no races are recognisable.

Pardalotus xanthopygus — Yellow-rumped Pardalote

Specimens. Western Australia. Eyre (16) 8 July 65, 1 ♂. Eucla (17) 1 – 5 July 65,

3 ♂, 1 ♀ imm. VICTORIA. Little Desert (91) 26 Jan 63, 1 fl.

Distribution & Abundance. Seen either singly or in pairs. Fairly common at Cocklebiddy (15) and common at Eucla (17). The western population had not previously been collected as far east as Eucla (17) and these specimens extend the range considerably, as already noted by Serventy & Whittell (1967).

Habitat. Coastal mallee (16, 17) and inland mallee scrub (91).

Behaviour. Goodwin notes the call of this species as of typical pardalote tone (insistent, somewhat ringing), a monosyllabic *teung* answered at once by a disyllabic *teung-teung*. Birds were usually observed while seeking food in the tops of mallee, but once seen feeding on the ground.

Food. Stomach contents were recorded as "grey-green mush", perhaps vegetable matter, once with insect fragments; and once as wax.

Annual Cycle. The gonads of all birds on the coast were small. At Eucla (17) one male was completing moult into adult plumage, while a female was in sub-adult dress with no apparent signs of moult. The bird from Victoria (91) had two innermost primaries new and the body was in moult.

Impermanent Colours. *Legs* light brown or dark grey; immature, flesh-grey. *Iris* olive-grey, dark olive-brown; immature, dark brown, olive grey. *Bill* black. *Mouth* flesh-pink, fleshy white or white.

Geographical Variation. There appears to be no appreciable difference between specimens of the eastern and western populations (Salomonsen, 1961a). Although the two populations seem isolated further collecting in the coastal mallee may show that the gap is not as wide as is supposed.

Pardalotus ornatus Eastern Striated Pardalote

Specimens. QUEENSLAND. Cape Gloucester (56) 5–7 Aug 64, 2 ♀ (11.5, 12.9 g). Banoona (58) 3 May 63, 1 ♂. Tyrone (59) 10 May 63, 1 ♀. NEW SOUTH WALES. Tenterfield (95) 15 Apr 63, 1 ♂.

Distribution & Abundance. Scarce, only recorded where collected. The specimens from Cape Gloucester (56) place this species 260 miles further north than previously recorded.

Habitat. Open forest and woodland. At Cape Gloucester (56) found in woodland scrub, consisting mainly of bushes and shrubs close to tropical layered woodland along the beach.

Behaviour. Usually found feeding high up in the foliage or canopy, but at Cape Gloucester (56) in the lower branches. Voice, a weak *cheep*.

Food. Seeds and insects.

Annual Cycle. Gonads were small, except at Cape Gloucester (56) where a female had a slightly enlarged ovary. This bird was found in close company with a male of *P. melanocephalus* and may have been paired (Macdonald 1969a: 42). All specimens were in body moult (56, 58, 59) or very fresh plumage (95).

Impermanent Colours. *Legs* dark purplish-grey to black. *Iris* light or dark brown. *Bill* black, basal part white. *Mouth* white.

Pardalotus substriatus Striated Pardalote

Specimens. WESTERN AUSTRALIA. Karratha (1) 29 May 66, 1 fl. Coolawanyah (4) 15 May 66, 1 ♂. Wonerup (8) 12 Mar 66, 1 ø. Scott River (11) 24 – 28 Feb 66, 1 ♂, 1 ø skel. Mount Barker (12) 24 Mar 66, 1 ♂ skel, 1 fl. QUEENSLAND. Tyrone (59) 10 – 12 May 63, 2 ♂. Monkira (61) 5 – 6 June 63, 1 ♂, 1 ♀. SOUTH AUSTRALIA. Innamincka (64) 29 Apr 65, 1 ♂. Etadunna (66) 15 – 21 May 65, 3 ♂, 2 ♀, 1 ♀, 1 ø, 1 fl. Kingoonya (70) 19 June 65, 1 ♂, 1 ♀. Sceale Bay (75) 14 Oct 63, 1 ♂, 1 ♀. Nonning (79) 3 Apr 65, 2 ♂ (11.6, 11.8 g). Aroona Valley (82) 8 – 9 May 65, 1 ♂ (10.0 g), 2 ♀ (10.0, 9.4 g), 1 ø (11.0 g). Angepena (84) 18 – 29 Nov 63, 3 ♂, 1 ♀. Furner (86) 7 Jan 63, 1 ø. VICTORIA. Coleraine (89) 19 Jan 63, 1 ♀. NEW SOUTH WALES. Albury (92) 10 – 12 Mar 63, 1 ♂, 1 ♀. Dunedoo (93) 18 – 19 Mar 63, 1 ♂, 1 ♀, 1 ♂ skel.

Specimens to other Museums. NMV 1 ♂, 1 ♀ Monkira (61), 1 ♂ Tyrone (59). AM 1 ♂ Albury (92).

Distribution & Abundance. Seen as single birds, two together or small flocks. Noted also at Thylungra (60) and at Newdegate (14). Common at Aroona Valley (82) and Etadunna (66), but uncommon at Kingoonya (70) and Nonning (79). Absent from the camps on the Australian Bight (15, 17) and in the arid centre of Western Australia (15-26). Specimens from Monkira (61) represent a 170 mile northward extension of the range in southern Queensland.

Habitat. Tall eucalypt woodland and trees near dry creeks. Jarrah forest (12). Mallee and Mulga shrubs, open woodland, and savanna woodland.

Behaviour. Feeds high in the foliage, sometimes in company with Weebills *Smicrornis brevirostris* (70) and Thornbills *Acanthiza* sp. (82). Voice noted as being both monosyllabic and disyllabic: "a liquid twittering and a more chirp-like single note; *will-oo-will, weeet-wit* or *whirra-whirra* with repeated trilling twitter both when flying and hopping" (DG).

Food. Mainly insects. "Seeds" also noted in stomachs may have been 'spangle galls' containing an insect, as noted by one collector.

Annual Cycle. No positive signs of breeding were found, but birds with slightly enlarged gonads were collected at Etadunna (66) in May. Gonads were slightly enlarged in June at Monkira (61) and Kingoonya (70).

In Western Australia a bird in immature plumage was collected in March (8) and in South Australia two specimens moulting from immature plumage to adult plumage were collected in April (79, 82) and one in May (66). Birds were in noticeably worn plumage in South Australia in October and November (75, 84). Other specimens completing either full or body moult were noted in January, March, April, May and June.

Impermanent Colours. *Legs* grey to light brown. *Iris* olive-brown to dark red-brown. *Bill* black, fleshy-white at base. *Mouth* pinkish-white; immatures yellow.

Geographical Variation. Apparently none, but some individual variation in size. The male and female from both Kingoonya (70) and Dunedoo (93) have exceptionally long wings, respectively ♂s 67, 68; ♀s 66.5, 66 mm against ♂ ♀ 63-65 mm elsewhere. Two specimens collected from Tyrone (59) (where *P. ornatus* was also collected) were described by Macdonald (1969a) as being hybrids, *substriatus* x *melanocephalus*, but both are males and are no different from typical *substriatus* males collected at Kingoonya (70) and Sceale Bay (75) (see below).

The specimens in fresh plumage from Western Australia confirm the opinions of Condon (1951a: 58) and Salomonsen (1961b) that it is not possible to recognise the subspecies *murchisoni*.

Other Notes. Adults in worn plumage are light grey rather than brown on the backs and the flanks are grey, not chestnut-brown as when in fresh plumage. In some specimens from Angepena (84) the plumage is very abraded with the white edges of the primaries worn almost entirely away. There is also apparently some sexual dimorphism. In the majority of males (13 out of 17 specimens), the white striations on the head are narrow, the forehead and crown unstreaked and glossy black: the lores are orange-yellow. In the females (10 out of 11 specimens), the white on the head extends further into the crown, the white centre of the striated feathers being broader towards the feather tip: the forehead is not glossy but a dull black and some feathers may be edged with buff, brown or yellow, similar to the immature plumage: the lores are yellow, not orange-yellow as in the males. Thus the males show a closer approach to *melanocephalus* than do the females. It is important to appreciate this when considering possible hybrids of *substriatus* x *melanocephalus* (Salomonsen, 1961b: 22).

Pardalotus melanocephalus — Black-headed Pardalote

Specimens. WESTERN AUSTRALIA. Derby (29) 22 – 25 May 68, 4 ♂ (8.1, 8.4, 8.4, 9.3 g), 1 ø (7.8 g). Geikie Gorge (31) 2 – 5 June 68, 2 ♂ (7.8 g), 2 ♀ (7.5, 8.3 g), 1 ø (9.1 g). Mount Bell (32) 14 – 20 June 68, 1 ♂, 1 ♀ (8.4 g), 3 fl (8.4, 8.5, 8.7 g), 1 imm fl (7.8 g). Mount Caroline (33) 24 June 68, 2 fl (8.7, 8.7 g). Lissadell Hill (38) 29 July 68, 1 ♂ (7.8 g). QUEENSLAND. Moonlight Creek (47) 1 – 14 May 64, 7 ♂ (8.1, 8.3, 8.4, 9.0, 9.3, 9.5, 10.0 g), 1 ♀ (7.8 g), 1 imm ♀ (8.5 g). Wernadinga (48) 19 – 21 May 64, 5 ♂ (8.2, 8.6, 9.4 g), 1 ♀. Glenore (49) 8 – 13 Apr 64, 5 ♂ (7.9, 8.6, 8.6, 8.6 g), 6 ♀ (8.6, 8.6, 8.8, 9.0, 9.1, 9.3 g), 1 ø (8.7 g). Inkerman (50) 1 – 9 June 64, 4 ♂ (7.8, 8.4, 8.5, 8.6 g), 1 ♀ (7.6 g): plus 1 ♀ hybrid (8.8 g). Tully (54) 17 – 23 June 64, 3 ♂ (9.2, 9.8, 9.8 g), 1 ♀ (12.7 g). Glendower (55) 14 Mar – 2 Apr 64, 4 ♂ (9.2, 9.3, 9.3, 9.3 g), 2 ♀ (8.9, 9.7 g), 2 ø (9.6, 9.7 g), plus 1 ø hybrid (10.3 g). Cape Gloucester (56) 6 – 8 Aug 64, 5 ♂ (9.7, 9.9, 9.9, 9.9, 10.4 g). Mount Dryander (57) 3 – 14 Aug 64, 5 ♂ (9.5, 9.6, 9.9, 10.9 g). NEW SOUTH WALES. Grafton (94) 3 Apr 63, 1 ♂, plus 1 ♂ hybrid.

Distribution & Abundance. Not found south of the Fitzroy River (31) in Western Australia or south of Grafton (94) in New South Wales. Encountered singly, in pairs or in small parties of up to six birds.

Habitat. Savanna woodland adjacent to water. At Tully (54), open rainforest. At Cape Gloucester (56), open forest. At Inkerman (50), trees at the edge of river mangroves.

Behaviour. Feeds high up in the foliage, sometimes in small flocks. Frequently seen calling from a high perch in tree tops. There is enough difference between descriptions of the voice from different localities to indicate geographical variation. In the Kimberleys the call was either two or three notes: at Glenore (49) often four rapid notes; at Inkerman (50) one long and two short notes; at Glendower (55) two double notes.

Food. Insects.

Annual Cycle. In June a male was collected as it flew from a nest site at Mount Bell (32). The nest, which was 3 ft from the ground and at the end of an 18-inch burrow in a sandy bank, was domed, made of grass and paperbark, and contained three eggs. Also in June the exceptionally heavy female from Tully (54) had a large oocyte, fully formed with yolk, in the ovary. Birds with enlarged gonads and in fresh plumage were found at Moonlight Creek (47) and Glenore (49) in April and May and Cape Gloucester (56) in August. Specimens in fresh plumage were also found at Grafton (94). Body and head moult was found in birds at Glendower (55). At other camps not already mentioned the birds were in worn plumage.

Impermanent Colours. *Legs* various shades of grey and brown. *Iris* yellow-olive to dark brown, birds from the Kimberleys and New South Wales apparently having rather browner and darker eyes than those from Queensland. *Bill* black, white at the margin of the gape. *Mouth* grey, whitish, flesh.

Geographical Variation. There is considerable variation in this species and Salomonsen (1961b) has divided it into two main "subspecies groups", the cinnamon/chestnut-rumped *melanocephalus* group in the east and the yellow-rumped *uropygialis* group in the north and west. This species is known to hybridise (Salomonsen) with the southern species *substriatus* (see below).

From the series of 72 skins, some collected at critical areas, it is evident that there is intergradation between *melanocephalus* and *uropygialis,* as Salomonsen has suggested, with a rough cline from pale yellow-rumped birds at Geikie Gorge (31) through orange-yellow on Melville Island to the yellow-and-chestnut-rumped birds at Glenore (49) on the Gulf of Carpentaria, and finally to the wholly chestnut-rumped birds at Grafton (94) on the east coast of New South Wales.

Among the yellow-rumped *uropygialis* group four of the five specimens collected at Geikie Gorge (31) are noticeably pale, the backs are a light grey, with the rump a pale yellow: the under-parts are more white and there is only a little pale cinnamon on the flanks: the amount of yellow on the breast and throat is reduced and pale. The remaining male is closer to the Derby (29) specimens. If it is found useful to recognise a paler race of the western Kimberleys the name *tormenti* is available.

Progressing eastwards, the rumps are a little more buff-yellow and a zone of intergradation between the *uropygialis* and *melanocephalus* groups is found at Glenore (49), where the yellow rumps of three specimens are tinged with chestnut, while ten others are wholly yellow. This population seems to fall within the description of *barroni.* North of Glenore (49), at Inkerman (50), the rumps are yellow tinged with buff, while to the southeast, at Glendower (55), intergradation is once more evident. Here the rumps are predominantly chestnut but two specimens have the chestnut distinctly tinged with yellow, while one is entirely yellow on the rump. At Tully (54) there is again intergradation, the rumps being cinnamon on two specimens and tinged with yellow on the two others. Specimens from these localities (54, 55) also appear to fit the race *barroni.* All other specimens from eastern Queensland (56, 57, 94) are typically *melanocephalus,* with darker birds found at Mount Dryander (57) and Grafton (94).

Other Notes. In this species the crown of the female is not as black as that of the male, the feathers being tipped or edged with grey or olive.

Hybrids, *P. melanocephalus* x *P. substriatus*

Salomonsen has recorded hybridisation between *P. m. melanocephalus* and *P. substriatus* and examples of this were collected at Glendower (55) and Grafton (94). Both specimens are mostly *melanocephalus* with very few white centred feathers of *substriatus* on the heads. The specimen from Grafton has already been described by Macdonald (1969a), together with three others said also to be hybrids from Tyrone (59) which have more the characters of *substriatus* and have been included under that species.

Hybridisation between *P. melanocephalus uropygialis* and *P. substriatus* is thought not to occur (Salomonsen) but a single specimen from a series of *uropygialis* collected at Inkerman (50) has some white striations on the head, the characters of a hybrid, although it comes from unexpectedly far north.

A possible pairing of *P. melanocephalus* and *P. ornatus* is recorded under the latter species.

Pardalotus rubricatus Red-browed Pardalote

Specimens. Western Australia. Karratha (1) 29 May 66, 1 ♂, 1 ♀, 1 ø skel, 1 fl. Peg's Beach (2) 1 June 66, 1 ♂, 1 ♀. Mungilli Claypan (20) 13 Sept 65, 1 ♂. Warburton (23) 12 Aug 65, 1 ♂. Mount Anderson (28) 28 Apr – 7 May 68, 1 ♀, 1 ♂ skel, 1 ø fl. Geikie Gorge (31) 2 June 68, 1 ♂ (10.0 g), 3 ♀ (10.1, 10.2, 11.2 g). Queensland. Glenore (49) 16 – 24 Apr 64, 1 ♂ (9.1 g), 2 ♀ (9.5, 9.9 g). Monkira (61) 2 – 6 June 63, 1 ♂, 1 ♀. South Australia. Innamincka (64) 27 – 29 Apr 65, 1 ♂, 2 ♀. Clifton Hills (65) 12 May 65, 1 ♂, 1 ♀. Etadunna (66) 15 – 21 May 65, 3 ♂, 1 ø, 1 ♂ fl., 1 ø fl.

Specimens to other Museums. NMV 1 ♂ Thylungra (60) 23 May 63. 1 ♀ Monkira (61).

Distribution & Abundance. Recorded only where collected; usually singly or in pairs.

Habitat. In grassland savanna, gibber plains, sandhills or spinifex on red sandstone, usually in association with tall trees and woodland bordering the banks of rivers, creeks and claypans, but also in coastal scrub near sand-dunes (2).

Behaviour. Active and usually feeding high in the foliage of tall eucalypts, also in low bush. Several calls were described – at Etadunna (66) a repetitive liquid phrase *pwur pwe-ti-ti-tit* was uttered with a swelling throat and was often answered at once by another bird: this call distinguishing the species immediately from *P. substriatus* which was also present (DG); also *dee-dee-dee-di-di-di* (23), a long first note followed by four short (49), and a triple sharp note (28).

Food. Stomachs contained mainly insects, also unidentified vegetable matter.

Annual Cycle. No positive signs of breeding were found, although many birds had slightly enlarged gonads. Fighting coupled with possible nest-site prospecting was noted at Mount Anderson (28). Fighting was also observed at Etadunna (66), where birds were completing body moult. Two females with slightly enlarged ovaries were noted in mid-April at Glenore (49) and both were in body moult. One female at

Geikie Gorge (31) had the ovary enlarged but there was no sign of moult. Specimens from Karratha (1) and Peg's Beach (2) had the gonads slightly enlarged but are in worn plumage with some body moult showing in those from Peg's Beach (2).

Impermanent Colours. *Legs* various shades of grey, except at Glenore (49) where the colours were olive-brown and pale mauve-brown. *Iris* greenish-yellow to orange. *Bill* upper mandible dark brown, black, or grey; lower mandible cream or off-white with a small dark brown spot towards the tip. *Mouth* light flesh or white.

Geographical Variation. There appears to be very little variation in wing length between either races or sexes. The plumage coloration is similar in both sexes.

Birds from northern Queensland (49) represent the subspecies *carpentariae*. They are darker than others with a noticeably deeper yellow rump and upper tail-coverts and have variegated feathers on the back, giving a rather scaly appearance.

Among the remaining specimens taken from widely-separated localities any possible geographical variation is obscured by seasonal variation, fresh-plumaged birds from South Australia (64-66) being darker than others, although they come from within 200 miles of Birdsville where Salomonsen (1961a: 18) found the palest birds.

An older specimen in the British Museum collected by Wilkins at Prairie, central Queensland (near Glendower (55)) outside the published range of the species, stands out as being quite different from all others. The back is tinged with rufous, there is a light wash of yellow on the upper tail-coverts, and the breast is heavily washed with buff-brown: wing 64 mm. Salomonsen (1961a: 20) noted an unusually dark bird from Winton, some 140 miles SW of Prairie and suggested that it might represent another subspecies inhabiting the verge of the semi-arid zone.

NECTARINIIDAE — Sunbirds
F. Carline

Nectarinia jugularis — Yellow-breasted Sunbird

Seen in or near mangroves at Inkerman (50) where three females were collected but not retained, and also on the edges of rainforest on the east coast of northern Queensland.

ZOSTEROPIDAE — Silvereyes
F. Carline

Zosterops lateralis (including *gouldi*) — Grey-breasted (and Western) Silvereye

Specimens. WESTERN AUSTRALIA. Perth (7) 19 – 20 Nov 62, 2 ø skel. Metricup (9) 17 Mar 66, 1 ♀. Scott River (11) 19 – 22 Feb, 1 Mar 66, 2 ♂ skel, 2 ø fl. Mount Barker (12) 26 Mar 66, 1 ø fl. Green Range (13) 31 Mar – 8 Apr 66, 1 ♀, 1 ♀ skel. Cocklebiddy (15) 8 July 65, 1 ♀. Eyre (16) 4 Dec 62, 1 ø. Eucla (17) 5 July 65, 1 ♂. QUEENSLAND. Big Tableland (51) 1 – 2 July 64, 1 ♂ (12.8 g), 1 ♀ (10.8 g). Ayton (52) 24 – 30 June 64, 6 ♂ (10.9, 11.1, 11.1, 11.6, 12.1, 12.4 g), 1 ♀ (10.8 g), 1 imm ♂ (11.6 g). Walter Hill

Range (53) 21 – 23 July 64, 2 ♂ (11.6, 11.8 g), 3 ♀ (10.8, 11.3, 11.5 g). Tully (54) 14 – 17 July 64, 1 ♂ (10.7 g), 3 ♀ (9.9, 10.8, 10.8 g). Mount Dryander (57) 1 – 13 Aug 64, 9 ♂ (10.4, 10.7, 12.0, 12.1, 12.5, 12.8, 13.0, 13.0, 13.3 g), 3 ♀ (11.1, 11.3, 13.0 g), 2 imm ♂ (13.0, 13.1 g), 1 imm ♀ (12.5 g), 1 imm ø (11.1 g). South Australia. Nullarbor HS (69) 6 – 7 Oct 63, 2 ♂. Streaky Bay (74) 10 Dec 62, 1 ♂. Elliston (76) 15 Oct 63, 2 ♂. Aroona Valley (82) 14 – 15 Apr 65, 2 ♂ (8.7, 10.3 g), 1 ø. Victoria. Nelson (88) 30 Dec 62, 1 ♀. Little Desert (91) 26 Jan 63, 1 ♂.

Specimens to other Museums. NMV 1 ♂ Nelson (88).

Distribution & Abundance. Found regularly in the camps of the extreme southwest (7-14) and along the Bight (15-17, 69, 74-76). Apparently common at coastal camps in Queensland (51-54, 57) and seen in New South Wales (92-94). Not found very far inland, Aroona Valley (82) in the Flinders Range being the furthest from the sea.

Habitat. Open forest on the Darling Range (7); in jarrah, peppermint and karri (11, 12); regularly in an orchard (13); along the south coast generally in mallee and other low scrub, also in acacias near grassland or sand-dunes. At Aroona Valley (82) in mixed open forest of pines, eucalypts and low bushes, also in saltbush near water. In northern Queensland, mainly at edge of rainforest, in tall grass clearings or other heavy undergrowth (including the introduced weed *Lantana*), occasionally mangroves. At Mount Dryander (57) in vine-covered rainforest, often on bouldery slopes.

Behaviour. Usually searching actively in undergrowth and thick foliage at low levels but also in trees 15-50 ft up. Sometimes found alone or in pairs especially in Queensland; in the south more frequently in loose flocks with varying numbers of its own species up to 20-40 and other birds such as honeyeaters and thornbills. At Walter Hill Range (53) notably with *Aegintha temporalis*. At Aroona Valley (82) seen drinking, also hopping about after food at base of eucalyptus trunk.

Its voice, not often heard, was a simple plaintive whistle or cheep, which Harrison (1968) found to be shriller than that of *Z. lutea*.

Food. Birds in the southwest were feeding extensively on ripe figs and grapes. Stomachs contained small seeds, fruits were often found, small beetle-like insect fragments and once "fibrous matter".

Annual Cycle. Little definite evidence of breeding seasons. Birds in moult were collected in the southwest (9, 13) in March: on the Bight (69) and in the Eyre Peninsula (76) in October; in the Flinders Range (82) in April: and in northern Queensland (51-54, 57) in June and July. The only gonad activity noted was in the majority of birds from Queensland and in a December female from Victoria (88) which is in very worn plumage. Among the Queensland series four birds were judged to be immature from their skull condition, small gonads and lack of moult, but no other differences are discernible.

Impermanent Colours. *Legs* light grey in various tints; dark flesh and brown once each from the southwest; Queensland birds show a tendency to be darker (light purplish-grey to dark brownish-grey). *Iris* brown; in Queensland light brown to ochraceous. *Bill* dark grey or grey-brown; basal two-thirds of lower mandible pale horn or greyish-white, the paler colour often extending along the cutting edges. *Mouth* inside bill usually grey with palate yellowish or pinkish flesh.

Geographical Variation. Mees (1969) has reviewed this species and we follow his arrangement, considering the green-backed *gouldi* as conspecific with the grey-backed *lateralis*. Our specimens extended the known range of *Z. l. gouldi* east to Nullarbor HS (69), though birds from the Bight (16–17) tend to be rather greyer below than those from the southwest (9, 11) showing an approach to the duller, grey-backed *Z. l. halmaturina* represented by birds from the Eyre Peninsula (74, 76), Flinders Range (82) and Victoria (88, 91). These vary, the Eyre birds having the sides of the throat yellowish green and the centre whitish, the Flinders birds have the throat almost wholly lacking yellow, and Victoria birds have the throat uniformly washed with pale yellow. Furthermore the Flinders birds have the flanks much more strongly washed with brown than others, the difference partly, but probably not wholly, due to their fresher plumage.

In Queensland the birds from furthest north (51–54) are typical *Z. l. ramsayi*, a short-tailed race with bright under tail-coverts. Those from Mount Dryander (57), in the area designated by Mees as intermediate between *ramsayi* and *familiaris*, are similar in colour but slightly longer-tailed, ♂ ♀ 39–44 against 37–41 mm for our *ramsayi*. There is some individual variation amongst them, one fully adult male from Mount Dryander (57) in particular having a lot of yellow on the belly.

Parasites. See Appendix 1.

Zosterops lutea — Yellow Silvereye

Specimens. Western Australia. Karratha (1) 29 May – 1 June 66, 1 ♂, 4 ♀ skel, 3 fl. Peg's Beach (2) 1 June 66, 1 ♂, 1 ♀, 1 ø, 3 fl. Carnarvon (5) 19 June 66, 1 fl. Derby (29) 26 May 68, 1 ♂ (8.6 g). Wyndham (39) 31 July 68, 1 fl (9.2 g). Northern Territory. Mount Roper (45) 1 Nov 68, 1 ♂ (8.6 g). Queensland. Moonlight Creek (47) 4 – 12 May 64, 4 ♂ (9.1, 9.9, 10.6, 10.7 g), 2 ♀ (9.3 g), 3 imm (?) ♂ (8.8, 9.4, 10.6 g), 3 imm ø (7.4, 10.1, 10.4 g). Inkerman (50) 28 May – 1 June 64, 2 ♂ (8.4, 11.3 g), 3 ♀ (9.0, 10.0, 11.0 g), 1 ø (9.4 g), 1 imm (?) ø (10.0 g).

Distribution & Abundance. Only found where collected, abundant at Moonlight Creek (50) but less so at Inkerman (50).

Habitat. Usually low down in mangroves, sometimes in neighbouring scrub or savanna woodland. At Derby (29) found among flowering *Melaleuca* near river bank.

Behaviour. Appears to be similar to *Z. lateralis* (see Harrison, 1968); loose flocks of up to 40 were found, and in association with *Gliciphila* (45) and with a party of *Malurus melanocephalus* in savanna woodland at Moonlight Creek (47). Voice described as *zeep* (29), a melodious chirrup, a high-pitched plaintive mew (47) and a shrill repeated piping while mobbing collector. See also notes on *Z. lateralis*.

Food. Stomachs contained insects including ants and caterpillar, fruit pulp and undigested seeds. Harrison (1968) showed that although *Z. lutea* in the wild fed less on fruit than *Z. lateralis*, birds of both species in captivity did not differ in their choice of food.

Annual Cycle. The one November bird from Mount Roper (45) is in fresh plumage, with testes slightly enlarged. The Moonlight Creek (47) May birds were all moulting, with little gonad activity, and include many which show signs of immaturity. In late

May the series from Inkerman (50) and the one bird from Derby (29) were mainly fresh or completing moult, two females having enlarged ovaries. From Western Australia (1, 2) 30 May – 1 June birds were worn or in an earlier stage of moult.
Impermanent Colours. *Legs* dark grey or blackish. *Iris* yellow or orange-brown. *Bill* greyish-black with the basal part of the lower mandible pale grey. *Mouth* tends to be light in throat and darker inside bill; immatures from Moonlight Creek (47) tend to have the mouth creamy distally, rather than grey as in adults, though descriptions are very variable.
Geographical Variation. The specimens from Western Australia (1, 2, 29) are *Z. l. balstoni,* the remainder *Z. l. lutea,* as indicated by Mees (1961b). The former average slightly smaller, with shorter bill; but the supposed brighter colour of *Z. l. lutea* is hardly discernible in our specimens.

MELIPHAGIDAE – Honeyeaters
P. R. Colston (F. Carline)

Lichmera indistincta — Brown Honeyeater

Specimens. WESTERN AUSTRALIA. Chichester Range (3) 16 May 66, 1 ⌀, 1 ♂ skel. Metricup (9) 17 Mar 66, 1 ♂. Scott River (11) 24 – 27 Feb 66, 1 ♂, 2 ♂ skel, 1 fl. Green Range (13) 2 Apr 66, 1 ♂. Newdegate (14) 13 Apr 66, 1 fl. Mount Anderson (28) 27 Apr – 13 May 68, 3 ♂, 2 ♂ skel, 1 ♀ skel, 1 ⌀ skel, 2 fl. Derby (29) 22 – 26 May 68, 2 ♂, 1 ♀ skel. Joint Hill (34) 1 July 68, 1 ♂. NORTHERN TERRITORY. Mount Roper (45) 27 – 30 Oct 68, 2 ♂ (11.2, 11.3 g), 1 ⌀ (9.9 g). QUEENSLAND. Glenore (49) 11 Apr 64, 1 ♀ (10.2 g). Glendower (55) 13 – 30 Mar 64, 3 ♂ (10.1, 11.4 g), 2 ♀ (8.7, 9.4 g), 2 fl (10.4, 11.5 g). Cape Gloucester (56) 5 – 7 Aug 64, 3 ♂ (10.9, 11.0, 11.2 g), 1 ♀ (9.8 g). Thylungra (60) 25 May 63, 1 ♀.
Distribution & Abundance. In Western Australia found in all camps in the southwest except Mount Barker (12). Not recorded in central Western areas (18-26) and coastal camps (15, 17). Common in the Kimberleys, especially at Mount Anderson (28), Geikie Gorge (31), and Joint Hill (34). Common at all main camps in Northern Territory, especially in thickets on the Roper River (45). In Queensland especially common at Cape Gloucester (56) and in New South Wales common at the mouth of the Clarence River (94).
Habitat. Gums along creeks (3), young flowering Marri in open forest and in high coastal scrub (11); low heath growth with patches of scattered mallee (13): mallee (14): savanna woodland (28, 29): dense riverside thickets in riverine forest (45): very low savanna woodland and scrub, mostly tea-trees (49): open forest or riverine forest (55): low mangroves (56) and savanna woodland (56): open woodland (60).
Behaviour. Feeds at all levels and on many types of flowering trees and shrubs. Flies out repeatedly from bushes and small trees to take insects on the wing in flycatcher fashion. Usually single or in pairs. Can be very aggressive; seen chasing other birds, including *Gerygone* sp. through trees into mangroves at Hope Inlet, Darwin (41) (DJF).
Food. Small insects and nectar.

Annual Cycle. Gonads of birds collected February to April in the southwest (9, 11, 13, 14) very small and inactive. Birds with enlarged gonads recorded in May (3, 28) and August (57). At Joint Hill (34) a nest with two eggs was found on 3 July and also a pair were observed feeding two young just out of the nest. Moult especially of wings and tail, was found in some birds (9, 11); many were in fresh plumage (13, 28, 45, 49, 55, 57).
Impermanent Colours. *Legs* blue grey to black, soles of feet sandy. *Iris* grey, greyish brown or brown. *Bill* black. *Mouth* yellow and black. *Bare skin below eyelid* pale grey.
Geographical Variation. Negligible. We agree with Mees (1961a) that eastern birds do not differ significantly from southwestern.

Myzomela obscura — Dusky Honeyeater

Specimens. Northern Territory. Darwin (41) 6 – 19 Sept 68, 2 ♂ skel (13.3, 13.8 g), 1 ♀ skel (11.6 g), 7 fl (8.5, 13.1, 13.2, 13.2, 13.5, 13.5, 13.8 g). Pine Creek (44) 27 – 30 Aug 68, 4 ♂ (12.3, 12.6, 13.3, 15.1 g), 1 ♀ (10.9 g). Queensland. Ayton (52) 21 June 64, 1 ♂ (14.4 g).
Distribution & Abundance. Noted as plentiful at Darwin (41). A few observed in the Oenpelli (43) area. Fairly common at Pine Creek (44) but not seen at Mount Roper (45). Common at Ayton (52).
Habitat. Pseudo-rainforest and edge of mangroves (41), savanna woodland by creek (44), grassy forest above rainforest (52).
Behaviour. Encountered singly or in pairs. Noted as being inquisitive and confiding, often remaining to feed close at hand. Tolerates other honeyeaters. Much of the time spent feeding in flowering eucalypts, often hanging upside down.
Food. Small insects.
Annual Cycle. Gonads small. One male at Pine Creek (44) was in full moult, the remainder completing moult.
Impermanent Colours. *Legs* slate grey or greyish brown. *Iris* brown. *Bill* black. *Mouth* yellow or orange yellow. *Gape* pale yellow.
Geographical Variation. Specimens belong to the two recognised races, *M. o. obscura* of Northern Territory (41, 44) and *M. o. harterti* of the northeast (52), which is darker above and browner below.

Myzomela erythrocephala — Red-headed Honeyeater

Specimens. Western Australia. Wyndham (39) 8 Aug 68, 1 ♂ (7.5 g). Northern Territory. Darwin (41) 10 – 12 Sept 68, 2 ♂ skel (8.6, 9.2 g), 2 fl (7.9, 8.2 g). Queensland. Moonlight Creek (47) 5 – 17 May 64, 2 ♂ (8.6, 9.1 g), 2 ♀ (7.7, 8.1 g), 4 imm ♂ (7.6, 7.8, 8.2, 9.4 g). Inkerman (50) 11 June 64, 6 ♂ (7.8, 8.1, 8.2, 8.2, 8.2, 8.6 g), 3 ♀ (7.1, 7.8, 7.9 g).
Distribution & Abundance. A few were seen at Point Torment (30) and Wyndham (39). Common at Darwin (41) and also Mount Roper (45), and where collected in Queensland.

Habitat. Mangroves and other scrub and trees in close vicinity. Riverine mangroves along muddy creeks and channels.

Behaviour. Singly or in pairs. Often in loose association with other species such as *Lichmera indistincta, Rhipidura fuliginosa, R. rufiventris, Zosterops lutea* and *Gerygone palpebrosa.* Very active and lively in upper foliage of trees. Inquisitive and readily "called up". Voice recorded as a shrill loud *cheep* and a buzzing chirp. Also a repetitive, metallic, jingling song noted.

Food. Small insects, including ants.

Annual Cycle. Four young males at Moonlight Creek (64) in May were moulting into adult plumage and acquiring red heads and scarlet feathering on rump. In June three of the males from Inkerman (50) had enlarged testes; the females may have been first-year birds as none had the skull fully pneumatised. At Mount Roper (45) a pair were seen with fledged young at the end of October.

Impermanent Colours. *Legs* black, soles of feet yellowish. *Iris* adults brown; immature males olive grey. *Bill* black or blackish brown. *Mouth* bright yellow.

Geographical Variation. All our specimens are *M. e. erythrocephala* including the most easterly ones from Inkerman (50), which might have been expected to approach the larger and darker *M. e. infuscata* of New Guinea and Cape York.

Myzomela sanguinolenta — Scarlet Honeyeater

Fairly common in rainforest at Ayton (52): high up in tea-trees of savanna woodland on the edge of swampy pools of Clarence River at Grafton (94) and in similar woodland at Mount Dryander (57). One male (8.6 g) collected but not retained at Ayton (52) 21 June had minute testes: two others (8.0, 9.0 g) from Mount Dryander (57) 8 August had enlarged testes. Stomachs contained minute insects and seeds.

Myzomela nigra — Black Honeyeater

Specimens. WESTERN AUSTRALIA. St George's Range (27) 8 – 9 June 68, 2 ♂ (9.1 g), 3 ♂ fl (7.5, 9.1, 9.2 g). QUEENSLAND. Monkira (61) 2 – 5 June 63, 2 ♂, 1 ♀, 1 ♂ fl.

Specimens to other Museums. NMV 1 ♂ Monkira (61).

Field Notes. Found in semi-arid country and savanna grassland near rocks on St George's Range (27) and at Monkira (61) on arid stony hillsides with scattered clumps and patches of low scrub, bushes and a few trees bordering the Diamantina River and its associated dry water courses. Only a few birds were seen at either place and they included males in song-flight display. Not seen elsewhere.

Food. Small insects including a small caterpillar and a few winged ants.

Annual Cycle. All adults examined had enlarged gonads. One female from Monkira (61) with a large yolk sac was presumed to be paired with a nearby male in display flight. This male is in moult, as are two from St George's Range (27).

Impermanent Colours. *Legs* males black; females grey. *Iris* dark brown. *Bill* males black; females brown. *Mouth* yellow or brownish yellow.

Myzomela pectoralis Banded Honeyeater

Specimens. Western Australia. Derby (29) 21 – 25 May 68, 4 ♂ (7.8, 11.4, 11.4 g), 3 ♀ (7.2, 8.3, 10.0 g), 1 imm ♂ (8.1 g), 2 imm ♀ (7.4, 7.8 g), 3 ♂ skel (9.9, 10.8, 10.9 g), 1 ♀ skel (7.9 g), 1 fl. Mount Bell (32) 14 – 17 June 68, 1 ♀ (9.4 g), 1 imm ♂, 3 imm ♀ (7.7, 8.5, 10.1 g), 5 juv ♂ (8.3, 9.7 g), 1 juv ♀(9.7 g), 5 fl (9.6, 11.2, 11.5 g). Mount Elizabeth (35) 13 – 14 July 68, 3 fl (8.1, 9.2 g).
Distribution & Abundance. Very common at Derby (29) and Mount Bell (32) where adults seemed to be in the ratio of one to forty immatures. Few noted at Mount Bell (35). Recorded as plentiful at Pine Creek (44), a few seen at Mount Roper (45) and fairly numerous on the Borroloola plain (46). Not encountered in north Queensland.
Habitat. Woodland savanna with flowering bloodwood trees and shrubs (29); particularly partial to flowering *Grevillea* bordering river banks (35).
Behaviour. Parties were seen feeding in tops of flowering bloodwoods; very active with quick erratic flights between trees, frequently chasing one another. Voice described as a loud (for size of bird) whistle, or a high-pitched double whistle. One juvenile in company with adults gave a plaintive *wheeze*.
Food. Insects and nectar.
Annual Cycle. The breeding season was apparently coming to an end, for most adults were in moult with inactive gonads and juveniles were plentiful.
Impermanent Colours. *Legs* adults black; young birds grey or greyish purple. *Iris* adults dark brown; young birds grey, grey-brown or brown. *Bill* adults black; young birds dark brown to black. *Mouth* adults black; young birds pinkish or yellow, black in center. *Sides of gape* yellow in adults.
Other Notes. Our series shows that juveniles of both sexes have the upperparts uniform pale sandy buff, a black line through the eye, pale yellow ear-coverts and a narrow buff breast band. This plumage is replaced by a complete post-juvenile moult into a first-year plumage in which the breast band is black and the feathers of the back browner with black centres: there are black feathers in the hind crown and brown wing and tail feathers which moult out slowly into the final black and white plumage of the adult. In the past this species has been considered sexually dimorphic, but our series shows only that the black pectoral band of the female is slightly narrower and the back is less glossy black than in the male, and the female is smaller, wings 61-63 against 66-70 mm (see also Officer, 1964, p. 22).

Certhionyx variegatus Pied Honeyeater

Specimens. Western Australia. Chichester Range (3) 13 May 66, 1 ♂ skel. Mount Shenton (18) 2 – 3 Aug 65, 2 ♂. South Australia. Nullarbor HS (69) 6 – 8 Oct 63, 2 ♂, 1 ♀. Angepena (84) 18 Nov 63, 1 ♀.
Field Notes. Normally encountered singly or in pairs, once (69) in a flock of trillers *Lalage sueurii*. At Mount Shenton (18) it was fairly common, though wary, but elsewhere it was scarce and only seen otherwise at Hunt Oil Camp (22) and Warburton (23). Always in Mulga in Western Australia.
Food. Caterpillars and nectar were each found once in stomachs.

Annual Cycle. A male collected at Mount Shenton (18) had enlarged gonads. Two males from Nullarbor HS (69) were in fresh plumage but still retaining very worn primaries. The November female (84) was in complete moult.
Impermanent Colours. *Legs* blue-grey. *Iris* brown or reddish brown. *Bill* male, upper mandible, blue-grey with a darker tip, lower mandible blue-grey, paler at base; female greyish brown paler at base of lower mandible. *Mouth* whitish, pale grey or mauve. *Small wattle under the eye of the male* light blue.

Meliphaga gracilis Graceful Honeyeater

Specimens. Queensland. Ayton (52) 20 – 27 June 64, 4 ♂ (16.3, 16.6, 17.9 g), 5 ♀ (14.1, 14.6, 15.2, 15.9, 16.0 g). Walter Hill Range (53) 27 July 64, 1 ♂ (14.7 g), 1 ♀ (14.1 g). Tully (54) 13 – 18 July 64, 2 ♂ (15.1, 15.6 g), 5 ♀ (13.8, 14.2, 15.2, 15.6 g).
Distribution & Abundance. Uncommon at Big Tableland (51) where one was collected but discarded. Very common at Ayton (52), common at Tully (54). The records from above 2,000 ft at Big Tableland (51) and Walter Hill Range (53) are interesting since this and *M. notata* are normally lowland species, their place being taken at higher altitudes by *M. lewinii*.
Habitat. Essentially a rainforest species going out into open forest and mangroves nearby.
Behaviour. Found singly or in pairs, rarely encountered in large groups; feeding at heights of 6-60 ft, in crown of substage trees or near tops and in crowns of mangroves. Voice described as a series of high ascending whistles (SAP); also a loud *click*.
Food. Mainly small insects, some beetles.
Annual Cycle. All specimens are in fresh plumage, with small or only slightly enlarged gonads.
Impermanent Colours. *Legs* slate purple with greenish yellow soles to feet. *Iris* dark olive grey. *Bill* black. *Mouth* bright orange-yellow. *Gape* bright-yellow.
Geographical Variation. Our material is referable to *M. g. imitatrix* which is darker green above and more dusky below than *M. g. gracilis* of Cape York.

Meliphaga notata Lesser Lewin Honeyeater

Specimens. Queensland. Ayton (52) 20 – 27 July 64, 5 ♂ (25.6, 29.9, 29.9, 30.2, 32.0 g), 4 ♀ (24.2, 25.1, 27.1, 27.2 g), 2 ø (28.9, 31.6 g). Tully (54) 14 – 25 July 64, 3 ♂ (21.2, 30.4, 32.3 g), 1 ♀ (23.1 g).
Distribution & Abundance. Also collected (but discarded) at Big Tableland above 2,000 ft; see notes on *M. gracilis* and *M. lewinii*.
Habitat. The same as for *M. gracilis*.
Behaviour. Usually encountered singly, feeding in foliage of trees and vines at heights of 10-40 ft. Voice, a series of urgent querulous whistles (SAP); also a clear piercing repeated whistle.
Food. Varied. Stomachs contained small seeds, berries and fruit remains: also insect fragments including a few ants.
Annual Cycle. Adults had completed moult and had small or slightly enlarged gonads.

Impermanent Colours. *Legs* dark greyish-brown, back of shanks ochraceous. *Iris* pale olive-grey. *Bill* black. *Mouth* orange-yellow. *Gape* yellow or lemon.
Geographical Variation. Our birds are *M. n. mixta* which is greener above and below than *M. n. notata* of the Cape York Peninsula.

Meliphaga lewinii Lewin Honeyeater

Specimens. QUEENSLAND. Big Tableland (51) 25 June – 1 July 64, 3 ♂ (35.2, 35.6, 38.3 g). Walter Hill Range (53) 21 – 23 July 64, 4 ♂ (30.4, 31.5, 35.0, 36.5 g), 2 ♀ (30.0, 31.3 g). Cape Gloucester (56) 7 Aug 64, 1 ♂ (31.8 g). Mount Dryander (57) 1 – 3 Aug 64, 5 ♂ (34.2, 34.5, 34.6, 34.7, 37.9 g), 8 ♀ (29.6, 29.8, 30.4, 30.4, 32.6, 33.7, 35.3, 36.5 g), 1 ø (38.5 g). NEW SOUTH WALES. Tenterfield (95) 12 Apr 63, 1 ♂ skel.
Field Notes. Within range of *M. gracilis* and *M. notata* found only above 2,000 ft: here common (51) or very common (53) in tropical rainforest with large stands of Johnstone Hardwood (51). Further south (57) in forest at lower altitudes and woodland on the edge of mangroves. Fairly common at Tenterfield (95) in rainforest with thick undergrowth near streams. Most birds were solitary but were twice in parties with *Zosterops lateralis.*
Food. Stomachs contained various insects including some ants and the remains of fruit pulp and berries, seeds and once a conical budleaf.
Annual Cycle. In Queensland (51, 53 and 57) gonads in June were very small, in July slightly enlarged, in August becoming active and moderately enlarged. The majority of specimens were in fresh plumage and several (56, 57) had the skulls not fully pneumatised but otherwise appear fully adult.
Impermanent Colours. *Legs* dark brown, greyish brown to slate purple. *Iris* olive grey to bluish grey. *Bill* black. *Mouth* yellow to orange. *Gape* white.
Geographical Variation. Our birds show little variation in plumage or size and are referable to *M. l. lewinii.*

Meliphaga flava Yellow Honeyeater

Specimens. QUEENSLAND. Glenore (49) 21 – 23 Apr 64, 4 ♂ (20.5, 22.2, 22.2, 23.7 g), 1 ♀ (18.6 g). Inkerman (50) 27 May – 2 June 64, 5 ♂ (23.6, 23.7, 24.3, 25.4, 26.2 g), 2 ♀ (17.4, 19.5 g), 1 ø (20.6 g). Tully (54) 14 July 64, 1 ♂ (26.2 g). Cape Gloucester (56) 7 Aug 64, 1 ♂ (29.2 g). Mount Dryander (57) 4 – 14 Aug 64, 3 ♂ (26.1, 27.4, 30.8 g), 5 ♀ (21.9, 23.8, 23.8, 23.9, 25.4 g), 1 ø (22.5 g).
Field Notes. Collected wherever found. In woodland or forest with flowering tea-trees, and in mangroves. Always solitary or in pairs; curious and readily called up. Voice, a loud single high-pitched whistle, a chattering like *Rhipidura leucophrys* but mellower, and a short squeaky chirrup.
Food. Spiders, beetles and other insects. Small number of seeds also in stomach.
Annual Cycle. Gonads of majority of birds were very small and all were completing moult.
Impermanent Colours. *Legs* yellowish brown to yellowish olive. *Iris* olive grey. *Bill* black. *Mouth* orange or bright yellow.

Geographical Variation. Birds from the Gulf camps (49, 50) have the underparts just fractionally more yellow, lacking the slight greenish wash to sides of breast and flanks of the east coastal birds (54, 57).

Meliphaga albilineata White-lined Honeyeater

Specimens. NORTHERN TERRITORY. Oenpelli (43) 28 Sept – 17 Oct 68, 7 ♂ (22.3, 25.1, 25.7, 29.9 g), 6 ♀ (22.3 g), 1 ♂ skel, 1 fl (25.1 g), 1 juv fl.

Distribution & Abundance. Locally common but not seen elsewhere.

Habitat. Restricted to the sandstone and granite gorges with thick tropical vegetation. Not infrequently seen out in the open, but rarely far from rocky cover.

Behaviour (contributed by D. J. Freeman). Generally a secretive bird, either singly or in pairs and difficult to locate in thick vegetation. Occasionally feeding in blossoming trees where it would remain fairly high and out of the way of other species, though once noted (AH) as pugnacious to other species and once found in association with *Meliphaga unicolor*. Collectors' notes, though varied, show much the same patterns of behaviour. "Flew into thick vegetation in gully; sat quietly in tree; a wary bird flying hard in or through vegetation close to the rock face" and "singing about forty feet up from small bush growing from out of rocky sandstone gorge" (DJF). "Hovering and feeding in flowering eucalypt" and "preening in tall trees in wooded gorge by lake (close to rock face)" (AH).

The bird is probably best identified in an area by its voice. This is a series of unmelodious whistles (loud whistling double call repeated twice, *too-ep, too-in* (AH) and *tue-ee-ee, tue-i-in,* the second and third syllables higher pitched (BB)) which are not only amplified greatly by the narrow gorges but even echo sufficiently to cause confusion as to where the bird is hiding. Any imitation of these calls, whether birds are singing locally or not, will usually bring at least one of them up to investigate. "To an imitation of its call it responded at once and within half a minute it appeared though remained well concealed about twenty feet up in the denser foliage . . . . once it has located the imposter it is off at once" (BB). They were either breeding or had just bred during September and October and probably held well defined territories while we were there.

Food. Varied. Stomachs contained pieces of pericarp from small brownish berries, soft green seeds up to half an inch long, small black seeds, general vegetable matter, small insects and remains of spiders and flies.

Annual Cycle. Nest with newly hatched young (see Freeman, 1971) found on 2 October. Seven males and three females all with skull recorded as fully pneumatised had enlarged or slightly enlarged gonads. Of four birds with skulls not fully pneumatised all appear adult, one female had the ovary slightly enlarged while the remainder showed no significant gonad activity. All skins show some moult. Plumage is mainly in fairly good condition with body moult either complete or well advanced. Primaries and rectrices were in various stages of replacement and in some cases were moulting asymmetrically.

Impermanent Colours. *Legs* grey, grey-brown or brown, feet darker. *Iris* pale grey or pale blue grey, once, grey-green. *Bill* black. *Mouth and Gape* yellow or orange-yellow.

Meliphaga virescens Singing Honeyeater

Specimens. WESTERN AUSTRALIA. Chichester Range (3) 13 May 66, 1 fl. Bush Bay (6) 17 June 66, 1 ♀ skel. Newdegate (14) 11 Apr 66, 1 ♂ skel, 1 ♀ skel, 1 fl. Mount Shenton (18) 30 July – 3 Aug 65, 1 ♂, 2 ♀. Carnegie (19) 16 Sept 65, 2 ♂, 1 ♀. Lake Gruszka (21) 11 Sept 65, 2 ♂ (25.0 g), 1 ♀. Hunt Oil Camp (22) 9 Sept 65, 1 ♂. Warburton (23) 10 Aug 65, 1 ♂, 1 ♀, 2 ø. Van der Linden Lakes (24) 6 Sept 65, 1 ♂, 1 ♀. Giles (25) 20 Aug – 1 Sept 65, 3 ♂, 5 ♀. Mount Anderson (28) 27 Apr – 8 May 68, 1 ♂, 4 ♂ skel, 3 ♀ skel (18.0 g), 3 fl. NORTHERN TERRITORY. Mount Roper (45) 27 – 1 Nov 68, 1 ♀ skel (16.7 g), 1 fl (19.3 g). QUEENSLAND. Glendower (55) 18 Mar 64, 2 ø (22.3, 22.8 g). Tyrone (59) 10 – 15 May 63, 1 ♂, 1 ♀, 3 juv ø. Thylungra (60) 25 May 63, 1 fl. Monkira (61) 4 June 63, 2 ♂, 1 ♀, 1 fl. SOUTH AUSTRALIA. Innamincka (64) 3 May 65, 1 ♀ (23.5 g). Etadunna (66) 20 May 65, 1 ♂. Granite Downs (68) 5 – 10 June 65, 2 ♂, 1 ♀. Nullarbor HS (69) 7 Oct 63, 1 ♀, 1 imm ♀; 23 Oct 64, 2 ♂, 1 imm ♀. Kingoonya (70) 16 – 25 June 65, 3 ♂, 4 ♀, 1 ♂ skel, 1 fl. Lake Acraman (71) 3 Nov 64, 1 ♂, 2 ♀, 1 imm ♀. Nonning (79) 29 – 31 Mar 65, 2 ♂, 1 ♀, 1 imm ♂. Parachilna (81) 12 Apr 65, 1 ♂ (28.8 g), 1 ♀ (23.5 g). Aroona Valley (82) 15 Apr 65, 1 ♀ skel (23.8 g). Angepena (84) 26 Nov 63, 1 imm ♂.

Specimens to other Museums. AM 1 ♂ Monkira (61). SAM 1 ♂ Nullarbor HS (69) 27 Oct 64.

Distribution & Abundance. In Western Australia fairly common in the south (14-17); in the centre noted at all main camps (18-25) in small numbers: in the Kimberleys common at Mount Anderson (28) and a few noted at several other camps (29-31, 33-35). In Northern Territory a few at Mount Roper (45) and a single sight record at Borroloola (46).

In Queensland, in the north seen on the Gulf (47, 48): common at Glendower (55): widespread in small numbers in the southwest (59-61). In South Australia uncommon in the north (64-66, 68), but common in southerly main camps (70, 79, 82). In the southeast seen only at Nelson (88).

Habitat. Present in most types of semi-arid country with bushes or trees; also in dwarf mangroves along beaches in Western Australia (1, 2, 6).

Behaviour. Normally encountered singly or in pairs, sometimes a family party, but never in flocks; often sitting still for long period on tops of mangroves or bushes in shrike-like attitude in early mornings (PRC). Very aggressive towards smaller birds including some unlikely to be food competitors (DG). Many calls heard, including a loud distinctive trilling whistle. Also a note like clucking of White-fronted Honeyeater, but louder and more mellow. Also suspected of vocal mimicry (DG).

Food. Stomachs contained insects, twice mainly beetles, also a few ants, a bee and a moth: also pulp and seeds, including mistletoe berries (64) and seeds of *Heterodendron oleifolium* (Desf.), Sapindaceae (79).

Annual Cycle. The gonads were inactive in all birds examined except for two males and five females (18, 19, 25) out of the series taken in the Gibson Desert in August and September. Three juveniles collected in May at Tyrone (59) had only recently left the nest.

Impermanent Colours. *Legs* dark grey to black. *Iris* brown or greyish brown. *Bill* black. *Mouth* yellow to orange. *Gape* immature pale yellow.

Geographical Variation. Variation in this species is clinal. The darkest and most heavily streaked birds (*M. v. virescens*) occur around the southern and peripheral areas of the range while birds from the interior and more northerly regions (18-25, 28, 55, 61) (*M. v. forresti*), are paler above, whiter below and with less pronounced or very reduced streaking on the breast, the palest being from Glendower (55): those from central South Australia (70, 81, 84) are intermediate between *forresti* and *virescens*. Salomonsen (in Peters' 12, 1967) expresses doubts as to whether birds from the central part of southern Queensland (*M. v. westwoodia*) are separable from *M. v. virescens:* our Tyrone (59) birds do not seem to be. The male from Etadunna (66) is interesting as it has golden or bronzy yellow edges to the primaries and rectrices instead of the normal green.

Parasites. See Appendix 1.

Meliphaga fasciogularis — Mangrove Honeyeater

Specimens. QUEENSLAND. Cape Gloucester (56) 7 Aug 64, 7 ♂ (26.5, 28.2, 28.4, 29.0, 29.0, 29.9, 33.1 g), 5 ♀ (24.2, 24.7, 25.0, 26.5, 28.6 g).

Field Notes. Very common along the beach of a small island which was connected to the mainland at low tide. A dozen or more birds were seen in loose association, flying amongst the mangroves, stunted trees and bushes, chasing one another and singing. Voice, a loud fluting song, mainly a series of voluble ascending whistles, with a flourish at the end (SAP). Not seen elsewhere.

Food. Stomachs contained insect remains, and a small shell.

Annual Cycle. The majority of birds were in full breeding condition with enlarged gonads, one female with three very large oocytes.

Impermanent Colours. *Legs* dark grey or blackish slate. *Iris* pale grey-blue tinged brown around pupil. *Bill* black. *Mouth* bright yellow-orange. *Gape* whitish yellow.

Meliphaga fusca — Fuscous Honeyeater

Specimens. QUEENSLAND. Mount Dryander (57) 13 – 14 Aug 64, 3 ♂ (16.5, 17.8, 18.0 g), 1 ♀ (15.6 g). NEW SOUTH WALES. Albury (92) 11 Mar 63, 1 ♀, 1 imm ø. Grafton (94) 30 Mar – 5 Apr 63, 1 ♂, 1 ♀ (10.0 g), 1 ø, 1 ♀ skel. 1 fl.

Specimens to other Museums. NMV 1 ♀ skel Grafton (94).

Distribution & Abundance. Common at Mount Dryander (57) but not encountered elsewhere in north Queensland. Small numbers at Albury (92) and a few scattered small parties at Grafton (94).

Habitat. Open grassed forest (57), lightly timbered grassed slopes of Table Top Mountain (92), open forest with tea-tree undergrowth (94).

Behaviour. A trusting and confiding honeyeater, feeding on or near eucalyptus blossom at all heights. Voice, a short high-pitched rising *twee-twee-twee* and also a longer *tew-tew-tew*.
Food. Insects.
Annual Cycle. In New South Wales young birds were with an adult female in March at Albury (92) and adults from Grafton (94) three weeks later were in moult. In Queensland (57) in August two males were in moult and had enlarged gonads but those of the third male and the female were small.
Impermanent Colours. *Legs* light horn to slate purple. *Iris* brown. *Bill* adult black; immature light horn with basal half yellowish flesh. *Mouth* yellow.
Geographical Variation. Our specimens belong to *M. f. fusca* (92, 94) and *M. f. subgermana* (57) which differs in having a greenish wash on the sides of face and forehead.

Meliphaga flavescens — Yellow-tinted Honeyeater

Specimens. WESTERN AUSTRALIA. Mount Anderson (28) 27 Apr – 1 May 68, 1 ♂, 1 ♀, 7 ♂ skel, 3 ♀ skel, 3 fl. Mount Elizabeth (35) 8 – 14 July 68, 3 ♂ (13.2, 13.6, 13.8 g). Lake Gladstone (36) 21 – 23 July 68, 2 ♂ (12.2 g). Mount Remarkable (37) 25 July 68, 1 ♂ (15.2 g), 1 fl (13.5 g). QUEENSLAND. Moonlight Creek (47) 9 – 14 May 64, 3 ♂ (12.6, 13.4, 14.0 g), 1 imm ♂. Glenore (49) 9 – 17 Apr 64, 3 ♂ (11.7, 13.0, 13.0 g), 2 ♀ (9.7, 11.3 g).
Distribution & Abundance. Common at Mount Anderson (28), elsewhere in the Kimberleys (27, 35, 36, 37) only seen in small numbers.
Habitat. Savanna woodland bordering creeks, water holes or rivers.
Behaviour. Noted feeding in flowering *Grevillea* and paperbark trees. Fairly active, although searching thoroughly in one tree before moving on, unlike *Myzomela pectoralis* that barely rests for a minute in the same tree. Seen chasing Grey-crowned Babblers (49). Voice similar to *M. penicillata,* but deeper and quicker (SAP).
Food. Insects and nectar.
Annual Cycle. In Western Australia gonads were inactive in April (28) but single males in the second half of July had them enlarged (35, 37): these birds were in worn plumage with some stage of moult, particularly wings and tail (35, 36, 37). In Queensland most birds were finishing moult though some still had the gonads enlarged including a female (49) which was replacing the primaries.
Impermanent Colours. *Legs* light grey, purplish grey or brownish purple. *Iris* yellowish brown, olive brown or dark brown. *Bill* adults black, immature ochre at base grading to black at tip.
Geographical Variation. The Kimberley birds *M. f. flavescens* (28, 35-37) do not differ significantly from those taken in the Gulf (47, 49), the amount of light grey streaking on the breast seeming to vary individually rather than geographically. It seems that *M. f. zanda,* type locality Normanton, is best synonymised with *M. f. flavescens.*

Meliphaga plumula — Yellow-fronted Honeyeater

Specimens. WESTERN AUSTRALIA. Mount Shenton (18) 31 July – 3 Aug 65, 3 ♂, 2 ♀,

1 ♂ skel, 1 fl. Warburton (23) 14 Aug 65, 1 ♀. Van der Linden Lakes (24) 6 Sept 65, 3 ♂. Giles (25) 27 Aug 65, 1 ♂. Mount Anderson (28) 27 Apr – 1 May 68, 2 ♂, 2 ♀, 1 ø, 2 ♀ skel, 3 fl. Derby (29) 29 May 68, 1 ♀ skel. Mount Elizabeth (35) 14 July 68, 1 ♂ (13.8 g), 1 ♀. Lake Gladstone (36) 21 July 68, 1 ♂. QUEENSLAND. Wernadinga (48) 20 May 64, 1 ♂ (15.0 g). SOUTH AUSTRALIA. Kingoonya (70) 19 June 65, 1 ♂, 3 ♀. Aroona Valley (82) 9 – 15 Apr 65, 3 ♂ (18.8, 20.7, 22.5 g). Angepena (84) 25 Nov 63, 1 ♂.

Distribution & Abundance. In central Western Australia, common at Mount Shenton (18) but less common further north (23, 24, 25). In northeastern Western Australia, fairly common at Mount Anderson (28) and smaller numbers recorded elsewhere in the Kimberleys (29, 31, 32, 33, 34, 35, 36). A few seen in Northern Territory on the Victoria River (40), and at Glendower (55) in northern Queensland. In South Australia fairly common at Kingoonya (70) and Aroona Valley (82).

Habitat. Open mulga woodland (18), mallee (23, 24), river gums (25), savanna woodland (28, 29, 35, 36), tropical savanna woodland (48): in mallee and acacia woodland on sandhills (70) and in stunted eucalypts and bushes bordering creek (82).

Behaviour. Usually in pairs or small flocks in flowering eucalypts. Very active, flitting and chasing each other. Readily called up. Voice, loud sharp calls similar to *M. penicillata* and *M. ornata* (DG).

Food. Stomachs contained insects mainly, also some seeds, whitish berry remnants and nectar.

Annual Cycle. Gonads inactive except in one female from Mount Shenton (18), one male out of three from Van der Linden Lakes (24) and two Kimberley males (35, 36).

Impermanent Colours. *Legs* grey or greyish brown. *Iris* dark brown. *Bill* black. *Mouth* yellow to orange-yellow.

Geographical Variation. Variation appears to be clinal. The darkest birds with most pronounced streaking on the underparts are from the Flinders Range in South Australia (82, 84) *M. p. ethelae,* and the palest, brightest birds with least streaking, *M. p. planasi,* are from the Kimberleys (28-36) and northwest Queensland (48), while those of the centre (18-25, 70) are intermediate, *M. p. plumula.* We differ from Salomonsen (in Peters' 12, 1967) only in considering the male from Wernadinga (48) closer to *ethelae* than to *plumula.*

Parasites. See Appendix 1.

Meliphaga chrysops — Yellow-faced Honeyeater

Specimens. SOUTH AUSTRALIA. Furner (86) 10 Jan 63, 1 fl. NEW SOUTH WALES. Tenterfield (95) 10 – 12 Apr 63, 1 ♂, 1 ♀.

Distribution & Abundance. Observed singly or in pairs at Furner (86) on several occasions. In Victoria small numbers at Nelson (88) and moderately common at Coleraine (89). In New South Wales seen at Dunedoo (93) and common at Tenterfield (95). These were believed to be migrant birds moving south (PRC).

Habitat. Mixed woodland (86) and rainforest (95).

Behaviour. Found in flocks of 20 or more (95). A very animated and sprightly bird constantly visiting blossoms. Call, a series of cheerful *chick-up, chick-up* notes (PRC).

Annual Cycle. Gonads were very small.
Impermanent Colours. *Legs* grey. *Iris* dark grey. *Bill* black. *Mouth* orange.
Geographical Variation. Our two skins are of the nominate form.

Meliphaga cratitia — Purple-gaped Honeyeater

Specimens. WESTERN AUSTRALIA. Green Range (13) 31 Mar – 4 Apr 66, 5 ♂, 1 ♀, 1 ♂ skel, 2 ♀ skel, 2 fl.
Distribution & Abundance. Locally common in suitable habitat at Green Range (13). Not encountered elsewhere.
Habitat. A mixture of mallee, scattered scrub, low bushes and dwarf acacias on open heath.
Behaviour. Encountered singly or in pairs. Confident and pugnacious to spinebills and other medium-sized honeyeaters. Usually found feeding low down, systematically searching foliage and blossoms of one or two bushes for some time, then flying to another patch of scrub some distance away. Chattering calls only noted.
Food. Insects.
Annual Cycle. Birds collected were in fresh plumage, little evidence of moult. No gonad activity.
Impermanent Colours. *Legs* purplish-grey or greyish brown, soles of feet buffish. *Iris* dark brown. *Bill* black. *Mouth* orange or yellow. *Wattle at gape* purple.

Meliphaga keartlandi — Grey-headed Honeyeater

Specimens. WESTERN AUSTRALIA. Chichester Range (3) 14 – 17 May 66, 1 ♂, 2 ♀, 1 ♀ skel, 2 fl. Van der Linden Lakes (24) 6 Sept 65, 2 ø. Giles (25) 20 – 30 Aug 65, 6 ♂, 4 ♀, 2 ø (15.0 g), 2 ♂ skel, 1 fl. Walter James Range (26) 26 Aug 65, 1 ♂. NORTHERN TERRITORY. Borroloola (46) 9 – 11 Nov 68, 1 ♂ (15.0 g), 1 imm ♂ (13.4 g), 1 imm ♀ (11.1 g).
Distribution & Abundance. In Western Australia locally common, singly or in loose groups of four to six birds, in the Chichester Range (3). Common at Giles (25) in groups of up to 12 birds, but not found on the Warburton Range (23) although the country appeared suitable. In Northern Territory seen in small numbers at Borroloola (46) and also on the sandstone escarpment at Mount Roper (45).
Habitat. Mulga and spinifex with richer vegetation (gums, Desert Oaks, thickets): in broken country, with gullies and creeks (3), in or near the foothills of the Rawlinson Range (Plate 3(b)) (24, 25) or in gorges of sandstone ridges (45, 46).
Behaviour. Very active and constantly darting about feeding and calling. Voice a loud and deep-toned *kwoyt-kwoyt-kwoyt* (DG), also *kew-kew-kew-kew* or *peep-peep-peep* (CBF). Some calls seemed very similar to *M. penicillata* and *M. flavescens.* A party of six were observed drinking at midday from a small pool with Painted Finches *Emblema picta* (3).
Food. Stomachs contained remains of hover-flies, small beetles, some ants and unidentifiable vegetable matter.

Annual Cycle. The young November birds from Borroloola (46) had the skulls incompletely pneumatised and the adult was in moult, as were those from the Chichester Range (3) in May. August birds from Giles (25) were very worn, especially the wings and tail, some with rectrices starting to moult. All adults examined had small gonads.

Impermanent Colours. *Legs* light fleshy grey to reddish brown. *Iris* brown or greyish brown. *Bill* adults black, immatures black with the basal half of the lower mandible yellow. *Mouth* yellow.

Geographical Variation. Our birds show no appreciable differentiation either in plumage or size.

Other Notes. The birds from Giles (25) had the crown and lores heavily stained reddish-brown. Obviously a sticky resin or nectar had been picked up whilst feeding and was subsequently coated with a light powdery red dust.

Meliphaga penicillata White-plumed Honeyeater

Specimens. WESTERN AUSTRALIA. Chichester Range (3) 16 May 66, 1 fl. Coolawanyah (4) 15 – 20 May 66, 1 ♀, 1 ♂ skel. Mungilli Claypan (20) 12 – 13 Sept 65, 3 ♂, 1 ♀. Lake Gruszka (21) 10 – 11 Sept 65, 4 ♂, 1 ♀. Warburton (23) 10 Aug 65, 3 ♂, 3 ♀, 1 imm ♂. Giles (25) 20 – 27 Aug 65, 5 ♂ (18.0 g), 4 ♀. Walter James Range (26) 26 Aug 66, 1 ♂, 1 ø. QUEENSLAND. Glendower (55) 17 – 30 Mar 64, 2 ♀ (14.7, 14.9 g), 1 ø (13.8 g). Banoona (58) 30 Apr – 4 May 63, 2 ♀. Tyrone (59) 10 – 12 May 63, 2 ♂. Thylungra (60) 26 May 63, 3 ♂. Monkira (61) 2 June 63, 2 ♂, 1 ♀. SOUTH AUSTRALIA. Innamincka (64) 26 Apr 65, 2 ♂, 1 ♀. Clifton Hills (65) 10 – 12 May 65, 2 ♂, 1 ♀. Etadunna (66) 17 – 22 May 65, 3 ♂, 1 ♀, 1 fl. Oodnadatta (67) 31 May 65, 1 ♂, 3 ♀, 1 ø. Aroona Valley (82) 8 Apr 65, 1 ♂ (18.7 g). Angepena (84) 22 – 29 Nov 63, 2 ♂. Furner (86) 8 Jan 63, 1 ♂. NEW SOUTH WALES. Albury (92) 9 – 12 Mar 63, 1 ♀, 1 ♂ skel, 1 ♀ skel. Dunedoo (93) 19 Mar 63, 1 ♂, 1 juv ♀, 1 ø skel, 2 fl. Tenterfield (95) 15 Apr 63, 1 ♂ skel.

Specimens to other Museums. NMV 1 fl Albury (92).

Distribution & Abundance. In Western Australia found only in the west centre (1-4) and centre (20-23, 25-26), where common. Common also in central South Australia (64-67, 70, 81-83), a few seen in the southeast (86), but not seen in the southwest (69, 71-80) or, surprisingly, at Granite Downs (68). Common in southern Queensland (58-61) north to Glendower (55) and at most camps in New South Wales (92, 93, 95).

Habitat. In drier areas only where eucalypts were found; usually along creeks or in mixed woodland with acacia, bloodwoods and corkwood trees. In the east also in savanna woodland or open forest (55, 86, 92, 93, 95) once with Yellow Box and ironbark association (93). Open woodland and low bushes (95).

Behaviour. A very active, noisy and inquisitive species. Also very aggressive, several birds at times fighting and chasing each other, while smaller birds such as pardalotes, Mistletoe Birds and small honeyeaters were regularly chased out of food trees. Persistently mobbed Boobok Owls and attacked other species in flight, especially the Grey Shrike-thrush *Colluricincla harmonica:* but was in turn sometimes driven out of food trees by the larger Yellow-throated Miner *Manorina flavigula*. It was often seen making aerial sallies after flying insects. Apart from the well known *chick-oo-wee* a variety of calls were noted: *chak-chak-chak;* scolding notes that rise and fall in pitch; a mewing call; short stuttering calls; squeaking notes while mobbing a Pallid Cuckoo; a ringing *feeu-feeu-feeu* given during display flight.

Food. Stomachs contained some beetle and ant remnants, seeds, unidentifiable vegetable matter, nectar and pollen.

Annual Cycle. A juvenile with half-grown rectrices collected Dunedoo (93) 19 March confirms breeding there had taken place in February: the other young bird, from Warburton (23) in August, was in post-juvenile moult. Nearly all adults from central Western Australia (20-26) had enlarged gonads August-September and so had the males, but not the female, from southern Queensland (59-61) in May – June. Elsewhere there was no indication of breeding.

Impermanent Colours. *Legs* light flesh brown, brown. *Iris* brown. *Bill* adults black; juvenile olive brown with area above nostrils yellow. *Mouth* yellow to orange.

Geographical Variation. Our series shows clinal variation with the brightest and palest birds occupying the drier interior regions (*M. p. centralia*) and becoming progressively duller and darker in the east and southeast, but a larger series is needed from the peripheral areas to establish how many races it is useful to recognise amongst them.

Meliphaga ornata — Yellow-plumed Honeyeater

Specimens. Western Australia. Cocklebiddy (15) 10 July 65, 1 ♂, 2 ♀, 1 ♀ skel. South Australia. Kingoonya (70) 19 June 65, 1 ♂. Nonning (79) 30 Mar – 3 Apr 65, 3 ♂ (19.0, 21.0 g), 1 ♀ (16.4 g).

Distribution & Abundance. Moderately common at Cocklebiddy (15), a few at Kingoonya (70) and at Streaky Bay (74). Very common at Nonning (79).

Habitat. In mallee (15, 79) or open Mulga woodland near mallee with some Black Oak and many small flowering shrubs (70).

Behaviour. Active and arboreal, often skirmishing with its own kind and other honeyeaters such as *M. penicillata*.

Food. Insects and nectar.

Annual Cycle. Gonads generally small. At Nonning (79) birds were moulting wings and tail feathers.

Impermanent Colours. *Legs* grey or greyish brown. *Iris* dark greyish-brown. *Bill* black. *Mouth* bright yellow.

Meliphaga leucotis — White-eared Honeyeater

Specimens. Western Australia. Newdegate (14) 14 – 19 Apr 66, 1 ♂, 1 imm ♂, 1 ø, 2 ♂ skel, 1 ♀ skel, 2 fl. South Australia. Kingoonya (70) 19 June 65, 2 ♀. Nonning (79) 1 Apr 65, 1 ♂ (22.2 g). New South Wales. Tenterfield (95) 10 – 15 Apr 63, 2 ♂, 1 ♀, 1 imm ♀, 1 ♀ skel.

Specimens to other Museums. NMV 1 ♀ skel Tenterfield (95).

Distribution & Abundance. Common at Tenterfield (95) and encountered in small groups of five or six; elsewhere moderately common (14) or scarce (70, 79).
Habitat. In Salmon Gums, Blackbutt, Gimlet and mallee associations (14). In mallee intergrading with Mulga shrub with shrubs (70); open forest (79) or dry sclerophyll forest, at 2,800 feet (95).
Behaviour. Once seen hopping about in leaf litter and also examining the roots and trunks of trees in the manner of a Rufous Tree-creeper *Climacteris rufa*. It has a clear voice, with a variety of far-carrying calls including a liquid *quilp* uttered by birds in a foraging party (95) (PRC).
Food. Beetles and other small insects, also probably some nectar.
Annual Cycle. Gonads of all birds examined were very small. In April birds at Tenterfield (95) had nearly completed moult and the young bird from Newdegate (14) was acquiring the black areas to the sides of the head, chin and throat of the adult.
Impermanent Colours. *Legs* black. *Iris* brown. *Bill* black. *Mouth* orange.
Geographical Variation. Specimens from Tenterfield (95) are a brighter green above with the crown of the head less intensely streaked than either the male from Western Australia (14), or the birds from South Australia (70, 79). Salomonsen (1967 in Peters' 12) refers Western Australian birds to *M. l. novaenorciae* and all others to *M. l. leucotis*. Our specimens suggest that South Australian birds (70, 79) are intermediate, closer to *M. l. novaenorciae*.

Meliphaga melanops — Yellow-tufted Honeyeater

Specimens. VICTORIA. Victoria Range (90) 22 Jan 63, 1 ♀, 2 imm ♂, 2 fl.
Field Notes. Not observed elsewhere, but at Victoria Range (90) fairly common in parties of five to eight birds. Usually seen high in the trees of open woodland consisting of stringy-bark, *Banksia*, saplings and scrub.
Annual Cycle. The female was in full moult with very small gonads.
Impermanent Colours. *Legs* black. *Iris* brown. *Bill* black. *Mouth* black. *Gape* yellow in immatures.

Meliphaga unicolor — White-gaped Honeyeater

Specimens. WESTERN AUSTRALIA. Mount Anderson (28) 18 May 68, 1 fl (33.8 g). Derby (29) 26 May 68, 1 ♂ imm? (31.4 g). Geikie Gorge (31) 2 – 4 June 68, 1 ♂ (31.6 g), 1 ♀, 2 ♂ skel (34.0 g). Mount Bell (32) 21 June 68, 1 ♀ (27.0 g). NORTHERN TERRITORY. Oenpelli (43) 4 – 7 Oct 68, 2 fl (30.9, 34.9 g), 1 ♂ skel. Pine Creek (44) 30 – 31 Aug 68, 1 ♂ (29.1 g), 1 ♀ (32.4 g). QUEENSLAND. Moonlight Creek (47) 4 – 12 May 64, 3 ♂ (31.3, 35.8, 40.4 g), 1 ♀ (32.1 g), 1 ø (33.3 g), 1 ø imm? (30.8 g). Inkerman (50) 27 – 30 May 64, 4 ♂ (27.8, 30.5, 30.9, 33.9 g), 1 ø (28.8 g).
Distribution & Abundance. Usually singly or in pairs though occasional parties of four or five birds were seen. Evenly distributed throughout the Kimberleys and collected where encountered. In Northern Territory not recorded at Darwin (41) or Mount Roper (45) possibly because collecting there was largely confined to the mangroves. In northern Queensland common where collected.

Habitat. Generally confined to the wetter areas within its range. In the Kimberleys rarely seen far from the thicker vegetation of water-courses or lagoons in savanna woodland. In Northern Territory it occupied this habitat at Pine Creek (44) and Oenpelli (43), though at the latter it was also found in the monsoon forests of the deeper gorges.

In Queensland found on the coast in thick vine-draped woodland "islands" surrounded by mud-flats (47), tall timber with dense undergrowth (50) and dry mangroves, including the area immediately landward, along a creek (50).

Behaviour. Usually skulking in thick vegetation. More often recorded singly though pairs and larger groups, with individuals well separated or concealed, may be more common than records suggest. The song, a "short rich fluting whistle of about seven notes" (ICJG) was recorded from a stationary bird at Inkerman (50) and similar sequences were noted in the Kimberleys (31). The more general contact calls of active or feeding birds were short harsh repetitive chirps likened (SAP) to those of sparrows *Passer* spp.

More secretive than most honeyeaters, though when feeding mixes more freely with other species. In Northern Territory (41) seen with *Meliphaga albilineata* and in Queensland (47) with *Zosterops lutea* in tropical savanna woodland and (50) with *Colluricincla megarhyncha* in dry mangroves.

Food. Stomach contained seeds, vegetable mash and insect remains. The yellow colouring in some stomachs probably attributable to nectar.

Annual Cycle. Apparent adults throughout the species' range are in various stages of moult. Those not so advanced tended to have the gonads slightly enlarged and were also the only birds with skulls recorded as fully pneumatised. One recorded immature (30.8 g) from Moonlight Creek (47) had the skull not fully pneumatised and gonads not found. This specimen does not differ noticeably from adult birds. Others with not fully pneumatised skulls and inactive gonads vary in weights and measurements and only one was recorded as having a yellow gape flange.

Impermanent Colours. *Legs* grey, blue-grey or black; soles, white. *Iris* olive-grey, brown or grey-brown. *Bill* upper mandible, black; lower mandible, black, once "tending to brown". *Mouth* yellow or orange. *Gape* with white or yellow (immature ?) patches.

Meliphaga macleayana — Macleay Honeyeater

Specimens. QUEENSLAND. Big Tableland (51) 2 July 64, 3 ♂ (31.5, 32.9, 34.2 g), 1 ♀ (29.5 g). Ayton (52) 22 – 30 June 64, 1 ♂ (37.4 g), 1 ♀ (29.2 g). Walter Hill Range (53) 21 July 64, 1 ♂ (30.3 g). Tully (54) 11 – 18 July 64, 1 ♂ (32.0 g), 2 ♀ (25.6, 27.5 g).

Distribution, Abundance & Habitat. Found only in the main rainforest block of northern Queensland at varying altitudes from 20 feet at Ayton (52) to 2,300 feet at Big Tableland (51).

Behaviour. Encountered singly or in pairs, sometimes sitting in vines only 3 ft above the ground at others feeding on vertical tree trunks and in vines up to 50 ft. A pair once seen feeding on ripe mandarins in company with *Ptiloris victoriae* and on another

occasion associating with *Myzomela obscura* and *Zosterops lateralis*. The call was a musical *chewit-che-wew* (SAP).
Food. Insects, including beetles, ants and once, a grasshopper.
Annual Cycle. Gonads were small.
Impermanent Colours. *Legs* bluish-grey to grey. *Iris* blackish brown. *Bill* black. *Mouth* pale yellowish-pink, whitish-grey distally or white grading to yellowish in throat. *Bare skin around eye* usually white or whitish, but one male (52) greenish-yellow.
Other Notes. Males are heavier than females and have longer wings (6 ♂ 85-92, 4 ♀ 80-85 mm).

Meliphaga frenata — Bridled Honeyeater

Specimens. QUEENSLAND. Big Tableland (51) 2 July 64, 1 ♀ (31.6 g). Ayton (52) 21 June 64, 1 ♀ (28.4 g).
Field Notes. The only birds seen were the two collected, one of which was in tropical rainforest at 2,200 feet (51) and the other in a small pocket of rainforest along the Bloomfield River (52) almost at sea level. Both were high in the trees.
Food. Small black beetles.
Annual Cycle. Gonads very small, plumage fairly fresh.
Impermanent Colours. *Legs* grey-blue. *Iris* pale blue or grey-blue. *Bill* distal half black, proximal half bright yellow. *Mouth* yellow. *Bare skin round eye* grey.

Melithreptus brevirostris — Brown-headed Honeyeater

Specimens. WESTERN AUSTRALIA. Green Range (13) 2 Apr 66, 1 ♀ skel. Newdegate (14) 14 – 16 Apr 66, 1 ♂, 1 ♀, 1 ♂ skel, 1 ♀ skel, 3 fl. SOUTH AUSTRALIA. Kingoonya (70) 19 June 65, 1 ♀. Aroona Valley (82) 10 Apr 65, 1 ♂ (15.0 g). NEW SOUTH WALES. Dunedoo (93) 20 Mar 63, 1 ♂.
Field Notes. In southwestern Australia (13, 14) groups of a dozen or so were encountered on several occasions but it was rare in South Australia (70, 82). Seen also at Streaky Bay (74) and Tenterfield (95). Found in mallee on the edge of scrub or heath (13, 14, 70), native pine (82) and forest (93). Generally a rather restless species moving from tree to tree. Call *breet-breet*.
Food. Insects and vegetable matter.
Annual Cycle. All March – April birds (13, 14, 18, 93) were in moult. The June female from Kingoonya (70) is in fairly fresh plumage. Gonads of all birds inactive.
Impermanent Colours. *Legs* mostly light flesh-brown. *Iris* brown. *Bill* black. *Mouth* yellow or orange. *Bare skin round eye* sandy yellow or deep dull yellow.
Geographical Variation. Of our five skins the male from New South Wales (93) stands out through having the feathers of the crown grey with dark brown centres, whereas in all others they are uniform dark brown: below it is slightly paler, less buffy than those from South Australia (70, 82), while those from Western Australia (14) are again slightly buffier. Apart from this slight difference the Western and South Australian birds are alike except that the Aroona Valley (82) male appears greyer on the mantle through the presence of some old, worn feathers. The condition of older specimens from all states suggests that plumage may fade after death as well as in life.

On locality our birds should belong to the three races recognised on the mainland, *M. b. brevirostris* (93), *M. b. augustus* (70, 82), *M. b. leucogenys* (14) but the material is inadequate for judging the validity of these forms.

Melithreptus lunatus — White-naped Honeyeater

Specimens. WESTERN AUSTRALIA. Scott River (11) 19 Feb – 3 Mar 66, 1 ♂, 1 ♀, 1 ø, 2 imm ♂, 1 imm ø, 2 ♂ skel, 3 fl. Mount Barker (12) 22 Mar 66, 1 ♀ skel. Green Range (13) 6 Apr 66, 1 ♂. NEW SOUTH WALES. Tenterfield (95) 10 Apr 63, 2 ♂, 1 ♀.
Specimens to other Museums. NMV 1 ♂ Tenterfield (95).
Distribution & Abundance. In the southwest common at Scott River (11) in loose flocks or family parties; in small numbers at Mount Barker (12) and at Green Range (13), either single birds or small nomadic parties.

In the southeast in small numbers at Coleraine (89) but common in the ranges at Tenterfield (95).
Habitat. Well established forested areas of the southwest, particularly in Karri, Red Gum and Jarrah (11, 12) and mixed and open woodland (13). In New South Wales (95) in heavily forested areas at 2,800 ft and in open second-growth forest and low scrub.
Behaviour. Observed singly or in loose flocks actively feeding on eucalypt blossoms and taking insects. At Tenterfield (95) large flocks of small to medium-sized honey-eaters were seen moving south over the canopy of forest, including some of this species and many *Meliphaga chrysops* (PRC).
Food. Stomachs contained small insects, beetles, some vegetable matter and pollen.
Annual Cycle. In Western Australia breeding was apparently over in February and March with immature birds about, some (fully-grown) being fed by parents, and all adults in moult with inactive gonads. In New South Wales (95) adults in April were in fresh plumage and showed no gonad activity.
Impermanent Colours. *Legs* light purplish-brown or purplish-grey with sandy or dull yellow soles to feet. *Iris* brown. *Bill* black. *Mouth* yellow to orange. *Bare skin round eye* either lower lid light blue, upper white (11, 12, 13) or upper and lower lid orange (95).
Geographical Variation. *M. l. chloropsis* (11-13) of the southwest is a longer billed form, having the bare skin around the eye blue and white, instead of orange as in *M. l. lunatus* (95).
Other Notes. Mathews (1924: 243) gives measurements which indicate that the female is larger than the male. Specimens in the British Museum show the reverse, wings 4 ♂ 80-85, 3 ♀ 73-75 (*chloropsis*) and 6 ♂ 77-80, 5 ♀ 68-75 (*lunatus*).

Melithreptus albogularis — White-throated Honeyeater

Specimens. WESTERN AUSTRALIA. Geikie Gorge (31) 3 June 68, 1 ♂. Mount Bell (32) 21 June 68, 1 ♂ (13.0 g). Mount Caroline (33) 25 June 68, 2 ♂ (11.3, 11.7 g), 1 ♀ (10.6 g), 1 fl (10.4 g). Joint Hill (34) 1 July 68, 1 ♂ (12.3 g). Mount Elizabeth (35) 13 July 68, 1 ♂ (12.2 g). NORTHERN TERRITORY. Pine Creek (44) 28 Aug – 2 Sept 68, 2 ♂ (11.4, 11.8 g), 1 imm ♂ (12.5 g), 1 ♂ skel, 3 ♀ skel (9.0, 10.3 g), 5 fl (9.6, 9.9, 10.2, 12.2, 12.8 g).

Queensland. Moonlight Creek (47) 13 May 64, 1 ♂ (11.1 g). Glenore (49) 10 – 17 Apr 64, 2 ♂ (10.5, 11.0 g), 1 ♀ (9.6 g). Glendower (55) 22 – 28 Mar 64, 1 ♂ (12.3 g), 1 ♀ (11.1 g), 1 ø (9.1 g), 1 juv fl (8.8 g). New South Wales. Grafton (94) 30 Mar – 3 Apr 63, 2 ♂, 1 ♀, 3 fl.

Specimens to other Museums. NMV 1 fl Grafton (94).

Distribution & Abundance. In the Kimberleys found in the central camps (31-35). In all Northern Territory camps except at Mount Roper (45). Also found at other camps in Queensland (50, 52, 54, 56, 57). In New South Wales common at Grafton (94).

Habitat. In the Kimberleys (31-35) riverside gums in rocky gorge country, river valleys and savanna woodland in fairly close proximity to fresh water, at altitudes between 1,000 and 2,000 ft. In Northern Territory (44) savanna woodland with wet creeks. In northern Queensland (47, 49) tropical savanna woodland with flowering tea-trees bordering creeks. Low arid woodland, savanna woodland, and open forest; up to 1,500 ft (55). Open grassy forest near rainforest (57). Lowland forest with tea-trees (94).

Behaviour. Single or in small loose parties. Frequently seen hanging upside down taking insects from leaves or inspecting eucalypt blossoms.

Food. Mainly small insects.

Annual Cycle. In the Kimberleys (31-35) in June and July specimens are mostly worn with slight body moult and gonads small or slightly enlarged. Northern Territory (44) August birds are similar, but without apparent moult. The young male was acquiring black plumage to the head; an older specimen in the British Museum collected at Katherine, 17 September, is in a slightly more advanced state of moult to adult plumage. In northern Queensland March birds (55) are fresh, completing body moult, with no gonad activity and a juvenile (fluid specimen) was collected. In April (49) one male had testes fully enlarged, and in May (47) one had them slightly enlarged and had some body moult. In July (54, 56, 57) there was some gonad enlargement. Further south birds from Grafton (94), March – April, were fresh or completing moult with gonads small.

Impermanent Colours. *Legs* dark flesh or brown. *Iris* light chestnut or reddish brown. *Bill* black; immature dark brown, paler at base. *Mouth* orange. *Bare skin round eye* whitish with a faint blue tint above the eye.

Geographical Variation. Our Kimberley (31-35), Northern Territory (44) and north Queensland (47. 49) birds together with older specimens from Darwin, Katherine, South Alligator River, Normanton and Cape York are similar and refer to *M. a. subalbogularis*. They have the upperparts especially the rump, a brighter yellower green than others from southeast Queensland and New South Wales (94), *M. a. albogularis*. The series from Glendower (55) is intermediate.

Melithreptus gularis — Black-chinned Honeyeater

Specimens. New South Wales. Grafton (94) 3 Mar 63, 1 ♀.

Field Notes. Found in open forest and never encountered in parties exceeding three or four birds, unlike most *Melithreptus* spp. Seen also at Dunedoo (93).

Annual Cycle. Ovary not enlarged. In fresh plumage.
Impermanent Colours. *Legs* sandy. *Iris* brown. *Bill* black. *Mouth* orange. *Bare skin above eye* pale blue.

Melithreptus laetior Golden-backed Honeyeater

Specimens. WESTERN AUSTRALIA. Chichester Range (3) 20 – 23 May 66, 5 ♂, 1 ♀ skel, 1 fl. Mount Anderson (28) 28 Apr – 18 May 68, 1 ♂, 1 ♀ (19.1 g), 1 juv ø (17.4 g), 1 ♂ skel (17.2 g), 1 imm ø skel, 2 fl. Derby (29) 21 – 29 May 68, 1 ♂, 2 ♂ skel, 2 ♀ skel (16.1 g), 1 ø skel (15.8 g). Mount Elizabeth (35) 8 – 13 July 68, 2 ♀ (20.2 g). Mount Remarkable (37) 26 July 68, 1 ♂ (19.3 g).
Distribution & Abundance. Local; several sight-records at Karratha (1) and seen at most of the Kimberley camps (28-37), and Victoria River (40) in Northern Territory but not elsewhere.
Habitat. Savanna or riverine woodland, often eucalypts. Also in other flowering trees, bushes and herbage including *Grevillea* (37) and *Crotalaria* (28).
Behaviour. In pairs, small groups or family parties. Occasionally in mixed bird parties, including honeyeaters such as *Meliphaga penicillata* and *Lichmera indistincta* and pardalotes. Mostly observed high up in foliage and often hanging upside down examining blossoms for considerably longer periods than most honeyeaters. But sometimes low down on flowering bushes when feeding and once searching for insects under bark of eucalypt (PRC).

Voice recorded variously as a loud, single or double, whistling or bell-like note. A juvenile gave a quick *peep-peep* similar to *Climacteris melanura* (DJF).
Food. Insects and nectar (liquid green matter).
Annual Cycle. May birds from the Chichester Range (3) show some wear or slight moult and the gonads were enlarged in two out of three males (see Harrison & Colston, 1969). The juvenile (28) collected on 13 May indicates breeding in April in the Kimberleys. The Derby (29) specimen, 29 May, shows slight body moult and all birds from there had small or only slightly enlarged gonads. July birds (35, 37) had similarly inactive gonads and were in partially worn plumage.
Impermanent Colours. *Legs* light brown with sandy or orange feet. *Iris* dark brown. *Bill* black; juvenile light brown, lower mandible orange basally. *Mouth* orange or yellow, brighter in juvenile. *Bare skin over eye* yellow-green, the same colour as the mantle; this changed to blue in half an hour, and to black one hour after death.
Geographical Variation. The Chichester Range (3) series does not differ significantly from the Kimberley series (28, 29, 35, 37) and all are of the nominate form.
Other Notes. The juvenile shows only a slight yellowish tint to the feathers of the lower back, although the rump is yellow. Otherwise the upper parts and head are a very pale brown with a dilute version of the adults' head pattern, the blackest area being the loral region extending below the eye to the ear coverts. Underparts pure white, very pale grey on chin.

Entomyzon cyanotis Blue-faced Honeyeater

Specimens. WESTERN AUSTRALIA. Mount Bell (32) 14 – 19 June 68, 3 ♂ (110.5, 121.2,

121.4 g), 3 ♂ skel (110.6, 112.1, 115.0 g), 2 fl (100.0, 116.6 g). NORTHERN TERRITORY. Darwin (41) 17 Sept 68, 1 ♂ (116.6 g). QUEENSLAND. Moonlight Creek (47) 2 – 11 May 64, 4 ♂ (110.9, 118.8, 121.8, 135.0 g), 5 ♀ (94.1, 98.6, 100.0, 106.2, 118.8 g), 4 imm ♂ (104.6, 111.5, 112.9, 113.2 g), 1 imm ♀ (91.3 g), 1 imm ø (90.8 g). Inkerman (50) 25 May – 3 June 64, 1 ♂ (111.8 g), 2 ♀ (89.1, 93.3 g), 1 imm ø (89.3 g). Glendower (55) 1 Apr 64, 1 ♀ (88.4 g). Banoona (58) 28 Mar – 9 Apr 63, 1 ♂, 1 ♀, 1 imm ♀. Tyrone (59) 15 May 63, 2 ♂ skel.

Specimens to other Museums. NMV 1 imm ♂, 1 ♂ skel Banoona (58).

Distribution & Abundance. Mount Bell (32) represents a western extension of the known range (Gannon, 1962). In Northern Territory fairly common at Darwin (41), Oenpelli (43), Pine Creek (44) and Borroloola (46). Common in suitable habitat in northern Queensland (47, 49, 50) but not found at Wernadinga (48). Further south locally numerous at Tyrone (59) and a few at Banoona (58) and Grafton (94).

Habitat. Generally in tall trees of riverine or open forest and woodland with a creek or some water source in the vicinity.

Behaviour. Sometimes in groups of twenty and thirty or more (47, 59), once in loose association with ten *Manorina flavigula* (32). Voice recorded as similar to that of *Manorina flavigula* but deeper; also a harsh mew.

Food. Stomachs contained medium-sized insects, including grasshoppers, nectar and fragments of leaves.

Annual Cycle. Individuals in varying degrees of moult were taken at all localities except Darwin (41) (and possibly Tyrone (59) – no skins retained), and at most places young birds were collected. The only adults with enlarged gonads were from Moonlight Creek (47) although birds were finishing or had finished moult.

Impermanent Colours. *Legs* blue grey to greyish brown; immature olive green. *Iris* straw or bright yellow. *Bill* black with the basal third pale blue or bluish white in the adult but dull yellow to yellowish green in the immature. *Mouth* black; immature yellowish, streaked black. *Bare skin on face* adults, cobalt blue grading to a paler blue above eye level; immatures yellow-green, usually dull blue around and below eye: the area of bare skin more extensive in Inkerman (50) birds.

Geographical Variation. There is clinal variation in several characters. Birds from southern Queensland (58, 59), *E. c. cyanotis,* have the least amount of white or pinkness beneath the wing; the throat-patch, wings and tail are not pure black, and the bills are comparatively short. The amount of white or pink beneath the wing and on the base of the primaries increases northwards, eventually forming a clear patch above and below the wing in all northern birds but most marked in those from the Kimberleys (32). Birds from northwestern Queensland (47), Northern Territory (41) and the Kimberleys (32) also have the throat-patch, wings and tail blacker than *E. c. cyanotis,* and the bill comparatively long: they can all be referred to *E. c. albipennis*. Cape York birds and ours from Inkerman (50), *E. c. harterti,* have larger areas of bare skin on the head, brighter green mantles and rumps and are slightly smaller than *albipennis* and *harterti*. The Glendower (55) female is close to *cyanotis* but shows an approach to *harterti* in the colour of the mantle, but there seems to be no intergradation across the Carpentaria Gap between *harterti* from Inkerman (50) and *albipennis* from Moonlight Creek (47).

Philemon citreogularis Little Friar-bird

Specimens. Western Australia. Mount Anderson (28) 27 Apr – 18 May 68, 1 ♂ (70.0 g), 1 ♀, 1 ø, 1 imm ø, 1 ♂ skel (78.6 g), 2 fl. Mount Bell (32) 15 – 22 June 68, 2 imm ♂ (57.6, 70.5 g), 1 imm ø (52.7 g), 1 ♂ skel (55.0 g), 2 fl (60.0, 71.1 g). Northern Territory. Darwin (41) 11 Sept 68, 1 imm fl (76.4 g). Pine Creek (44) 2 Sept 68, 2 ♂ skel (50.0, 50.0 g). Mount Roper (45) 30 Oct 68, 1 ♀ (46.4 g), Borroloola (46) 8 – 11 Nov 68, 3 fl (48.1, 53.3, 55.8 g). New South Wales. Albury (92) 13 Mar 63, 1 fl.
Distribution & Abundance. In the Kimberleys (30-38) found at most camps. In Northern Territory mostly rather scarce (41, 44, 45) but common at Borroloola (46). In Queensland very common at Glendower (55) and found also at Moonlight Creek (47) and Glenore (49). In New South Wales found at Grafton (94) as well as Albury (92).
Habitat. At Mount Anderson (28) and Mount Bell (32), savanna woodland, near water, also wooded gorges with tall grass (32). Similar habitats in Northern Territory (44, 46) and Moonlight Creek (47), again near creeks on lagoons. Mangroves at Darwin (41) and stony scrub at Pine Creek (44). At Glendower (55) a large party found in a patch of stunted or dead woodland. At Albury (92) cultivated savanna.
Behaviour. Appeared singly, in pairs, or parties of six to 15 which usually kept up a continuous discordant chatter. At Borroloola (46) a flock was seen feeding on eucalypts. One bird came to collector's whistling (28). An adult observed feeding a juvenile Koel *Eudynamys scolopacea* at Glenore (49) on 16 April was presumably its foster parent. Voice, an abrupt alarm-note.
Food. Insects including ants, and beetles.
Annual Cycle. All Kimberley (28, 32) birds examined were in moult, including those with imperfectly pneumatised skulls, believed to be immatures moulting into adult plumage. The October female from Mount Roper (45) was mainly worn with a few fresh-looking feathers in the wing coverts: the ovary was small.
Impermanent Colours. *Legs* slate-grey. *Iris* brown. *Bill* black. *Mouth* variable, usually purplish-pink orange or whitish individually. *Bare skin on side of face* blue-black.
Geographical Variation. Weights and the measurements of the females from Mount Anderson (28) and Mount Roper (45) indicate that Kimberley birds *P. c. occidentalis* are larger than those from Northern Territory *P. c. sordidus* (wing 1 ♀ 128 against 117 mm; bill 32 against 30 mm) while a male in the British Museum from Roebuck Bay south of Derby (29) has a wing 139 mm against 131 for the male from Mount Anderson (28). With the exception of the Roebuck Bay bird the northwestern birds appear generally whiter below with brighter yellow on the breast, but our material is inadequate for proper comparison.

Philemon buceroides (including *P. gordoni*) Melville Island Friar-bird

Specimens. Northern Territory. Darwin (41) 14 Sept 68, 1 ♂ (98.8 g), 2 ♀ (84.8, 91.2 g), 1 imm ♂ (86.1 g). Oenpelli (43) 14 Oct 68, 1 ♂, 1 ♀, 2 juv ♂.
Distribution & Abundance. Found in small numbers in the Noogoo Swamp and Hope Inlet (41) and 9 miles SSW of Oenpelli (43).

Habitat. Mangroves in the tidal stretch between swamp and 'wet' mangrove; monsoon forest and thickets within one mile of the sea (41). Thick vegetation at foot of sandstone gorge near savanna woodland (43).
Behaviour. Seen actively feeding at all levels in eucalypts. Voice, a harsh-warbling; "louder than *P. argenticeps* from which it differs in having something of a repetitive theme" (BB).
Food. Stomachs of adults contained insects (ladybirds), a spider, seeds, berries and flowers, and of the juveniles insects and a small lizard.
Annual Cycle. At Darwin (41) two adults had enlarged gonads and were in rather worn plumage with most rectrices very abraded: the second (heavier) female was completing moult from immature to adult, a few whitish fringing edges to mantle remaining. The immature male has distinct whitish crescents to sides of neck and back together with a slight touch of yellow on upper chest. Adults at Oenpelli (43) had small gonads and were rather worn, commencing moult. The male has several rectrices and most of throat and upperbreast feathers fresh, whereas they are exceedingly worn in the female. The two juveniles, which were being fed by the adults, were capable of short flights.
Impermanent Colours. *Legs* bluish grey to dark grey. *Iris* adult and immature deep red, juvenile dull brownish red. *Bill* black. *Mouth* adult purplish pink or whitish, immature male (41) and juveniles (43) yellow. *Bare skin on face* black; juvenile dark grey.
Geographical Variation. There appear to be two discrete populations, one from Melville Island and the coastal mangroves on the adjacent mainland at Darwin (41) to which the name *gordoni* is applicable, and one from the inland sandstone gorges around Oenpelli (43). The male from Oenpelli (43) differs from those from Darwin (41) in having a white, silvery wash on the head, while the feathers of the ear-coverts and below the eye are brown rather than black and broader, less silky and hair-like, differing from those of all other specimens examined including material from Timor and Lesser Sunda islands. The Oenpelli (43) birds are also larger with stouter feet than those from Darwin (41).

Since this report was prepared Parker (1971) has discussed these populations and concluded that the mangrove populations should be treated as a distinct species, *P. gordoni*. He discussed differences in size and head-colour finding, on more material than available to us, the Oenpelli (43) birds indistinguishable from *P. b. buceroides* of Timor and Savu. He did not however comment on the texture of the facial feathers.

Philemon argenticeps — Silver-crowned Friar-bird

Specimens. Western Australia. Mount Bell (32) 16 – 17 June 68, 1 ♀ (69.0 g), 2 fl (73.1, 75.7 g). Mount Elizabeth (35) 13 – 14 July 68, 1 ♂ (89.6 g), 1 fl (91.6 g). Northern Territory. Darwin (41) 12 – 22 Sept 68, 2 ♂ (76.0, 76.1 g), 1 fl (73.1 g). Pine Creek (44) 29 Aug 68, 2 ♂ (92.0, 95.3 g), 1 ♂ skel (90.1 g). Mount Roper (45) 30 Oct 68, 1 ♂ skel (86.6 g). Borroloola (46) 8 Nov 68, 1 fl (66.2 g).
Field Notes. Found at most camps in the Kimberleys and Northern Territory, mostly in rather small numbers. In Queensland, recorded at Moonlight Creek (47), Tully (54)

and Glendower (55). Usually in savanna woodland, sometimes near water (32, 35, 46) in thick vegetation, often in dead or stunted trees (41, 45) or scrub country (45). Active when feeding, sometimes hanging upside down; often in company with *P. citreogularis* (32, 45, 47), but in smaller numbers and less noisy.

Voice, harsh and discordant; a guttural croaking.

Food. Insects, including grasshoppers (47) and beetles.

Annual Cycle. June, July and August birds (32, 35, 44) were in moult with small gonads, the August birds having fresh secondaries and rectrices. Of two September birds (41), one was in breeding condition, with testes fully enlarged, and moult completed. The other had smaller, unequal testes and still retained some worn feathers.

Impermanent Colours. *Legs* slate-grey or black. *Iris* deep red or red-brown (the red fades rapidly to brown or white after death). *Bill* black. *Mouth* pink washed purple or white, twice greyish white. *Bare skin on head* black.

Geographical Variation. Northern Territory birds are currently recognised as *P. a. alexis* and Kimberley birds as *P. a. argenticeps,* but no significant difference is apparent in our series.

Philemon corniculatus — Noisy Friar-bird

Specimens. QUEENSLAND. Glendower (55) 20 Mar 64, 1 ♀ (91.6 g). Tyrone (59) 15 May 63, 1 ♂. NEW SOUTH WALES. Dunedoo (93) 18 Mar 63, 1 ♂ skel.

Field Notes. A few recorded in New South Wales (92-94), and in southern Queensland (58-60) north to Glendower (55) and Cape Gloucester (56).

In open woodland; tea-tree (56) sandalwood, Mulga and gidgee (59), tall eucalypts (93). At Glendower (55) near a water-hole in river bed. Usually singly or in pairs, ten birds being the greatest number seen together (92). Voice variously described as *chok-chok, four-o-clock* and a noisy double call.

Food. Insects, including grasshoppers (55), beetles (93) and soft-bodied kinds.

Annual Cycle. The March male (55) is in general moult, especially heavy on back of neck, but in the May male (59) moult was completed except for some inner secondaries. Gonads in both birds small. The Tyrone (59) bird had a large amount of body and subcutaneous fat.

Impermanent Colours. *Legs* dark grey (once buffy down front ridge). *Iris* brown. sometimes mottled with grey. *Bill* black. *Mouth* dark grey to black (sometimes mottled). *Bare skin on head and neck* dull black.

Geographical Variation. Salomonsen 1966 recognises three races on size. The large *P. c. corniculatus* from New South Wales, wing ♂ 159-166: the intermediate *P. c. clamans* from southeastern Queensland, wing ♂ 154-156: and the small *P. c. ellioti* from northern Queensland, wing ♂ 137-147 mm.

The Glendower female, wing 145 mm, is clearly *ellioti* on both locality and size. The Tyrone (59) male, wing 159 mm, on locality should be *clamans* but is as large as *P. c. corniculatus.*

Phylidonyris pyrrhoptera — Crescent Honeyeater

Specimens. VICTORIA. Victoria Range (90) 31 Jan 63, 1 ♂, 1 ♀ (?imm), 1 ø (?imm ♂).

Field Notes. Fairly common in sclerophyll forest with a scrubby understory, on rocky slopes.

Noisy and active, decidedly more arboreal than other *Phylidonyris* species. Observed coming to streams to drink and bathe mid day. Voice a loud distinctive *egypt* frequently repeated (PRC).

Annual Cycle. The adult male was in breeding condition; the other two birds had small or indistinct gonads, but are not in moult.

Impermanent Colours. *Legs* grey. *Iris* adult male red; immatures (?) brown. *Bill* males black; immature female brown. *Mouth* adult male black; immature flesh; immature female yellow.

Phylidonyris novaehollandiae — New Holland Honeyeater

Specimens. WESTERN AUSTRALIA. Scott River (11) 19 Feb – 1 Mar 66, 4 ♂, 2 ø, 1 ♂ skel. Mount Barker (12) 21–28 Mar 66, 3 ♂, 2 ♀, 1 imm ♂, 1 juv ♂, 2 ♀ skel, 1 fl. Green Range (13) 31 Mar–4 Apr 66, 1 ø skel, 1 fl. SOUTH AUSTRALIA. Port Lincoln (77) 16 Oct 63, 1 ♂. Furner (86) 7 Jan 63, 1 ♂ skel. VICTORIA. Victoria Range (90 31 Jan 63, 3 fl.

Distribution & Abundance. Seen regularly in the southwest (8-13). Elsewhere seen also at Aroona Valley (82) once, Nelson (88) and Little Desert (91). Common in the central Victoria Ranges (90).

Habitat. In Western Australia found in a variety of flowering trees and scrub – tea-tree, peppermint, Jarrah, Karri, apple; also on bracken-covered hillside (12) and often open heathland (11, 13), the characteristic habitat of *P. niger*. Elsewhere, in high thick scrub (77), scrub with light timber (86) or on sclerophyll-covered slopes (90).

Behaviour. Most birds from Scott River (11) were in twos or threes but others were in feeding parties, sometimes with *Zosterops* (11) or *Acanthorhynchus* (13). A large flock at Mount Barker (12) was feeding among the fallen fruit in an apple orchard, apparently searching for insects.

Annual Cycle. A pair of birds at Scott River (11) were thought to be nest-building and one male was seen carrying a large feather: the specimens were in moult with little or no gonad activity. The fledgling collected at Mount Barker (12) later in March appears to be 12 – 16 days old: most adults there showed some moult. The Port Lincoln (77) October male is in wing moult while the January specimen from Furner (86) (now a skeleton) was apparently in fresh plumage, with enlarged gonads.

Impermanent Colours. *Legs* black or dark grey; juvenile pinkish-grey. *Iris* white; one immature grey-brown; juvenile dark brown. *Bill* black, immature paler; juvenile flesh-brown with yellow cutting edge. *Mouth* yellow with black throat; juvenile yellow.

Geographical Variation. There is little geographical variation in our series but birds from the southwest have rather less white on the "whiskers" than others and may justify separation as *P. n. longirostris*.

Phylidonyris niger — White-cheeked Honeyeater

Specimens. WESTERN AUSTRALIA. Scott River (11) 22 Feb 66, 1 ♂, 1 ø juv. Green Range (13) 2 – 6 Apr 66, 3 ♂, 1 ø, 1 imm ♀ skel, 1 fl.

Field Notes. Locally common where collected and at Newdegate (14). Typically in open heathland with clumps of mallee nearby (13, 14). Some pairs seen also in New South Wales in coastal heathland at the mouth of the Clarence River (94).
Food. Small insects and vegetable matter.
Annual Cycle. Breeding was apparently over, with young birds about and the adults in moult, those from Green Range (13) being mainly in fresh plumage.
Impermanent Colours. *Legs* black, immatures greyish-black or greyish-brown. *Iris* dark brown. *Bill* black. *Mouth* adults black, immature yellow, juvenile black with yellow gape. *Bare skin* grey.
Geographical Variation. Our birds, *P. n. gouldi,* are clearly separable from material representing *P. n. niger* from the east in barely showing any white across the forehead.
Other Notes – Identification. Apart from differences in pattern, *niger* differs from the closely related *P. novaehollandiae* in having a brown instead of a white or bluish white iris; also it lacks the white hairy filaments to the throat which are present in *novaehollandiae.*

Phylidonyris albifrons — White-fronted Honeyeater

Specimens. Western Australia. Cocklebiddy (15) 10 July 65, 1 ø. Eucla (17) 1 – 5 July 65, 3 ♂. Mount Shenton (18) 2 – 3 Aug 65, 2 ♂, 1 ♂ skel. Warburton (23) 14 Aug 65, 1 ♀. South Australia. Granite Downs (68) 7 June 65, 1 ♀. Kingoonya (70) 19 June 65, 2 ♂. 1 ♀, 1 ø.
Distribution & Abundance. Common or very common where collected except at Granite Downs (68) where only one was found. Not seen elsewhere.
Habitat. Scrubland or mulga in sandy or rocky country, usually with large trees nearby (18, 23, 68, 70). Mixed woodland with mallee (15) and Bluebush (17).
Behaviour. Often in loose parties; alert and shy of man. At Eucla (17) seen in early evening to leave eucalyptus flowers on which it was feeding in order to hawk for flying insects from the tree top (DG).

Voice: the alarm note was a chuckling or squawking monosyllable given at variable rate. The Kingoonya (70) birds were very active and vocal, giving the alarm note and canary-like calls and song-phrases (DG). Also heard to mimic the Crested Bell-bird and Port Lincoln Parrot.
Food. Stomachs contained insects and nectar, once green buds (18).
Annual Cycle. Most June and July birds had small gonads and all were in fairly fresh plumage. August specimens (18, 23) were more worn but not in moult, and their gonads were enlarged.
Impermanent Colours. *Legs* black. *Iris* dark rich brown. *Bill* black. *Mouth* orange or yellow; tongue pinkish or whitish, once yellowish. *Bare skin behind eye* bright red (3 ♂, 1 ♀, 1 ø Western Australia) or pink (1 ♂ Western Australia, 1 ♂, 2 ♀ South Australia), the colour fading entirely within a few hours of death.

Gliciphila melanops — Tawny-crowned Honeyeater

Specimens. Western Australia. Scott River (11) 22 Feb 66, 1 ♀ skel. Green Range (13) 2 – 6 Apr 66, 1 ♂ skel, 2 fl.

Field Notes. Moderately common where collected and at Newdegate (14). Occupying open heathlike country on sea coast, low bush and mallee (14). Seen also in the Victoria Range (90) and at the mouth of the Clarence River (94).
Behaviour. Seen singly, in pairs, or occasionally up to five together. Feeds in low bushes, flies swiftly and strongly often for considerable distances: perches prominently on bare branches or stumps. Aggressive to spinebills *Acanthorhynchus* and other small honeyeaters. Voice a long clear whistle and a mournful long drawn out note (PRC).
Annual Cycle. The Scott River (11) female was in worn plumage with the tail in moult. Gonads were small.
Impermanent Colours. *Legs* dark grey. *Iris* brown. *Bill* black. *Mouth* dark grey.

Ramsayornis fasciatus — Bar-breasted Honeyeater

Specimens. WESTERN AUSTRALIA. Mount Bell (32) 16 – 20 June 68, 1 ♂ (15.1 g), 1 ♀ (14.4 g), 1 imm ♂ (13.2 g). Mount Caroline (33) 25 June 68, 1 ♂ (14.5 g), 1 ♀ (12.7 g). Joint Hill (34) 1 – 3 July 68, 1 ♂ (14.6 g), 2 ♀ (13.4, 13.4 g), 1 imm ♂, 1 ♂ skel (12.4 g). Mount Elizabeth (35), 13 – 15 July 68, 2 ♀ (11.8 g), 1 fl (11.9 g). NORTHERN TERRITORY. Darwin (41) 18 Sept 68, 1 ♂ skel (12.4 g), 1 fl (12.6 g). QUEENSLAND. Moonlight Creek (47) 1 May 64, 1 ♂ (13.8 g), 1 imm ♂ (11.8 g). Glenore (49) 11 Apr 64, 1 ø (11.8 g).
Distribution & Abundance. Found throughout the central Kimberleys (32-36) being numerous at Joint Hill (34). In Northern Territory found at most camps (41, 43-46) common at Pine Creek (44) and Borroloola (46). Only found where collected in Queensland.
Habitat. Mostly savanna woodland near water, but at Darwin (41) at Hope Inlet far closer to salt water than to fresh.
Behaviour. Usually encountered singly sometimes two to four, or forming part of a mixed flock feeding among blossoms. Freeman noted (32) flycatcher-like habits, hawking over water for insects in jerky, irregular fluttering fashion. Voice, soft purr or burring chirrup resembling the Crimson Finch *Neochmia phaeton*.
Food. Stomachs contained insects, and some nectar.
Annual Cycle. Of the Queensland specimens the April adult (49) shows some body moult, the May adult (47) is in wing and tail moult, and the immature is worn. Of the Kimberley series June, July adults, were acquiring fresh plumage and moult was almost complete in mid July. All gonads were small.
Impermanent Colours. *Legs* mostly pinkish-brown. *Iris* light brown. *Bill* dark brown, lighter below. *Mouth* whitish, sometimes with purplish tinge.
Geographical Variation. Our Kimberley birds (32-35) do not differ significantly from those of Northern Territory and Queensland, including, from older collections in the British Museum, three from Victoria River (see Macdonald & Colston, 1966), one from Melville Island (*R. f. apsleyi* of Mathews) and three from Normanton, although three races are usually recognised.

Plectorhyncha lanceolata — Striped Honeyeater

Specimens. QUEENSLAND. Glendower (55) 12 – 25 Mar 64, 1 ♀ (36.3 g), 1 imm ♂ (32.7 g). Cape Gloucester (56) 7 Aug 64, 1 " ♂ " (♀ on size) (33.5 g). Banoona (58)

6 May 63, 1 ♂ skel. Tyrone (59) 10 May 63, 1 ♀ skel. New South Wales. Dunedoo (93) 20 Mar 63, 1 ♂.

Specimens to other Museums. NMV 1 ♂ Banoona (58).

Distribution & Abundance. The specimen from Sinclair Bay Inlet, Cape Gloucester (56) represents an extension of range in Queensland (Gannon, 1962) where it has not previously been found on the coast. It is sometimes coastal in New South Wales and single birds were seen at the mouth of the Clarence River (94) in April 1963 (PRC).

Habitat. Open woodland, lightly timbered or with low bushes; also mulga (59) and mangroves (56).

Behaviour. Nearly always alone, though once in company with miners (58), and once eight together drinking at a dam (55). Melodious song heard once delivered from the crown of a river gum (55).

Food. Insects, including ladybird and small caterpillars. One stomach contained 10-12 black seeds, 4 mm in diameter (58).

Annual Cycle. Of the March specimens, two adults (93, 55) are in moult; one immature (distinguished by being more spotted on the crown, less striped and having brownish tips to wings and tail-feathers and coverts) is fairly fresh. The August specimen (56) is rather worn, especially on the wings. Gonads all small.

Impermanent Colours. *Legs* slate-grey. *Iris* brown. *Bill* grey, tending to darken towards tip. *Mouth* whitish or flesh; immature pinkish-yellow.

Geographical Variation. Our birds and others in the British Museum show that Queensland birds are smaller than those of New South Wales. Wings, Queensland 6 ♂ 109-116, 6 ♀ (103) 107-110; New South Wales 2 ♂ 117-120, 4 ♀ 108-115.

The smallest bird, wing 103 mm, is the "male" from Cape Gloucester (56), presumed to be mis-sexed.

Conopophila whitei — Grey Honeyeater

Specimens. South Australia. Granite Downs (68) 7 June 65, 1 ♂.

Field Notes. The first record from South Australia, though known from southern Northern Territory 200 miles to the north. Perhaps overlooked owing to its inconspicuous plumage and habits (Cowles, 1967). Found on stunted mulga on red sandy soil, where many trees were parasitised by mistletoe; with a flock of Yellow-rumped Thornbills.

Food. Stomach contained mistletoe berries.

Annual Cycle. In fairly fresh plumage, with small testes.

Impermanent Colours. *Legs* black. *Iris* mid-brown. *Bill* black. *Mouth* black.

Conopophila albogularis — Rufous-banded Honeyeater

Specimens. Northern Territory. Oenpelli (43) 28 Sept – 11 Oct 68, 6 ♂ (10.3, 11.0, 12.0 g), 2 ♀ (11.7 g), 1 imm ♀ (10.6 g), 1 ♂ skel (11.1 g), 1 ♀ skel (9.4 g), 2 fl (10.3, 12.1 g). Queensland. Inkerman (50) 27 May – 9 June 64, 2 ♂ (13.3, 13.6 g), 1 imm ♂ (12.1 g), 2 ♀ (11.7, 13.2 g), 1 ø (13.5 g).

Distribution & Abundance. Common but local at Oenpelli (43); possibly one seen at Stokes Hill (41) but not elsewhere in Northern Territory. Fairly common locally at Inkerman (50) the only place it was seen in Queensland.

Habitat. At Oenpelli (43) in flowering trees of savanna woodland in black-soil plain with lakes and sandstone escarpments nearby. At Inkerman (50) only in tropical layered woodland, tea-tree swamps, and mangrove.

Behaviour. Birds in loose flocks of 10-30 individuals, sometimes including *C. rufogularis,* followed one another through flowering trees (43). Also pairs chased each other through foliage very quickly, twisting, turning, in and out of branches, stopping only for a very short time and calling continuously. Voice, a continuous melodious twitter with occasional cheeps; a complicated squeaky song, comparable with *Lichmera indistincta.*

Food. Small insects, spiders, and caterpillars. Nectar also noted in the throat.

Annual Cycle. The Inkerman (50) birds, end of May to June, were in full moult with small gonads. At Oenpelli (43), September – October, moult was almost complete and gonads were small or slightly enlarged.

Impermanent Colours. *Legs* blue-grey or dark grey. *Iris* light brown. *Bill* blackish; base of lower mandible light grey. *Mouth* greyish or pale flesh; immatures cream or dull yellowish.

Conopophila rufogularis Rufous-throated Honeyeater

Specimens. WESTERN AUSTRALIA. Mount Anderson (28) 29 Apr – 7 May 68, 1 ♂, 2 ♀, 1 imm ♂, 1 imm ø, 3 ♀ skel, 1 imm ø skel, 3 fl. Joint Hill (34) 2 July 68, 1 ♀ (10.9 g). NORTHERN TERRITORY. Mount Roper (45) 30 Oct 68, 1 ♂ (9.9 g). QUEENSLAND. Glenore (49) 9 – 11 Apr 64, 1 ♂ (9.1 g), 1 ♀ (10.7 g). Inkerman (50) 30 May – 6 June 64, 3 ♂ (11.0, 13.1, 13.9 g), 3 ♀ (11.2, 11.6, 12.1 g). Glendower (55) 20 Mar 64, 1 ♂ (13.0 g), 1 imm ♀ (11.1 g).

Distribution & Abundance. Seen throughout the Kimberleys (28-35) where it was netted in fair numbers in mixed flocks at Mount Anderson (28). In Northern Territory fairly numerous at Oenpelli (43) and Mount Roper (45) and in smaller numbers at Borroloola (46). In Queensland fairly common in denser habitats where collected, and also at Moonlight Creek (55).

Habitat. Mostly savanna woodland, sometimes along river-beds or gorges in sandstone country (28, 34), and in thicker growth at Inkerman (50). It was found feeding in eucalypts (34), tea-trees (28, 49, 50) and mistletoe (50).

Behaviour. Often found in association with other honeyeaters, particularly *Lichmera indistincta* (34, 50) and *Ramsayornis fasciatus* (49), also lorikeets (34). Otherwise not more than five or six together. Voice, "rasping" note (50).

Food. Small insects including ants, and nectar.

Annual Cycle. In Western Australia the April birds were in very worn plumage (two females) or in full moult (one male). There is no red on the chin or throats of either of the immatures. The July bird (34) had almost completed moult. The October male (45) from Northern Territory was in fresh plumage with small gonads.

In north Queensland, March adults (55) are in worn plumage, both April birds (49) were in full moult, and May/June birds (50) are worn and in varying stages of moult with gonads small. The March immature bird (55) appears older than those from the Kimberleys (28) in April for it is in worn plumage and replacing the wing quills: also the base of the bill in the skin is grey as in adults.
Impermanent Colours. *Legs* grey. *Iris* olive grey. *Bill* very dark brown, base of lower mandible light-grey in adults, yellowish in immatures. *Mouth* grey, throat pinkish, yellower in immatures.
Geographical Variation. Allowing for plumage wear and individual variation the Kimberley birds are only marginally separable from those of Queensland in being paler brown on the upperparts. The difference hardly seems sufficient to justify the recognition of *C. r. queenslandica.*

Conopophila picta — Painted Honeyeater

Specimens. QUEENSLAND. Thylungra (60) 25 May 63, 1 ♂, 1 ♀, 2 imm ⌀.
Specimens to other Museums. NMV 1 ♂, 1 imm ♀ Thylungra (60).
Field Notes. Only found at Thylungra (60) where about 30 birds were present in the area for a week. This is the most westerly record for this nomadic honeyeater (Colston, 1968). Found in pairs or groups up to six together, active and noisy, feeding on mistletoe in Mulga. Voice a loud, liquid *georgie* repeated.
Food. Stomachs contained mistletoe berries and seeds. Apparently mistletoe is the main if not the only food.
Annual Cycle. Groups seen included adults with juveniles, probably less than four to five weeks old, and it was thought that the birds had been nesting nearby. The adult male specimen is completing moult on head and body. The female and an unsexed bird, probably male, are in general moult, the latter probably from immature to adult plumage. The other immature bird is worn. Gonads small.
Impermanent Colours. *Legs* grey to black. *Iris* male red brown, female and immature light brown. *Bill* pinkish, dark at tip. *Mouth* pink, yellow in one immature.

Acanthorhynchus tenuirostris — Eastern Spinebill

Specimens. QUEENSLAND. Walter Hill Range (53) 23 July 64, 1 ♂ (13.3 g). VICTORIA. Victoria Range (90) 31 Jan 63, 1 imm ⌀, 2 imm fl. NEW SOUTH WALES. Tenterfield (95) 12 Apr 63, 1 ⌀.
Specimens to other Museums. NMV 1 imm ⌀ Victoria Range (90).
Field Notes. Common in Central Victoria Ranges (90) and Tenterfield (95); also found at Big Tableland (51).

In tropical rainforest above 2,000 ft (51, 53), mixed woodland and dry sclerophyll forest on slopes of ranges (90). Edges and more open parts of rainforest at Tenterfield (95) at 3,300 ft.

In loose groups of three to five feeding on flowering eucalypts (90), sometimes with mixed bird parties of *Zosterops,* other honeyeaters or small warblers. An active searcher of blossoms from low bushes to tree tops, usually at 10–30 feet.

Food. Nectar and insects.
Annual Cycle. The Tenterfield (95) bird in April was in fresh plumage, completing wing moult.

The Victoria Range (90) bird is in fresh immature plumage; the upperparts resemble those of adults, but it has no pattern on the head and throat, and the abdomen is duller orange. The Queensland male (53) has the skull recorded as not fully pneumatised but is in full adult plumage.
Impermanent Colours. *Legs* adults dark grey or dark brown; immatures grey. *Iris* adults red, immatures brown. *Bill* black, immatures black with gape and base of lower mandible yellowish. *Mouth* adult light grey with yellow throat; immatures yellow throughout.
Geographical Variation. The north Queensland (53) male belongs to the isolated race of the Atherton Tableland *A. t. cairnsensis*. It differs from birds of eastern Australia, *A. t. tenuirostris,* in having a paler throat with no contrasting chestnut throat-band.

Acanthorhynchus superciliosus — Western Spinebill

Specimens. WESTERN AUSTRALIA. Metricup (9) 6 – 11 Mar 66, 2 ♂, 1 imm ♀, 1 ♀ skel, 3 fl. Scott River (11) 19 – 28 Feb 66, 1 ♂, 2 imm ♂, 1 ♂ skel, 1 fl. Mount Barker (12) 25 Mar 66, 1 fl.
Distribution & Abundance. Common at all the southwest camps (8-14), especially between Wonerup (8) and Scott River (11). A few seen in the Sterling Range (7).
Habitat. Sclerophyll forest (9); open Karri forest and clearings near river (11); apple-trees near tea-tree thicket (12): blossoming shrubs, bushes and *Banksia* on open heath (14).
Behaviour. Singly, in pairs or occasionally in small groups. Sometimes a member of mixed bird parties. Frequently seen probing tubular flowers for nectar, sometimes only 3 – 4 inches off the ground. Darts about from perches with a rapid zig-zag.
Annual Cycle. The three adults were in moult (wings and tail) with small gonads. Immatures were seen with adults on numerous occasions, and a nest with a sitting bird was found at Green Range (13) in early April.
Impermanent Colours. *Legs* black; immatures paler grey. *Iris* red; immatures brown. *Bill* black; immatures dark horn, yellowish at base of lower mandible. *Mouth* black becoming yellow in throat; immatures yellow throughout.
Other Notes. The immature bird is dull buffish below lacking the frontal gorget of the adult.

Manorina melanocephala — Noisy Miner

Specimens. QUEENSLAND. Banoona (58), 28 Apr – 6 May 63, 3 ♀, 1 fl. SOUTH AUSTRALIA. Furner (86) 8 Jan 63, 1 ♀. NEW SOUTH WALES. Albury (92) 9 – 12 Mar 63, 1 ♂, 2 ♂ skel. Dunedoo (93) 18 Mar 63, 1 fl. Grafton (94) 3 Apr 63, 1 fl. Tenterfield (95) 10 Apr 63, 1 ♀, 1 fl.
Specimens to other Museums. NMV 1 ♂ skel Albury (92).

Field Notes. Common where found, and also recorded from Furner (86). Mostly in open country, but sometimes thicker woodland.

Sometimes single but usually in groups of three or four, or loose flocks of 15-20; one individual was shot when walking along and up a branch, "creeper-like" (58). Noisy birds, with variable screeches, whistles and scoldings.

Food. Stomachs contained insects, including a medium-sized beetle, vegetable matter and pepper seeds (58).

Annual Cycle. One January specimen from Furner (86) is mostly worn, with moult in tail and body. March and April birds (92, 95) were in moult, the later one completing wing moult. The three Banoona (58) birds, April – May are in fresh plumage. Gonad activity was insignificant; the largest were in the January bird.

Impermanent Colours. *Legs* sandy yellow, frontal surface usually brown. *Iris* brown. *Bill* yellow. *Mouth* yellow. *Bare skin round eye* yellow, not necessarily the same shade as the bill.

Geographical Variation. Queensland birds have been separated as *M. m. crassirostris:* ours and others in the British Museum are smaller than *M. m. melanocephala* from New South Wales, Victoria and South Australia; wings 5 ♂s 129-136, 4 ♀s 126-130 against 7 ♂s 138-145, 10 ♀s 134-142.

Manorina flavigula — Yellow-throated Miner

Specimens. WESTERN AUSTRALIA. Chichester Range (3) 16 May 66, 2 ♂, 1 ♀, 1 ø, 2 ♂ skel, 1 fl. Newdegate (14) 13 April 66, 1 ♀ skel, 3 fl. Cocklebiddy (15) 8 July 65, 1 ♂. Eucla (17) 1 July 65, 1 ♂, 1 ♀. Mount Shenton (18) 30 July 65, 1 ♂. Warburton (23) 10 Aug 65, 2 ♂. Van der Linden Lakes (24) 6 Sept 65, 1 ♀. Giles (25) 25 Aug – 1 Sept 65, 2 ♂ (48.0 g), 3 ♀, 1 ♂ skel. Mount Anderson (28) 28 Apr – 17 May 68, 1 ♂, 3 ♂ skel, 1 ♀ skel, 1 ø skel, 2 fl. Mount Elizabeth (35) 15 July 68, 1 imm ♀ (57.8 g). QUEENSLAND. Tyrone (59) 12 – 15 May 63, 1 ♂, 1 ♀, 1 juv, 1 ♂ skel. Thylungra (60) 22 May 63, 2 ♂, 1 ♂ skel. SOUTH AUSTRALIA. Innamincka (64) 1 May 65, 1 ♂. Etadunna (66) 17 May 65, 1 " ♂" (♀ on size). Oodnadatta (67) 31 May 65, 1 ♀. Granite Downs (68) 7 June 65, 1 ♀. Yardea (72), 1 Nov 64, 1 ♀. Nonning (79) 31 Mar 65, 1 ♂ (50.6 g). Aroona Valley (82) 8 Apr 65, 1 ♀. Angepena (84) 27 Nov 63, 1 ♀.

Specimens to other Museums. NMV 1 ø skel Eucla (17), 1 ♂ Tyrone (59).

Distribution & Abundance. Found throughout Western Australia, except in the extreme southwest (8-13). Found at most Northern Territory camps but was noticeably absent at Pine Creek (44) and 'doubtful' at Borroloola (46). In Queensland it was recorded from all the southern camps (58-61) and at Glendower (55) but not further north. Generally distributed in central South Australia, particularly common at Etadunna (66).

Habitat. In general, lightly wooded to dry open country (Plate 3(b)): mallee and coastal scrub. In particular: scattered trees on spinifex plain (3), mulga and river gums in creekbeds (18-25), Desert Oaks (24), Gidgee and corkwood flowers (25), *Melaleuca* copse (28), native pine (82).

Behaviour. In most areas found in small groups, five to 10, or parties up to 20, and only occasionally single. However in the Kimberleys (28) and southeast South Australia (72, 79, 82), birds were mostly single or in pairs. Sometimes associated with other smaller honeyeaters such as *Meliphaga penicillata;* once when feeding on the ground (67).

Goodwin (1967c: 247) describes how it searches for food in crevices and under debris, and he gives an account of an apparently submissive display by one individual when chased by three others.

Noisy and wary, warning others of the presence of intruder, such as observer or bird of prey. Alarm notes are varied monosyllabic screams and squawks. Also heard were a call like that of *Cracticus* – given by a single bird, possibly mimicry (18); a low guttural *aarh* from six birds returning to position where one of their number had been shot, and a melodious chuckle from a group near nest (35).

Food. Stomachs contained insects including beetles, 'small black bugs' and a small white caterpillar (17); also seeds (64, 79, 82), nectar and pollen (once) and grit (once).

Annual Cycle. In Western Australia the mid-May birds from Chichester Range (3) were in a slightly less advanced stage of moult than the April male from Mount Anderson (28); gonads not enlarged. The young July bird from Mount Elizabeth (35) is in transitional juvenile to immature plumage (see below). In the central camps (18, 23-25) in late July to early September all specimens were worn and in varying degrees of moult, but some adults from all camps had enlarged gonads. Further south, three early July specimens from Cocklebiddy (15) and Eucla (17) show varying degrees of wear to plumage with gonads small.

In South Australia the Nonning (79) male has primaries in pin with the outer tail-feathers half grown. The Aroona Valley (82) female had almost completed tail and wing moult. The May – June birds (64, 67, 68) are very worn, and the Angepena (84) female has some primaries half grown. Gonads all small. In southern Queensland birds were nesting in May. At one nest in Tyrone (59) adults were watched feeding young in nest, while from another 20 ft up, a young bird had fallen out. This had sprouting flight feathers, and the complete tail in pin, and is of the same age as a juvenile from an earlier collection taken at Prairie, near Glendower (55) on 27 March. One of the Thylungra (60) males had very enlarged testes.

Impermanent Colours. *Legs* dull yellow, slightly darker in front. *Iris* brown or grey brown. *Bill, Mouth* and *Bare skin around eye* yellow.

Geographical Variation. Our series shows some geographical variation in both size and colour. Birds from the Kimberleys and Chichester Range (3, 28, 35) are the largest and palest (wing 4 ♂ 132-134 mm), with a buff wash on the mantle, an orange-yellow forehead and light scaling on the breast. Those of the arid centre east to Granite Downs (18, 23-25, 66-68) are smaller (wing 5 ♂ 121-128), darker and greyer above, more heavily patterned on the breast, and have the forehead greener (though most birds have the colour obscured by pollen-staining). Those from Innamincka (64), southern Queensland (59, 60) and southern South Australia (72, 79, 82, 84) are slightly larger (5 ♂ 126-132) and more heavily patterned below, but while the eastern birds (59, 60, 64) are also darker above, the southern ones, especially the one from Nonning (79) and the two from the Flinders Range (82, 84) are rather pale and grey, possibly due mainly to their fresh plumage.

The three birds from the Bight (15, 17) are the smallest (wing 1 ♂ 118), darkest and most heavily patterned of our series.

A number of races are recognised by Salomonsen (in Peters', 1967) but it is not easy to fit our birds into these. Probably the pale northwestern birds are all *M. f. lutea;* those of the centre (18, 23-25, 66-68) may be *M. f. pallida;* those of the south and west (59, 60, 64, 72-84), though variable, may be *M. f. flavigula;* and the Bight birds (15, 17) intermediate between *flavigula* and the dark *M. f. clelandi* of the southwest.

Other Notes. White tips to the feathers of the mantle are not a sign of immaturity, for they are not found in either the juvenile from Tyrone (59) or the immature bird from Mount Elizabeth (35). The juvenile has few feathers on the chin but the remainder of the throat and upper breast are pale grey with vestigial barring. The immature has the centre of the throat white, the sides grey and the upper breast buff with no scaling. Otherwise the young birds in colour and pattern resemble the adults.

Parasites. See Appendix 1.

Anthochaera rufogularis — Spiny-cheeked Honeyeater

Specimens. WESTERN AUSTRALIA. Bush Bay (6) 18 June 66, 1 ♂. Newdegate (14) 16 Apr 66, 1 ♂ skel. Mount Shenton (18) 3 Aug 65, 1 ♂. Mungilli Claypan (20) 13 Sept 65, 1 ♂. Warburton (23) 10 Aug 65, 1 ♂. Giles (25) 30 – 31 Aug 65, 1 ♂, 1 fl. QUEENSLAND. Glendower (55) 12 Mar 64, 1 ♀ (42.7 g). Banoona (58) 6 May 63, 1 fl. Tyrone (59) 12 – 15 May 63, 1 ♂, 1 ♀, 1 juv skel, 1 fl. SOUTH AUSTRALIA. Etadunna (66) 18 May 65, 1 ♀. Nullarbor HS (69) 4 Oct 63, 1 imm ♂, 30 Oct 64, 1 imm ♂. Kingoonya (70) 17 – 23 June 65, 3 ♂, 1 ♀, 2 fl. Streaky Bay (74) 9 Dec 62, 1 ♂ skel. Nonning (79) 29 Mar 65, 2 ♂, 1 ♀ (41.8 g). Aroona Valley (82) 10 – 14 Apr 65, 1 ♂ (51.9 g), 1 ♀ (39.1 g).

Specimens to other Museums. NMV 1 ♀, 1 ♂ skel Tyrone (59). AM 1 ♂ Tyrone (59).

Distribution & Abundance. In Western Australia scarce at Bush Bay (6), fairly common in the main camps of the Gibson Desert (23, 25) but along the south coast only a few seen at Cocklebiddy (15) and none at Eucla (17). In South Australia very common in the south (70, 79, 82), less so at Kingoonya (68), a few at Etadunna (66) and only one at Innamincka (64). More common again in Queensland (58, 60, 64) especially at the limit of its range at Glendower (55).

Habitat. Avoids heavy forest and more usually found in trees or bushes in open country, e.g. coastal scrub (6, 69), roadside bushes and mallee (14), mainly Mulga woodland and creekbeds in the interior of Western Australia (18-25) and particularly in *Curara* (23) and flowering *Hakea* (25); also Black Oaks (70), eucalypts (74, 82), casuarinas (79) and native pines (82) in South Australia; in Queensland, Mulga, sandalwood bushes and Gidgee (59), and once in open riverine forest (55).

Behaviour. Usually seen singly or in twos and threes feeding or calling, and flying swiftly from tree to tree. The Glendower (55) bird was one of five. It associated with other honeyeaters including *Meliphaga virescens, Entomyzon cyanotis* and *Philemon corniculatus* and with babblers (79). Has various pleasant loud and liquid whistling phrases; also a *tok-tok,* used as a contact note between pairs (PRC). At Eyre (16) a song flight was seen (PRC) in which a male soared high into air, 70 ft above mallee and sand dunes, calling as it descended. The full song was later given from a perch, a liquid and guttural gurgling jumble lasting for five or six seconds, repeated at intervals. It was delivered with neck outstretched in semi-upright posture, and head tilted back.

Food. Stomachs contained mostly seeds, including those of *Heterodendron olefolium* (79) and mistletoe (70); also fruit, nectar and insects, including ants, small beetles, and once a caterpillar.

Annual Cycle. Breeding at Nullarbor HS (69) appears to have taken place about September/October two years in succession for both young birds are similar, distinguished from most adults by some bright yellow feathers in the moustachial streak, pale throats and rather smaller size. The iris (noted in one only) was white. They are not in moult, and wings and tail appear fresh, except that the tips of the rectrices are somewhat abraded. In March and April (55, 70, 82) all birds were moulting and had small gonads: the females from Nonning (70) and Aroona Valley (82) have some yellow "whiskers" and may be subadult (see below). May birds are in fresh plumage; the female from Etadunna (66) and the male, but not the female, from Tyrone (59) had quite large gonads. June birds from South and Western Australia (70, 6) were fairly fresh and, except for one male, had slightly enlarged gonads. The August males from Mount Shenton (18) and Warburton (23) are in mixed worn and fresh plumage, but were not apparently in moult, and had very enlarged testes. A few weeks later the males from Giles (25) and Mungilli Claypan (20) were in very worn plumage with small testes.

Impermanent Colours. *Legs* dark grey. *Iris* pale blue or blue-grey; sometimes described as dark brown ringed with pale blue-grey: white in one immature. *Bill* pink or mauve with black tip. *Mouth* orange or yellow. *Bare skin* pinkish or bluish mauve round gape and under eye, fading rapidly after death.

Geographical Variation. There is a certain amount of individual variation in size and colour but we agree with Salomonsen (in Peters', 1967) that subspecies should not be recognised.

Other Notes. A number of our specimens have yellow in the moustachial streak, which seems to be a character of young birds but is also found in some apparent adults, including a male from Bush Bay (3) and some females (66, 70, 82) and these may be first year birds.

Anthochaera chrysoptera — Little Wattle-bird

Specimens. WESTERN AUSTRALIA. Scott River (11) 27 Feb 66, 1 ♂ skel. Green Range (13) 6 Apr 66, 1 ♂ skel, 1 fl. SOUTH AUSTRALIA. Furner (86) 7 Jan 63, 1 fl.

Distribution & Abundance. Seen only where collected in the west but in the east found also at Nelson (88) and Albury (92).

Habitat. In Western Australia, among scrub or sandhills in clumps of taller trees such as tea-trees. In open forest at Furner (86).

Behaviour. Never more than three or four birds observed together. Pugnacious to other birds feeding in same tree. Noisy, particularly at daybreak: at Furner (86) one male regularly for two weeks took up its stance in a tree four feet from a tent and

emitted a series of loud short notes *wak, wak, keik-keik, keik-kewick* while holding up its head and depressing its tail.
Annual Cycle. The February bird from Scott River (11) was noted as having some moult on breast and mantle and the January bird from Furner (86) as being in good plumage except for worn rectrices. Gonads were very small.
Impermanent Colours. *Legs* dark grey (11, 13), light grey (86). *Iris* brown (11, 13), pinkish grey (86). *Bill* black, light tip to lower mandible (13). *Mouth* yellow.

Anthochaera carunculata — Red Wattle-bird

Specimens. WESTERN AUSTRALIA. Perth (7) 15 Feb 66, 1 imm ♀. Wonerup (8) 12 Mar 66, 2 ♂, 1 imm ♀, 1 fl. Scott River (11) 19 – 24 Feb 66, 1 ♀, 1 ♂ skel, 1 fl. Mount Barker (12) 21 – 26 Mar 66, 1 ♂, 1 juv ♂, 1 ♀ skel, 1 fl. Eucla (17) 1 – 5 July 65, 1 ♂, 1 ♀. NEW SOUTH WALES. Dunedoo (93) 23 – 25 Mar 63, 1 ♂, 1 ♀, 1 ♂ skel.
Distribution & Abundance. In Western Australia found everywhere in the southwest camps (7-14) and also common at Cocklebiddy (15) and Eucla (17). Further east it was seen at Streaky Bay (74), Nonning (79) and Furner (86). Also at Tenterfield (95) at the northern end of its range.
Habitat. Mainly open forest of various kinds; also mallee scrub (17).
Behaviour. Singly, in pairs or groups. Noisy, uttering "single, guttural croaks" and other loud harsh calls.
Food. Stomachs contained beetles (8, 93), insect remains, nectar and an unidentified berry.
Annual Cycle. In February and March breeding was over in Western Australia, adults were mostly in moult, parents were seen feeding young and the juvenile collected was not long out of the nest. In the east, the March bird from Dunedoo (93) had completed moult. The July birds from Eucla (17) are worn and had fairly small gonads.
Impermanent Colours. *Legs* pinkish-brown; juvenile, flesh-grey. *Iris* red or red-brown; young birds brown or grey-brown. *Bill* black. *Mouth* yellow. *Wattle* adults, pinkish (7-17) or red (93); immature (vestigial) pink (7) or whitish (8).
Geographical Variation. The two Dunedoo (93) birds, nominate *carunculata,* are considerably larger than the rest. These fall within the range of *woodwardi,* which has proportionately longer wattles; the throat and breast also appear to be darker brown, as characterised by Mathews (1925). Wings, *carunculata* ♂160, ♀ 150; *woodwardi* 3 ♂ 145-150, 1 ♀ 141.
Parasites. See Appendix 1.

ESTRILDIDAE — Grass-finches
C. J. O. Harrison

Zonaeginthus oculatus — Red-eared Firetail

Specimens. WESTERN AUSTRALIA. Green Range (13) 6 Apr 66, 1 ♂ skel, 1 ♂ fl.
Field Notes. Found only in the Green Range (13) region; in heath vegetation with scattered mallee bushes on coastal dunes at Warriup, further inland in similar vegetation

on a low scarp at Beckon. Usually in pairs feeding on bare ground between shrubby growths; when alarmed would perch on shrubs. They could be called up by imitation of the rather plaintive piping call-note.
Annual Cycle. Both birds showed some evidence of moult. Gonads were small.
Impermanent Colours. *Legs* pale pinkish-buff. *Iris* dark brown. *Bill* bright scarlet. *Mouth* pink. *Bare skin around the eyes* bright blue.
Nomenclature. It is considered that the differences between *Emblema picta* and the three species placed in *Zonaeginthus* justify the recognition of the latter genus.

Zonaeginthus guttatus — Diamond Firetail

Specimens. New South Wales. Dunedoo (93) 18 – 25 Mar 63, 2 ♂, 1 ♀, 1 fl.
Specimens to other Museums. NMV 1 ♂ Dunedoo (93).
Field Notes. Only found at Dunedoo (93). On edge of woodland and in lightly timbered savanna woodland.
Food. Seeds and insects, including beetles.
Annual Cycle. Gonads were rather enlarged.
Impermanent Colours. *Legs* grey, drab sandy soles to feet. *Iris* deep red. *Bill* red, pinker at base of upper mandible. *Mouth* yellow to orange. *Orbital ring* deep red.

Emblema picta — Painted Finch

Specimens. Western Australia. Chichester Range (3) 16 – 17 May 66, 1 ♂, 1 imm ♂ 2 ♂ skel, 2 ♀ fl. Coolawanyah (4) 19 – 20 May 66, 2 ♂, 1 imm ♂, 1 ♂ fl. Mount Anderson (28) 7 May 68, 2 ♀. Geikie Gorge (31) 2 June 68, 1 ♂, 1 fl.
Field Notes. Only seen at the above camps and at Karratha (1), in spinifex grassland with some trees and rocky outcrops, usually near water. Usually in pairs or small parties but sometimes with other grass-finch flocks at water.
Food. Spinifex seed (3, 4), grass seed (28).
Annual Cycle. Immature birds were collected at Chichester Range (3) and Coolawanyah (4) in May when adults showed no signs of breeding, unlike other grassfinch species also present. A female from Mount Anderson (28) in May had the ovary slightly enlarged.
Impermanent Colours. *Legs* males dull brown, buffish-brown, light grey-brown, or light grey, with whitish soles noted once; females dull brown, greyish-brown, pinkish-buff; immature males, light fleshy grey. *Iris* creamy white or white, with a very narrow black outer rim. *Bill* adults upper mandible black with red or orange tip, lower mandible red or orange and pinker towards the base, with a light blue patch on the lower part of the base; immature males upper mandible blackish with pinkish culmen and tip, lower mandible light pinkish orange with pale blue at base. *Mouth* adults pale blue-grey or bluish-white; immature males fleshy-grey, or white spotted with black.

Neochmia phaeton — Crimson Finch

Specimens. Western Australia. Mount Bell (32) 15 June 68, 1 ♂, 1 imm ♂. Joint Hill (34) 28 June 68, 1 ♂ (10.2 g), 1 imm ♂ (11.2 g). Mount Elizabeth (35) 14 – 15 July 68, 1 ♂ (9.9 g). Queensland. Tully (54) 18 July 64, 1 ♂ (9.9 g), 1 ♀ (9.6 g).

Field Notes. Irregularly distributed where suitable habitat was present. Also recorded at Geikie Gorge (31), Darwin (41), Pine Creek (44) and Borroloola (46), and near Ayr in Queensland. In Western Australia occurred in *Pandanus* by rivers in savanna woodland; in Queensland in a grassy clearing in rainforest. Usually seen in small parties.
Food. Seeds, of *Pandanus* and grasses.
Annual Cycle. In males in Western Australia, from mid-June to mid-July, gonads were small. A male on 15 June was in worn plumage, one on 28 June had begun to moult, and one on 15 July was almost fully moulted. An immature male on 15 June showed the beginning of moult to adult plumage, and birds taken on the other dates showed more extensive moult. Birds from Queensland in mid-July appeared to be recently moulted and gonads were small.
Impermanent Colours. *Legs* males ochraceous, light brown or light fawn; females ochraceous; immature males, light brown. *Iris* light brown. *Bill* males dull red with whitish base, red, crimson; females deep orange with whitish base; immature males brown turning crimson, brown with reddish tint, or red with greyish base. *Mouth* dull yellow to yellow.
Geographical Variation. Only a single black-bellied race is recognised for most of Australia other than northern Cape York where a white-bellied form occurs. In worn specimens the deep crimson tends to change to bright vermilion.

Neochmia ruficauda — Star Finch

Specimens. Western Australia. Karratha (1) 28 May – 6 June 66, 3 ♂, 1 ♀, 1 ♀ skel, 1 ♂ fl, 1 ♀ fl. Derby (29) 26 May 68, 1 ♀ (10.6 g), 1 imm ♂ (10.5 g). Queensland. Inkerman (50) 30 May 64, 1 ♂ (9.9 g), 1 ♀ (9.5 g), 1 imm ♂ (9.5 g), 1 imm ø (8.0 g).
Field Notes. Distribution is limited by the need for low thick plant growth by water. In addition to the above localities a small flock was seen on the shore at Onslow, Western Australia in early June. They usually occurred in small flocks even when breeding. At Karratha (1) spent much of the day in small flocks in long rushes and reeds growing in water, moving among the plants with agility, and coming at times to feed, or drink at shallow pools, in sparse grass and bushes bordering the creek. At Derby (29) they were in company with a party of *Lonchura castaneothorax*.
Food. Grass seeds (50), spinifex seeds (1).
Annual Cycle. Birds at Karratha (1) in late May and early June were beginning to breed after late rains. Gonads were enlarged and one female had a full-sized yolk. At Derby (29) and Inkerman (50) gonads were small; immature birds had not yet commenced moult into adult plumage.
Impermanent Colours. *Legs* males sandy yellow, buff, deep orange-yellow; females deep yellow or pale sandy-yellow; immatures ochraceous or buff. *Iris* orange to orange-red in adults, brown to olive-brown in immatures. *Bill* males scarlet, red, orange or bright pinkish-red; females scarlet or pinkish-orange with browner culmen and light orange lower mandible; immatures deep chestnut-brown, dark slate grey, or black. *Mouth* adults orange, flesh-coloured, or yellow with purplish marks; immatures white or greyish-white.

Geographical Variation. Keast (1958c) recognised two races, nominate *ruficauda* in the east north to Rockhampton, Queensland, and *clarescens* elsewhere. Our material agrees with this save that a male from Inkerman (50) on the Gulf of Carpentaria is dull enough dorsally to be intermediate and suggests that the colour variation may be clinal.

Aegintha temporalis — Red-browed Finch

Frequently found in company with *Zosterops lateralis* on the Walter Hill Range (53).

Erythrura trichroa — Blue-faced Finch

Specimens. Queensland. Ayton (52) 22 – 25 June 64, 1 imm ♂ (12.5 g), 1 imm ♀ (16.0 g).
Field Notes. Only encountered at Ayton (52) in tropical rainforest, once on a steep slope, once beside a fast stream; seen feeding among vines. The female was one of a party of four.
Food. One stomach contained white, grain-like seeds; the other, insect remains and grit.
Annual Cycle. Both birds were in immature plumage, with unpneumatised skulls. The ovary of the female appeared slightly enlarged.
Impermanent Colours. *Legs* immature male deep salmon-pink; immature female, pale buff. *Iris* very dark brown. *Bill* black, paler on the gonys of the lower mandible. *Mouth* white, the palate of the immature male showing a half-circle of five small black spots.

Chloebia gouldiae — Gouldian Finch

Specimens. Western Australia. Mount Bell (32) 21 June 68, 1 ♂, 1 ♀, 1 imm ♀. Mount Elizabeth (35) 10 – 15 July 68, 3 ♂ (14.5, 14.8, 15.1 g), 1 imm ♂ (15.5 g), 1 ♀ (16.5 g), 1 imm ♀ (15.9 g).
Field Notes. Local, elsewhere recorded only at Wandie Creek (Pine Creek (44)); in savanna woodland with long grass, near water. In flocks of up to 150 birds, once in association with Long-tailed Finches *Poephila acuticauda*.
Food. Small seeds, including grass seeds.
Annual Cycle. Testes were slightly enlarged in a male at Mount Bell (32) in late June, but otherwise gonads were small, and body plumage generally was worn and appeared blue. Birds in immature plumage were collected at both localities and one at Mount Elizabeth (35) on 15 July was showing one or two scattered adult feathers.
Impermanent Colours. *Legs* males dull yellow, yellow-buff, pale yellowish-brown, straw; females straw, and yellow-buff; immatures yellow-buff. *Iris* dark brown. *Bill* males pale pink with red tip; white upper mandible, rose pink lower mandible and pink tip; and whitish suffused red, with dark tip; females, whitish with dark tip. Some apparent adults had bills similar to those of immatures, with blackish upper mandibles with greyer base, lower mandible paler and suffused with red and with dark tip. *Mouth* whitish, with immature and some adults showing a pattern of black spots on the palate. *Eye-rim* a thin pale blue eye-rim was present on adults and young.

Poephila acuticauda — Long-tailed Finch

Specimens. WESTERN AUSTRALIA. Mount Anderson (28) 14 – 18 May 68, 1 ♂ (16.2 g), 1 ♀ (13.3 g), 2 fl (14.1, 16.0 g). Derby (29) 21 – 24 May 68, 3 ♂ (14.7, 14.8, 15.2 g), 2 ♀ (14.3, 17.6 g), 1 imm ø, 1 juv ♂ (11.7 g), 3 ♀ skel (13.0, 15.1, 15.7 g), 1 ø skel. Mount Bell (32) 17 June 68, 1 ♂ (15.4 g). NORTHERN TERRITORY. Pine Creek (44) 28 Aug 68, 1 ♂ (14.7 g). QUEENSLAND. Moonlight Creek (47) 2 – 7 May 64, 4 ♂ (14.0, 14.8, 15.0, 15.1 g), 4 ♀ (13.1, 13.5, 14.6, 15.8 g).
Field Notes. Also noted at Borroloola (46). The birds from Moonlight Creek (47) are apparently the first to be found in Queensland. Found in savanna woodland, usually near water and in shrubby growth along a river. Recorded feeding on ground; once in association with *Chloebia gouldiae*.
Food. Seeds, grass seed.
Annual Cycle. Birds were in breeding condition at Moonlight Creek (47) and Mount Anderson (28) in early May, and breeding in late May at Derby (29), where a half-grown young bird was collected. A male from Pine Creek (44) had enlarged testes and very worn plumage.
Impermanent Colours. *Legs* adults bright red, rarely orange-red; immatures flesh-pink to pink. *Iris* adults deep red to chestnut; immatures dark brown or brown. *Bill* nominate form orange-yellow to yellow; *hecki* bright red to orange-red, occasionally noted as paler towards the base of the upper mandible; immatures black, or grey with black tip. *Mouth* nominate form whitish to yellow and orange-yellow; in *hecki* the bill was noted as orange-red internally, the palate yellowish or grey and white, while tongue and throat were pale pink: immatures, white patterned with black.
Geographical Variation. *P. a. hecki,* type locality Port Essington, has a red bill instead of deep yellow (a character which is lost in museum specimens) and generally darker body plumage. Specimens from Pine Creek (44) and Queensland are this form, and red-billed birds were seen at Oenpelli (43); but at Pine Creek (44) some family parties were seen with "washed-out yellow" bills (RAW). Birds from Western Australia are *P. a. acuticauda.* There is no evidence of the clinal variation suggested by Immelmann (1965).

Poephila cincta — Black-throated Finch

Specimens. QUEENSLAND. Wernadinga (48) 19 May 64, 4 ♂ (14.1, 15.0, 16.7 g), 4 ♀ (15.5, 15.8, 17.1 g). Glenore (49) 15 – 20 Apr 64, 4 ♂ (12.9, 13.3, 13.9, 14.3 g), 3 ♀ (14.9, 15.0, 15.1 g). NEW SOUTH WALES. Tenterfield (95) 15 Apr 63, 1 ♂.
Field Notes. Encountered only where collected; in pairs or small parties, in cultivation in woodland (95) and in tropical savanna woodland near water (48, 49). Usually noted either perched in trees, or feeding on bare ground or in tall grass, or drinking. In one instance associating with a small flock of *Lonchura pectoralis* and *Taenopygia guttata.*
Food. Seeds, including grass seeds.
Annual Cycle. The Tenterfield (95) male had testes a little enlarged. Birds at Glenore (49) and Wernadinga (48) in April and May were breeding.

Impermanent Colours. *Legs* males orange-pink to red; females orange, orange-pink. *Iris* dark brown, deep purplish brown, medium brown, light chestnut, dull chestnut, creamy brown, light yellow brown (paler eye colour was associated with dark markings on palates and possibly indicated younger birds). *Bill* black, sometimes paler at base. *Mouth* white, greyish-white, or light grey. Black patches present on the palates of some apparently adult birds.

Geographical Variation. There are two distinct forms. The Tenterfield (95) bird is of the browner, white-rumped nominate form, and the others of the paler, black-rumped form, *P. c. atropygialis,* of the peninsula east of the Gulf of Carpentaria.

Parasites. See Appendix 1.

Stictoptera bichenovii Banded Finch

Specimens. WESTERN AUSTRALIA. Mount Anderson (28) 17 – 18 May 68, 1 ♂, 3 fl (8.3, 9.0, 9.9 g). Derby (29) 25 – 27 May 68, 1 ♂ (8.4 g), 5 fl (8.2, 8.2, 8.4, 8.8, 9.1 g). QUEENSLAND. Moonlight Creek (47) 1 – 9 May 64, 2 ♂ (8.5 g), 2 ♀ (7.9, 10.3 g), 1 imm ø (7.9 g). Banoona (58) 29 Apr – 6 May 63, 2 ♂, 1 fl. Tyrone (59) 14 May 63, 1 ♀. NEW SOUTH WALES. Dunedoo (93) 23 – 25 Mar 63, 2 ♂ ,1 ø, 1 fl. Grafton (94) 1 – 3 Apr 63, 1 ø, 2 skel, 2 fl.

Specimens to other Museums. NMV 1 skel Grafton (94).

Field Notes. Also noted at Mount Roper (45) in Western Australia; Darwin (41), Oenpelli (43), Pine Creek (44), and Borroloola (46) in Northern Territory; Inkerman (50) and Glendower (55) in northern Queensland. In savanna woodland sometimes with paperbarks, and usually by water; usually in parties or small flocks of up to 30, at times associating with *Taenopygia guttata.*

Food. Small seeds.

Annual Cycle. In Western Australia in May and in southern Queensland and New South Wales in March to May birds had small or only slightly enlarged gonads; with the exception of one male at Banoona (58) on 6 May with enlarged testes. At Moonlight Creek (47) in northern Queensland in early May gonads were enlarged. At Dunedoo (93) in late March and Moonlight Creek (47) in early May immatures were commencing moult into adult plumage.

Impermanent Colours. *Legs* males slate blue to black; females deep blue, grey, black; immatures dark grey with ochraceous soles, blue-grey, slate-grey. *Iris* reddish-brown, dark brown, chocolate brown. *Bill* adults pale blue to light grey; immatures similar with some brown on tip and culmen. *Mouth* pale blue-grey to greyish-white, or cream; immatures white with pale grey pattern.

Geographical Variation. There are two well-marked races, the white-rumped nominate form, and a dark-rumped western form, *annulosa,* the two interbreeding in eastern Northern Territory. Queensland and New South Wales birds are referable to the former, Western Australian birds to the latter. Black-rumped and white-rumped individuals were seen at Borroloola (46). This species differs markedly in plumage pattern, colour, and behaviour, from the typical grass-finches of the genus *Poephila,* and it seems preferable to retain it in the monotypic genus *Stictoptera.*

Taenopygia guttata Zebra Finch

Specimens. WESTERN AUSTRALIA. Chichester Range (3) 25 May 66, 1 imm ♀. Giles (25) 31 Aug 65, 1 ♂. QUEENSLAND. Tyrone (59) 15 May 63, 1 ♀ skel, 1 ♂ fl, 1 imm fl. Thylungra (60) 23 – 25 May 63, 1 ♂, 1 ♀, 1 ♂ skel, 1 ♀ skel, 2 fl. SOUTH AUSTRALIA. Clifton Hills (65) 10 – 12 May 65, 5 ♂, 2 ♀. Etadunna (66) 26 May 65, 1 ♀. Nullarbor HS (69) 4 Oct 63, 1 ♀. Kingoonya (70) 17 June 65, 1 ♂, 1 ø. Nonning (79) 31 Mar 65, 1 ♂ (11.0 g), 1 ♀ (11.7 g), 1 ♂ skel (11.9 g). Aroona Valley (82) 8 Apr 65, 1 ♀ (12.0 g). Wirrealpa (83) 15 Apr 65, 1 ♂ (11.8 g), 1 ♀ (13.8 g), 1 ♂ skel (12.9 g). Angepena (84) 29 Nov – 3 Dec 63, 1 ♂, 1 ♀, 2 imm ♂.

Distribution & Abundance. Widely distributed in suitable country and usually abundant when encountered. Usually in flocks or family parties. In addition to the above localities it was also noted in South Australia at Innamincka (64), Oodnadatta (67), Granite Downs (68); in Queensland at Glendower (55); in Western Australia at Mungilli Claypan (20), also plentiful along the inland road between Paynes Find and Roy Hill in 1966 after rains, and was encountered right through the Kimberleys in 1968.

Habitat. Occurred mainly in arid or semi-arid areas, usually where some shrubby growth was present. Recorded from salt bush flats, low mulga scrub, woodland edge adjoining salt bush, savanna woodland, spinifex with scattered mulga, and patches of trees and bushes in open Bluebush country. Often in the vicinity of a source of water such as a creek, dam, waterhole or well.

Food. Small seeds, including grass seeds. Insects in one instance (83).

Annual Cycle. This species comes rapidly into breeding condition after rain at any time of year. Birds were beginning to nest at Mount Anderson (28) in May, and a very young bird was collected at Chichester Range (3) in late May. A male at Thylungra (60) in late May had enlarged testes. Two immature males from Angepena (84) in late November were moulting into adult plumage, and females from Nullarbor HS (69) in early October, and Aroona Valley in early April (82), and a male from Nonning in late March show evidence of body moult.

Impermanent Colours. *Legs* adults light orange, orange, orange-pink; immatures flesh or orange. *Iris* adults orange-brown to red; immatures brown to orange-red. *Bill* adults orange-pink to red, slightly paler lower mandible noted once; immatures black, dirty brick-red with blackish tip, orange with orange-yellow lower mandible, orange. *Mouth* adults flesh, light yellow, orange with yellowish palate, orange; immatures yellow with black markings.

Geographical Variation. Our material confirms that the Australian specimens represent a single population *castanotis*. In view of the differences in plumage pattern, colour and behaviour between this species and the typical grass-finches of the genus *Poephila*, it seems preferable to retain the monotypic genus *Taenopygia*.

Aidemosyne modesta Plum-headed Finch

Specimens. QUEENSLAND. Tyrone (59) 12 May 63, 1 ♂, 1 ♀, 1 imm ♂.

Specimens to other Museums. NMV 1 imm ♂ Tyrone.

Field Notes. Only recorded at Tyrone (59). In small parties in open woodland and

groups of trees near a creek with water; feeding on the ground, but restless, moving frequently.
Food. Seed.
Annual Cycle. Gonads were minute. The male showed some body and head moult.
Impermanent Colours. *Legs* males lilac-grey; female pinkish. *Iris* brown. *Bill* black with blue-grey on the sides of the basal half of the lower mandible. *Mouth* male black with yellow in throat; female grey with yellow on tongue and roof of mouth.

Lonchura castaneothorax — Chestnut-breasted Finch

Specimens. WESTERN AUSTRALIA. Mount Anderson (28) 7 May 68, 2 imm ♀ skel. QUEENSLAND. Moonlight Creek (47) 2 May 64, 1 ♂ (14.9 g).
Field Notes. Apparently local, and elsewhere only seen at Mount Dryander (57) where it was very common; found in tropical savanna woodland (47, 57) and grassland savanna with spinifex (28). Recorded feeding in small flocks in grass or on open ground.
Food. Seeds.
Annual Cycle. Immature birds were collected on 7 May at Mount Anderson (28). A male at Moonlight Creek (47) on 2 May had enlarged testes. At Mount Dryander (57) in late August birds (collected but not retained) were not in breeding condition.
Impermanent Colours. *Legs* male black; female dark purplish-grey; immatures purplish-grey. *Iris* dark brown. *Bill* light grey or light grey-blue, darker in immatures. *Mouth* male yellowish-white; others pale blue-grey, pinkish-purple in the throat.
Geographical Variation. The only skin collected was typical of the eastern nominate race.

Lonchura pectoralis — Pictorella

Specimens. WESTERN AUSTRALIA. Mount Anderson (28) 29 Apr – 18 May 68, 2 ♂ (16.6 g), 1 ♀ (16.6 g), 6 fl, 1 imm fl. Derby (29) 25 May 68, 2 fl (15.5, 16.8 g). Mount Bell (32) 10 July 68, 1 ♀ (16.5 g). Mount Elizabeth (35) 10 July 68, 1 ♂ (14.0 g), 1 imm ♂ (13.8 g), 1 ø (14.0 g). Lissadell Hill (38) 28 – 29 June 68, 1 imm ♂ (14.0 g). QUEENSLAND. Glenore (49) 10 – 20 Apr 68, 7 ♂ (13.2, 13.8, 14.0, 14.3, 14.3, 14.5, 15.7 g), 5 ♀ (14.3, 15.1, 15.7, 16.3, 16.5 g).
Field Notes. Recorded throughout the Kimberleys (28-38) but not in Northern Terri-fory: apparently limited to grassland within easy reach of water. Size of groups varied, trom parties of three to fifteen when apparently breeding, to large flocks of mixed sexes and ages when not breeding, sometimes associated with *Poephila cincta, P. acuticanda* and *Taenopygia guttata;* recorded feeding on ground. Flocks frequently noted at pools where they drank, birds sucking water without intermittently raising the head. At Mount Anderson (28) a nest was found in a spinifex clump about 10 inches off the ground, two eggs and three young on 30 April and all eggs hatched on 1 May.
Food. Seeds, frequently noted as grass seed, once as of two different sizes.
Annual Cycle. Birds at Glenore (49) in early April had enlarged testes and ovaries, one female, 10 April, having two fully developed yolks on the ovary, and one, 20 April, with an egg in the oviduct: another female showed no signs of gonad activity. In

Western Australia, birds from Mount Anderson (28), where the nest was found, in early May had enlarged ovaries: immatures were collected (35, 38) in July in plain plumage, two showing a few feathers of adult plumage on breasts, wing-coverts and mantle edges. An adult male from Mount Elizabeth (35) showed enlarged testes but also body moult.
Impermanent Colours. *Legs* adults and immatures pinkish-buff to dark brown. *Iris* dark brown. *Bill* upper mandible blue-grey, dark grey, or olive-brown, paler on lower mandible. *Mouth* light grey or blue-grey; immature yellow with bluish gape.

FRINGILLIDAE — Typical Finches
C. J. O. Harrison

Carduelis carduelis — Gold Finch

Specimens. VICTORIA. Coleraine (89) 31 Jan 63, 1 ø skel.
Found in the stockyard of a station. Partly decomposed.

ORIOLIDAE — Orioles
B. P. Hall (F. Carline)

Oriolus sagittatus — Olive-backed Oriole

Specimens. WESTERN AUSTRALIA. Mount Bell (32) 15 – 18 June 68, 3 ♂ (93.5, 99.0 g), 1 imm ♂, 1 imm ♀ (99.9 g). Mount Caroline (33) 24 – 26 June 68, 1 ♂ (100.5 g), 1 imm ♂ (102.9 g), 1 imm ♀ (99.1 g). Mount Elizabeth (35) 15 July 68, 1 ♂ (94.0 g). Lake Gladstone (36) 21 July 68, 1 imm ♀ (82.9 g). Wyndham (39) 8 Aug 68, 1 fl (85.0 g). NORTHERN TERRITORY. Darwin (41) 6 Sept 68, 2 fl (90.0, 104.7 g). Borroloola (46) 9 Nov 68, 1 fl (87.7 g). QUEENSLAND. Moonlight Creek (47) 1 May 64, 1 imm ♂ (85.0 g). Glendower (55) 16 – 31 Mar 64, 1 imm ♂ (101.7 g), 2 imm ♀ (95.8, 98.4 g). Banoona (58) 28 – 29 Apr 63, 1 ♂, 1 ♂ skel. NEW SOUTH WALES. Dunedoo (93) 20 – 25 Mar 63, 2 imm ♂.
Specimens to other Museums. NMV 1 fl Dunedoo (58).
Distribution & Abundance. Also at Pine Creek (44), Inkerman (50), Tully (54), and Grafton (94).
Habitat. Nearly always near water: in the Kimberleys (32-36) by wooded streams and freshwater mangroves; at Wyndham (39) on edge of mangroves and woodland of river estuary; at Darwin (41) in belt of pseudo-rainforest bordering mangroves, and in savanna woodland. In Queensland in eucalypt woodland (47), open forest of flowering trees (54), edge of riverine forest and open woodland (55). In New South Wales (93) in woodland.
Behaviour. Usually alone, occasionally with one or two others, once with *Chlamydera maculata.* Most birds were sitting quietly or feeding above 15 ft in the trees. One was seen frequently extending its head and neck in a curious snake-like attitude (BB).

Usually silent but a melodious whistle was heard once and on another occasion parrot-like squeaks and screechings from a bird crouched in a tree, calling continuously.

Food. Insects, including caterpillars, beetles and ladybirds, and vegetable matter were found almost equally among the thirteen stomachs examined, though the Kimberley birds showed a preference for insects.

Annual Cycle. No gonad activity. All the adults skinned, regardless of locality or date, are in fresh plumage or just completing moult, while all the immatures are moulting from immature to sub-adult plumage.

Impermanent Colours. *Legs* dark grey to black. *Iris* adults orange, immatures red-brown. *Bill* adults maroon to dark brown; immatures dark brown to black. *Mouth* white, pinkish or orange (once).

Geographical Variation. Keast (1956: 20, 23) recognises that northwestern birds, east to the Norman River on the Gulf of Carpentaria (*O. s. affinis*) are paler above, both in the adult and sub-adult plumages, than those of the east (*O. s. sagittatus*) and have a cream or buff-coloured breast, narrow stripes on the abdomen, and less white in the tail. These characters are well marked in our Kimberley adults as compared with the male from Banoona (58), and, as far as can be judged in their unsatisfactory mixed plumage, the immature birds from the Kimberleys (32, 33, 36) and Moonlight Creek (47) when compared also conform with the two from New South Wales (93). However the three from Glendower (55), which, on locality, should be *O. s. sagittatus*, seem to be closer to *affinis* in the colour of the back and intermediate between the two races in the amount of streaking on the underparts and white on the tail. Adults are needed from this area.

Oriolus flavocinctus Yellow Oriole

Specimens. Northern Territory. Darwin (41) 7 – 14 Sept 68, 1 ♂ (91.0 g), 1 imm ♀ (83.7 g), 2 fl (82.1, 99.4 g).

Distribution & Abundance. Fairly common at Darwin (41), less so at Oenpelli (43). Not seen elsewhere in Northern Territory but in Queensland specimens were collected but not retained at Inkerman (50), as follows: 1 ♂ (117.1 g), 1 ♀ (108.3 g), 1 imm ♀ (101.4 g).

Habitat. Pseudo-rainforest with mangrove underlay at Hope Inlet (41). At Inkerman (50) in dense, low trees over swamp near mouth of Salt Arm Creek.

Behaviour. Usually single, once a pair, twice three birds together, once with *Chlamydera nuchalis* (50). Once in thick foliage 10 ft above the ground but more usually in the upper canopy. A variety of calls, some very guttural and harsh, also tuneful whistling and chortles.

Food. Stomachs contained seeds including *Strychnos lucida* (41) and various berries (50), including some purple (wild currant ?), black and yellow.

Annual Cycle. Neither of the immature birds had the skull fully pneumatised: the one skinned (41) is in fresh immature body plumage (duller and heavily marked with black) but with primaries and rectrices fairly worn. The Inkerman (50) adults had the gonads slightly enlarged, but the Darwin (41) male did not.

Impermanent Colours. *Legs* blue-grey. *Iris* male vermilion, immature female dark brown. *Bill* male dark flesh, immature female dark brown. *Mouth* flesh.
Geographical Variation. Northern Territory birds are *O. f. flavocinctus;* on the limited material examined we agree with Keast (1956: 24) that there is no reason to separate Queensland birds.

Sphecotheres vieilloti — Southern Figbird

Specimens. QUEENSLAND. Mount Dryander (57) 3 – 10 Aug 64, 2 ♂ (128.0, 131.2 g), 1 ♀ (146.2 g).
Field Notes. The two males were collected from flocks of 10 and 30 in low, dry rainforest, the smaller party feeding and calling from a tall fig-tree. The female was feeding on reddish berries about 30 ft up in a tree in savanna woodland close to the rainforest. "A soft mimicking variety of calls" were noted from the party (IM).
Food. Stomachs contained small seeds; various green fruit; purple berries.
Annual Cycle. All had small gonads and are finishing moult, the males more advanced than the female.
Impermanent Colours (noted 10 minutes after death). *Legs* males buff; female light grey. *Iris* males red-brown, female blackish-brown. *Bill* males blackish-brown or black; female blackish-brown. *Orbital skin* males white or yellowish white, heavily spotted with dull red; female, deep slate-blue.
Geographical Variation. Keast (1956: 22) provisionally recognises only two subspecies, the dark olive-green southern form *S. v. vieilloti* and the lighter and brighter *S. v. salvadorii* ranging from Port Moresby, New Guinea, "south to about Townsville". Our birds clearly belong with the northern form, extending its range southwards, but the evidence of specimens in the British Museum suggests that a further sub-division should be made, for the type of *salvadorii* and one other New Guinea male have considerably more yellow on the abdomen than six males from Queensland, from Cape York to Mount Dryander (57), which are reasonably uniform. The name *stalkeri,* given to a bird from Townsville, is the oldest name available for the north Queensland birds but may have been mislocalised (Keast, 1956: 24).
Parasites. See Appendix 1.

Sphecotheres flaviventris — Yellow Figbird

Specimens. NORTHERN TERRITORY. Darwin (41) 18 – 20 Sept 68, 3 ♂ (97.1, 101.6, 111.2 g). Oenpelli (43) 15 – 16 Oct 68, 1 ♀ (113.8 g), 1 imm ♂. QUEENSLAND. Inkerman (50) 1 June 64, 1 ♂ (131.3 g).
Distribution & Abundance. Fairly common in small parties at Hope Inlet, northeast of Darwin (41), but less common at Oenpelli (43) where the immature collected was solitary.
Habitat. Pseudo-rainforest at Hope Inlet (41), with eucalypts and abundant fruit trees; in thick vegetation at the edge of the paperbarks on the rocky sandstone escarpment at Oenpelli (43), and in spreading trees with orange berries in tall woodland near the homestead at Inkerman (50).

Behaviour. Usually high in the tree tops, the solitary immature (43) chattering and squeaking very much like *Trichoglossus* (CBF) of which there were a few around. In company with *Oriolus sagittatus* and *flavocinctus* at Inkerman (50).
Food. Stomachs contained red berries; green and red seeds; vegetable matter.
Annual Cycle. The Darwin (41) males are in worn plumage, one just starting moult on body, and had large testes. The Oenpelli (43) female is also worn, just starting moult, and with a small ovary, while the young bird is in full moult apparently to adult plumage. The Inkerman (50) male is fresh.
Impermanent Colours. *Legs* males flesh pink, female light green-grey. *Iris* red-brown or dark brown. *Bill* males black, female brown. *Mouth* deep pink. *Orbital skin* males crimson (including immature), female greyish.
Geographical Variation. Mathews separated Northern Territory birds as *S. f. ashbyi* on their larger size. This is not borne out by our birds, the Inkerman (50) male being both heavier and longer in the wing than Darwin (41) males (149 against 144-147 mm) confirming Keasts' view (1956: 22) that no races should be recognised in Australia.

DICRURIDAE — Drongos
B. P. Hall (F. Carline)

Dicrurus bracteatus — Spangled Drongo

Specimens. NORTHERN TERRITORY. Darwin (41) 6 – 19 Sept 68, 1 ♂ (73.9 g), 4 ♀ (68.6, 68.7, 71.2, 74.8 g), 1 ♀ skel (70.5 g), 2 fl (71.7, 75.2 g). NEW SOUTH WALES. Grafton (94) 3 – 5 Apr 63, 1 imm ♂ skel, 1 fl.
Distribution & Abundance. Fairly common around Hope Inlet, northeast of Darwin (41); a few were seen around Alligator Crossing at Oenpelli (43); a male was collected (but not kept) near the mouth of Salt Arm Creek, Inkerman (50).
Habitat. Coastal belt of monsoon forest and mangroves at Darwin (41), dense forest at Inkerman (50) and in open woodland on the edge of a swamp at Grafton (94).
Behaviour. Single or in pairs, usually perched 5-10 ft from the ground. A hovering display flight from the tops of dead trees was seen just before dusk, and one bird was seen mobbing a Barking Owl *Ninox connivens*. Voice a harsh repetitive chatter.
Food. Insects, including grasshoppers and beetles.
Annual Cycle. Of the five Darwin (41) birds the male had enlarged testes and is in fairly fresh plumage; none of the females had enlarged ovaries, two are in worn plumage (one just starting moult) and two are fairly fresh. The immature at Grafton (94) was in moult to adult plumage.
Impermanent Colours. *Legs* black. *Iris* adults orange or vermilion; immature brown. *Bill* black. *Mouth* black with pinkish palate.
Parasites. See Appendix 1.

GRALLINIDAE — Mud-nest Builders
C. J. O. Harrison

Grallina cyanoleuca — Magpie Lark

Specimens. WESTERN AUSTRALIA. Metricup (9) 12 – 14 Mar 66, 1 ♂, 1 ♂ skel. Newdegate (14) 18 Apr 66, 1 imm ♀ skel. Derby (29) 29 May 68, 1 ♂ (82.9 g), 1 ♀ (80.3 g). Lissadell Hill (38) 29 July 68, 1 ♂ fl (84.2 g). Wyndham (39) 3 Aug 68, 2 fl (87.5, 87.5 g). NORTHERN TERRITORY. Mount Roper (45) 31 Oct 68, 1 ♀ fl (65.8 g). Borroloola (46) 8 Nov 68, 1 fl (98.3 g). QUEENSLAND. Glendower (55) 25 Mar 64, 2 imm fl (75.6, 75.6 g). Banoona (58) 3 May, 63 1 ♀. Monkira (61) 7 June 63, 1 ♀. SOUTH AUSTRALIA. Lake Acraman (71) 3 Nov 63, 1 juv ♀. NEW SOUTH WALES. Albury (92) 10 – 11 Mar 63, 1 ♀ skel, 1 fl. Dunedoo (93) 20 Mar 63, 2 juv ♂, 2 juv fl. Grafton (94) 5 Apr 63, 1 ♂.

Distribution & Abundance. Very widely distributed, in areas where some surface water was available, but absent from the Gibson Desert (20-25) and southeast (15-17) of Western Australia and the west centre (68, 70, 79) of South Australia. Usually in pairs or singly, once in a flock of 14 at Wyndham (39) in August.

Habitat. Varied. In Tuart and peppermint forest (9), open pasture (14), savanna woodland (29, 46, 55), grassland (45), low scrub (71) and pastures and trees by creek (92).

Behaviour. Usually seen feeding by water, or resting perched fairly high in trees. Calls, a metallic *pee-pee,* and the antiphonal *pee-pee, pee-o-wee.*

Food. Insects, including grasshoppers and beetles; seeds and vegetable matter.

Annual Cycle. Females at Albury (92) and Banoona (58) in March had slightly enlarged ovaries; otherwise there was little evidence of sexual activity. Birds in juvenile plumage were collected at Albury (92) and Dunedoo (93) in March, and at Lake Acraman (71) in November.

Impermanent Colours. *Legs* adults black; immatures greyish-black. *Iris* adults, pale creamy white, light yellow or light grey, immatures light grey to dark brown. *Bill* adults, whitish or pale pinkish grey, with blackish on culmen and tip of upper mandible; immatures blackish or dark grey with pale tip; intermediate birds have slate grey bills with light buff on lower mandible and cutting edges. *Mouth* adults whitish, pinkish-grey, grey, blue-grey, black; the whiter birds are probably sub-adult; immatures bright yellow, lemon yellow.

Geographical Variation. There is a general but not necessarily smooth southwards cline of increase in size, wings varying from about 190 mm in the south to 160 mm in the north. Our birds include two extreme examples, a single male from Metricup (9) with the wing 196 mm and a female from Derby (29) with the wing 157 mm. Mees (1961a) records two females from the Pilbara region, collected near each other, with wings of 164 and 184 mm. Mathews proposed the name *neglecta* for one of the small northwestern individuals.

Corcorax melanorhamphus — White-winged Chough

Specimens. QUEENSLAND. Tyrone (59) 15 May 63, 1 ♂ skel. SOUTH AUSTRALIA. Nonning (79) 3 Apr 65, 2 ♂ (312.9, 324.6 g). Furner (86) 7 Jan 63, 1 ♂ skel, 1 imm ♂ skel. NEW SOUTH WALES. Albury (92) 12 Mar 63, 4 fl. Dunedoo (93) 18 Mar 63, 2 ♂ skel.

Field Notes. Also seen at Eucla (17) and Streaky Bay (74). Occurred in parties or small flocks of 8-30 birds. In mallee and savanna woodland. Once noted actively feeding in trees and on the ground.
Food. Insects, including beetles, caterpillars, grasshoppers, and possibly ants; also spiders.
Annual Cycle. The immature from Furner (86) in January was fully-grown and in moult but still being fed by its parents, of which the male was also in moult with slightly enlarged testes. The Nonning (79) bird was in heavy moult in April and one specimen from Albury (92) was noted as being in partial moult.
Impermanent Colours. *Legs* black. *Iris* adults red, bright reddish-orange, or with yellow inner ring and scarlet outer ring; immature pale red and pink (paler than parent) fading to white after death. *Bill* black; immature with pale tip. *Mouth* adult cream or pinkish; immature pinky-white (whiter than parent).

Struthidea cinerea Apostle Bird

Specimens. QUEENSLAND. Banoona (58) 28 Apr 63, 3 ♂ skel, 3 fl. Tyrone (59) 10 – 15 May 63, 1 ♀, 1 ♀ skel. Glendower (55) 24 – 31 Mar 64, 2 ♂ (123.8, 129.7 g), 4 fl (124.5, 128.7, 134.3, 136.5 g).
Field Notes. Also seen at Moonlight Creek (47), Glenore (49), Inkerman (50) and Thylungra (60). Occurred in parties or flocks of up to 40-50 birds in open woodland and pine forest (58), open areas in mulga (59) and low arid woodland (85). Seen feeding on the ground or moving through the trees in noisy flocks. Very wary and still noisier when disturbed, uttering loud incessant screeches "like bullets ricochetting off metal".
Food. Seeds, insects, grasshoppers.
Annual Cycle. Gonads generally small, slightly enlarged in one male at Glendower (55) in March. The female from Tyrone (59) in May showed slight body moult.
Impermanent Colours. *Legs* black. *Iris* olive-grey to dark brown. *Bill* black. *Mouth* pale yellow or yellow, once orange.

ARTAMIDAE — Wood-Swallows

B. P. Hall (F. Carline)

Artamus leucorhynchus White-breasted Wood-swallow

Specimens. WESTERN AUSTRALIA. Peg's Beach (2) 3 June 66, 2 ♂ skel, 1 fl. Mount Anderson (28) 15 – 18 May 68, 3 ♂ (42.5, 45.8 g), 4 ♂ skel (40.7, 46.5 g), 10 fl (44.7 g). Mount Elizabeth (35) 15 July 68, 1 ♂ skel. Lake Gladstone (36) 15 July 68, 1 ♂. SOUTH AUSTRALIA. Innamincka (64) 2 May 65, 1 ♂ (43.0 g), 1 ♀ (39.4 g), 1 ♀ skel (42.4 g). Etadunna (66) 15 May 65, 1 ♀.
Distribution & Abundance. Fairly common locally in the Kimberleys (28-36). In Northern Territory a few over the lagoons at Oenpelli (43) and Borroloola (46), and a few at Pine Creek (44). In South Australia a party of 12 at Innamincka (64), a few birds at Etadunna (66) and a single bird seen at Clifton Hills (65). Encountered widely in

north Queensland; in New South Wales a few over the Clarence River at Grafton (94).
Habitat. Always associated with water: in mangroves at Peg's Beach (2), lake-sides or rivers elsewhere. At Etadunna (66) in open sandy woodland or small eucalypts in the vicinity of Lake Kopperamanna. At Mount Anderson (28) parties came to the lake at dusk and circled over at about 60 ft hawking insects, then went to roost huddled in the top branches of trees fringing the shallows. Some birds had been feeding in the flowering *Grevillea* trees (35). Once after a bird had been collected from a party others stayed in the trees scolding with a harsh *chyek*! *cheyk*! (DG). At other times single soft barking notes were heard.
Food. Insects, including beetles, and one dragonfly.
Annual Cycle. All gonads were small: skinned birds are in fresh plumage, or just completing moult. The male from Innamincka (64) has white tips to the primaries and may not be fully adult.
Impermanent Colours. *Legs* dark grey. *Iris* dark brown. *Bill* grey-blue with black tip. *Mouth* usually grey sometimes with back of palate fleshy, and three times yellowish.
Geographical Variation. Keast (1958b: 209) found that breeding birds from northern Australia, *A. l. parvirostris*, were smaller than southern *A. l. leucopygialis*. There are too few dated and localised specimens in the British Museum on which to judge this, but if it is so then the South Australian (64, 66) specimens are wintering birds of *parvirostris* for their wing lengths are 1 ♂ 123, 2 ♀ 127 mm, compared with 124-131 mm for four Kimberley males.

Artamus personatus — Masked Wood-swallow

Specimens. WESTERN AUSTRALIA. Coolawanyah (4) 20 May 66, 1 ♂ skel. Carnegie (19) 16 Sept 65, 1 ♂ (32.0 g). Mount Anderson (28) 5 May 68, 1 imm ♂, 2 ♂ skel. QUEENSLAND. Glenore (49) 17 Apr 64, 1 ♂ (35.6 g). SOUTH AUSTRALIA. Nullarbor HS (69) 6 Oct 63, 2 ♂, 1 ♀.
Distribution & Abundance. Common at Mount Anderson (28) and Giles (25). Encountered widely in north Queensland, elsewhere also seen at Warburton (23) (one flock), Van der Linden Lakes (24), Tyrone (59), at Nullarbor HS (69) in December 1962 as well as the following year, and on the Eyre Highway 50 miles east of Norseman in Western Australia.
Habitat. Savanna woodland, grassland and mulga.
Behaviour. Sometimes single (19) or in pairs (28) but more usually in flocks of up to 100 birds, or with *A. cinereus* and *A. cyanopterus*. Once (49) a large flock feeding on the ground but at Mount Anderson (28) noted as flying very high. Perches on tops of trees and bushs. A nest 50 miles east of Norseman was in a broken hollow branch of a dead tree eight feet from the ground.
Food. Insects, including grasshoppers, ichneumon flies and many small winged insects.
Annual Cycle. The nest near Norseman had two eggs, with the female sitting, on 4 December. In May the immature bird from Mount Anderson (28) had the skull not fully pneumatised and a few white-tipped feathers of the immature plumage on the body; the adults there were in moult. The male from Carnegie (19) had slightly enlarged testes and from Coolawanyah (4) had small testes.

Impermanent Colours. *Legs* grey. *Iris* brown. *Bill* grey-blue with black tip. *Mouth* whitish, yellower in immature (28), or black (19), or greyish pink (49).

Artamus superciliosus White-browed Wood-swallow

Specimens. QUEENSLAND. Glendower (55) 17 Mar 64, 1 imm ♂ (41.1 g). Tyrone (59) 10 May 63, 1 fl.

Field Notes. Not encountered north of Glendower (55) where it was in flocks of 150-200, including a few *A. personatus*. At Tyrone (59) in flocks of up to 30 birds. Also seen at Thylungra (60) and Monkira (61), and at Little Desert (91), Victoria.

In open woodland, either wheeling high in rather open flocks or massed in trees, descending to the ground to feed. A chirping call in flight and a high, querulous short call when disturbed.

Food. Grasshoppers.

Annual Cycle. The young male is in plumage resembling the female and with the skull partly pneumatised.

Impermanent Colours. *Legs* adult black; immature greyish-horn. *Iris* adult brown; immature golden-yellow. *Bill* grey-blue with black tip. *Mouth* greyish-pink, yellower in immature.

Artamus cinereus Black-faced Wood-swallow

Specimens. WESTERN AUSTRALIA. Chichester Range (3) 13 – 24 May 66, 1 ♂, 2 ♀, 2 imm ♂, 2 imm ♂ skel, 3 fl. Cocklebiddy (15) 8 July 65, 1 ♂. Mount Shenton (18) 30 July – 3 Aug 65, 1 ♂, 1 ♀, 1 imm ♀. Giles (25) 27 Aug 65, 1 ♂. Walter James Range (26) 27 Aug 65, 1 ♀, 1 ø. Mount Anderson (28) 28 Apr – 18 May 68, 1 ♂, 6 ♂ skel, 2 ♀ skel, 1 ø skel, 1 imm ♀ skel, 2 fl (28.2 g). Joint Hill (34) 27 June 68, 1 ♀ skel. Mount Elizabeth (35) 9 July 68, 2 ♂ (38.4, 40.0 g), 1 imm ♀ (37.2 g). Mount Remarkable (37) 26 July 68, 1 ♂ (36.7 g). Lissadell Hill (38) 29 July 68, 1 fl (33.0 g). QUEENSLAND. Moonlight Creek (47) 9 May 64, 1 ♂ (37.4 g), 2 ♀ (37.8, 39.6 g). Glenore (49) 10 – 17 Apr 64, 5 ♂ (32.2, 33.3, 33.4, 34.9, 35.6 g), 3 ♀ (30.9, 31.8, 36.9 g), 3 ø (32.2, 32.3, 35.8 g), 2 imm (32.2, 36.7 g). Inkerman (50) 4 June 64, 3 ♂, 3 ♀. Glendower (55) 14 Mar – 1 Apr 64, 6 ♂ (35.0, 35.9, 36.0, 36.3, 36.4, 36.7 g), 4 ♀ (33.7, 34.6, 35.6, 40.0 g), 2 ø (34.1, 36.0 g), 1 imm ♀ (35.4 g). Thylungra (60) 22 May 63, 1 ♂, 1 ♀, 1 ♀ skel. SOUTH AUSTRALIA. Innamincka (64) 27 Apr – 3 May 65, 1 ♂, 1 ♀. Clifton Hills (65) 12 May 65, 1 ♀. Etadunna (66) 21 May 65, 1 ♂. Oodnadatta (67) 31 May 65, 1 ♂. Nullarbor HS (69) 7 Oct 63, 1 ♂, 1 ♀. Kingoonya (70) 24 June 65, 1 ♂. Nonning (79) 31 Mar 65, 1 ♀ (37.9 g), 1 ♂ skel (35.6 g). Parachilna (81) 12 Apr 65, 1 ♀ (32.0 g), 1 ø (38.8 g).

Distribution & Abundance. Present and usually common or very common at all camps where records were kept except in the southwest (7-13) and southeast (58, 86-95). Certainly the most omnipresent bird in the central area. In north Queensland not encountered at eastern localities (51-54 and 56-57) but very common at Glendower (55) and fairly common at Glenore (49).

Habitat & Behaviour. Any open country with some trees. Birds were collected from flocks of up to 30, sometimes with other *Artamus* species (*cyanopterus, personatus*), *Lalage* *sueurii, L. leucomela* or *Ephthianura tricolor*. Seen hawking insects as well as feeding on the ground, once pecking at a large termite mound. Once seen mobbing a Black-faced Cuckoo-shrike *Coracina novaehollandiae*. Birds coming to a communal roost began to cluster and then dispersed about three times before settling together in a variety of sleeping postures along a sheltered branch (DG).

A male prior to copulation was seen in a crouching posture on a horizontal bough, wings and tail widely spread; it waved its wings and made pivoting movements of the body before flying to the female 30 yards off. While displaying it uttered a husky *tch-tch-tch-tch*. A scolding bird uttered a throaty *tchat-tchat-tchat-tchat* while making lateral movements of its partly-spread tail. This call was similar to that made when coming to roost; other calls were longer and higher pitched (DG).

Food. Insects, including beetles, grasshoppers, bees, dragonflies, grubs, ants, flies; twice spiders.

Annual Cycle. Breeding appears to be synchronised throughout the country for the youngest of the immature birds, with immature feathers on back as well as head, is from Glendower (55) in March; those from Chichester Range (3), Mount Anderson (28) and Glenore (49) in April/May have immature feathers on the head, while those from Mount Shenton (18) and Mount Elizabeth (35) in July have only a few light-edged wing-coverts to indicate their immaturity. Moult appears to be prolonged; most March/April birds were in full moult (49, 79, 81) but others in August (25, 26) had some feathers still coming through and the rest of the plumage was fairly worn. These birds show some enlargement of the gonads.

Impermanent Colours. *Legs* dark grey to black. *Iris* dark brown. *Bill* adults, grey-blue with black tip, the blue paler in birds from Glendower (55), *A. c. hypoleucus* (see below); immatures whitish with darker tip. *Mouth* adults variable yellowish to black, sometimes mottled; immatures yellowish. *Lower eyelid* grey.

Geographical Variation. Variation in these specimens, except for those from the Kimberleys (28-37), has already been discussed by Macdonald (1967). He found that the majority (18-26, 47, 55-81) belonged to the black-vented *melanops*, with western birds (3) possibly distinguishable as more buffy (*A. c. tregellasi*), while the populations of the northeast with white under tail-coverts could be divided into *A. c. normani* of the Gulf coast (49, 50) in which the belly is grey and the vent black or dusky, and *A. c. hypoleucos* from eastern Queensland (55) south of Townsville which is paler on the belly and vent.

Kimberley birds match well enough with typical *melanops* in comparable plumage (worn birds are more buffy) and size. Indeed, although Keast (1958b: 213-4) found considerable variation in wing length in the specimens examined, with a cline of increasing size from north to south, our series is remarkably uniform, wings ♂ 119-123, ♀ 116-122 in all birds except those from Glenore (49), which are in moult, and those from Inkerman (50), *A. c. normani*, which have wings ♂ 115-119, ♀ 113-116.

Parasites. See Appendix 1.

Artamus cyanopterus Dusky Wood-swallow

Specimens. WESTERN AUSTRALIA. Scott River (11) 10 – 20 Feb 66, 1 ♀, 1ø, 1 imm ♂,

1 juv ♂. Mount Barker (12) 21 – 25 Mar 66, 1 ♀, 1 ♀ skel. SOUTH AUSTRALIA. Minippa (73) 1 Nov 64, 1 ♀. Kimba (78) 3 Oct 63, 1 ♀. Nonning (79) 5 Apr 65, 1 imm ♂. VICTORIA. Little Desert (91) 26 Jan 63, 2 juv ø. NEW SOUTH WALES. Albury (92) 10 Mar 63, 1 ♂ skel. Dunedoo (93) 18 Mar 63, 1 ♂, 1 ♀, 1 fl.

Specimens to other Museums. NMV 1 fl Little Desert (91), 1 ♂ Dunedoo (93). AM 1 ♀ Albury (92).

Field Notes. Seen also at Metricup (9), Newdegate (14), Furner (86), Glendower (55) and Cape Gloucester (56) where one male (36.0 g) and two females (30.0, 33.3 g) were collected but not retained from a flock of about 300. Found in open forest and heathland (11), eucalypts by swamp (12), mallee (79, 91), savanna (92, 93, 56). In small parties of up to ten birds everywhere except at Cape Gloucester (56), feeding either on the wing or on the ground.

Food. Insects, including grasshoppers, beetles, Hymenoptera.

Annual Cycle. Juveniles, unable to fly, at Scott River (11) and Little Desert (91) in January/February and an immature bird, moulting to adult plumage, at Nonning (79) in March establish summer breeding in these areas. Adults were mostly in moult (11, 12, 73, 93).

Impermanent Colours. *Legs* adults black or blackish-grey, once blue-grey; young birds usually grey, once black. *Iris* dark reddish-brown. *Bill* adults grey-blue with black tip; young birds, light horn with yellow gape (the youngest) to blackish with grey gape. *Mouth* adults grey or yellow, young birds yellow.

Geographical Variation. Fresh-plumaged adults from Western Australia (12) and New South Wales (93) match in colour, but more worn birds (73, 78) are browner above and below.

Our series confirms Keast's view (1958b: 211) that birds of Western and South Australia, *A. c. perthi,* differ from eastern *A. c. cyanoptera* in having less white on the tail, and having two and not three primaries broadly edged with white, the 4th primary having no white or only a narrow edge.

Artamus minor — Little Wood-swallow

Specimens. WESTERN AUSTRALIA. Coolawanyah (4) 20 May 66, 1 ♂, 1 ♀, 1 skel, 1 fl. Mount Anderson (28) 28 Apr 68, 2 ♂, 1 ♀ (15.9 g), 1 ø (15.3 g), 4 ♀ skel (16.2, 16.2 g), 1 ø skel, 5 fl. Mount Elizabeth (35) 14 July 68, 1 ♂ (13.9 g), 1 fl (14.8 g). QUEENSLAND. Moonlight Creek (47) 1 – 14 May 64, 2 ♂ (15.1, 15.7 g), 1 imm ♂ (12.4 g), 1 imm ♀ (15.6 g), 1 imm ø. Wernadinga (48) 19 May 64, 1 ♂ (13.1 g). Glenore (49) 20 – 23 Apr 64, 2 ♂ (14.2, 14.3 g), 1 ♀ (15.5 g). Inkerman (50) 26 May – 8 June 64, 3 ♂ (14.2, 14.7, 16.9 g), 2 ♀ (15.4, 15.4 g), 1 ø (15.4 g), 1 imm ♂ (14.4 g), 1 imm ø (13.6 g). Glendower (55) 12 Mar 64, 1 ♂ (15.4 g). Mount Dryander (57) 11 Aug 64, 3 ♂ (18.0, 18.6, 18.8 g), 1 ♀ (20.9 g), 1 ø (18.1 g). Tyrone (59) 12 – 14 May 63, 1 ♂, 1 fl, 1 juv fl.

Specimens to other Museums. NMV 1 fl Tyrone (59).

Distribution & Abundance. Seen also at Karratha (1) and all camps in the Kimberleys (28-36). In Northern Territory breeding on the sandstone plateau southwest of the mission at Oenpelli (43), possibly breeding on the sandstone hills at Borroloola (46);

plentiful at Pine Creek (44). Common at Glendower (55) and fairly common at Moonlight Creek (47).
Habitat. Always in Western Australia and twice in Northern Territory (43, 46) associated with sandstone outcrops, but at Pine Creek (44) and in Queensland in savanna woodland mainly in flat open country. At Tyrone (59), where a fledgling was collected, there were no big sandstone ranges similar to those where the species bred at Coolawanyah (4) but only low ridges a few miles away rising just above tree-level, and where no birds were observed (PRC).
Behaviour. Unlike other wood-swallows found mainly in flocks of under 20 or even singly, but it is evident that they also roost communally for at Coolawanyah (3) on three successive mornings flocks of over 200 were found in a tight assembly in dead trees beside a permanent creek. After 08.00 hrs they dispersed to hawk high above the area and over the neighbouring low ranges. The flight was noted as more effortless than that of other wood-swallows, and in the Kimberleys they did not seem to mix with other species in flight, though a party were found with some *A. cinereus* feeding amongst the flower clusters of a *Grevillea* (BB). The voice was seldom heard but was noted as a high-pitched call, or a single chipping note.
Food. Insects, including beetles and some ground-living ants (though birds were not actually noted as feeding on the ground). One of the birds which had been feeding in the *Grevillea* (see above) had insect remains and some yellow fluid in the stomach.
Annual Cycle. No gonad activity. In the Kimberleys, April (28) birds are in moult and the July (35) bird in fresh plumage. In Northern Territory birds appeared to be breeding in early October (43) and November (46). In Queensland breeding was over, except possibly at Tyrone (59) where the juvenile was collected in May: elsewhere adults from March to June were in moult, suggesting that the season of moult may be prolonged, and most of the immature birds were moulting into adult plumage.
Impermanent Colours. *Legs* dark grey or black. *Iris* brown. *Bill* adults grey-blue, with black tip: young birds orange-pink with black tip. *Mouth* variable, mainly blackish and flesh, with a tendency to be yellower in young birds. *Lower eye-lid* whitish.
Geographical Variation. None in colour. Weights indicate that birds from Mount Dryander (57) are larger, but there is no overall difference in measurement, wings ♂ 105-106, ♀ 103-104, against ♂ 99-106, ♀ 99-105 elsewhere.

CRACTICIDAE — Australian Magpies
B. P. Hall (F. Carline)

Strepera graculina — Pied Currawong

Specimens. QUEENSLAND. Banoona (58), 30 Apr 63, 1 ♂ skel.
Field Notes. Seen also at Big Tableland (51), Glendower (55) and Tyrone (59) and in New South Wales (93-95). Collected from high up in a tree in woodland. Very wary.
Impermanent Colours. *Legs* black. *Iris* yellow. *Bill* black. *Mouth* black.

Strepera versicolor Grey Currawong

Specimens. WESTERN AUSTRALIA. Mount Barker (12) 21 – 23 Mar 66, 1 skel, 1 imm fl. Green Range (13) 31 Mar – 4 Apr 66, 1 ♂ skel, 1 ♀ skel, 2 fl.

Field Notes. Common at Green Range (13). Seen also at Newdegate (14); one party at Cocklebiddy (15); common at Eucla (17); a few at Mount Shenton (18), Warburton (23) and Nonning (79).

Found in pines bordering swamp (12), edge of Jarrah forest bordering heathland (13), mallee (13, 17, 18, 79), mulga (18, 23). Usually in parties of 3 – 11 birds, sometimes feeding on ground.

Food. Grasshoppers.

Impermanent Colours. *Legs* black. *Iris* yellow. *Bill* black. *Mouth* black with yellow gape and yellow markings on sides of upper palate and in centre of lower.

Cracticus nigrogularis Pied Butcher-bird

Specimens. WESTERN AUSTRALIA. Mount Shenton (18) 3 Aug 65, 1 imm ♀ (130 g). Mungilli Claypan (20) 12 Sept 65, 1 ♂ (155 g). Mount Anderson (28) 8 – 11 May 68, 1 ♀, 1 imm ♀. Derby (29) 24 May 68, 1 imm ♂. Mount Bell (32) 14 – 18 June 68, 1 ♀, 1 imm ♂ (105.5 g). Mount Elizabeth (35) 14 July 68, 1 ♂ (125.5 g), 1 ♀ (123.8 g), 2 fl (121.2, 124.4 g). Mount Remarkable (37) 25 – 26 July 68, 1 ♂ (105.5 g), 1 fl (109.0 g). Lissadell Hill (38) 28 – 29 July 68, 1 ♂ (102.8 g), 1 fl. NORTHERN TERRITORY. Darwin (41) 7 Sept 68, 1 imm fl. Pine Creek (44) 30 Aug 68, 1 imm ♀ (82.5 g). Borroloola (46) 13 Nov 68, 1 ♂, 1 ♀ (97.0 g), 1 fl (96.1 g). QUEENSLAND. Banoona (58) 30 Apr – 3 May 63, 1 ♂, 2 ♀. NEW SOUTH WALES. Grafton (94) 3 Apr 63, 1 ♂ skel, 1 fl.

Distribution & Abundance. Common at Mount Shenton (18) and one or two at most camps in the Gibson Desert (19-21), at Warburton (23), Giles (25), Karratha (1) and the Chichester Range (3). Seen at most camps in the Kimberleys and Northern Territory but absent from Oenpelli (43) and the Alligator River (42). In Queensland common at Glendower (55), seen also at Tyrone (59) and Thylungra (60), and at Tenterfield (95), but not further south.

Habitat. Mulga with corkwoods and river gums (18, 20); savanna woodland, sometimes near creeks in the Kimberleys and Northern Territory; open woodland (58, 94).

Behaviour. Usually single or in pairs, once four birds together. Usually perched in trees but once feeding on ground. One bird had been killed and was being carried by a hawk, which dropped it when alarmed. On another occasion a bird when shot flew wildly around in circles: a second bird appeared and the two flew round together touching, one above the other. The song was heard more often before it was fully light than later in the day.

Food. Lizards (twice) and large insects, including grasshoppers and large weevils (once).

Annual Cycle. In the Kimberleys May birds were in moult; June and July birds have the plumage in various stages of wear but show no sign of moult, and some gonads were starting to enlarge (37-38). The immature May bird from Derby (29) was moulting into adult plumage and appears older than those taken later (32, 41, 44). From central

Australia the Mount Shenton (18) immature taken in August was similarly moulting into adult plumage. The Mungilli Claypan (20) male had enlarged testes and worn plumage.

The southern Queensland May birds (58) showed no gonad activity and are in fairly fresh plumage.

Impermanent Colours. *Legs* dark grey. *Iris* dark brown. *Bill* pale grey to slate grey, anterior third blackish. *Mouth* adults mainly black, some mottled or fleshy; immatures whitish. *Skin behind eye* dark grey.

Geographical Variation. The weights and the measurements of our series show that southern birds are larger than northern, and that the northern birds *C. n. picatus* decrease in size eastwards from Derby, confirming the pattern found by Amadon (1951: 8-11) and Mees (1961a: 125-126). Amadon also divided the southern birds into a long-billed western race, *C. n. kalgoorli,* and the shorter-billed *C. n. nigrogularis,* but Mees considered bill size too variable to admit this. Skins in the British Museum support the view that western birds have long bills, 3 ♂ 52-53, 2 ♀ 45-50, against 7 ♂ 45-47, 4 ♀ 43-45 in birds from Queensland.

Cracticus quoyi — Black Butcher-bird

Specimens. NORTHERN TERRITORY. Darwin (41) 7 – 20 Sept 68, 1 ♂ (215.5 g), 2 fl (184.9, 194.3 g). QUEENSLAND. Ayton (52) 25 June – 3 July 64, 4 ♂ (159.1, 164.6, 172.0 g).

Field Notes. Fairly common at Hope Inlet, northeast of Darwin (41) in mangroves and eucalypts in pseudo-rainforest. At Ayton (52) in tropical rainforest and once in palms. Two birds came to squeaking. Voice a rich yodelling call. Not seen elsewhere.

Food. Stomachs contained insects and once (41) a few tiny pieces of crab-shell.

Annual Cycle. No moult is apparent in any skin, and only the male from Darwin (41) showed gonad activity. In this species most immature birds are all black but a rufous phase occurs among immatures in *C. q. rufescens* (see below) of Queensland, and it seems likely that one of the Ayton (52) males which has a few brown-edged feathers on the bend of the wing is a first-year bird. One of the Darwin (41) birds put into spirit (the heavier) was also believed to be young when collected.

Impermanent Colours. *Legs* dark grey or black. *Iris* very dark brown. *Bill* grey-blue at base, remainder black; individuals vary in having between one-third and one-half grey-blue. *Mouth* black, or once whitish (41) – possibly immature. *Lower eye-lid* grey.

Geographical Variation. Amadon (1951: 11-14) recognises a larger race *C. q. spaldingi* from Northern Territory and Cape York (wing 184-207 mm) and the smaller *C. q. rufescens* from the rest of Queensland (wing 171-183) in which there is a rufous phase in immature birds. Our birds show that *rufescens* reaches as far north as Ayton (52) (wings 175-179). The weights of the Darwin (41) birds also demonstrate the greater size of *spaldingi*.

Cracticus torquatus — Grey Butcher-bird

Specimens. WESTERN AUSTRALIA. Metricup (9) 7 – 15 Mar 66, 1 ♂ skel, 2 fl. Scott

River (11) 22 Feb 66, 1 fl. Newdegate (14) 11 Apr 66, 1 imm ♂, 1 imm ø. Eucla (17) 5 July 65, 1 imm ♂. Mount Bell (32) 20 June 68, 1 ♀ (82.7 g). Mount Elizabeth (35) 15 July 68, 1 ♀ (91.1 g). NORTHERN TERRITORY. Darwin (41) 6 – 17 Sept 68, 2 ♂ (80.9 g), 2 ♀ (74.7, 84.0 g), 1 ♂ skel (83.4 g). QUEENSLAND. Glendower (55) 16 Mar – 2 Apr 64, 4 ♂ (86.3, 87.2, 88.4, 96.7 g), 4 ♀ (75.8, 84.2, 85.5, 87.5 g), 3 imm ♂ (85.7, 88.7, 91.4 g), 1 imm ♀ (77.9 g). Banoona (58) 30 Apr 63, 1 ♂, 1 ♀ skel. SOUTH AUSTRALIA. Kingoonya (70) 20 June 65, 1 ♂. Lake Acraman (71) 3 Nov 64, 1 imm ♀. Sceale Bay (75) 13 Oct 63, 1 juv ♂. Nonning (79) 3 Apr 65, 1 imm ♂. Aroona Valley (82) 9 Apr 65, 1 ♂ (86.5 g). Angepena (84) 22 Nov 63, 1 imm ♀. NEW SOUTH WALES. Dunedoo (93) 18 Mar 63, 1 ♀ skel. Grafton (94) 5 Apr 63, 1 ♂, 1 imm ♀. Tenterfield (95) 10 Apr 63, 1 ♀.

Specimens to other Museums. AM 1 imm ♀ Dunedoo (93).

Distribution & Abundance. In Western Australia also seen at Chichester Range (3), Perth (7), Green Range (13) and Cocklebiddy (15) where it was very common. Elsewhere a few at Oodnadatta (67), Granite Downs (68) and Furner (86). Uncommon at Glendower (55) where it was always found near the much commoner *C. nigrogularis*. Not found elsewhere in northern Queensland.

Habitat. Any type of wooded country or light forest: mulga (70), mallee (14, 17, 79), native pines (82), sub-monsoon forest (41) and riverine forest (55).

Behaviour. Usually single or in pairs, though sometimes several birds would be heard within a half a mile of each other. At Eucla (17) it was found to sit habitually in any patch of sun just after dawn, when it was very lethargic and allowed a closer approach than at other times. Seen to fly straight from its look-out to take insects from the ground and trees, and also to chase *Acanthiza uropygialis* without success (DG). Many melodious phrases of song; at Glendower (55) described as a sonorous *wik-wok-wurrawoo* as distinct from the *wik-ik-ik, eou* of *C. nigrogularis* (ICJG).

Food. Stomachs contained mainly large insects, particularly grasshoppers, but also vegetable matter, seeds and berries.

Annual Cycle. Among the adults, all those collected in March and April are in moult (9, 55, 82, 94) except for one from Tenterfield (95) in which moult is completed: some from Glendower (55), especially the females, have the gonads slightly enlarged (probably receding). Those collected in June and July are worn (35, 70) or starting moult (32). September birds from Darwin (41) are also worn and the gonads were slightly enlarged. Plumages of young birds are complex and may be individually variable. As far as I can understand them from our series they are as follows: (a) *Juvenile* above, dark brown with chestnut shaft-streaks on head and mantle; below, fluffy white with some dark markings on the breast; wings brown with light brown edges to coverts and secondaries; tail brown with small light brown patches at tips – as in the October juvenile from Sceale Bay (75) in which the wings and tail are not yet fully grown. (b) *Immature* head, dark brown or dull black with very faint light shaft-streaks; mantle, dark brown, faintly scalloped, with black shaft-streaks; underparts, whitish grey with a buff wash on the breast when fresh, vestigial grey barring, and black shaft-streaks; wings brown, coverts and secondaries with light brown or white edges; juvenile tail – April birds from Newdegate (14) and Nonning (79) are in this plumage, the latter having a number of feathers coming through on the body: one from Grafton (94) in April, and from Kingoonya (70) in July are plainer on the back and wings and only faintly washed with buff below, and appear to represent a worn stage of this plumage. The Lake Acraman (71) November bird is exceedingly worn, but just starting to moult into sub-adult plumage, having a new full-grown black rectrix. (c) *Sub-adult* as adult but with the head duller black; mantle darker and plainer; underparts with usually more vestigial grey barring; wings very dark brown, fading to pale brown; tail, almost black when fresh, fading to brown, with white patches on tips not so extensive as in adults – the November bird from Angepena (84) is in this plumage and several from Glendower (55) are apparently moulting from sub-adult to adult.

On such a small series from scattered localities it is impossible to postulate how long each plumage may be retained, especially in the puzzling absence of any juvenile or immature birds among the series of sub-adults and adults in presumably post-breeding moult from Glendower (55) in April. A full study of this problem is desirable.

Impermanent Colours. *Legs* dark grey or black. *Iris* brown. *Bill* anterior half black, base, whitish-grey to slate blue. *Mouth* whitish, pink, mauvish, whitish-blue blotched black, blackish. *Eyelid* pale blue-grey. No colours were noted for the juvenile bird, and from the labels there seem to be no consistent differences between immatures, sub-adults and adults, but descriptions suggest that the bills of young birds are usually greyer at the base, and in the skin they usually appear paler and browner.

Geographical Variation. Amadon (1951: 4-8) recognises only three races on the mainland, the darkest *C. t. torquatus* in the southeast, the palest *C. t. argenteus* in the northwest, and *C. t. leucopterus* elsewhere. Our birds confirm this in that those from the Kimberleys (32, 35) and Darwin (41) are clearly *argenteus,* and adults from Queensland (55, 58) and northern New South Wales (94, 95) are similar to some *leucopterus* in the British Museum taken at Southern Cross, W.A., when allowance is made for the very fresh plumage of the eastern birds. The two worn birds from South Australia (70, 82) appear darker, and may indicate an approach to *C. t. torquatus,* but fresh material is needed to judge this.

Cracticus torquatus argenteus x *C. nigrogularis picatus*

Specimen. Western Australia. Mount Bell (32) 18 June 68, 1 ♂. Wing 160, bill (from base) 44, tail 107 mm.

This specimen is in *torquatus*-type plumage, differing only in having more conspicuous patches of black at the sides of the breast and more black at the bases of the breast feathers than others of the species, though individuals show these characters to a lesser extent. It is its greater size, particularly the wing and bill, which makes us believe that this is a hybrid with *nigrogularis,* for the highest measurements for five *C. t. argenteus* are wing 146, bill 40 and tail 105 mm.

Furthermore, comparison with specimens of both species taken at Mount Bell (32) shows that its plumage condition (fairly fresh) corresponds with the annual cycle of *nigrogularis* rather than *torquatus,* in which moult had only just started.

Gymnorhina tibicen Black-backed Magpie

Specimens. Western Australia. Mount Elizabeth (35) 6 – 9 July 68, 1 ♀ (250.1 g), 1 imm ♀ (205.4 g). Queensland. Banoona (58) 3 – 5 May 63, 1 imm ♀, 1 ♂ skel. South Australia. Innamincka (64) 29 Apr 65, 1 ♂ (255 g). Granite Downs (68) 11 June 65, 1 ♂. Aroona Valley (82) 16 – 17 Apr 65, 2 ♂ (288.9, 295.5 g). New South Wales. Dunedoo (93) 23 Mar 63, 1 imm ♂. Grafton (94) 5 Apr 63, 1 ♀ skel.

Distribution & Abundance. Seen also at Karratha (1), Chichester Range (3) and most Kimberley camps (33-36). A few at Borroloola (46) and at several camps in central Australia (23, 25, 65-68, 70). Common at Aroona Valley (82) and at Glendower (55).

Habitat. Usually in gums along creeks (35, 64, 82, 93) or in more open country.

Behaviour. Birds collected were single or from parties of not more than five birds. Some were feeding on the ground, one was perched 100 ft up in a gum and when disturbed flew to open country looking back over its shoulder. Other birds in flight were noted as turning their heads alternately left to right.

Food. Stomachs contained insects, including beetles, seeds, and once a piece of plastic tape used for markers by surveyors.

Annual Cycle. The Kimberley (35) female was in fairly fresh plumage while the immature bird was rather worn. All adults from South Australia were in worn plumage with no gonad activity. The May immature bird from Banoona (58) was starting to moult into adult plumage.

Impermanent Colours. *Legs* dark brown, dark grey or black. *Iris* brown. *Bill* adults blue-grey or greyish white, black at tip; young birds, dark horn or dark grey. *Mouth* adults black or grey, sometimes with some pink, young birds whitish.

Geographical Variation. The two adults from Mount Elizabeth (35) and Granite Downs (68) have longer bills (61, 59 mm respectively) than the three males from Aroona Valley (82) and Innamincka (64) (52-54 mm), showing that if *G. t. longirostris* of Western Australia is recognised they should be referred to it. The only other noteworthy variation in this small series is the paleness of the immature from the Kimberleys (35) in comparison with those from Queensland (58) and New South Wales (93), but not enough young birds have been examined to know whether this variation is individual or geographical.

Parasites. See Appendix 1.

Gymnorhina dorsalis Western Magpie

Specimens. Western Australia. Metricup (9) 12 – 17 Mar 66, 1 ♂, 1 imm fl. Margaret River (10) 9 Mar 66, 1 fl. Scott River (11) 20 Feb 66, 1 ø skel, 1 imm ♂ skel.

Field Notes. All birds were collected from parties, in open paddocks with gums (9), heathland (11) and forest of Tuart and peppermint (9) or Marri and Jarrah (10). Seen also at Perth (7) and other camps in the southwest (7-14).

Food. Large seeds and insects.

Annual Cycle. The immature and adult birds from Scott River (11) were in moult. The Metricup (9) male was in fresh plumage.

Impermanent Colours. *Legs* dark grey or black. *Iris* usually brown, once dark red

(11). *Bill* dark grey or black, with steel blue base; one immature light horn. *Mouth* grey or flesh.

Gymnorhina hypoleuca White-backed Magpie

Specimens. VICTORIA. Nelson (88) 4 Jan 63, 1 imm skel.
Field Notes. Seen also at Furner (86) and at Streaky Bay (74).
Impermanent Colours (of immature). *Legs* black. *Iris* brown. *Bill* light grey.

Gymnorhina hypoleuca x tibicen

Specimens. NEW SOUTH WALES. Albury (92) 9 – 12 Mar 63, 3 ♀.
Field Notes. Common in paddocks and savanna woodland; two birds were single, the third was from a party of five.
Food. Stomachs contained grasshoppers and beetles; also grass.
Annual Cycle. All are in fresh plumage, completing moult on the head. Ovaries were small.
Impermanent Colours. *Legs* black. *Iris* brown. *Bill* blue-grey with anterior third black. *Mouth* flesh or blue-grey.
Geographical Variation. These birds come from a known hybrid zone between the White- and Black-backed Magpies. They vary in the amount of grey and black on the mantle, one having the centre of the back black with wide borders of grey on the hind neck and lower back; another has the same area grey with a few black feathers; the third has an asymmetrical mixture of black and grey feathers.

PTILONORHYNCHIDAE — Bower-birds

B. P. Hall (F. Carline)

Ailuroedus melanotis Catbird

Specimens. QUEENSLAND. Big Tableland (51) 1 – 2 July 64, 3 ♂ (163.5, 181.3, 182.4 g), 1 ♀ (170.8 g). Ayton (52) 21 – 29 June 64, 2 ♂ (174.3, 179.3 g), 4 ♀ (142.5, 161.5, 163.5, 165.8 g), 1 imm ♂ (185.9 g). Walter Hill Range (53) 23 July 64, 2 ♂ (186.0, 195.2 g), 1 ♀ (162.6 g).
Field Notes. Found in rainforest on steep slopes from sea level to at least 2,200 ft. Usually single or in pairs, once in flock of 20. Perched birds were within 10 ft of the ground but those feeding were up to 50 ft high in the trees. Once in Johnstone Hardwoods. Came to collector's squeaking and to the call of a wounded *Megapodius freycinet*.
Food. Stomachs contained fruit, berries (many purplish-black), seeds and once some insect matter.
Annual Cycle. The immature male from Ayton (52) did not have the skull fully pneumatised; it is in fresh plumage resembling the adult female (see below). Of the adults, those from the Walter Hill Range (53) have some enlargement of the gonads, as has one female from Big Tableland (51). July birds are in fresh plumage and June birds mainly in moult.

Impermanent Colours. *Legs* grey sometimes with purple or bluish tinge. *Iris* males olive or yellow-brown to bright red (2 ♂, 53); females olive to orange (♀ 53) birds in moult having usually duller-coloured eyes than others; immature olive green. *Bill* whitish or yellowish-grey. *Mouth* greyish-white, palate black.
Geographical Variation. We follow Gilliard (1969) in regarding *A. melanotis* and *A. crassirostris* as distinct species, our birds belonging therefore to the Queensland race of *melanotis, A. melanotis maculosus.* These, and six specimens in the British Museum, show that there is a tendency for populations from the Ayton area (51, 52) southwards to be larger with less heavy brown markings on the breast and head than more northern birds, showing an approach to *crassirostris.*
Other Notes. Both Gilliard (1969: 267) and Mayr & Jennings (1952: 3) deny that there is a sexual difference other than size but in our series females are clearly recognisable from all but one of the adult males of the same population by being greener on the crown and forehead and less heavily marked on the breast. Confusion may have arisen because the immature plumage of the males, as shown by our bird with the skull not fully pneumatised, resembles the female. The single other male which does not conform is the least heavy of the two Ayton (52) birds and is shorter in the wing than others, so is probably also not fully adult.

Scenopoeetes dentirostris Tooth-billed Bower-bird

Specimens. Queensland. Big Tableland (51) 1 July 64, 1 ♂ (165.1 g). Walter Hill Range (53) 21 July 64, 1 ♂ (199.2 g).
Field Notes. In tropical rainforest at 2,200 ft. Both birds came to squeaking and perched fairly low ("in a small tree", "8 ft up") to investigate.
Annual Cycle. Gonads small. Both birds just completing moult. The bird from Big Tableland (51) may be sub-adult (see below).
Food. Stomachs contained large purple fruit (53) and light green fruit paste (51).
Impermanent Colours. *Legs* deep olive-green (51) or shiny greyish-green (53). *Iris* pale brownish-grey (51) or dark brown (53). *Bill* blackish-brown with edge of mandibles and tip of lower mandible whitish (51); or black (53) – the differences visible in the skins. *Mouth* "white dorsally, interior yellow-pink" (51) or "mottled white, pink and grey" (53); gape orange in both.
Geographical Variation. The male from Walter Hill Range (53) is longer in the wing (149 against 147 mm) and more heavily streaked on the upper breast as well as being heavier than the Big Tableland (51) bird. If the latter is correctly sexed, these differences and the differences in the impermanent colours suggest either that there is some geographical variation or that it is sub-adult, which is more likely.

Ptilonorhynchus violaceus Satin Bower-bird

Specimens. Queensland. Big Tableland (51) 2 July 64, 2 imm ♂ (208.5, 231.3 g).
Field Notes. Rainforest at 2,300 ft. Both in a party of about six birds feeding high in a hardwood tree, with some *Ailuroedus*. Not seen elsewhere.
Food. Soft purplish-red berries like those of *Rubus* sp.

Annual Cycle. One bird is in the immature plumage described by Gilliard (1969: 346) as "like adult female but with tail feathers more pointed, wings more brownish and upper wing coverts with rufous edges". The heavier (sub-adult) bird is in Gilliard's male intermediate "(a)" plumage with browner ear-coverts, less spotting on the lower neck, greener on the back, and without light edges to the wing coverts. Both are in fresh plumage but showing some wear on the tail.
Impermanent Colours. *Legs* sub-adult grey-green, immature whitish-buff. *Iris* bright blue. *Bill* blackish-brown. *Mouth* creamy-white.
Geographical Variation. These belong to the smaller northern race *P. v. minor*.

Chlamydera maculata — Spotted Bower-bird

Specimens. WESTERN AUSTRALIA. Chichester Range (3) 16 May 66, 1 ♀ skel. Giles (25) 23 Aug 65, 1 ♀ (135 g). QUEENSLAND. Glendower (55) 16 – 26 Mar 64, 6 ♂ (135.0, 140.5, 143.0, 143.9, 151.3, 154.1 g), 2 ♀ (139.9, 140.3 g), 1 imm ø (138.1 g), 1 imm ♀ (head only) fl. Banoona (58) 30 Apr – 6 May 63, 1 ♂, 1 ♀ skel, 1 fl.
Specimens to other Museums. NMV 1 ♂ Banoona (58), 1 ♀ Thylungra (60) 28 May 63.
Distribution & Abundance. Only the one bird from Giles (25) was seen by the 3rd Expedition. At Glendower (55) very common and seen also at Tyrone (59), but not elsewhere in Queensland or in New South Wales.
Habitat. Corkwood trees on spinifex plain (3): a clump of fig trees *Ficus platypoda* at the head of steep, dry valley on the south face of the Rawlinson Range (25) where *Amytornis textilis* was found; both low arid woodland and riverine forest with fig trees, casuarinas, gums (55). Grassy paddocks with Pepper Trees (58).
Behaviour. Usually single or at Glendower (55) in mixed parties with honeyeaters, and once *Platycercus adscitus*. Came to squeaking and distress calls.
Food. Stomachs contained figs, berries, various seeds, spiders, grasshoppers.
Annual Cycle. The Giles (25) female had the ovary slightly enlarged and is in fresh plumage. In Queensland the March adult birds from Glendower (55) were in full moult; the unsexed immature was moulting into adult body plumage with a nuchal patch. The Banoona (58) male in early May was completing moult.
Impermanent Colours. *Legs* brown horn (25) or light olive-green to brownish green (55-60). *Iris* olive-brown to dark brown. *Bill* adults black, immature dark brown. *Mouth* mainly yellow.
Geographical Variation. The Giles (25) bird clearly belongs to the western *C. m. guttata* but differs from the only other western bird in the British Museum (from the northwest Cape) in having the head more spotted and less rufous, the spots of the mantle slightly less rufous, and the underparts mainly pale yellow with no rufous on the breast or flanks.

No races are currently recognised among eastern birds, *C. m. maculata*.

Chlamydera nuchalis — Great Bower-bird

Specimens. WESTERN AUSTRALIA. Mount Anderson (28) 14 May 68, 1 ♀ skel (208.5 g).

Geikie Gorge (31) 5 June 68, 2 ♀ (189.0, 200.5 g). Joint Hill (34) 4 July 68, 1 fl. Mount Remarkable (37) 26 July 68, 1 imm fl. Lissadell Hill (38) 28 July 68, 1 fl (192.4 g). Wyndham (39) 9 Aug 68, 1 fl (211.6 g). NORTHERN TERRITORY. Darwin (41) 16 Sept 68, 1 imm ♂. Borroloola (46) 8 – 16 Nov 68, 1 ♂ skel (166.4 g), 3 ♀ skel, 1 ♂ fl (187.7 g), 1 fl (250.7 g!). QUEENSLAND. Moonlight Creek (47) 4 – 6 May 64, 2 ♂ (216.8, 226.1 g). Glenore (49) 24 Apr 64, 1 ♂ (188.3 g). Inkerman (50) 27 May – 6 June 64, 1 ♂ (233.2 g), 5 ♀ (191.3, 196.2, 197.1, 202.6, 203.9 g), 1 imm ♂ (213.4 g), 1 imm ø (201.7 g).

Distribution & Abundance. Encountered at all main camps in northwestern Australia and Northern Territory (28-46) but less common at Pine Creek (44) and Mount Roper (45) than elsewhere. One bird seen at Glendower (55) (Parker, 1968) but none in the northeastern Queensland camps (51-54, 56-57).

Habitat. In Kimberleys (28-31) in riverine eucalypts or spinifex on rocky outcrops. In Northern Territory, savanna woodland close to monsoon forest (41), wooded swamps with paperbarks near lagoons (46).

In Queensland, mixed savanna woodland by gravel banks or sand ridges near coast (47, 64) – tall trees with low undergrowth.

Behaviour. Single or in small parties of up to eight birds; twice in same trees as an *Oriolus sagittatus.* Seen at all levels in the trees. Secretive but came to squeaking and distress calls.

Food. Stomachs contained fruit pulp and seeds of various sizes and shapes including those of passion fruit (31, 47). Also a spider (once) and insects including grasshoppers, larvae, ants.

Annual Cycle. The young September bird from Darwin (41) was just moulting out of juvenile into immature body plumage. The other Inkerman (50) young May birds were older, being in very worn immature plumage. The Geikie Gorge (31) June females were completing moult and had small ovaries. One of the Borroloola (46) females in November was on the point of lay but others had small gonads and some were in moult.

In Queensland the Glenore (49) and Moonlight Creek (47) birds were in moult, and of the Inkerman (50) series the male and three females were in mainly fresh plumage; two other females were possibly sub-adult moulting into adult plumage, but the very worn condition of the tails makes it difficult to be sure; the ovaries of all were slightly enlarged. In this species pneumatisation of the skull apparently takes some time to complete as two of the males with nuchal patches (47, 50) and one of the apparently fully adult females (31) have some "windows" in the skull.

Bowers. In the Kimberleys unroofed bowers were seen at Mount Anderson (28), Derby (29) and Geikie Gorge (31), and roofed bowers at Mount Bell (32) and Mount Caroline (33). All contained stones and bones, and one (33) also snail shells and green fruits.

Impermanent Colours. *Legs* greyish-green to greyish-brown; immature (41) green-yellow. *Iris* greyish-brown to dark brown. *Bill* black or dark sepia, sometimes with lighter tip in females or young birds. *Mouth and inside of nostrils* bright yellow or orange.

Geographical Variation. In *C. nuchalis* Mayr & Jennings (1952: 12-15), followed by Gilliard (1969: 372) recognised a western group in which males are less patterned above and females less liable to be barred below in comparison with eastern birds, the division between the two groups lying probably at the head of the Gulf of Carpentaria (on the evidence of a single male from the Gregory River which appeared similar to western birds on the upper parts). Within each group two subspecies were recognised: in the west *C. n. nuchalis* of Northern Territory (and possibly northwestern Queensland) being smaller than *C. n. oweni* of the Kimberleys but otherwise similar; in the east *C. n. yorki* of Cape York being smaller and paler than *C. n. orientalis* from further south, intermediates occurring at Cairns. Gilliard inadvertently reversed the characters of these subspecies. (Note that the Inkerman of Mayr & Jennings is on the east coast near Townsville and is not the same as our locality, on the Gulf of Carpentaria.) Our series from the Gulf should help resolve the boundaries between the races but unfortunately lack of good comparative material prevents positive conclusions.

The Kimberley (31) females represent *C. n. oweni* and stand out from the rest as less patterned on the back. The Darwin (41) bird is the only specimen available in this stage of immature plumage and others of *C. n. nuchalis* from Northern Territory are too worn for useful comparison, so that birds from the Gulf have to be compared direct with *C. n. oweni*. The most western Gulf birds (47) are very slightly less patterned than those from Inkerman (50) but are much closer to them than they are to *C. n. oweni* (or presumably *C. n. nuchalis*). On the other hand the adult males appear to have a smaller nuchal patch than a representative of *C. n. orientalis* from the east coast, though they match it in colour and size (wing 175-182). Most females, including the one from Moonlight Creek (47), have some barring on the underparts; one is unique in having a nuchal patch like the males. This bird had the ovary somewhat enlarged so there is little likelihood of mis-sexing; furthermore in size (wing 168 mm) and weight (202.6 g) it matches other females. For the present it is perhaps best to regard the Gulf birds as near *orientalis,* showing a slight approach to *nuchalis.*

PARADISAEIDAE — Birds of Paradise
B. P. Hall (F. Carline)

Ptiloris victoriae — Victoria Rifle-bird

Specimens. QUEENSLAND. Big Tableland (51) 1 – 2 July 64, 2 ♀ (82.9, 84.3 g), 1 imm ♂ (92.5 g). Ayton (52) 24 – 30 June 64, 1 ♀ (83.0 g), 5 imm ♂ (87.2, 88.7, 91.1, 92.7, 100.4 g), 1 imm ♀ (78.0 g). Walter Hill Range (53) 21 – 23 July 64, 2 ♂ (99.3, 106.8 g), 1 ♀ (88.4 g), 1 imm ♂ (104.2 g). Tully (54) 11 – 15 July 64, 1 ♀ (76.7 g), 2 imm ♂ (93.2, 94.3 g), 1 imm ♀ (89.5 g).
Distribution & Abundance. Common where found, but not seen elsewhere.
Habitat. Rainforest between sea level and 2,200 ft, and on the edge of forest and mangroves on islands in the mouth of the Bloomfield River, Ayton (52).
Behaviour. Several birds were seen pecking at trunks or branches, another was feeding on mandarin oranges. The two full-plumaged males were high in the trees (above 30 ft), other birds were at various heights from ground level to 30 ft. Five members of this species were the most seen together; these were in company with several honeyeater species, including *Acanthorhynchus tenuirostris*. At other times birds were with a Rufous

Shrike-thrush *Colluricincla megarhyncha* and an oriole. One young male was scolding a snake. Several birds came to squeaking or distress calls.

Food. Most commonly insects, including beetles, wood-grubs, a centipede, a grasshopper. Also spiders, fruit, including oranges (see above) and small pea-sized berries.

Annual Cycle. All birds were in fresh or fairly fresh plumage having completed moult, and showed no appreciable gonad activity. Only the two males from Walter Hill Range (53) were in full adult plumage; of the males listed as immature three had the skull not fully pneumatised but they do not differ in any other respect from the other five. Similarly the two females listed as immature were diagnosed on the skull character alone. It may be that pneumatisation in this species, as in some bower-birds, is prolonged.

Impermanent Colours. *Legs* adult males black, others dark grey. *Iris* dark brown. *Bill* blackish-brown or black. *Mouth and gape* bright yellow, sometime with greenish tinge.

Geographical Variation. We follow Gilliard (1969: 112) in regarding *P. victoriae* as a species distinct from the larger *P. paradiseus* from southern Queensland in which the females and young males are more heavily marked and less rufous below. Our specimens come from close to the northern (51, 52) and southern (53, 54) limits of the range of *victoriae;* they show no geographical variation in size but the Tully area (53, 54) females and young males show an approach to *paradiseus* in being, in series, more heavily marked than those from Ayton area (51, 52), and the females are also slightly paler. Specimens from older collections taken around Cairns (in between our two areas) are individually more variable than our series but on the whole closer to those from the Tully area (53, 54).

CORVIDAE — Crows & Ravens
D. Goodwin

Corvus coronoides — Australian Raven

Specimens. Western Australia. Scott River (11) 21 – 24 Feb 66, 1 imm skel, 1 fl. Mount Barker (12) 24 Mar 66, 1 imm. Newdegate (14) 12 – 17 Apr 66, 1 imm ♂, 1 ♂ skel, 1 imm ♀. Eucla (17) 1 July 65, 1 ♂, 1 skel. Queensland. Moonlight Creek (47) 1 – 12 May 64, 3 ♂ (640.4, 646.1, 685.1 g), 2 ♀ (572.6, 633.5 g). Glenore (49) 13 – 24 Apr 64, 1 ♂ (576.2 g), 2 imm ♂ (547.1, 606.4 g), 2 imm ø (495.0, 504.2 g). Glendower (55) 14 – 31 Mar 64, 2 ♂ (677.5, 695.3 g), 1 ♀ (577.0 g), 2 imm ♂ (571.3, 626.6 g), 2 imm ø (560.8, 583.4 g). Banoona (58) 6 May 63, 1 ♂ skel. Monkira (61) 5 June 63, 1 ♂ skel. South Australia. Innamincka (64) 29 Apr 65, 1 ♀ (582.6 g). Clifton Hills (65) 11 May 65, 1 ♀. Nullarbor HS (69) 25 – 30 Oct 64, 2 imm ♂. Aroona Valley (82) 12 – 14 Apr 65, 1 ♂, 1 skel.

Distribution & Abundance. In Western Australia also at Metricup (9), Green Range (13), and Cocklebiddy (15). In the southeast (15, 17) it was noted as very common but it is possible that some of the birds seen may have been *C. mellori*.

In Queensland common at Glendower (55). In South Australia common in the northeast (64-66) and south centre (70, 79, 82) but less so in the northwest (67, 68). In New South Wales seen at Tenterfield (95).

Habitat. Usually in or fairly near tall trees but often feeding in open paddocks, around homesteads or even perching in small trees on gibber plain (65). Recorded in Jarrah-Marri Forest (11), open Jarrah forest (12), mallee (14), (17), tropical savanna woodland (47, 49), riverine forest (55), open woodland (58), gibber and sand plains (65), low scrub (69) and woodland (82). Rowley (1967, 1970) has found this species apparently dependent on large trees for nesting and that breeding pairs feed within their extensive territories.

Behaviour. Usually in pairs, considerable numbers often gathering at food sources such as the cattle stockade at Clifton Hills (65), dead kangaroos and the shallow water in the creek at Aroona Valley (82). It is not certain that all the birds concerned were this species but many certainly were.

Wary and nervous of man where shot at but at other times not markedly shy. In northern South Australia was seen to alight and investigate camp sites as soon as the party left.

On 8 May two were seen to fly into a tree in which there was a nest. Two others then flew into it and all four perched fairly near each other. Two (probably males) which uttered the long moaning territorial call and had the throat feathers raised into a conspicuous "dewlap", erected the feathers above the eyes to form two "ears", in the same manner as do the Raven *Corvus corax,* Carrion Crow *C. corone* and Rook *C. frugilegus* in some agonistic situations. Birds collected at Glendower (55) had calls which differed so markedly from those with which McCallum and Pink were familiar in western Victoria, that they were at first thought to be a different species.

Food. Grasshoppers and other insects, berries and carrion.

Annual Cycle. The adult taken at Scott River (11) in Western Australia was in moult. In northern Queensland adults taken at Glenore (49) in April and at Glendower (55) in late March were in moult, those taken at Moonlight Creek (47) in early May were in fresh new plumage, one female still moulting its tail. Immatures taken at Glenore (49) and Glendower (55) were in body moult but showed no moult of wing or tail quills, except for one bird from Glendower (55) which was moulting tail and inner secondaries. These immatures mostly had the wing and tail quills very bleached and worn and in some cases almost certainly retained juvenile quills, but they had no trace of juvenile contour feathers. In southern Queensland, two immatures taken at Banoona (58) and Monkira (61) in May and June were similarly in moult. Two immatures from Nullarbor HS (69), South Australia in late October were moulting the juvenile contour feathers.

Impermanent Colours. *Legs* black. *Iris* adults white, often with faint blue inner ring. *Bill* black. *Mouth* adults black or purplish black; juvenile pinkish, immatures with varying amounts of pinkish and black. *Gular skin* black or very dark grey.

Corvus mellori Little Raven

Specimens. SOUTH AUSTRALIA. Elliston (76) 15 Oct 63, 1 imm ♂. Aroona Valley (82) 8 Apr 65, 1 ♀ (521.0 g).

Field Notes. Found at Aroona Valley (82) in and near gum creeks and also on more open ground. Almost certainly abundant but not at the time distinguished in the field from *C. coronoides* which was also present (see Rowley, 1967). No observations on behaviour were made that can with certainty be ascribed to this species.
Food. No information.
Annual Cycle. The immature taken at Elliston (76) in mid-October was still in juvenile plumage. The female taken at Aroona Valley (82) was in fresh-looking plumage with no sign of moult and very few signs of wear.
Impermanent Colours (adult female only). *Legs* black. *Iris* white. *Bill* dark grey. *Mouth* very dark grey.

Corvus bennetti Little Crow

Specimens. WESTERN AUSTRALIA. Carnarvon (5) 18 June 66, 1 ♀ skel. QUEENSLAND. Moonlight Creek (47) 15 – 16 May 66, 1 ♂ (461.8 g), 1 imm ♀ (352.0 g). SOUTH AUSTRALIA. Clifton Hills (65) 12 May 65, 1 ♂ (397.9 g), 1 ♀ (396.5 g). Etadunna (66) 18 May 65, 1 skel.
Distribution & Abundance. A breeding colony of crows, assumed to be this species as no other Australian crow is known to nest colonially, was seen at Burketown, northwestern Queensland on 28 April 64. No specimens were collected. Crows believed to have been this species were seen in South Australia also at Granite Downs (68) and Kingoonya (70), and in Western Australia at Onslow.
Habitat. Specimens collected in northern Queensland (47) were in tropical savanna woodland. In South Australia in various types of arid, sandy country with shrubs and small trees with larger eucalypts in the vicinity.
Behaviour. Usually in flocks but one of the males collected in northern Queensland, was with one other crow. At Clifton Hills (65) large numbers of this species gathered to feed around dying and dead cattle in and near stockades on open desert. Fed most actively in early morning and late afternoons. In the middle hours of the day the crows (*C. bennetti* and *C. coronoides*) spent most of their time in the shade, in or under small trees and shrubs, in a small shallow creek about a quarter of a mile or less from the stockades. At Onslow in Western Australia birds scavenged on the local municipal camping site.
Food. Carrion (65), small red berries and orthoptera (47).
Annual Cycle. The May female from Moonlight Creek (47) was in full body moult but all its wing and tail quills were old and, like the old contour feathers, much worn and faded to blackish brown. Both the South Australian (65) May specimens were in the final stages of a complete moult.
Impermanent Colours. *Legs* black. *Iris* white with light blue inner ring; brown in immature. *Bill* black. *Mouth* black, pinkish grey in throat; immature pinkish with black on insides of tips of mandibles.
Other Notes. Rowley (1970) states that there is no morphological character that will distinguish all specimens of this species from all of *Corvus orru* (q.v.), although the latter is usually larger and heavier. The series of Australian *Corvus* in the British Museum

confirm this and it seems that, in the present state of knowledge, the identification of some specimens of *orru* or *bennetti*, other than those collected by people thoroughly competent to distinguish them in the field, must be regarded as provisional.

In the present series the two specimens from Moonlight Creek (47) are provisionally referred to *bennetti* for the following reasons: the immature female not only agrees best in size with *bennetti* but was one of a flock of about thirty individuals that flew and foraged on the ground as a flock. Both it and the male agreed better in minor, and possibly irrelevant, details of bill shape and texture and gloss of plumage with specimens of *bennetti* than with specimens of *orru* from northern Queensland. The male is larger than normal for *bennetti* (wing 333 mm, culmen 52 mm) but both S. A. Parker (who examined this and all other Australian *Corvus* specimens here) and I independently came to the conclusion that it appeared to be an unusually large individual of *C. bennetti*.

The male (but not the female) taken at Clifton Hills (65) has the feather bases considerably tinged with grey instead of the usual clear white, so that in this character it is intermediate between typical *bennetti* (or *orru*) and the *mellori-coronoides-tasmanicus* group. This does not, I think, indicate hybridisation or introgression between "crows" and "ravens" in Australia but does emphasise that this convenient character may be of little phylogenetic significance. In some other species, such as the Brown-necked Raven *Corvus ruficollis*, there may be considerable geographical or individual variation in the colour (white or grey) of the feather bases.

Corvus orru — Australian Crow

Specimens. WESTERN AUSTRALIA. Warburton (23) 7 Sept 65, 1 skel. St George's Range (27) 8 – 9 June 68, 2 ♂ (474.5 g). Joint Hill (34) 4 July 68, 1 imm ♀ skel. QUEENSLAND. Moonlight Creek (47) 10 – 16 May 64, 1 ♂ (566.9 g), 1 ♀ (458.3 g). Wernadinga (48) 20 May 64, 1 imm ♀ (347.2 g). Inkerman (50) 30 May – 11 June 64, 1 ♂ (597.9 g), 1 ♀ (512.8 g), 1 imm ♀ (420.7 g), 1 ø (467.9 g), 1 skel (482.0 g). SOUTH AUSTRALIA. Granite Downs (68) 4 June 65, 1 skel.

Distribution & Abundance. Crows believed to have been this species were also seen in Western Australia western camps (1-4) and in the centre (19, 20, 22); it was abundant at Warburton (23) but only a few, in family (?) parties were seen at Giles (25). A few individuals only were seen at all main camps in Northern Territory (41, 43-46) except Pine Creek (44) where it was present in fair numbers. In South Australia a bird picked up dead 32 miles south of Granite Downs (68) and preserved as a skeleton appears to be this species on bill and tarsus measurements.

Habitat. In central and northeastern Western Australia in arid country but in or near gum creeks, although one pair were nesting on a windmill in spinifex flatlands (27). In northern Queensland in tropical savanna woodland and tall timber at the edge of grassland.

Behaviour. Usually in pairs or small parties, sometimes singly; but once about thirty were seen around a carcass (50).

At Paradise Bore, St George's Range (27), a pair had a nest of thorn twigs, lined with rootlets, on the moving platform of a windmill about 30 ft high. The male, which was collected, was on the nest, probably building as there were no eggs in the nest when it was examined next day and the male does not have a brood patch.

Calls heard presumably from this species (as all collected specimens were *orru*) were a short sharp *caa-caa* of apparent alarm and, from birds apparently at ease, varying deep guttural calls and higher-pitched warblings "fading away" with a deep gargling sound lasting for two or three seconds, this latter type of (apparently *coronoides*-like) call was heard throughout the western and central Kimberleys (DJF).

In most places this species appeared very wary of man.

Food. Grasshoppers and other insects. One (50) also had black hair and fragments of bivalve shell in stomach.

Annual Cycle. The male collected on 9 June from a nearly or quite completed nest at St George's Range (27), was in the final stage of moult (all plumage new, outermost primaries still growing). An adult female taken on 16 May at Moonlight Creek (47) and an adult male from Inkerman (50) on 30 May were also in final stages of moult. An unsexed immature from Inkerman on 31 May was also in a moult.

Impermanent Colours. *Legs* black. *Iris* adults white or (once) pearl grey with narrow blue inner ring; immatures brownish or mottled brownish and bluish. *Bill* black. *Mouth* black, mottled black and pink, or purplish in throat of some specimens, both adult and immature; juveniles have predominantly pink or reddish mouths. *Eyelids* pinkish to pale grey.

Other Notes. Rowley (1970) who has made and is making intensive studies of Australian *Corvus* species, can find no morphological character that will serve to separate all specimens of *bennetti* from all specimens of *orru,* although the former is normally smaller. In the field they can be distinguished by their behaviour and calls, but we do not feel that any of the members of the Harold Hall Expeditions had sufficient comparative experience to be able to do this with certainty. Rowley (1970) has shown that small individuals of *orru* may, when fully adult, "key out" as *bennetti* on measurements.

Two specimens from northern Queensland, an immature female from Wernadinga (48) and another from Inkerman (50), have wing measurements of 306 and 311 mm which are below the range given by Rowley for *orru*. Their bill (47, 52 mm) and tarsus (55, 62 mm) measurements would, however, fit either species. They agree perfectly in every nuance of plumage and bill shape with other normally-sized specimens of *orru* taken in the same areas and are, I think, merely small individuals of *C. orru ceciliae*. Nothing recorded in their data sheets as to behaviour or circumstances under which they were found would fit *bennetti* rather than *orru*.

Parasites. See Appendix 1.

APPENDIX 1

Parasites identified from birds collected

G. S. Cowles

During the Expeditions both ecto- and endoparasites were collected from the birds: they were determined in the Departments of Entomology (Dr T. Clay) and Zoology (Dr W. G. Inglis, C. G. Ogden, S. Prudhoe) of the British Museum (Natural History) and the Commonwealth Institute of Entomology (D. Macfarlane). Several rare nematodes were found, and one new species was described and named in honour of Major Hall (Ogden, 1967). One species was redescribed from the new material (Inglis, 1967). A single tick identified by Dr H. Hoogstraal, United States Naval Medical Unit No. 3, is the only specimen of *Haemaphysalis doenitzi* known from Australia. Details are given in Table 3.

Table 3

List of parasites collected by the Harold Hall Expeditions

	Host	Camp No.	Identification	Site
Trematoda Flukes	*Vanellus miles*	49	*Cyclocoelum (C.) obscurum* (Leidy, 1887)	Coelom
Cestoda Flat worms	*Sphecotheres vieilloti*	57	*Raillietina* sp.	Not noted
Nematoda Round worms	*Chrysococcyx plagosus*	55	*Ascaridia galli* (Schrank, 1788)	Coelom
	Ninox novaeseelandiae	55	*Hamatospiculum mcnelli* (Johnston & Mawson, 1941)	Eye socket
			Aprocta brevicaudata (Chow, 1939). Not determined	Between stomach and body wall
	Podargus strigoides	55	*Allodapa suctoria* (Molin, 1860)	Outside of the liver
			Excisa excisiformis (Yamaguti, 1935)	Stomach
	Coracina papuensis	50	*Diplotriaena tricuspis* (Fedtschenko, 1874)	Between stomach and body wall
	Anthus novaeseelandiae	50	Not determined	Eye socket
	Pomatostomus temporalis	50	*Seuratinema brevicaudatum* Johnston & Mawson, 1941	Body cavity
	Cisticola exilis	50	*Diplotriaena* sp. (♀ only)	Between thigh muscle and the skin
	Malurus lamberti	55	*Diplotriaena delta* Johnston & Mawson, 1941	Coelom
			**Diplotriaena halli* Ogden, 1967	Abdomen
	Peneoenanthe pulverulenta	50	*Seuratinema* sp. (♀ only)	Coelom
	Arses kaupi	52	*Diplotriaena* sp. (♀ only)	Amongst intestines
	Monarcha trivirgata	54	**Diplotrianea halli* Ogden, 1967	Abdomen
	Colluricincla harmonica	26	*Diplotriaena golvani* Anderson, 1959	Body cavity
	Zosterops lateralis	54	*Diplotriaena* sp. (♀ only)	Coelom
	Corvus orru	47	*Diplotriaena flabellata* Linstow, 1888	
Acanthocephala Spiny-worms	*Turnix velox*	49	*Oligacanthorhynchus pomatostomi* (Johnston & Cleland, 1912)	Shoulders and neck
	Anthus novaeseelandiae	18	,, ,,	Subcutaneous region of neck and breast
	Cinclosoma cinnamomeum	25	,, ,,	,, ,,
	Pomatostomus superciliosus	16	,, ,,	,, ,,
		22	,, ,,	,, ,,
		23	,, ,,	,, ,,
		70	,, ,,	,, ,,
	Pomatostomus temporalis	47	,, ,,	Neck, shoulder and head
		49	,, ,,	Hindneck, throat, roof of mouth

* = new species.

	Host	Camp No.	Identification	Site
	Acanthiza chrysorrhoa	15	*Oligacanthorhynchus pomatostomi*	Subcutaneous region of neck and breast
	Pyrrholaemus brunneus	17	,, ,,	,, ,,
		25	,, ,,	,, ,,
	Calamanthus fuliginosus	17	,, ,,	,, ,,
	Amytornis textilis	25	,, ,,	,, ,,
	Microeca leucophaea	49	,, ,,	Shoulders and neck
	Oreoica gutturalis	67	,, ,,	Subcutaneous region of neck and breast
	Neositta chrysoptera	25	,, ,,	,, ,,
	Meliphaga virescens	24	,, ,,	,, ,,
	M. plumula	24	,, ,,	,, ,,
	Manorina flavigula	25	,, ,,	,, ,,
	Anthochaera carunculata	17	,, ,,	,, ,,
	Poephila cincta	49	,, ,,	Under the skin of the neck
	Artamus cinereus	49	,, ,,	Shoulders and neck
	Gymnorhina tibicen	68	,, ,,	Subcutaneous region of neck and breast
Acarina	*Sericornis magnirostris*	95	*Ixodes holocyclus* Neumann	
Mites & Ticks	*Dacelo gigas*	57	*Haemaphysalis doenitzi* (♀ only) Warburton & Nuttall 1905	
Phthiraptera Feather lice	*Platalea regia*	28	*Ibidoecus plataleae* (Denny, 1842)	
	Haliastur sphenurus	28	*Laemobothrion maximum* (Scopoli, 1763)	
	Hieraaetus morphnoides	28	*Degeeriella* sp.	
	Falco subniger	29	*Degeeriella* sp.	
	Porphyrio porphyrio	29	*Eulaemobothrion* sp.	
	Jacana gallinacea	43	*Saemundssonia platygaster theresae* (Timmerman, 1971) *Rallicola* sp.	
	Recurvirostra novaehollandiae	45	*Quadraceps decipiens* (Denny, 1842)	
		45	*Quadraceps recurvirostrae* (Linné, 1758)	
		45	*Quadraceps signatus* (Piaget, 1880)	
	Ninox novaeseelandiae	44	*Strigiphilus cursitans* group	
		66	*Kurodaia cryptostigmatia* (Nitzsch, 1861)	
	Orthonyx spaldingi	52–54	Not identifiable	
	Myiagra alecto	41	*Ricinus* sp.	
	Rhipidura rufiventris	41	*Ricinus* sp.	
	Pachycephala simplex	41	*Ricinus* sp.	
	Colluricincla woodwardi	43	*Brueelia* sp.	
	Struthidea cinerea	58	*Myrsidea* sp.	
	Corcorax melanorhampus	86	*Myrsidea* sp.	
	Dicrurus bracteatus	41	*Menacanthus* sp.	

PUBLISHED RESULTS OF THE HAROLD HALL AUSTRALIAN EXPEDITIONS

Publication numbers 1–32

No.

1. G. S. Cowles, 1964. A new Australian babbler. *Emu*, **64** : 1–5.
2. J. D. Macdonald, 1966. The status of some Australian treecreepers. *Emu*, **66** : 113–115.
3. I. C. J. Galbraith, 1967. The Black-tailed and Robust Whistlers *Pachycephala melamura* as a species distinct from the Golden Whistler *P. pectoralis*. *Emu*, **66** : 289–294.
4. D. Goodwin, 1967. Notes on behaviour of some Australian birds. *Emu*, **66** : 237–251.
5. J. D. Macdonald, 1967. Variation in the Black-faced Wood-swallow, *Artamus cinereus*. *Emu*, **66** : 341–345.
6. W. G. Inghs, 1967. The relationships of the nematode superfamily Scuratoidea. *J. Helminth.*, **41** : 115–136.
7. C. G. Ogden, 1967. Some parasitic nematodes from Australian birds. *Ann. Mag. nat. Hist.*, **9** : 505–518.
8. J. D. Macdonald, 1968. Speciation in the *Colluricincla harmonica* complex. *Emu*, **67** : 215–224.
9. J. D. Macdonald, 1968. Notes on the genus *Cinclosoma*. *Emu*, **67** : 283–289.
10. I. C. J. Galbraith, 1969. The Papuan and Little Cuckoo-shrikes, *Coracina papuensis* and *robusta*, as races of a single species. *Emu*, **69** : 9–29.
11. G. S. Cowles, 1967. The Grey Honeyeater (*Lacustroica whitei*) a new record for South Australia. *S. Aust. Orn.*, **24** : 147.
12. A. R. McEvey & W. G. Middleton, 1968. Birds and vegetation between Perth and Adelaide. *Emu*, **68** : 161–212.
13. P. R. Colston & C. J. O. Harrison, 1968. Some records of Palacarctic waders in Western Australia. *West. Aust. Nat.*, **11** : 45–46.
14. P. R. Colston, 1968. The occurrence of the Painted Honeyeater (*Grantiella picta*) in S.W. Queensland. *Qd Nat.*, **18** : 116.
15. J. D. Macdonald, 1969. Hybridisation in *Pardalotus*. *Emu*, **69** : 41–44.
16. I. C. J. Galbraith & S. A. Parker, 1969. The Atherton Scrub-wren *Sericornis keri* Mathews a neglected Australian species. *Emu*, **69** : 212–232.
17. J. D. Macdonald, 1969. Variation in the White-browned Treecreeper. *Emu*, **69** : 110–111.
18. J. D. Macdonald, 1969. Notes on the taxonomy of *Neositta*. *Emu*, **69** : 169–174.
19. C. J. O. Harrison & P. R. Colston, 1969. Some records of breeding after late rains in the Hamersley region. *West. Aust. Nat.*, **11** : 49.
20. C. J. O. Horrison, 1969. Emu-wrens and the forest zone in south-western Australia. *West. Aust. Nat.*, **11** : 66.
21. C. J. O. Horrison, 1969. Helpers at the nest in Australian passerine birds. *Emu*, **69** : 30–39.
22. C. J. O. Harrison, 1969. The affinities of the blue wrens genus *Malurus* and related genera, with special reference to the grass-wren genus *Amytornis*. *Emu*, **69** : 1–8.

23. D. Goodwin, 1969. A new sub-species of the White-quilled Rock Pigeon. *Bull. Br. Ofn. Club.*, **89** : 131–133.
24. C. J. O. Harrison, 1970. Relationship of the Brown and Black-backed Tree-creepers. *Emu*, **70** : 9–11.
25. C. B. Frith, 1970. Nests and eggs of the Australian rock pigeons *Petrophassa* spp. *Emu*, **70** : 29–30.
26. C. J. O. Harrison, 1969. The possible affinities of the Australian Treecreepers of the genus *Climacteris*. *Emu*, **69** : 161–138.
27. D. J. Freeman, 1970. The rediscovery of the Black Grass Wren *Amytornis housei* with additional notes on this species. *Emu*, **70** : 193–195.
28. D. J. Freeman, 1971. The nest of the White-lined Honeyeater. *Emu*, **71** : 86–87.
29. S. A. Parker, 1971. Taxonomy of the Northern Territory friarbirds known as *Philemon buceroides gordoni*. *Emu*, **71** : 54–56.
30. D. Goodwin, 1967. Australian pigeons: their affinities and status. *Emu*, **66** : 319–336.
31. S. A. Parker, 1973. The tongues of *Ephthiamura* and *Ashbyia*. *Emu*, **73** : 19–20.
32. C. J. O. Harrison, 1972. A re-examination of the chestnut-shouldered wren complex of Australia. *Bull. Br. Mus. nat. Hist.* (Zool.), **21** : 313–328.

APPENDIX 3

Supplementary expedition to northern Queensland

C. B. Frith

Introduction

This short additional Harold Hall Expedition was made possible by a number of factors. Mr Cyril Walker of the Palaeontology Department, British Museum (Natural History), had made arrangements to collect fossils in the Hughenden area late in 1970. Having left the Bird Room staff I was at this time living in Sydney, N.S.W., and suggested to Walker that I should join him to make a collection of birds if funds could be obtained. Subsequently equipment was provided by the Bird Room and financial help by the Bird Exploration Fund Committee.

While preparations were being made Mr William Naylor, at that time working at Sydney Zoo, offered to join the party to assist with the collecting and the domestic arrangements, and later Mr Anthony Hiller, who had been with me on the 5th Expedition, decided to spend his annual leave collecting with us.

Thanks are due to these two men who both took part voluntarily, receiving only their subsistence in the field. Particular thanks must go to Mr Hiller who helped to cut costs by freely accommodating all members of the party at his home and by making much of his personal, specialised equipment available for use.

Thanks also go to Dr Bartholomai, Director of the Queensland Museum, and Mr Vernon, taxidermist, for their assistance with official matters and a number of practical problems. Obviously a collecting trip of this kind would not be possible without the understanding and co-operation of property owners and land overseers, and we are greatly indebted to all at the homesteads we visited.

Diary of Expedition

See map opposite.

2 – 4 October, 1970. Sydney to Brisbane by New England Highway. Some collecting was done along the road.
4 – 8 October. Preparations in Brisbane.
8 – 31 October. **Soda Valley Bore (A1),** 20 miles northwest of Hughenden. Camp was made in an old building about two hundred yards from the bore. This very productive bore was in a shallow valley the floor of which was stony, sparsely covered with low scrub and some poor eucalypt trees but no grass. The surrounding hills were covered with eucalypt woodland with large volcanic basalt boulders and dry grasses beneath.

Vegetation about the bore was relatively lush as the water formed a small marshy area before flowing, in a very thin gully, southward down the valley. A few large flowering eucalypts dominated the scene directly over the bore with lush grass and small reed-beds beneath. The Reed-warbler was found to be fairly common in these reeds. About the perimeter of this vegetation was a cattle-proof fence just outside which low scrub was plentiful, offering cover for smaller passerines.

Map of C. B. Frith's supplementary expedition to Northern Queensland, 1970

Two miles south of the bore the water formed two shallow lakes covering approximately two acres. These had good local vegetation, but little grass or reeds. A number of dead trees stood in the water providing good perches for White Ibis, White-necked Heron, Darter, Royal Spoonbill and other water-frequenting species, as well as cockatoos. Three nests of Little Grebe were found, all containing eggs.

It was on and around these areas that most of the collecting from this camp took place. Few species were encountered in the drier country that were not to be seen at the waters. However one such bird was the elusive Ground Cuckoo-shrike, four individuals being seen but none collected.

From Soda Valley Bore visits were made to other localities including **Pelican Bore (A2),** 46 miles northwest of Hughenden, a small bore consisting of an open tank for cattle, in hill country similar to that surrounding Soda Valley but with more scrub, grass and bushes. A number of species not found in Soda Valley were encountered here, including Singing Bushlark, Western Warbler, Yellow-rumped Thornbill, Weebill and Red-capped Robin. It was here that Walker found the majority of his fossils including several ichthyosaurs and fish.

Another excursion was to **Mount Sturgeon (A3),** 50 miles north of Hughenden, off the Hann Highway. Only one day could be spent here. This country consisted of extremely dry, sparse savanna woodland with some low shrubs. The grass stood about ten inches high and dead, rotting trees were not uncommon. Mount Sturgeon itself was a mound of broken rock covered with grass and eucalypts about which most of our collecting took place. Only twelve species were seen during the day: Laughing Kookaburra, Squatter Pigeon, Fairy Martin, Black-faced Cuckoo-shrike, Red-backed Wren, Variegated Wren, Weebill, Brown Tree-creeper (see Appendix 4), Noisy Miner, Magpie Lark, Black-faced Wood-swallow and Dusky Wood-swallow.

2 – 6 November. **Edington Homestead (A4),** 12 miles west of Julia Creek. The drive from Hughenden to Julia Creek was through dry, flat country, consisting of open clay plains with very stunted, sparse grass and almost no trees. Next to no birds were seen during this drive other than at water, which was very scarce. The Blue-and-white Wren was encountered at a small **water hole 20 miles east of Julia Creek (A5)** and a family party was collected.

Edington Homestead was the obvious collecting locality having a large, deep lake (about 900 x 100 feet) from a bore, giving rise to a fair amount of vegetation in otherwise arid country. Eucalypts and much low bush lined the banks and tall grass was to be found in dry shallow 'run-off' gulleys. Wild pigs were common here, particularly in the grass beds.

Fork-tailed Swifts were feeding over the lake, and Australian Pratincoles were both at the lake edge and by the road between camp and Julia Creek. We found only 35 species at this camp and as it was obvious few more would be found we left after three days.

6 – 7 November. **Dalgonelley Homestead (A6),** 54 miles north of Julia Creek. North of Julia Creek the country was like that on the journey from Hughenden but with more grass. This homestead is situated close to a very extensive, long, shallow lake, part of which is used for water skiing. We made camp at the opposite end. Despite the extensive area of water, vegetation is confined to a thin strip at the lake edge, due to the nature of the soils which are predominantly clays. This vegetational strip consisted of tall eucalypts with low bushes beneath and much long grass. Outside the strip were dry open stony plains with very little shrub and sparse short grass.

Due to rain this camp had to be left prematurely. Unlike the previous hot, dry weather, days were becoming humid and heavy clouds were building up in the afternoons. After a very impressive rain-storm it was decided to get quickly to the tarmac road for fear of getting stuck in the clay area. Our only consolation was that, after particularly watching out for pardalotes previously, we managed to obtain a number here.

8 – 10 November. **Norman River (A7).** 15 miles east of Normanton. On the road north to Norman River we stopped for a few hours at a shallow lake 12 miles north of **Donor's Hill Homestead (A8)** in some rocky country consisting of open stony plains and escarpments with stunted eucalypts and spinifex grass. Here we collected the Plumed Pigeon at the extreme north of its range.

Camp was pitched close to the point the Normanton – Croydon road crosses the Norman River which at this point is tidal, rising and falling about four to six feet. It is about 30 yards wide, muddy but only slightly salted, and lined with quite lush woodland of eucalypt and paper-bark trees with some creepers and tendrils. In the muddy shallows stood a few mangroves which were more common as one followed the river seawards.

Away from the river, grass and shrubs were plentiful with eucalypts here and there. Seventy-four species of birds were recorded here, including our first encounters with White-winged Black Tern, Leaden Flycatcher, Yellow Honeyeater and the Banded Finch. Owing to imminent rains it was decided to move from this camp and cover the remaining area of un-tarmaced road between this point and Croydon.

11 – 12 November. **Alehvale Homestead (A9),** 15 miles southeast of Croydon. Camp was made in a small rocky gully between two escarpments about one mile east of the homestead, at the site of a small running spring. Most of the collecting took place in the gully, around the spring, in dense eucalypt woodland with low scrub and grasses. On the scree and escarpment eucalypts were sparse and on top of the hills woodland was sparse with broken sandstone beneath. A small flock of Rainbow Lorikeets were seen to fly over. Other species not seen before were Channel-billed Cuckoo, Great Bowerbird and Banded Honeyeater.

Again this camp was abandoned due to the rains, which were becoming more frequent and heavier, and it was decided to drive direct to Brisbane and finish the expedition.

List of specimens collected

Podiceps novaehollandiae. Norman River (A7), 1 fl.
Falco berigora. Pelican Bore (A1), 1 fl.
Charadrius cinctus. Soda Valley Bore (A1), 1 ♀ skel.
Charadrius melanops. Soda Valley Bore (A1), 1 ♂ skel. Edington Homestead (A4), 2 ♂ skel.

Himantopus himantopus. Norman River (A7), 1 ♂ skel.
Stiltia isabella. Edington Homestead (A4), 1 ♂ skel, 2 ♀ skel.
Geopelia humeralis. Norman River (A7), 1 ♂ skel, 1 ♀ skel.
Phaps chalcoptera. 3 October. New England Highway, 14 miles south of Ipswich, Qld, 1 ♂ skel.
Lophophaps plumifera. 12 miles north of Donor's Hill (A8), 1 ♀ skel, 2 fl.
Geophaps scripta. Mount Sturgeon (A3). 1 ♀ skel, 1 fl, Alehvale Homestead (A9), 1 ♀ skel.
Cacatua galerita. Soda Valley Bore (A1), 1 ♂ skel, 1 ♀ skel. Pelican Bore (A2), 1 ♀ skel.
Nymphicus hollandicus. 3 October, New England Highway, 23 miles south of Ipswich, Qld, 1 ♂ skel.
Platycercus eximius. 3 October, New England Highway, 8 miles south of Deepwater, N.S.W., 1 ♂ skel.
Platycercus adscitus. 11 October, on road 6 miles north of Hughenden, 1 ♂ skel, 1 ♀ skel, 1 ♀ skel, 1 fl.
Cacomantis variolosus. Norman River (A7), 3 ♂ skel.
Chrysococcyx basalis. Soda Valley Bore (A1), 1 ♂. Pelican Bore (A2), 1 ♂ skel. Dalgonelly Homestead (A6), 1 ø.
Ninox novaeseelandiae. Soda Valley Bore (A1), 1 fl.
Podargus strigoides. 3 October, New England Highway, 3 miles south of Deepwater, N.S.W., 1 ♀ skel. Soda Valley Bore (A1), 1 ♀ skel. Edington Homestead (A4), 1 ♂ skel. Alehvale Homestead (A9), 1 ♂ skel.
Apus pacificus. Edington Homestead (A4), 2 ♂ skel.
Dacelo leachii. Soda Valley Bore (A1), 1 ♀ skel. Norman River (A7), 1 ♂ skel. Alehvale Homestead (A9), 1 ♂ skel.
Halcyon pyrrhopygia. Soda Valley Bore (A1), 1 ♂ skel, 1 fl. Pelican Bore (A2), 1 ♀ skel.
Halcyon sancta. Soda Valley Bore (A1), 2 ♂ skel, 1 ♀ skel.
Eurystomus orientalis. Pelican Bore (A2), 1 ø skel. Norman River (A7), 2 ♂ skel.
Mirafra javanica. Pelican Bore (A2), 3 ♂.
Petrochelidon ariel. Soda Valley Bore (A1), 1 ♂ skel. Edington Homestead (A4), 1 ♂ skel.
Coracina novaehollandiae. Soda Valley Bore (A1), 2 ♂, 1 ♀ skel.
Lalage sueurii. Soda Valley Bore (A1), 1 ♂, 4 fl.
Pomatostomus temporalis. 3 October, New England Highway, 14 miles south of Ipswich, Qld, 1 ♂. Soda Valley Bore (A1), 1 ♀ skel, 1 ø skel.
Acrocephalus stentoreus. Soda Valley Bore (A1), 1 fl.
Malurus leucopterus. 20 miles east of Julia Creek (A5), 1 ♂, 1 ♀ skel, 2 fl.
Malurus lamberti. Soda Valley Bore (A1), 5 ♂, 1 ♀, 1 ø, 1 ♀ skel, 5 fl. Edington Homestead (A4), 1 ♂, 1 ♀. Dalgonelly Homestead (A6), 2 ♂. Alehvale Homestead (A9), 3 ♂, 1 ♀, 1 ♂ skel.
Malurus melanocephalus. Mount Sturgeon (A3), 1 ♂. Norman River (A7), 1 ♂, 1 ♀ skel.
Gerygone olivacea. Alehvale Homestead (A9), 3 ♂, 1 ♀, 1 ♂ skel.
Gerygone fusca. Pelican Bore (A2), 1 ♂, 2 fl.
Smicrornis brevirostris. Soda Valley Bore (A1), 1 fl. Pelican Bore (A2), 1 ♂, 1 ♀ skel, 1 fl.
Acanthiza chrysorrhoa. Pelican Bore (A2), 2 ♂, 1 ♀, 2 fl.

Microeca leucophaea. Soda Valley Bore (A1), 1 fl. Pelican Bore (A2), 1 ø. Norman River (A7), 1 fl.
Petroica goodenovii. Pelican Bore (A2), 1 ♀ skel.
Myiagra rubecula. Norman River (A7), 1 ♂ skel.
Seisura inquieta. Soda Valley Bore (A1), 1 ♀ skel. Edington Homestead (A4), 2 ♂ Norman River (A7), 2 ♀.
Pachycephala rufiventris. Soda Valley Bore (A1), 1 ♂, 1 ♀ skel, 2 fl. Pelican Bore (A2), 1 fl.
Colluricincla harmonica. Pelican Bore (A2), 1 ♂.
Neositta chrysoptera. Soda Valley Bore (A1), 6 ♂, 1 ♀, 2 fl. Pelican Bore (A2), 1 fl.
Climacteris picumnus. Mount Sturgeon (A3), 3 ♂. (See Appendix 4).
Dicaeum hirundinaceum. Soda Valley Bore (A1), 1 ♂ skel, 1 fl.
Pardalotus melanocephalus. Pelican Bore (A2), 1 ♀. Dalgonelly Homestead (A6), 2 ♂, 4 ♀.
Zosterops lutea. Norman River (A7), 1 ♂.
Lichmera indistincta. Soda Valley Bore (A1), 1 ø skel. Alehvale Homestead (A9), 2 ♂ skel, 1 fl.
Meliphaga flava. Norman River (A7), 1 ♀ skel. Alehvale Homestead (A9), 1 ♂ skel.
Meliphaga virescens. Pelican Bore (A2), 1 ♀ skel, 1 ø skel, 1 fl.
Melithreptus sp. Norman River (A7), 2 fl.
Entomyzon cyanotis. Soda Valley Bore (A1), 1 fl. Norman River (A7), 1 ♂ skel.
Philemon citreogularis. Soda Valley Bore (A1), 1 ♀ skel, 2 fl.
Philemon sp. Pelican Bore (A2), 2 ♂ skel. 26 October, Hann Highway, 30 miles north of Hughenden, 1 ♀ skel, 1 fl.
Plectorhyncha lanceolata. Pelican Bore (A2), 1 ♂ skel, 1 fl.
Conopophila rufogularis. Soda Valley Bore (A1), 1 ♂, 1 ♂ skel.
Manorina melanocephala. Soda Valley Bore (A1), 1 fl.
Anthochaera rufogularis. Soda Valley Bore (A1), 1 ♂. Pelican Bore (A2), 1 ♂ skel, 1 ♀ skel.
Poephila cincta. Norman River (A7), 2 ♀.
Stictoptera bichenovii. Norman River (A7), 3 ♀ skel.
Oriolus sagittatus. 26 October, Hann Highway, 30 miles north of Hughenden, 1 ♂ skel. Norman River (A7), 1 ♂ skel, 1 ♀ skel.
Grallina cyanoleuca. Soda Valley Bore (A1), 2 ♂ skel, 2 ♀ skel.
Struthidea cinerea. Soda Valley Bore (A1), 1 ♂ skel, 1 ♀ skel, 1 fl.
Artamus superciliosus. Soda Valley Bore (A1), 2 ♂ skel, 1 ♀ skel, 1 fl.
Artamus cinereus. Soda Valley Bore (A1), 1 ♂, 1 ♀, 1 ♂ skel, 1 fl. Dalgonelly Homestead (A6), 1 ♂. Norman River (A7), 1 ♂.
Artamus cyanopterus. Mount Sturgeon (A3), 1 ♂ skel, 1 ø skel.
Cracticus nigrogularis. 26 October, Hann Highway, 30 miles north of Hughenden, 1 ♂ skel, 1 ♀ skel, 1 fl.
Cracticus torquatus. Alehvale Homestead (A9), 1 ♂, 1 ♀.
Gymnorhina tibicen. 12 miles north of Donor's Hill (A8), 1 ♀ skel, 1 fl. Edington Homestead (A4), 1 ♂ skel.
Chlamydera maculata. Pelican Bore (A2), 1 ♂ skel, 2 ♀ skel.
Chlamydera nuchalis. Alehvale Homestead (A9), 1 ♂.

APPENDIX 4

Note on some treecreepers collected by C. B. Frith

C. J. O. Harrison

On 28 October, 1970, in northern Queensland, C. B. Frith collected for the British Museum (Natural History) three male specimens of the Brown Tree-creeper, *Climacteris picumnus*. Two were collected at Mount Sturgeon, 50 miles north of Hughenden, and the other ten miles further north. These three show further evidence of intermediacy between typical *C. p. picumnus* from further south and *C. p. melanota* to the north, in addition to that already cited (Harrison, 1970). They are comparable with the intermediate specimens in the Queensland Museum, and darker than specimens of *C. p. picumnus* from Glendower (55) which also show some degree of intermediacy (Harrison, *op. cit.*).

The specimens were moulting into fresh plumage but still show a considerable amount of worn plumage. Even their worn dorsal plumage is darker, duller brown, lacking the slightly warmer tint of the lighter brown dorsal plumage of typical *C. p. picumnus;* and new feathers are dark brown. The plumage of wings, tail and crown of the head is darker than that of the typical form. They agree with *C. p. melanota* in possessing a light, cream-coloured superciliary stripe which contrasts strongly with the dark crown and a dark brown streak through the eye. The mid-throat is almost white and the foreparts of the ear coverts are cream-coloured, unlike those of the typical form.

These specimens were collected in the zone where interbreeding between the two named forms might be expected.

GLOSSARY OF BOTANICAL TERMS

Collectors on the Expeditions were largely dependent on local information for the identification of trees, plants and vegetation types. It is the names they were given in the field which appear on labels and data sheets and have been transcribed into this report, except in the special instances when botanical samples were brought back for identification by Dr R. Melville at the Royal Botanic Gardens, Kew.

It was not appreciated until the expeditions were completed that many of the names in common usage would be insufficient for proper identification later, and indeed the same common name might be used for different species in different states. Therefore in compiling the report no attempt has been made to apply scientific names to common names except where positive identification was established at the time, or specimens were collected. The mangrove plants encountered by the 2nd Expedition were identified by Dr W. Macnae from colour transparencies.

As a guide to the terms used the following glossary has been compiled with the help of Dr Melville, Dr D. L. Serventy and Mr R. D. Royce, Curator of the Western Australian Herbarium.

acacia scrub — *see* Gidgee, Mulga, wattle.
Apple Gum — *Eucalyptus clavigera* Myrtaceae.

Banksia Proteacea — many species.
banyan tree — used imprecisely for a number of species of wild fig *Ficus* Moraceae.
bird shrub — *Crotalaria* sp. Papilionaceae: commonly *C. cunninghamii*.
blackboys — *Xanthorrhoea* sp. Palmae (Plate 1(a)).
Blackbutt — *Eucalyptus pilularis*.
Black Oak/Black She Oak — probably *Casuarina littoralis* Casuarinaceae (NSW, Vic, Qld).
bloodwood — common name for a group of about 30 eucalypt species with red sap, plentiful in northern districts and the eastern coastal strip.
Bluebush — *Kochia sedifolia* Chenopodiaceae: *see* salt bush.
blue gum — used imprecisely for various eucalypts: *E. globulus* the Tasmanian Blue Gum, *E. bicostata* the Southern Blue Gum, *E. leucoxylon* the Yellow Gum or South Australian Blue Gum.
box (Yellow Box — *Eucalyptus melliodora*).
bracken/brackenfern — probably *Pteridium esculentum* Dicksoniaceae.
Buffel Grass — *Cenchrus ciliaris* Graminae: *see* pindan.
Bull Oak — *Cardwellia sublimis* Proteaceae in north Queensland: *Casuarina luehmanni* in inland areas.
Button Grass — *Dactyloctenium radulans* Graminae.

corkwood — applied imprecisely to many plants including *Hakea* sp. Proteaceae, particularly *H. lorea* in Western and central Australia.

Crotalaria Papilionaceae — *see* bird shrub.
Curara (Dead Finish) — *Acacia tetragonophylla* Leguminosae.
cypress pine, native pine — *Callitris* sp. Cupressaceae.

Dead Finish — *see* Curara.
Desert Oak — *Casuarina decaisneana.*

fig tree — *Ficus* sp.

gallery forest — forest bordering watercourses.
Ghost Gum — *Eucalyptus papuana.*
gibber desert — stony desert, found chiefly in South Australia.
Gidgee (Gidyea) — *Acacia pruinocarpa* in Western Australia: elsewhere *A. cambagei,* often confused with Yarran.
Gimlet Gum — *Eucalyptus salubris.*
Grevillea Proteaceae — several species, some similar to *Hakea* sp.

Hakea Proteaceae — *see* corkwood.
hardwood tree — used imprecisely but mainly for myrtaceous trees.
hummock grass — used imprecisely, often for *Triodia* sp.

ironbark — used imprecisely for various eucalypts.

Jarrah — *Eucalyptus marginata.*
Johnstone Hardwood/Johnstone River Hardwood — *Backhousia bancroftii* Myrtaceae.

Karri — *Eucalyptus diversicolor.*

Lantana camara Verbenaceae — an introduced weedy shrub.
lawyer vine — *Calamus* sp. Liliaceae. (Plate 9(a)).
Lignum scrub — *Muehlenbeckia cunninghamii* Polygonaceae.

mallee — many-stemmed eucalypts which grow from an underground or partly underground woody swelling (lignotuber) which contains numerous bark buds and functions as a root (Worrell & Sourry, 1968). (Plates 1(b), 3(a)).
mangroves — shrubs or trees in littoral zone, growing in formation that may be relatively open or of dense shady consistency. The word mangrove may be applied to any association of tree that grows in salt water or to individual species within that formation (Macnae, 1966). The four genera identified on the expeditions were *Avicennia* Verbenaceae, *Bruguiera, Ceriops* and *Rhizophora* Rhizoporaceae.
Marri — *Eucalyptus calophylla.*
mistletoe — Loranthaceae: several genera in Australia.

Mulga — *Acacia aneura,* but the word mulga is used broadly for the vegetation areas in which this is the dominant tree (Plate 2(b)).

native pine — *Callitris* sp. (Plate 4(b)).

Pandanus Pandanaceae (Screw palms): several species in tropical Australia.
paperbark tree (tea-tree) — used imprecisely for various *Melaleuca* sp. Myrtaceae.
passion fruit — Passifloraceae: several genera in tropical Australia.
peppermint tree — used imprecisely for various eucalypts and other myrtaceous trees.
Pepper Tree — *Schinus molle* Anacardiaceae: introduced.
pindan — generally open savanna country: in the Kimberleys in this formation several species of *Acacia* are associated with a large number of both annual and perennial grasses, although in recent years the introduced Buffel Grass is dominating the grass cover in some areas.
pine — *see* cypress pine and native pine.
Plum Bush — *Santalum esculentum* Santalaceae: a bush of the inland which reaches tree proportions and has small purplish fruit. But "plum" may have been used imprecisely for similar fruits.
Prickle Weed/Prickly Moses — *Acacia uliciphia.*
pseudo-rainforest (sub-monsoon forest) — narrow strips of dense softwoods along banks, etc., in north (Keast, 1961). The pseudo-rainforest at Darwin may in parts be tropical layered woodland, *q.v.*

quandong (1) Quandong — *Santalum acuminatum:* a small tree with bright red fruit widely distributed on sandy and clay soils, encountered by 3rd Expedition in W.A.
(2) correctly the Blue Quandong — *Elaeocarpus grandis* Eleocarpaceae: a tree associated mainly with forested areas in coastal districts (NSW, Qld).

Red Cap Gum — *Eucalyptus erythrocorys:* also called Illyarrie in its native habitat in Western Australia.
red gum — used imprecisely for various eucalypts including Red Cap Gums, River Gums, Marri.
River Gum/River Red Gum — *Eucalyptus camaldulensis:* but the term may have been used imprecisely for gums along rivers.

Salmon Gum — *Eucalyptus salmonophloia.*
salt bush — *Atriplex* sp. Chenopodiaceae: similar in size and formation to Bluebush and very often found together as the main vegetation in open plains and paddocks (Plate 4(a)).
samphire — used imprecisely for several halophytes like *Salicornis* sp. and *Arthrocnemum* sp. Chenopodiaceae.
sandal wood — *Santalum* sp. Santalaceae.

sclerophyll forest — trees of forest form in closed community; dense undergrowth of hard-leaved shrubs; grass rare. Classed as "wet" with one or two tree layers, shrubs and herbs; and as "dry" with sclerophyllous shrubs with ground stratum absent or sparse (Keast, 1961).
spinifex — used commonly but incorrectly for *Triodia* sp. Graminae.
Spotted Gum — *Eucalyptus maculata.*
stringy-bark — used imprecisely for various eucalypts of which the bark peels in long strips.
sub-monsoon forest — see pseudo-rainforest.
tea-tree (also paperbark) — species of *Melaleuca* and other myrtaceous genera.
tropical layered woodland — part of sclerophyll forest formation, dominated by *Eucalyptus* sp. (Williams, 1955): possibly confused with pseudo-rainforest at Darwin.
tropical rainforest — dense, mainly evergreen tree community with continuous canopy. Smaller trees, shrubs, ferns and herbs present (Williams, 1955) (Plate 9(a)). Impoverished 'rainforest' at Mount Dryander (Plate 9(b)) may have been softwood forest (CSIRO 1953).
Tuart — *Eucalyptus gomphocephalus.*

Water Hyacinth — *Eichornia crassipes* Pontederiaceae: introduced.
Water Lily — *Nymphea gigantea* Nymphaeaceae: the endemic blue form.
wattle — *Acacia* sp., especially *A. bakeri* (NSW, Qld).

Yarran — *Acacia homalophylla:* often confused with Gidgee in Queensland and Northern Territory.
Yellow Box — *Eucalyptus melliodora.*
York Gum — *Eucalyptus loxophleba.*

GAZETTEER OF CAMPS AND COLLECTING LOCALITIES

Throughout the Expeditions, a base was chosen in the field for each main collecting area, from which distances and directions were quoted on labels and data sheets. This base was sometimes the homestead (HS), although on a large station the actual camp may have been some distance from it; sometimes a peak, sometimes a nearby town; and sometimes the actual camp-site if it was identifiable on the map. The name of the base has not always been selected as the one by which that camp, which is numbered and mapped, is identified in this report. This gazetteer therefore contains: (a) the names and co-ordinates of each base, followed by the names of the mapped localities which refer to that base; (b) the names (in capitals) of the mapped and numbered localities with their co-ordinates or distance from base, followed by the position of the actual camp-site; (c) other collecting localities and features, with co-ordinates or distance from either a base or a mapped locality.

ACRAMAN LAKE (71), 32.02S: 135.26E.
Albany, 35.02S: 117.53E.
ALBURY (92), 36.04S: 146.55E (camp Kywanna HS).
Amelira Rockhole, 30m SW Kingoonya (70).
ANGEPENA (84), HS 30.34S: 138.51E.
AROONA VALLEY (82), 31.19S: 138.35E.
AYTON (52), 15.56S: 145.22E (camp Granite Creek).

BANOONA (58), HS 26m SE Roma.
Beckon HS, GREEN RANGE (13).
BIG TABLELAND (51), 15m NNW Ayton (52).
Birdsville Track, running through Clifton Hills (65) and Etadunna (66).
Blanchewater, 29.32S: 139.26E.
Bloomfield River, Ayton (52).
BOOL LAGOON (87), 3m W Struan.
BORROLOOLA (46), 16.04S: 136.17E (camps at Caranbirini Waterhole and McArthur R.).
Bow River camp, 23m SW Lissadell Hill (38).
Brachina Creek, Aroona Valley (82).
Brachina HS, 14m S Parachilna (81).
Brigham Creek, 30m SW Joint Hill (34).
Brokera Station, Moonlight Creek (47).
Bunyaroo Gorge, Aroona Valley (82).
BUSH BAY (6) 25.13S: 113.49E.
Bussleton, 5m SW Wonerup (8).

CADELGA (62) HS 26.05S: 140.25E.
Cambridge Gulf, Wyndham (39).
CAPE GLOUCESTER (56), 22m NNW Proserpine.
Caranbirini Waterhole, 21m SW Borroloola (46).
CARNARVON (5) 24.53S: 113.39E.
CARNEGIE (19) HS 25.42S: 123.00E (camp 4m W of HS).
Cherubim Station, St George's Range (27).
Cheyne Beach, SE of Green Range (13).
CHICHESTER RANGE (3) 21.42S: 117.44E (camp at Tampatha Well).
CLIFTON HILLS (65) HS 27.03S: 138.55E.
COCKLEBIDDY (15) 32.02S: 126.06E (camp 3m S).
COLERAINE (89) 37.36S: 141.42E.
COOLAWANYAH (4) 21.48S: 117.50E.
Coolgardie 31.01S: 121.12E.
Coopers Creek, Innamincka (64).
Cooper River (dry), also at Etadunna (66).
COORONG, THE (85) 36.00S: 139.30E.
Copley 30.32S: 138.26E.

CORDILLO DOWNS (63) HS 26.43S: 140.38E (camp 20m W of HS).
Cramps Camp, 13m SW Oodnadatta (67).

Darling Range, Perth (7).
DARWIN (41) 12.29S: 130.50E (camp at Hope Inlet).
Deltana HS, 7m NE Furner (86).
DERBY (29) 17.18S: 23.38E (camp 15m SE).
Diamantina R. flowing through Monkira (61) and Clifton Hills (65).
Dingo Beach, Cape Gloucester (56).
Dryander Camp, Mount Dryander (57).
Dudyun Flats, Mount Shenton (18).
DUNEDOO (93) 32.01S: 149.24E (camp Wheogo HS).
Durack Range, Mount Remarkable (37).

EAST ALLIGATOR RIVER MOUTH (42) 16m NW Oenpelli (43).
Elgie Cliffs, 26m WSW Mount Remarkable (37).
ELLISTON (76) 33.39S: 134.55E.
Escott HS 17.44S: 139.24E – base for Moonlight Creek (47).
ETADUNNA (66) HS 28.43S: 138.38E (camp 9m N of HS).
EUCLA (17) 31.43S: 128.52E.
EYRE (16) 32.15S: 126.18E.

Fitzroy R., Langey Crossing, Derby (29) and Geikie Gorge (31).
Flinders Range, North, Aroona Valley (82).
Forrest R. mouth, 12m NNW Wyndham (39).
Fortescue R. Crossing, 37m SE Coolawanyah (4).
Fortescue R. Flats, Coolawanyah (4).
FURNER (86) 37.25S: 140.20E (camp Deltana HS).

Gahnda Rockhole, 80m SW Warburton (23).
Gandys Hill, trig point for Pine Creek (44).
Gap Creek, St George's Range (27).
Gawler Range, Nonning (79).
GEIKIE GORGE (31) 18.08S: 125.42E.
Gibraltar Range, 19m E Tenterfield (95).
Gibson Desert, W & N of Warburton (23).
GILES (25) 25.02S: 128.18E (camp at Pass of Abencerrages).
GLADSTONE, LAKE (36) 17.12S: 126.14E.
GLENDOWER (55), 10m SE Wongalee.
Glenelg Estuary, Nelson (88).
GLENORE (49) HS 17.51S: 141.08E (camp at The Lakes).
GRAFTON (94) 29.41S: 152.56E (camp at Southgate).
Grampians, Victoria Range (90).
Granite Creek, Ayton (52).

Granite Downs (68) HS 26.57S: 133.30E (camp at Larry's Well).
Granite Range, Mount Anderson (28).
GREEN RANGE (13) 34.40S: 118.25E (camp at Beckon HS).
Gregory R., Mount Dryander (57).
GRUSZKA, LAKE (21) 25.19S: 125.34E.
Gunbarrel Highway, Carnegie (19) to Giles (25).

Hann R. Crossing, 9m SSE Mount Elizabeth (35).
Hann R., Upper, 21m NW Mount Elizabeth (35).
Hardy Inlet, 6m SW Scott River camp (11).
Harrow, 22m N Coleraine (89).
Helena Valley, 15m E Perth (7).
Hooley Creek, 13m ESE Coolawanyah (4).
Hope Inlet, Beagle Gulf, 11m NE Darwin (41).
Hume Reservoir, E of Albury (92).
HUNT OIL CAMP (22) 26.04S: 125.50E.

INKERMAN (50) HS 16.14S: 141.25E.
INNAMINCKA (64) HS 27.45S: 140.44E (camp 3m SE of HS).

JOINT HILL (34) 16.27S: 125.56E (camp Mount Barnett HS).

Kalamunda, 12m E of Perth (7).
KARRATHA (1) HS 20.53: 116.40E (camp at Marie's Pool).
Killalpaninna, L., 12m NW Etadunna (66).
KIMBA (78) 33.09E: 136.26S.
King R., Cambridge Gulf, Wyndham (39).
KINGOONYA (70) 30.54S: 135.19E (camp at North Well HS).
Kiriwana HS, 6m E Mount Barker (12).
Koolan Is. 16.08S: 123.45E.
Kopperamanna, L., 10m N Etadunna (66).
Kopperamanna Bore, 8m N Etadunna (66).
Kywanna HS, 10m NW Albury (92).

Labyrinth, L., 18m NW Kingoonya (70).
Laing's Rock, Mount Shenton (18).
Lakes, The, 15m SSW of Glenore HS (49).
Langey Crossing, Fitzroy R., Derby (29).
Langlo Crossing 26.08S: 145.40E, base for Tyrone (59).
Larry's Well, 32m S Granite Downs HS (68).
Laverton 28.44S: 122.29E.
Leopold Range, Mount Bell (32).
Lily Creek, 9m SSE Inkerman (50).
Linwood HS, 12m E of Newdegate (14).

LISSADELL HILL (38) 16.49S: 128.27E (camp at Bow River).
LITTLE DESERT (91), 10m N Nurcoung.

Maitland R. mouth, 12m NW Karratha (1).
MARGARET RIVER (10) 33.57S: 115.04E.
Margaret R., Geikie Gorge (31).
Maries Pool, 3m NW Karratha (1).
McArthur R., camp 28m SW Borroloola (46).
METRICUP (9) 33.47S: 115.07E.
MINIPPA (73) 32.50S: 135.10E.
Moana Bore, St George's Range (27).
MONKIRA STATION (61) 24.49S: 140.34E.
MOONLIGHT CREEK (47) 33m NNW Escott HS.
MOUNT ANDERSON (28) 18.01S: 123.54E.
MOUNT BARKER (12) 34.03S: 117.37E (camp at Kiriwana HS).
Mount Barnett HS, 14m S Joint Hill (34).
MOUNT BELL (32) 18.08S: 125.42E.
MOUNT CAROLINE (33) 17.00S: 125.59E (camp 1m N).
MOUNT DRYANDER (57) camp 8m NNW Proserpine.
Mount Eba, 16m NNW Kingoonya (70).
MOUNT ELIZABETH (35) 16.23S: 126.16E (camp on Hann River).
Mount Eveline, 38m E Warburton Mission (23).
Mount House, 20m SW Mount Caroline (33).
MOUNT REMARKABLE (37) 17.13S: 127.37E (camp at Elgie Cliffs).
MOUNT ROPER (45) 14.50S: 135.03E (camp 8m WNW).
Mount Russell, 6m N of Giles (25).
Mount Searle, Angepena (84).
MOUNT SHENTON (18) 28.00S: 123.22E.
MUNGILLI CLAYPAN (20) 25.23S: 124.15E.

Nammen, L., 70m NNW Perth (7).
Nareen, 17m NNW Coleraine (89).
Neales Creek, The, 13m SW Oodnadatta (67).
NELSON (88) 38.03S: 141.01E.
NEWDEGATE (14) 33.06S: 119.01E (camp at Linwood HS).
Nickol Bay, Peg's Beach (2).
NONNING (79) HS 32.31S: 136.30E (camp 10m S of HS).
Noogoo Swamp, 11m NE Darwin (41).
Norseman 32.8S: 121.43E.
North Well HS, 4m N Kingoonya (70).
NULLARBOR HS (69) 31.26S: 130.55E.
Nullarbor Plain, N Eucla (17) & Cocklebiddy (15).
Nurcoung 36.44S: 141.41E, base for Little Desert (91).

Occabalima, 15m E Angepena (84).
OENPELLI MISSION (43) 12.20S: 133.04E (camp 10m SSW of mission).
OODNADATTA (67) 27.33S: 135.28E (camp at Cramps Camp).
Oyster Creek, CARNARVON (5).

PARACHILNA (81) 31.08S: 138.23E.
Pass of the Abencerrages, 5m N Giles (25).
PEG'S BEACH (2) 20.45S: 116.50E.
PERTH (7) 31.56S: 115.50E.
PINE CREEK (44) 13.49S: 131.49E (camp at Wandie Creek).
POINT TORMENT (30) 17.06S: 123.40E.
Policeman's Soak, 19m SSW Granite Downs HS (68).
PORT LINCOLN (77) 34.44S: 135.52E.
Proserpine 20.25S: 148.35E – base for Cape Gloucester (56) & Mount Dryander (57).
Punchbowl Camp, Wernadinga (48).

Rawlinson Range, N & NW Giles (25).
Rocklands Dam, 28m NE Coleraine (89).
Roma 26.35S: 148.47E – base for Banoona (58).
Roper River, Mount Roper (45).

Sailor's Well, 28m S Granite Downs (68).
Salt Arm Creek, 5m NNW Inkerman HS (50).
SCEALE BAY (75) 33.01S: 134.12E.
SCOTT RIVER (11) camp 34.16S: 115.14E.
Sinclair Bay, Cape Gloucester (56).
Southgate, 8m NE Grafton (94).
Staaten R., Inkerman Station (50).
St George's Range (27) 18.39S: 125.27E.
Stokes Hill, trig point for Darwin (41).
STREAKY BAY (74) 32.48S: 134.13E.
Struan 37.07S: 140.47E, base for Bool Lagoon (87).
Strzelecki Track, Blanchewater to Innamincka (64).
Swan Lake, Hardy Inlet, SCOTT RIVER (11).
Swan R., Perth (7).
Surface Hill, 14m ESE Tenterfield (95).

Table Top Mt, 20m E of Albury (92).
Tablelands 17.16S: 126.51E.
Tambrey HS, W Chichester Range (3).
Tampatha Well, Chichester Range (3).
TENTERFIELD (95) 29.01S: 152.04E (camp at Timbarra HS).
Throssel, L., 160m SW Warburton Mission (23).
THYLUNGRA STATION (60) 26.04E: 143.28S.

Timbarra HS, 9m E Tenterfield (95).
TULLY (54) 17.56S: 145.56E (camp 10m NNW).
TYRONE (59) HS, 3m E Langlo Crossing.

VAN DER LINDEN LAKES (24) 25.00S: 127.00E.
Vassey, 20m NE Coleraine (89).
VICTORIA RANGE (90) 37.32S: 142.17E.
VICTORIA RIVER (40) camp 15.36S: 131.06E.
Vine Creek, Mount Dryander (57).

Wallatina, 32m S Granite Downs (68).
WALTER HILL RANGE (53), 24m NW Tully (54).
WALTER JAMES RANGE (26) 24.40S: 128.45E.
Wandie Creek camp, 20m ESE Pine Creek (44).
WARBURTON (23) Mission 26.05S: 126.34E (camp 18m E of mission).
Warriup, 5m SE Green Range (15).
Weary Bay, Ayton (52).
WERNADINGA (48) HS 18.07S: 139.58E (camp 12m ESE of HS).
Wertaloona, 32m E Angepena (84).
West Bastion, Wyndham (39).
Wheogo HS, 10m NNW Dunedoo (93).
WHYALLA (80) 33.02S: 137.35E.
WIRREALPA (83) 31.08S: 138.57E.
WONERUP (8) 33.38S: 115.26E.
Wongalee 20.37S: 144.25E – base for Glendower (55).
WYNDHAM (39) 15.27S: 128.07E.

YARDEA (72) 32.23S: 135.32E.
Yeeda Creek, 23m SSE Derby (29).

REFERENCES

Amadon, D., 1941. Notes on some Australian birds of prey. *Emu* **40:** 365–384.

Amadon, D., 1951. Taxonomic notes on the Australian butcher-birds (family Cracticidae). *Am. Mus. Novit.* **1504.**

Austin, C. N., Brown, A. G., Officer, H. R., & Wettenhal, H. N. B., 1963. Notes on the Black-banded Pigeon. *Emu* **62:** 238–240.

Brown, L., & Amadon, D., 1968. *Eagles, hawks, and falcons of the world.* London.

Cabanis, J. L., & Heine, F., 1863. *Museum Heineanum* **4,** 1862–63, 1:27, note.

Cain, A. J., 1954. Subdivisions of the genus *Ptilonopus* (Aves, Columbae). *Bull. Br. Mus. nat. Hist.* (Zool.) **2:** 267–284.

Cain, A. J., 1955. A revision of *Trichoglossus haematodus* and of the Australian Platycercine parrots. *Ibis* **97:** 432–479.

Campbell, A. J. C., 1901. Stray feathers. *Emu* **1:** 25.

Colston, P. R., 1968. The occurrence of the Painted Honeyeater (*Grantiella picta*) in S.W. Queensland. *Qd Nat.* **18:** 116.

Colston, P. R., & Harrison, C. J. O., 1968. Some records of Palaearctic waders in Western Australia. *West Aust. Nat.* **11:** 45-46.

Condon, H. T., 1941. The Australian broadtailed parrots (subfamily Platycercinae) *Rec. Aust. Mus.* **7:** 117-144.

Condon, H. T., 1951a. Notes on the birds of South Australia: occurrence, distribution and taxonomy. *S. Aust. Orn.* **20:** 26-68.

Condon, H. T., 1951b. Variation in the Brown Hawk. *Emu* **50:** 152-174.

Condon, H. T., 1962a. A handlist of the birds of South Australia. *S. Aust. Orn.* **23:** 85-151.

Condon, H. T., 1962b. Australian quail-thrushes of the genus *Cinclosoma. Rec. S. Aust. Mus.* **14:** 337-370.

Condon, H. T., 1968. *A handlist of the birds of South Australia.* 2nd ed. Adelaide: South Australian Ornithological Association.

Cowles, G. S., 1964. A new Australian babbler. *Emu* **64:** 1-5.

Cowles, G. S., 1967. The Grey Honeyeater (*Lacustroica whitei*). A new record for South Australia. *S. Aust. Orn.* **24:** 147.

C.S.I.R.O., 1953. (C. S. Christian *et al.*). Survey of the Townsville-Bowen region, north Queensland, 1950. *Land Res. Ser. C.S.I.R.O. Aust.* **2.**

C.S.I.R.O., 1964. (R. A. Penny *et al.*). General report on lands of the Leichhardt-Gilbert area, Queensland, 1953-54. *Land Res. Ser. C.S.I.R.O. Aust.* **11.**

C.S.I.R.O., 1969. An index of Australian bird names. *Tech. Pap. Div. Wildl. Res. C.S.I.R.O. Aust.* **20.**

C.S.I.R.O., 1970 (R. W. Galloway *et al.*). Lands of the Mitchell-Normanby area, Queensland. *Land Res. Ser. C.S.I.R.O. Aust.* **26.**

Deignan, H. G., 1950. Two new races of the Spotted Nightjar, *Eurostopodus guttatus* (Vigors & Horsfield). *Emu* **50:** 21-23.

Deignan, H. G., 1964. Birds of the Arnhem Land expedition. *Records of the American-Australian scientific expedition to Arnhem Land* 4, *Zoology* (ed. R. L. Specht): 345-425. Melbourne Univ. Press.

Delacour, J., 1956. *The waterfowl of the world,* vol. 2. London.

Disney, H. J. de S., 1969. Bird in the hand. Robins of the genus *Petroica. Aust. Bird Bander* **7:** 18-19, 38-39.

Fleay, D. H., 1926. Habits of the Boobook Owl. *Emu* **26:** 97-104.

Ford, J., 1970. Distribution of quail-thrushes in the Northern Territory, and their taxonomic relations. *Emu* **70:** 135-139.

Ford, J., 1971. Distribution, ecology and taxonomy of some Western Australian passerine birds. *Emu* **71:** 103-120.

Forshaw, J. M., 1969. *Australian parrots.* Melbourne.

Freeman, D. J., 1970. The rediscovery of the Black Grass Wren *Amytornis housei* with additional notes on the species. *Emu* **70:** 193-195.

Freeman, D. J., 1971. The nest of the White-lined Honeyeater. *Emu* **71:** 86-87.

Frith, C. B., 1970. Nest and eggs of the Australian rock pigeons *Petrophassa* spp. *Emu* **70:** 29-30, Plate 1.

Frith, H. J., 1952. Notes on the pigeons of Richmond River, N.S.W. *Emu* **52:** 89-99.

Galbraith, I. C. J., 1956. Variation, relationships and evolution in the *Pachycephala pectoralis* superspecies (Aves, Muscicapidae). *Bull. Br. Mus. nat. Hist.* (Zool.) **4:** 131-222.

Galbraith, I. C. J., 1967. The Black-tailed and Robust Whistlers *Pachycephala melanura* as a species distinct from the Golden Whistler *P. pectoralis. Emu* **66:** 289-294.

Galbraith, I. C. J., 1969. The Papuan and Little Cuckoo-shrikes, *Coracina papuensis* and *robusta,* as races of a single species. *Emu* **69:** 9-29.

Galbraith, I. C. J., & Parker, S. A., 1969. The Atherton Scrub-Wren *Sericornis keri* Mathews a neglected Australian species. *Emu* **69:** 212-232.

Gilliard, E. T., 1969. *Birds of paradise and bower birds.* London.

Givens, T. V., & Hitchcock, W. B., 1953. *Cisticola juncidis* (Raf.) in the Northern Territory. *Emu* **53:** 193-200.

Goodwin, D., 1956. Observations on the voice and some displays of certain pigeons. *Avicult. Mag.* **62:** 17-33, 62-70.

Goodwin, D., 1959. The validity of the genus *Megaloprepia* Reichenbach. *Bull. Br. Orn. Club* **79:** 52.

Goodwin, D., 1960. Taxonomy of the genus *Ducula. Ibis* **102:** 526-535.

Goodwin, D., 1967a. Australian pigeons: their affinities and status. *Emu* **66:** 319-336.

Goodwin, D., 1967b. *Pigeons and doves of the world.* London (Brit. Mus. (Nat. Hist.)).

Goodwin, D., 1967c. Notes on behaviour of some Australian birds. *Emu* **66:** 237-251.

Goodwin, D., 1969. A new sub-species of the White-quilled Rock Pigeon. *Bull. Br. Orn. Club* **89:** 131-133.

Gould, J., 1865. *Handbook to the birds of Australia,* vols 1 & 2. London.

Gould, J., 1868. On two new Australian birds. *Proc. zool. Soc. Lond.* 1868: 74-76.

Harrison, C. J. O., 1968. Some comparative notes on the Common and Yellow Silver-eyes. *Emu* **68:** 127-131.

Harrison, C. J. O., 1969a. The affinities of the blue wren genus *Malurus* and related genera; with special reference to the grass-wren genus *Amytornis. Emu* **69:** 1-8.

Harrison, C. J. O., 1969b. Emu-wrens and the forest zone in south-western Australia. *West Aust. Nat.* **11:** 66-67.

Harrison, C. J. O., 1970. The relationship of the Brown and Black-backed Tree-creepers. *Emu* **70:** 9-11.

Harrison, C. J. O., & Colston, P. R., 1969. Some records of breeding after late rains in the Hamersley region. *West. Aust Nat.* **11:** 49-50.

Harrison, C. J. O., & Parker, S. A., 1965. The behavioural affinities of the blue wrens of the genus *Malurus. Emu* **65:** 103-113.

Hartert, E., 1929. Types of birds in the Tring Museum. *Novit. zool.* **35:** 42-58.

Hill, G. F., 1911. Field notes on the birds of Kimberley, north-west Australia. *Emu* **10:** 258-290.

Hindwood, K. A., 1948. The Blue-faced Finch in Australia. *Emu* **48:** 53.

Hindwood, K. A., 1965. The "Sydney" bird paintings. *Aust. Zool.* **13:** 83-92.

Horton, W., 1973. Halls Babbler near Opalton, Queensland. *Aust. Bird Bander* **11:** 33-34.

Immelmann, K., 1963. Drought adaptation in Australian desert birds. *Proc. Int. Orn. Congr.* **13,** vol. 2: 649-657.

Immelman, K., 1965. *Australian finches in bush and aviary*. Sydney.

Inglis, W. G., 1967. The relationships of the nematode superfamily Seuratoidea. *J. Helminth.* **41:** 115-163.

Ingram, C., 1908a. Description of two new species from north Queensland. *Bull. Br. Orn. Club* **21:** 99-100.

Ingram, C., 1908b. On the birds of Inkerman Station, north Queensland. *Ibis* **(9)2:** 458-481, Plate 9.

Keast, A., 1956. Variation in the Australian Oriolidae. *Proc. R. zool. Soc. N.S.W.* 1954-55: 19-24.

Keast, A., 1957a. Variation in the Australian whitefaces (Aves, genus *Aphelocephala,* Oberholser, 1899). *Proc. R. zool. Soc. N.S.W.* 1955-56: 38-42.

Keast, A., 1957b. Variation in the Australian kingfishers (Aves: Alcedinidae). *Rec. Aust. Mus.* **24:** 61-72.

Keast, A., 1957c. Variation and speciation in the genus *Climacteris* Temminck (Aves, Sittidae). *Aust. J. Zool.* **5:** 474-495.

Keast, A., 1958a. The influence of ecology on variation in the Mistletoe-bird (*Dicaeum hirundinaceum*). *Emu* **58:** 195-206.

Keast, A., 1958b. Seasonal movements and geographic variations in the Australian wood-swallows (Artamidae). *Emu* **58:** 207-218.

Keast, A., 1958c. Infraspecific variation in the Australian finches. *Emu* **58:** 219-246.

Keast, A., 1958d. The genus *Psophodes* Vigors and Horsfield, and its significance in demonstrating a possible pathway for the origin of Eyrean species from Bassian ones. *Emu* **58:** 247-255.

Keast, A., 1958e. Speciation in the genus *Amytornis* Stejneger (Passeres: Muscicapidae, Malurinae) in Australia. *Aust. J. Zool.* **6:** 33-52.

Keast, A., 1958f. The relationship between seasonal movements and the development of geographical variation in the Australian chats (*Epthianura* Gould and *Ashbyia* North (Passeres: Muscicapidae, Malurinae)). *Aust. J. Zool.* **6:** 53-68.

Keast, A., 1958g. Geographic variation in the Weebill, *Smicrornis brevirostris* (Gould) (Passeres: Muscicapidae, Malurinae), a sedentary species with a continuous range. *Aust. J. Zool.* **6:** 152-161.

Keast, A., 1958h. Variation and speciation in the Australian Campephagidae (Passeres). *Aust. J. Zool.* **6:** 248-267.

Keast, A., 1958i. Variation and speciation in the Australian flycatchers (Aves: Muscicapinae). *Rec. Aust. Mus.* **24:** 73-108.

Keast, A., 1961. Bird speciation on the Australian continent. *Bull. Mus. comp. Zool. Harv.* **123:** 305-495.

Lane, S. G., 1968. Age/plumage relationship in Rufous Whistlers. *Aust. Bird Bander* **6:** 75-77.

Lynes, H., 1930. Review of the genus *Cisticola. Ibis* **(12) 6,** suppl.

Macdonald, J. D., 1966. The status of some Australian treecreepers. *Emu* **66:** 113–115.
Macdonald, J. D., 1967. Variation in the Black-faced Wood-swallow *Artamus cinereus*. *Emu* **66:** 341–345.
Macdonald, J. D., 1968a. Speciation in the *Colluricincla harmonica* complex. *Emu* **67:** 215–224.
Macdonald, J. D., 1968b. Notes on the genus *Cinclosoma*. *Emu* **67:** 283–287.
Macdonald, J. D., 1969a. Hybridisation in pardalotes. *Emu* **69:** 41–44.
Macdonald, J. D., 1969b. Notes on the taxonomy of *Neositta*. *Emu* **69:** 169–174.
Macdonald, J. D., & Colston, P. R., 1966. J. R. Elsey and his bird observations on Gregory's overland expedition, 1856. *Emu* **65:** 255–278.
Mack, G., 1934. A revision of the genus *Malurus*. *Mem. natn. Mus. Melb.* **8:** 100–125.
Mack, G., 1936. A systematic revision of the Australian thornbills. *Mem. natn. Mus. Melb.* **10:** 86–118.
Macnae, W., 1966. Mangroves in eastern and southern Australia. *Aust. J. Bot.* **14:** 67–104.
Mathews, G. M., 1911. *The birds of Australia,* vol. 1. London.
Mathews, G. M., 1912a. A reference-list to the birds of Australia. *Novit. zool.* **18:** 171–455.
Mathews, G. M., 1912b. Additions and corrections to my reference list to the birds of Australia. *Austral. avian Rec.* **1:** 28.
Mathews, G. M., 1923–24. *The birds of Australia,* vol. 11. London.
Mathews, G. M., 1925–27. *The birds of Australia,* vol. 12. London.
Mathews, G. M., 1927, 1930. *Systema avium Australasianarum*. London: Brit. Orn. Union.
Mayr, E., 1932a. Notes on thickheads (*Pachycephala*) from the Solomon Islands. *Am. Mus. Novit.* **522.**
Mayr, E., 1932b. Notes on thickheads (*Pachycephala*) from Polynesia. *Am. Mus. Novit.* **531.**
Mayr, E., 1933. Notes on the variation of immature and adult plumages in birds and a physiological explanation of abnormal plumages. *Am. Mus. Novit.* **666.**
Mayr, E., 1934. Notes on the genus *Petroica*. *Am. Mus. Novit.* **714.**
Mayr, E., 1937a. Notes on the genus *Sericornis* Gould. *Am. Mus. Novit.* **904.**
Mayr, E., 1937b. Birds collected during the Whitney South Sea Expedition 35. Notes on New Guinea birds II. *Am. Mus. Novit.* **939.**
Mayr, E., 1938. Birds collected during the Whitney South Sea Expedition 39. Notes on New Guinea birds IV. *Am. Mus. Novit.* **1006.**
Mayr, E., 1940a. Birds collected during the Whitney South Sea Expedition 43. Notes on New Guinea birds VI. *Am. Mus. Novit.* **1056.**
Mayr, E., 1940b. Notes on Australian birds. *Emu* **40:** 111–117.
Mayr, E., 1941. *List of New Guinea birds*. New York: Am. Mus. Nat. Hist.
Mayr, E., 1943. Notes on Australian birds. *Emu* **43:** 3–17.
Mayr, E., 1944. Timor and the colonization of Australia by birds. *Emu* **44:** 113–130.
Mayr, E., 1945. Birds collected during the Whitney South Sea Expedition 55. Notes on the birds of northern Melanesia 1. *Am. Mus. Novit.* **1294.**

Mayr, E., 1948. Geographic variation in the Reed Warbler. *Emu* **47:** 205-210.

Mayr, E., 1951. Notes on some pigeons and parrots from Western Australia. *Emu* **51:** 137-145.

Mayr, E., 1953a. Geographic and individual variation in the Shrike-Tit (*Falcunculus frontatus*). *Emu* **53:** 249-252.

Mayr, E., 1953b. Taxonomic notes on *Oreoica gutturalis*. *Emu* **53:** 252-253.

Mayr, E., 1954. Notes on Australian whistlers (Aves, *Pachycephala*). *Am. Mus. Novit.* **1653.**

Mayr, E., & Jennings, K., 1952. Geographic variation and plumages in Australian bowerbirds (Ptilonorhynchidae). *Am. Mus. Novit.* **1602.**

Mayr, E., & McEvey, A. R., 1960. The distribution and variation of *Mirafra javanica* in Australia. *Emu* **60:** 152-192.

Mayr, E. & Moynihan, M., 1946. Evolution in the *Rhipidura rufifrons* group. *Am. Mus. Novit.* **1321.**

Mayr, E., & Rand, A. L., 1937. Results of the Archbold Expeditions. No. 14. The birds of the 1933-1934 Papuan Expedition. *Bull. Am. Mus. nat. Hist.* **73:** 1-248.

Mayr, E., & Serventy, D. L., 1938. A review of the genus *Acanthiza* Vigors and Horsfield. *Emu* **38:** 245-292.

Mayr, E., & Wolk, R., 1953. The south-west Australian races of the Spotted Scrub-wren *Sericornis maculatus*. *W. Aust. Nat.* **4:** 66-70.

McEvey, A. R., & Middleton, W. G., 1968. Birds and vegetation between Perth and Adelaide. *Emu* **68:** 161-212.

Mees, G. F., 1961a. An annotated catalogue of bird-skins from West Pilbara, Western Australia. *J. Proc. R. Soc. West. Aust.* **44:** 97-143.

Mees, G. F., 1961b. A systematic review of the Indo-Australian Zosteropidae, Part II. *Zool. Verh. Leiden* **50:** 1-168, 4 pls.

Mees, G. F., 1964a. Geographical variation and distribution of some birds from Western Australia. *J. Proc. R. Soc. West. Aust.* **47:** 91-96.

Mees, G. F., 1964b. A revision of the Australian owls (Strigidae and Tytonidae). *Zool. Verh. Leiden* **65:** 3-62.

Mees, G. F., 1968. Colour of orbital ring in Partridge-Pigeon *Geophaps smithii*. *Emu* **67:** 294.

Mees, G. F., 1969. A systematic review of the Indo-Australian Zosteropidae. Part III. *Zool. Verh. Leiden* **102:** 1-390.

Meise, W., 1931. Zur Systematik der Gattung *Gerygone*. *Novit. zool.* **36:** 317-379.

Nicolai, J., 1964. Der Brutparasitismus der Viduinae als ethologisches Problem. *Z. Tierpsychol.* **21:** 129-204.

Officer, H. R., 1964. *Australian honeyeaters*. Melbourne: Bird Observers Club.

Ogden, C. G., 1967. Some parasitic nematodes from Australian birds. *Ann. Mag. nat. Hist.* **(13) 9:** 505-518.

Parker, S. A., 1968a. An instance of apparent sympatry between the Great and Spotted Bowerbirds. *Bull. Br. Orn. Club* **88:** 56.

Parker, S. A., 1968b. The type locality of the White-quilled Rock Pigeon, *Petrophassa albipennis* Gould. *Bull. Br. Orn. Club* **88:** 57-58.

Parker, S. A., 1970. Critical notes on the status of some Northern Territory birds. *S. Aust. Orn.* **25:** 115–125.

Parker, S. A., 1971. Taxonomy of the Northern Territory friarbirds known as *Philemon buceroides gordoni. Emu* **71:** 54–56.

Parker, S. A., 1973. The identity of *Microeca brunneicauda* Campbell, 1902. *Emu* **73:** 23–25.

Peters, J. L., 1931. *Check-list of birds of the world,* vol. 1. Cambridge, Mass.: Mus. Comp. Zool.

1934. *op. cit.,* vol. 2.

1937. *op cit.,* vol. 3.

1940. *op. cit.,* vol. 4.

1943. *op. cit.,* vol. 5.

1960. *op. cit.,* vol. 9.

1964. *op. cit.,* vol. 10 (ed. Mayr, E., & Paynter, R. A.).

1967. *op. cit.,* vol. 12 (ed. Paynter, R. A.).

Piechocki, R., 1961. Uber die Grossgefieder-Mauser von Schleiereule und Waldkauz. *J. Orn., Lpz.* **102:** 220–225.

Rand, A. L., 1940. Results of the Archbold Expeditions. No. 27. Ten new birds from New Guinea. *Am. Mus. Novit.* **1074.**

Rand, A. L., 1941. Results of the Archbold Expeditions. No. 32. *Am. Mus. Novit.* **1102.**

Rand, A. L., & Gilliard, E. T., 1967. *Handbook of New Guinea birds.* London.

Rowley, I., 1967. A fourth species of Australian corvid. *Emu* **66:** 191–210.

Rowley, I., 1970. The genus *Corvus* (Aves: Corvidae) in Australia. *C.S.I.R.O. Wildl. Res.* **15:** 27–71.

Salomonsen, F., 1961a. Notes on flowerpeckers (Aves, Dicaeidae) 5. The genera *Oreocharis, Paramythia* and *Pardalotus* (except the superspecies *Pardalotus striatus*). *Am. Mus. Novit.* **2067.**

Salomonsen, F. 1961b. Notes on flowerpeckers (Aves, Dicaeidae) 6. The superspecies *Pardalotus striatus. Am. Mus. Novit.* **2068.**

Salomonsen, F., 1966. Preliminary descriptions of new honeyeaters (Aves, Meliphagidae). *Breviora* **254:** 1–12.

Schodde, R., & Hitchcock, W. B., 1968. Contributions to Papuasian ornithology. 1. Report on the birds of the Lake Kutubu area, Territory of Papua and New Guinea. *Tech. Pap. Wildl. Res. C.S.I.R.O. Aust.* **13:** 1–73.

Serventy, D. L., & Whittell, H. M., 1967. *Birds of Western Australia,* 4th ed. Perth, W. A.

Sibley, C. G., 1970. A comparative study of the egg-white proteins of passerine birds. *Bull. Peabody Mus. nat. Hist.* **32.**

Simmons, K. E. L., 1953. Some studies on the Little Ringed Plover. *Avicult. Mag.* **59:** 191–207.

Storr, G. M., 1953. Birds of the Cooktown and Laura districts, north Queensland. *Emu* **53:** 225–248.

Storr, G. M., 1967. List of Northern Territory birds. *Spec. Publs West. Aust. Mus.* **4.**

Storr, G. M., 1973. List of Queensland birds. *Spec. Publs. West. Aust. Mus.* **5.**

Terrill, S. E., & Rix, C. E., 1950. The birds of South Australia their distribution and habitat. *S. Aust. Orn.* **19:** 53-100.

Thomson, D. F., 1935. *Birds of Cape York Peninsula.* Melbourne: Government Printer.

Timmerman, G., 1971. Mallophagologische Kollektaneen 2. *Senckenbergiana Biol.* **52:** 41-47.

Van Tets, G. F., D'Andria, A. H., & Slater, E., 1967. Nesting distribution and nomenclature of Australasian vanelline plovers. *Emu* **67:** 85-93.

Warham, J., 1962. Bird islands within the Barrier Reef. *Emu* **62:** 99-111.

Wheeler, R., 1959. The R.A.O.U. camp-out at Noosa Heads, Queensland, 1958. *Emu* **59:** 230-249.

White, H. L., 1922. Two new subspecies of birds. *Emu* **21:** 163.

Whitlock, F. L., 1924. Journey to Central Australia in search of the Night Parrot. *Emu* **23:** 254-255.

Wilkins, G. H., 1928. *Undiscovered Australia, being an account of an expedition to tropical Australia to collect specimens of the rarer native fauna for the British Museum, 1923-1925.* London.

Williams, R. T., 1955. *Atlas of Australian resources, vegetation regions.* Canberra: Dept. of Nat. Development.

Wilson, S. J., in press. Plumage succession in the male Golden Whistler, *Pachycephala pectoralis pectoralis. Aust. Bird Bander.*

Worrell, E., & Sourry, L., 1968. *Trees of the Australian bush.* Sydney.

VERNACULAR NAMES WITH SCIENTIFIC EQUIVALENTS

A name is listed under its final substantive whether hyphened or not, except where this is 'bird' or 'hen', or where only the double name (e.g. cuckoo-shrike) is meaningful. All the names in a few groups such as the herons, bitterns and egrets, to which substantives have been applied illogically and inconsistently, are listed under one substantive, with cross-references under the others. A generic name is represented by its initial in subsequent entries, where these are consecutive or there is no ambiguity.

ALBATROSS, Royal *epomophora, Diomedea*
Wandering *exulans, D.*
APOSTLE BIRD, *cinerea, Struthidea*
AVOCET, Australian *novaehollandiae, Recurvirostra*
BABBLER, Chestnut-crowned *ruficeps, Pomatostomus*
Grey-crowned *temporalis, P.*
Hall's *halli, P.*
White-browed *superciliosus, P.*
BEE-EATER, Rainbow *ornatus, Merops*
BELL-BIRD, Crested *gutturalis, Oreoica*
BITTERN—see Heron
BOWER-BIRD Great *nuchalis, Chlamydera*
Satin *violaceus, Ptilonorhynchus*
Spotted *maculata, C.*
Tooth-billed *dentirostris, Scenopoeetes*
BROLGA, *rubicunda, Grus*
BRONZEWING, Brush *elegans, Phaps*
Common *chalcoptera, P.*
BUDGERYGAH, *undulatus, Melopsittacus*
BUSHLARK, Singing *javanica, Mirafra*
BUSTARD, Australian *australis, Eupodotis*
BUTCHER-BIRD, Black *quoyi, Cracticus*
Grey *torquatus, C.*
Pied *nigrogularis, C.*
BUZZARD, Black-breasted *melanosterna, Hamirostra*
CATBIRD, *melanotis, Ailuroedus*
CHAT, Crimson *tricolor, Ephthianura*
Orange *aurifrons, E.*
White-fronted *albifrons, E.*
Yellow *crocea, E.*
CHOUGH, White-winged *melanorhamphus, Corcorax*
CHOWCHILLA, Northern *spaldingi, Orthonyx*
CICADA-BIRD, *tenuirostre, Edoliioma*
COCKATIEL *hollandicus, Nymphicus*
COCKATOO, Gang-gang *fimbriatum, Callocephalon*
Glossy Black *lathami, Calyptorhynchus*
Red-tailed Black *banksi, Calyptorhynchus*
Suphur-crested *galerita, Cacatua*
White-tailed Black *baudini, Calyptorhynchus*
Yellow-tailed Black *funereus, Calyptorhynchus*
COOT *atra, Fulica*
CORELLA, Little *sanguinea, Cacatua*
Long-billed *tenuirostris, C.*
CORMORANT, Black *carbo, Phalacrocorax*
Little Black *sulcirostris, P.*
Little Pied *melanoleucos, P.*
Pied *varius, P.*
COUCAL, Pheasant *phasianinus, Centropus*
CRAKE, Australian Spotted *fluminea, Porzana*
White-browed *cinerea, P.*
CROW, Australian *orru, Corvus*
Little *bennetti, C.*
CUCKOO, Black-eared *osculans, Chrysococcyx*
Brush *variolosus, Cacomantis*
Channel-billed *novaehollandiae, Scythrops*
Chestnut-breasted *castaneiventris, Cacomantis*
Fan-tailed *pyrrhophanus, Cacomantis*
Golden Bronze *plagosus, Chrysococcyx*
Horsfield Bronze *basalis, Chrysococcyx*
Little Bronze *minutillus, Chrysococcyx*
Pallid *pallidus, Cuculus*
CUCKOO-SHRIKE, Barred *lineata, Coracina*
Black-faced *novaehollandiae, C.*
Ground *maxima, Pteropodocys*
Little *robusta, C.*
(Papuan—see Little)

CURLEW, Eastern *madagascariensis, Numenius*
CURRAWONG, Grey *versicolor, Strepera*
Pied *graculina, S.*
DARTER *rufa, Anhinga*
DOLLAR-BIRD *orientalis, Eurystomus*
DOTTEREL—see Plover
DOVE, Bar-shouldered *humeralis, Geopelia*
Diamond *cuneata, G.*
Peaceful *striata, G.*
Senegal Turtle- *senegalensis, Streptopelia*
Spotted Turtle- *chinensis, S.*
DRONGO, Spangled *bracteata, Dicrurus*
DUCK, Black *superciliosa, Anas*
Blue-billed *australis, Oxyura*
Burdekin *radjah, Tadorna*
Grass Whistling *eytoni, Dendrocygna*
Mountain *tadornoides, T.*
Musk *lobata, Biziura*
Pink-eared *membranaceus, Malacorhynchus*
Water Whistling *arcuata, D.*
White-eyed *australis, Aythya*
Wood *jubata, Chenonetta*
EAGLE, Australian Little *morphnoides, Hieraaetus*
Red-backed Sea *indus, Haliastur*
Wedge-tailed *audax, Aquila*
Whistling *sphenurus, Haliastur*
White-breasted Sea *leucogaster, Haliaeetus*
EGRET—see Heron
EMU *novaehollandiae, Dromaius*
FALCON, KESTREL, HAWK
FALCON, Black *subniger, Falco*
Grey *hypoleucos, F.*
Little *longipennis, F.*
Peregrine *peregrinus, F.*
HAWK, Brown *berigora, F.*
KESTREL, Nankeen *cenchroides, F.*
FANTAIL, Grey *fuliginosa, Rhipidura*
Northern *rufiventris, R.*
Rufous *rufifrons, R.*
FIGBIRD, Southern *vieilloti, Sphecotheres*
Yellow *flaviventris, S.*
FINCH, Banded *bichenovii, Stictoptera*
Black-throated *cincta, Poephila*
Blue-faced *trichroa, Erythrura*
Chestnut-breasted *castaneothorax, Lonchura*
Crimson *phaeton, Neochmia*
Gold- *carduelis, Carduelis*
Gouldian *gouldiae, Chloebia*
Long-tailed *acuticauda, P.*
Painted *picta, Emblema*
Plum-headed *modesta, Aidemosyne*
Red-browed *temporalis, Aegintha*
Star *ruficauda, N.*
Zebra *guttata, Taenopygia*
FIRETAIL, Diamond *guttatus, Zonaeginthus*
Red-eared *oculatus, Z.*
FLYCATCHER, Australian Pied *kaupi, Arses*
Black-faced *melanopsis, Monarcha*
Boat-billed *flaviventer, Machaerirhynchus*
Broad-billed *ruficollis, Myiagra*
(Brown-tailed—see Lemon-breasted)
Leaden *rubecula, Myiagra*
Lemon-breasted *flavigaster, Microeca*
Restless *inquieta, Seisura*
Shining *alecto, Myiagra*
Spectacled *trivirgata, Monarcha*
White-eared *leucotis, Monarcha*
FRIAR-BIRD, Little *citreogularis, Philemon*
Melville Island *buceroides, P.*
Noisy *corniculatus, P.*
Silver-crowned *argenticeps, P.*
FROGMOUTH Papuan *papuensis, Podargus*
Tawny *strigoides, P.*
GALAH *roseicapilla, Cacatua*
GIBBER-BIRD *lovensis, Ashbyia*
GODWIT, Bar-tailed *lapponica, Limosa*
Black-tailed *limosa, L.*
GOOSE, Green Pigmy *pulchellus, Nettapus*
Magpie *semipalmatus, Anseranas*
GOSHAWK, Australian *fasciatus, Accipiter*
Grey *novaeseelandiae, A.*
Red *radiatus, Erythrotriorchis*
GRASSBIRD, Tawny *timoriensis, Megalurus*
GREBE, Hoary-headed *poliocephalus, Podiceps*
Little *novaehollandiae, P.*
GREENSHANK *nebularia, Tringa*
Little *stagnatalis, T.*
GROUND-THRUSH, Australian *dauma, Zoothera*

Bushlark, Skylark, Songlark under those names
LORIKEET, Little *pusilla, Glossopsitta*
Musk *concinna, G.*
Purple-crowned *porphyrocephala, G.*
Rainbow *haematodus, Trichoglossus*
Scaly-breasted *chlorolepidotus, T.*
Varied *versicolor, Psitteuteles*
LOTUS BIRD *gallinacea, Jacana*
LYREBIRD, Albert *alberti, Menura*
MAGPIE, Black-backed *tibicen, Gymnorhina*
Western *dorsalis, G.*
White-backed *hypoleuca, G.*
MAJOR MITCHELL *leadbeateri, Cacatua*
MALLEE FOWL *ocellata, Leipoa*
MARTIN, Fairy *ariel, Petrochelidon*
Tree- *nigricans, P.*
MINER, Noisy *melanocephala, Manorina*
Yellow-throated *flavigula, M.*
MISTLETOE BIRD *hirundinaceum, Dicaeum*
MOORHEN, Dusky *tenebrosa, Gallinula*
NATIVE HEN, Black-tailed *ventralis, Tribonyx*
NIGHTJAR, Owlet- *cristatus, Aegotheles*
Spotted *guttatus, Eurostopodus*
White-tailed *macrurus, Caprimulgus*
ORIOLE, Olive-backed *sagittatus, Oriolus*
Yellow *flavocinctus, O.*
OSPREY *haliaetus, Pandion*
OWL, Barking *connivens, Ninox*
Barn *alba, Tyto*
Boobook *novaeseelandiae, N.*
Masked *novaehollandiae, T.*
OYSTERCATCHER, Pied *ostralegus, Haematopus*
Sooty *fulginosus, H.*
PARDALOTE, Black-headed *melanocephalus, Pardalotus*
Eastern Striated *ornatus, P.*
Red-browed *rubricatus, P.*
Spotted *punctatus, P.*
Striated *substriatus, P.*
Yellow-rumped *xanthopygus, P.*
PARROT, Blue-bonnet *haematonotus, Psephotus*
Blue-winged *chryostoma, Neophema*
Bourke *bourki, Neopsephotus*
Elegant *elegans, Neophema*
Fig *diophthalma, Opopsitta*
Golden-winged *chrysopterygius, Psephotus*
Ground *wallicus, Pezoporus*
King *scapularis, Aprosmictus*
Mulga *varius, Psephotus*
Night *occidentalis, Pezoporus*
Port Lincoln *zonarius, Barnardius*
Red-capped *spurius, Purpureicephalus*
Red-rumped *haematogaster, Psephotus*
Red-winged *erythropterus, A.*
Regent *anthopeplus, Polytelis*
Ringneck *barnardi, B.*
Turquoise *pulchella, Neophema*
PELICAN, Australian *conspicillatus, Pelecanus*
PENGUIN, Little *minor, Eudyptula*
PETREL, Giant *gigantea, Macronectes*
PICTORELLA *pectoralis, Lonchura*
PIGEON, Black-banded *alligator, Ptilinopus*
Brown *amboinensis, Macropygia*
Chestnut-quilled Rock *rufipennis, Petrophassa*
Crested *lophotes, Ocyphaps*
Flock *histrionica, Histriophaps*
Green-winged *indica, Chalcophaps*
Partridge *smithii, Geophaps*
Plumed *plumifera, Lophophaps*
Purple-crowned *superbus, Ptilinopus*
Red-crowned *regina, Ptilinopus*
Red Plumed *ferruginea, L.*
Squatter *scripta, G.*
Torres Strait *spillorrhoa, Ducula*
White-quilled Rock *albipennis, Petrophassa*
Wompoo *magnificus, Ptilinopus*
PIPIT, Australian *novaeseelandiae, Anthus*
PITTA, Noisy *versicolor, Pitta*
Rainbow *iris, P.*
PLOVER AND DOTTEREL
DOTTEREL, Australian *australis, Peltohyas*
Black-fronted *melanops, Charadius*
Hooded *cucullatus, C.*
Large Sand- *leschenaultii, C.*
Mongolian Sand- *mongolus, C.*
Red-capped *alexandrinus, C.*
Red-kneed *cinctus, C.*
PLOVER, Banded *tricolor, Vanellus*
Eastern Golden *apricaria, Pluvialis*
Grey *squatarola, Pluvialis*
Masked *miles, V.*

THORNBILL, Brown *pusilla, Acanthiza*
Buff-rumped *reguloides, A.*
Chestnut-rumped *uropygialis, A.*
Little *nana, A.*
(Mountain—see Brown)
Robust *robustirostris, A.*
Samphire *iredalei, A.*
Striated *lineata, A.*
Western *inornata, A.*
Yellow-rumped *chrysorrhoa, A.*
THRUSH, Ground-, Quail-, Shrike- —see under those names
TREE-CREEPER, Black—see Brown
Black-tailed *melanura, Climacteris*
Brown *picumnus, C.*
Red-browed *erythrops, C.*
Rufous *rufa, C.*
White-browed *affinis, C.*
White-throated *leucophaea, C.*
TRILLER, Varied *leucomela, Lalage*
White-winged *sueurii, L.*
TURKEY, Brush *lathami, Alectura*
TURNSTONE *interpres, Arenaria*
WARBLER, Brown *mouki, Gerygone*
Buff-breasted *levigaster, G.*
Dusky *tenebrosa, G.*
Fairy *palpebrosa, G.*
Golden-headed Fantail *exilis, Cisticola*
Green-backed *chloronota, G.*
Large-billed *magnirostris, G.*
Reed- *stentoreus, Acrocephalus*
Speckled *sagitatta, Chthonicola*
Streaked Grass- *juncidis, Cisticola*
Western *fusca, G.*
White-throated *olivacea, G.*
WATTLE-BIRD, Little *chrysoptera, Anthochaera*
Red *carunculata, A.*
WEDGEBILL *cristatum, Sphenostoma*
WEEBILL *brevirostris, Smicrornis*
WHIMBREL *phaeopus, Numenius*
Little *minutus, N.*
WHIPBIRD, Eastern *olivaceus, Psophodes*
WHISTLER, Brown *simplex, Pachycephala*
Gilbert *inornata, P.*
Golden *pectoralis, P.*
Mangrove Golden *melanura, P.*
Rufous *rufiventris, P.*
White-breasted *lanioides, P.*
WHITEFACE *leucopsis, Aphelocephala*
Banded *nigricincta, A.*
WILLIE WAGTAIL *leucophyrs, Rhipidura*
WOOD-SWALLOW, Black-faced *cinereus, Artamus*
Dusky *cyanopterus, A.*
Little *minor, A.*
Masked *personatus, A.*
White-breasted *leucorhynchus, A.*
White-browed *superciliosus, A.*
WREN, Atherton Scrub- —see Bellenden Ker Scrub-
Bellenden Ker Scrub- *keri, Sericornis*
(Black-and-white—see Blue-and-white)
Black-backed Blue *melanotus, Malurus*
Black Grass *housei, Amytornis*
Blue-and-white *leucopterus, M.*
Blue-breasted *pulcherrimus, M.*
Eyrean Grass *goyderi, A.*
Fern- *gutturalis, Oreoscopus*
Field- *fuliginosus, Calamanthus*
Heath- *pyrrhopygia, Hylacola*
Large-billed Scrub- *magnirostris, S.*
(Lavender-flanked—see White-flanked)
Lovely *amabilis, M.*
Mallee Heath- *cauta, H.*
Purple-crowned *coronatus, M.*
Red-backed *melanocephalus, M.*
Red-winged *elegans, M.*
Southern Scrub- *frontalis, Sericornis*
Southern Emu- *malachurus, Stipiturus*
Splendid Blue, *splendens, M.*
(Spotted Scrub- —see Southern Scrub-)
Striated Grass- *striatus, A.*
Superb Blue *cyaneus, M.*
Turquoise *callainus, M.*
Variegated *lamberti, M.*
Western Grass- *textilis, A.*
(White-browed Scrub- —see Southern Scrub-)
White-flanked *dulcis, Malurus*
White-throated Grass- *woodwardi, A.*
Yellow-throated Scrub- *citreogularis, Sericornis*

INDEX OF GENERIC AND SPECIFIC NAMES

Only those names are indexed here which are accepted for Australian genera and species in the C.S.I.R.O. Index (1969) or the present work. Extralimital species, cited in the text for comparison or discussion, are not indexed.

Page numbers against a genus indicate the appropriate part of the systematic section. Those against a species include all references, by both scientific and vernacular names, with the first page of the systematic entry in **bold** type.

Names of genera and species used in the C.S.I.R.O. Index but not here, are in brackets.

The scientific equivalents of the vernacular names used are given in the list which precedes this index.

(a) Scott River (11), W.A. Blackboys and flowers (P. R. Colston)

(b) Newdegate (14), W.A. Mallee association (P. R. Colston)

Plate 1

(a) Lake Gruszka (21), W.A. Eucalypts along dry creek (G. S. Cowles)

(b) Warburton (23), W.A. Mulga near camp, 23 miles east of the Mission (G. S. Cowles)

Plate 2

(a) Eucla (17), W.A. Mallee near camp (G. S. Cowles)

(b) Giles (25), W.A. Lower slopes of north side of Rawlinson Range. Typical habitat of *Meliphaga keartlandi* (G. S. Cowles)

Plate 3

(a) Nonning (79), S.A. Salt bush/Bluebush pastures near camp (G. S. Cowles)

(b) Aroona Valley (82), S.A. Flinders Range with pines on slopes and gums in creek (D. G. Nichols)

Plate 4

(a) Albury (92), N.S.W. View from Table Top Mountain (P. R. Colston)

(b) Dunedoo (93), N.S.W. Mixed woodland on laterite ridges. Habitat of *Pachycephala rufiventris, Chthonicola sagittata* and *Gerygone olivacea* (P. R. Colston)

Plate 5

(a) Monkira (61), Q'd. General view (P. R. Colston)

(b) Thylungra (60), Q'd. General view (P. R. Colston)

Plate 6

(a) Moonlight Creek (47), Q'd. Woodland near camp (I. C. J. Galbraith)

(b) Glenore (49), Q'd. Woodland near camp (I. C. J. Galbraith)

Plate 7

(a) Ayton (52), Q'd. Mouth of the Bloomfield River from the north bank. Slopes with areas of rainforest and open forest, characteristic of the collecting area. Islets near the far bank were largely covered with *Bruguiera* mangroves in which *Peneoenanthe pulverulenta* was common (I. C. J. Galbraith)

(b) Moonlight Creek (47), Q'd. Varied woodland on an island of sandy ridges surrounded by mud-flats, near the mouth of the creek (I. C. J. Galbraith)

Plate 8

(a) Walter Hill Range (53), Q'd. Gully in rainforest with lawyer vine. *Sericornis keri* was collected in nearby undergrowth, where *Orthonyx spaldingi* and *Psophodes olivaceus* were common (I. C. J. Galbraith)

(b) Mount Dryander (57), Q'd. Dry rainforest on bouldery lower slopes, above Deep Creek (I. C. J. Galbraith)

Plate 9

(a) Mount Bell (32), W.A. General view. Habitat of *Petrophassa albipennis* on rocks behind, and of *Psitteuteles versicolor* and *Trichoglossus haematodus* in foreground (C. B. Frith)

(b) Mount Elizabeth (35), W.A. The Hann River in savanna woodland (D. J. Freeman)

Plate 10

MAP AND NUMERICAL LIST OF COLLECTING LOCALITIES, WITH DATES

Explanatory note

The map shows the camps and collecting localities of Harold Hall Australian Expeditions. The camps are listed below in numerical order, with dates. Main camps, where thorough collections were made over four or five days or more, are shown on the map in white numerals on black. Minor camps and some places visited during day excursions are in black numerals.

Main camps in capitals. South Australian Museum localities marked ★.

Western Australia

1 KARRATHA May/June 66
2 Peg's Beach May/June 66
3 CHICHESTER RANGE May 66
4 Coolawanyah May 66
5 Carnarvon June 66
6 BUSH BAY June 66
7 Perth Nov 62, Apr/May 66
8 Wonerup Mar 66
9 METRICUP Mar 66
10 Margaret River Mar 66
11 SCOTT RIVER Feb/Mar 66
12 MOUNT BARKER Mar 66
13 GREEN RANGE Mar/Apr 66
14 NEWDEGATE Apr 66
15 COCKLEBIDDY July 65
16 Eyre Dec 62
17 EUCLA Dec 72, July 65
18 MOUNT SHENTON July/Aug 65
19 Carnegie Sept 65
20 Mungilli Claypan Sept 65
21 Lake Gruszka Sept 65
22 Hunt Oil Camp Sept 65
23 WARBURTON Aug 65
24 Van der Linden Lakes Sept 65
25 GILES Aug/Sept 65
26 Walter James Range Aug 65
27 St George's Range June 68
28 MOUNT ANDERSON Apr/May 68
29 DERBY May/June 68
30 Point Torment May/June 68
31 GEIKIE GORGE June 68
32 MOUNT BELL June 68
33 MOUNT CAROLINE June 68
34 JOINT HILL June/July 68
35 MOUNT ELIZABETH July 68
36 LAKE GLADSTONE July 68
37 Mount Remarkable July 68
38 Lissadell Hill July 68
39 WYNDHAM July/Aug 68

Northern Territory

40 Victoria River Aug 68
41 DARWIN Sept 68
42 East Alligator R. mouth Oct 68
43 OENPELLI Sept/Oct 68
44 PINE CREEK Aug/Sept 68
45 MOUNT ROPER Oct/Nov 68
46 BORROLOOLA Nov 68

Queensland

47 MOONLIGHT CREEK Apr/May 64
48 Wernadinga May 64
49 GLENORE Apr 64
50 INKERMAN May/June 64†
51 Big Tableland June/July 64
52 AYTON June/July 64
53 Walter Hill Range July 64
54 TULLY July 64
55 GLENDOWER Mar/Apr 64
56 Cape Gloucester Aug 64
57 MOUNT DRYANDER Aug 64
58 BANOONA Apr/May 63
59 TYRONE May 63
60 THYLUNGRA May 63
61 MONKIRA June 63

† Not the Inkerman where Stalker collected.